Science, Democracy, and the American University

This book reinterprets the rise of the natural and social sciences as sources of political authority in modern America. Andrew Jewett demonstrates the remarkable persistence of a belief that the scientific enterprise carried with it a set of ethical values capable of grounding a democratic culture – a political function widely assigned to religion. The book traces the shifting formulations of this belief from the creation of the research universities in the Civil War era to the early Cold War years. It examines hundreds of leading scholars who viewed science not merely as a source of technical knowledge, but also as a resource for fostering cultural change. This vision generated surprisingly nuanced portraits of science in the years before the military-industrial complex and has much to teach us today about the relationship between science and democracy.

Andrew Jewett is Associate Professor of History and Social Studies at Harvard University, where he also participates in the History of American Civilization and Science, Technology, and Society graduate programs. He received his PhD from the University of California at Berkeley and previously taught at Yale University, Vanderbilt University, and New York University. He has held fellowships from the American Academy of Arts and Sciences, the National Academy of Education/ Spencer Foundation, and the Cornell Society for the Humanities.

Science, Democracy, and the American University

From the Civil War to the Cold War

ANDREW JEWETT

Harvard University

CAMBRIDGE
UNIVERSITY PRESS

CAMBRIDGE UNIVERSITY PRESS
Cambridge, New York, Melbourne, Madrid, Cape Town,
Singapore, São Paulo, Delhi, Mexico City

Cambridge University Press
32 Avenue of the Americas, New York, NY 10013-2473, USA

www.cambridge.org
Information on this title: www.cambridge.org/9781107027268

First published 2012

Printed in the United States of America

A catalog record for this publication is available from the British Library.

Library of Congress Cataloging in Publication data
Jewett, Andrew, 1970–
    Science, democracy, and the American university: from the Civil War
    to the Cold War / Andrew Jewett.
       pages cm
    Includes bibliographical references and index.
    ISBN 978-1-107-02726-8
    1. Democracy and science – United States.   2. Science and state – United States.
    3. Science – United States – History.   4. Social sciences – United States – History.
    I. Title.
    Q127.U6J49   2012
    303.48'309730904–dc23   2012015679

ISBN 978-1-107-02726-8 Hardback

Contents

Preface   *page* vii

Introduction: Relating Science and Democracy   I

PART I. THE SCIENTIFIC SPIRIT   21

1. Founding Hopes   28
    Against the Classical Model   29
    Antecedents   34
    The Scientific Spirit   39
    Defending a Modern Curriculum   46

2. Internal Divisions   55
    Stepping Back   56
    Science and Speculation   59
    Small-State Science   65
    The Ethical Economists   74

3. Science and Philosophy   83
    Positivism   84
    Pragmatism   88
    Science and Disciplinarity   98

PART II. THE SCIENTIFIC ATTITUDE   109

4. Scientific Citizenship   117
    Big-State Science?   120
    Social Selves   124
    Toward Culture   132
    Culture and the State   137

5. The Biology of Culture   148
    Determinism and Emergence   149
    Psychobiology   159
    Sciences of Subjectivity   162

6. The Problem of Cultural Change   171
   Participation and Expertise   172
   Foundations and Value-Neutrality   176
   Culture and Governance   183

7. Making Scientific Citizens   196
   Curricular Reform   197
   Engaging the Public   209
   The Critics   216

PART III. SCIENCE AND POLITICS   225

8. Science and Its Contexts   235
   Philosophy of Science   237
   Sociology of Science   243
   Scientific Histories   250
   Science and Language   258

9. The Problem of Values   272
   The Sociologists Divided   273
   Culture and Personality   282
   Consensus Liberalism   290

10. Two Cultures   302
   The Physical Scientists   303
   New Alliances   310
   Science and Values Again   320

11. Accommodation   335
   A House Divided   336
   Causes and Cohorts   342
   Expressions   348

Conclusion: Science and Democracy in a New Century   365

Index   375

Preface

This book seeks to explain what several generations of thinkers had in mind when they devoted their lives to the project of making America scientific. In writing it, I benefited from many important studies of science and values that have appeared in recent decades. This theme has weighed heavily on the minds of American scholars since the 1960s, when the mobilization of technical expertise in the service of war, counterinsurgency, and domestic surveillance generated widespread criticism of putatively value-neutral knowledge. Yet as I waded through primary sources from the late nineteenth and early twentieth centuries, I noticed that questions of democratic theory and practice frequently loomed behind debates about religious belief, scientific authority, and epistemological frameworks in a way that scholars had not recognized.

In that context, it struck me that the question "What should we believe?" often referred to a democratic "we" – the citizens of the United States. And the answers given seemed to be powerfully shaped by what the askers believed to be the cultural, social, and political effects of particular truth claims and methodological approaches. Moreover, remarkably few of the arguments I found prefigured today's understanding of science as a purely technical, value-neutral practice. In fact, from the 1860s to the 1960s, scientific thinkers in the United States repeatedly insisted that science *did* imply certain values – in fact, exactly those values needed to sustain the cultural foundations of American democracy.

At the heart of my argument is a claim of complexity: *Many more scholars addressed the question of science and democracy, and many more understandings of those key terms and their relationship flourished among them, than historians have realized.* The reader may believe that one of the voices I present here is right. My own sympathies, such as they are, will be fairly clear. Yet I am not prepared to choose once and for all between the competing arguments – only to listen with care and respect.

During this book's long gestation, many colleagues, friends, and family members listened to me in such a manner. Without them, the project would never have come to fruition. As a graduate student in the Department of History at Berkeley, financial support from the Jacob K. Javits Fellowship Program kept me enrolled and department administrator Mabel Lee kept us all sane. The wise counsel of several influential teachers, especially Paula Fass, Robin Einhorn, and Don McQuade, helped point me in the right direction, both intellectually and professionally. So did Cathryn Carson and Tom Leonard, who served on my dissertation committee and provided welcome support on the job market. David Hollinger deserves special thanks for his many contributions to this project and to my career over the years. Anyone who knows his work will recognize its profound impact on my own. Less obvious but equally important has been David's expert guidance through the ins and outs of the profession, his friendship, and, most of all, his example. As a source of inspiration for young intellectual historians, David is second to none.

During my first week of graduate school, David also introduced me to Healan Gaston, a fellow student and, as fate would have it, my future wife. Other students and friends offered camaraderie and compassion during those years: Dee Bielenberg, Molly and Daryl Oshatz, Dan Geary and Jennie Sutton, Andy Lakoff, Guian McKee, Julia Svihra, Buzzy Jackson, Susan Haskell Khan and Rehan Khan, Line Schjolden, Laura Mihailoff, Debbie Kang, Samantha Barbas, Jennifer Burns, Kevin Schultz, Paddy Riley, Rachel Hope Cleves, Amanda Littauer, Heather McCarty, Sarah Carriger, Fred Shoucair, Cole Ruth, Tim and Jenn David-Lang, Erbin Crowell and Kristin Howard, Ann Pycha and Ivan Ascher, Dave and Lyssa Gilson, Jason Grunebaum, Dana Ingersoll, Heather Blurton, Homay King, Jesse Gora, Yoonah Lee, Raoul Bhavnani, Rebecca Lemov, Helen Tilley, Jennifer Gold, Louise Nelson Dyble, Eric Klinenberg, J. P. Daughton, Chad Bryant, Charles Postel, Justin Suran, Ania Wertz, Ben Lazier, Sam Moyn, Ajay Mehrotra, Abe Levin, Lorien Redmond, Kristina Egan, and Brian Austerman. Subsequent endeavors have brought us closer to additional members of the remarkable mid-1990s cadre of Berkeley students, especially Jason Smith, David Engerman, Diana Selig, and Julian Bourg.

During graduate school, I also enjoyed the company of older friends, many of whom I met during my undergraduate days at Berkeley: Chris Welbon, Nathaniel Gordon-Clark, Jeremy Wallach, Eric Volkman, Dan Callahan, Danae Vu, Teresa Nero-Wirth, Melissa Gutierrez, Preeti Ramac, David Schuster, Alex Farr, Chuck and Cathy Brotman, Dafna Elrad and Brandon Sahlin, Chris von Pohl, Steve von Pohl, Shalene Valenzuela, Andy Clay, Kennedy Greenrod, Jim Alumbaugh and Rima Kulikauskas, Mary Cosola, Ken Stockwell and Anne Heindel, and Anne Eickelberg and Rick Weldon. And then there is Joh Humphreys, with whom I started to hash out some of the vexing questions raised in this book before it was even a glimmer in my eye.

After receiving my PhD, I embarked on what can only be described as an epic academic journey. During the first year, at the American Academy of Arts and Sciences, Leslie Berlowitz, Pat Spacks, and James Carroll provided a rousing introduction to the world of Big Thought. I learned a great deal from the rest of my cohort as well: Anne Mikkelsen and Dan Sharfstein, Joseph Entin and Sophie Bell, Jay Grossman, Page Fortna and Pete Beeman, Eric Bettinger, and David Greenberg and Suzanne Nossel. Alexandra Oleson, Malcolm Richardson, and James Miller also deserve thanks for ensuring that the year would be productive, as do Charlie Hogg and Jonna Meyler (and Lucky) for making us feel at home in Cambridge. For a visiting position at Yale the following year, I have Jon Butler and Glenda Gilmore to thank. The friendship of David Greenberg, Mark Brilliant, Dietmar Bauer and Susi Schwarzl (and Sebastian), and Jennifer Klein made our time in New Haven especially memorable. A generous fellowship from the Spencer Foundation and the National Academy of Education supported my work in New Haven and then in Tennessee, where Dan Usner gave me a chance to teach American intellectual history at Vanderbilt. Devin Fergus, Diana Selig, Laura DeSimone, Adam Nelson, Benita Blessing, John Rudolph, Norman and Cassie Fahrney, Michael Blick, and Corey Blick enriched my experience in those years.

On yet another fellowship, at the Cornell Society for the Humanities, Brett de Bary provided warm encouragement, and the cold winter days also seemed brighter for the presence of Heidi Voskuhl, Kevin Lambert, Suman Seth, Angela Naimou, Aaron Sachs, Walter Cohen, Petar Bojanic, and Jason Smith. Then it was on from Ithaca to Princeton, where I commuted to a position at NYU's John W. Draper Interdisciplinary Master's Program in Humanities and Social Thought. Robin Nagle and Robert Dimit showed me the ropes, while an extraordinary group of master's students taught me as much as I taught them. On the train to New York, I had the great privilege of stopping in to join Carla Nappi, the late Phil Pauly, and others in a discussion group led by Jackson Lears and Ann Fabian at Rutgers. Back in Princeton, I got to know Jason Josephson, Rebecca Davis and Mark Hoffman, Alan Petigny, Jim McCartin, and Leigh Schmidt and Marie Griffith through Healan's fellowship at the Center for the Study of Religion, and Larry Glickman and Jill Frank through the Center for Human Values.

At Harvard, many colleagues in the History Department and the Social Studies program helped me keep the project moving forward as I learned to balance the roles of junior faculty member and new father. Liz Cohen and Jim Kloppenberg expertly guided the History Department, and many other colleagues – especially Nancy Cott, Lisa McGirr, Sven Beckert, Vince Brown, Laurel Ulrich, David Armitage, Joyce Chaplin, Jill Lepore, Walter Johnson, Evelyn Higginbotham, Carrie Elkins, Erez Manela, Kelly O'Neill, Andy Gordon, Alison Frank, Ann Blair, Dan Smail, Charlie Maier, Roger Owen, and Emma Rothschild – shared their insights and good humor. Peter Gordon, Maya Jasanoff, Ian Miller and Crate Herbert, Mary Lewis and Peter Dizikes, Tryg Throntveit, and Rachel St. John deserve special thanks for their support and friendship. In Social Studies, Richard Tuck, Anya Bernstein, Michael and Coral Frazer, Verity Smith, Bo-Mi Choi, Thomas Ponniah, Darra Mulderry, and Jona Hansen helped me wrap my brain around the canon of Western social theory. Elsewhere in the university, Heidi Voskuhl, Sindhu Revuluri and Nina Moe, Rebecca Lemov and Palo Coleman, Jeremy Greene, Jeanne Haffner, Alex Wellerstein, Chris Phillips, and Robin Bernstein provided welcome companionship too. I also benefited enormously from conversations with colleagues in History of Science, especially Charles Rosenberg, Everett Mendelsohn, and Janet Browne; and the Divinity School, especially Dan McKanan, David Hall, and David Hempton. At the Kennedy School, Sheila Jasanoff was a particularly valuable source of support and inspiration. The annual cycle of Harvard visitors also brought faces old and new to town: Sam Haselby, Andy Lakoff and Daniela Bleichmar, Paul Kramer, Jamie Cohen-Cole, Dan Geary and Jennie Sutton, Mike Pettit, Robert Adcock, and Liz Lunbeck. In and around Cambridge, David Engerman, Julian Bourg and Jessie Yamas, Brooke Blower, Jason Josephson and Dalena Frost, Eleanor Goodman, Vanessa Ruget and Prabal Chakrabarti, Claire Rowberry and Tim McLucas, Patty Nolan and David Rabkin, Craig Malkin, and Martine Gorlier have been sources of camaraderie and sanity.

Back at the AAAS for a leave year, Dan Amsterdam, Debbie Becher, Angus Burgin, Dawn Coleman, Jason Petrulis, Jamie Pietruska, and Pat Spacks dissected my project and taught me worlds about much else, while John Tessitore kept the cookies coming. Harvard's Charles Warren Center for Studies in American History deserves thanks twice over. After providing much-needed financial support for my research, the CWC sponsored a postdoctoral program on "The Politics of Knowledge in Universities and the State" that I co-led with Julie Reuben. The project fostered vibrant conversations with Kornel Chang, Jamie Cohen-Cole, Jal Mehta, Gregg Mitman, Adam Nelson, Mark Solovey, Lisa Stampnitzky, Jeffrey Stewart, Jessica Wang, and many others. At the CWC, Arthur Patton-Hock and Larissa Kennedy made easy much that at first seemed hard. So did Janet Hatch and Mary McConnell in History; Sarah Champlin-Scharff, Kate Anable, and Katie Greene in Social Studies; and Phyllis Strimling at the Radcliffe Institute for Advanced Study, which funded a productive workshop on the social sciences and American liberalism in April 2010.

Good friends and colleagues I met on the conference circuit warrant prominent mention as well. Sarah Igo, Jessica Wang, John Carson, Ted Porter, Amy Kittelstrom, Daniel Immerwahr, Tom Stapleford, Joel Isaac, Jennifer Ratner-Rosenhagen, Andrew Hartman, Tim Lacy, Chris Loss, Chris Nichols, Gabe Rosenberg, Ethan Schrum, Martin Woessner, Joy Rohde, Michael Kimmage, Patrick Slaney, Nicole Sackley, and Guy Ortolano have taught me much about my project as well as theirs. Danielle Allen, Rob Reich, Philippe Fontaine, Susan Lindee, John Beatty, Keith Benson, and Alan Richardson offered opportunities to discuss my research in congenial settings and added their own insights.

And then – at long last – there is the book itself. Many people read all or part of the manuscript: Howard Brick, Jamie Cohen-Cole, Nancy Cott, Henry Cowles, Katrina Forrester, Healan Gaston, Joe and Kay Gaston, Dan Geary, David Hollinger, Joel Isaac, Jim Kloppenberg, Susan Lindee, Adam Nelson, Ron Numbers, Julie Reuben, Mitchell Stevens, Tryg Throntveit, and Jessica Wang. Dee Bielenberg and Eleanor Goodman deserve special mention for going beyond the call of duty. I also want to thank the anonymous readers of the manuscript and the ever-patient Lew Bateman at Cambridge University Press for their crucial contributions to the process. Howard Brick warrants extra thanks for shepherding this project through several iterations and serving as a valuable adviser over the years. Dorothy Ross, Ken Alder, Charlie Capper, Mary Jacobus, Jeff Sklansky, George Cotkin, Mary Furner, and Richard Teichgraeber also offered wise counsel. And of course, there would be no book at all without the research support of Marisa Egerstrom, Alexa Howitt, Elsa Kim, Bronwen O'Herin, Robert Owens, Eva Payne, Julian Petri, Arjun Ramamurti, and Kip Richardson.

Although printed sources proved most useful for capturing the tone of American public debates on science and democracy, archival work also shaped this project in important ways. I am grateful to the staffs of the Bancroft Library at the University of California-Berkeley; the Manuscript Reading Room at the Library of Congress; the Houghton, Pusey, and Andover-Harvard Theological Libraries at Harvard; Manuscripts and Archives at Yale; the Manuscripts Department of the American Philosophical Society; the Smithsonian Institution Archives; the National Archives and Records Administration in College Park, Maryland; the Hoover Institution at Stanford; the Wisconsin Historical Society; the University Archives and Jacques Maritain Center at Notre Dame; the Burke Library at Union Theological Seminary; and especially the Rare Book and Manuscript Library at Columbia.

I grew up in a family full of creative, dedicated teachers of science, mathematics, and engineering. Their commitments to pedagogical innovation and the value of careful thought overlapped with the Deweyan tenor of The School in Rose Valley, where I spent eight formative years. I want to single out my parents, grandparents, and in-laws for their enduring love and unflagging support: Mary and Howie Ditkof, John and Lisa Jewett, Joe and Kay Gaston, Sam and the late Debbie Mercer, Gretchen MacDonald, the late Dutch and Jo Schober, and the late Healan Baker. My deep thanks also to the rest of the Mercer, Jewett, Ditkof, Schober, Gaston, and Baker clans. This unusually large extended family was a source of great joy as I sought to balance work and play – and of forgiveness as I repeatedly erred on the side of work. It was a particular treat to have Cass and John Bing and Bob and Eileen Mercer nearby while in Princeton. Louise and Dave DeNight, Tom and Barb Mercer, Allyson Mercer and the late Paul Shunskis, and Barb Carr provided homes away from home in Pennsylvania. Carey Shunskis, Stephen Mercer and Melissa Estrella, and Tom and Lynne Baker (and Kate Sonderegger) represented the New England outposts of the families. In California, Charlie and Laura Jewett, Sarah and Chris Candela, John Powers and Kimmie Burgandine, Josh and Shannon Powers, and Debbie Schober and Mike Long shared their families and our tribulations. And in Tennessee, Josephine and Andrew Larson and their children helped out in countless ways. I am also grateful to a sadly departed menagerie – a loyal gecko, a flock of plucky parakeets, and a soulful rabbit – for their company during the long years in the academic wilderness.

Finally, I owe almost boundless thanks to Healan Gaston and Joseph Jewett, the heart and soul of my family. They have loved me through thick and thin, celebrated my successes and helped shoulder my burdens, and been the light at the end of each long workday. Healan and I were best friends and intellectual companions before we became life partners; her influence on this book and on my understanding and appreciation of the scholarly life are simply inestimable. Joseph reminds me daily that many, many things are more important than books, or at least grown-up books. Together, they never let me forget the power of love and laughter in this world.

Introduction

Relating Science and Democracy

For a full century, from the 1860s to the 1960s, American intellectual life took its central dynamic from a powerful impulse to build up the scientific disciplines. What accounted for this enormous investment of energy in science? Some interpreters invoke professional or class interests, arguing that the proponents of science sought to increase the status of their disciplines or of the professional middle class more generally. Others view the period's scientific advocacy in religious terms, as an outgrowth of either a theologically liberal form of Protestantism or a thoroughly secular worldview. This book augments and in many respects challenges such interpretations by offering a broadly political reading of the push to make America scientific. It excavates one of the most important and least examined dynamics in American intellectual and political history: a massive effort to mobilize science, so successful in its industrial applications, as a resource for strengthening American democratic practices. The book traces the origin of the campaign to turn the science-centered university into a tool for building a new culture. It explores where and how this project unfolded and explains why it largely failed to achieve its political goals, even as it powerfully aided the growth of scientific authority in general.

The campaign to bring science to bear on American public culture initially aimed at inspiring citizens to protect their autonomy from the state, but it soon became aligned with an emerging Progressive or "social liberal" attempt to strengthen the state as a counterpoint to big business. Many of science's advocates concluded that every modern society would feature both massive corporations and a regulatory apparatus to tame their ill effects. They then turned to the creation of citizens who could bring business and the state into line with their collective needs. These figures believed that science, properly understood and internalized, could protect Americans from a stultifying existence in the German theorist Max Weber's "iron cage" of bureaucracy.

Of course, Weber himself was responding to the longstanding claim of many European thinkers that the spread of science among the public would make possible the modes of self-government endorsed by republicans, liberals, and democrats. This political hope has constituted a central strand of Western thought since the Scientific Revolution, although it has come under fire in many precincts of the academic left since the 1960s. A century earlier, in the 1860s, a broad, science-centered political vision became a major component of American thought as well. Yet American political culture featured an egalitarian, populist edge that rarely registered among intellectual elites in other Western nations. In the post-Civil War United States, more than anywhere else, the advocates of a scientific culture felt obliged to actively reconcile the claims of scientific research with the requirement of democratic legitimacy. Never doubting that science and democracy would prove harmonious and even mutually reinforcing, they worked to transform European conceptions of science in keeping with American understandings of politics.

That work of intellectual reconciliation, undertaken during a crucial period in American economic, social, and political development, provides the subject matter for this book. The chapters trace the diverse and ever-shifting formulations of the commonly heard assertion that science embodied and inculcated a set of personal virtues, skills, beliefs, and values that could ground a modern, democratic public culture. They highlight the impact of this broad discourse about science and democracy on the career choices and knowledge claims of scores of scholars as it wound its way through American intellectual life from the creation of the modern universities in the 1860s to the postwar era.

To be sure, few advocates of science in this period thought it offered a comprehensive worldview equivalent to that of Protestant Christianity, which in one form or another had grounded American public culture since the Revolution. In fact, the late nineteenth century brought vigorous efforts to distance science from "ultimate" questions of theology and metaphysics. But even many of the thinkers who banished such ultimates still deemed science sufficiently robust *ethically* to take over from religious authorities the crucial political role of forming democratic citizens. In their view, science could largely replace what historians call the "pan-Protestant establishment": that cluster of mainline Protestant denominations whose leaders controlled the nation's major cultural institutions and acted as an informal religious establishment in the mid-nineteenth century.

After the Civil War, many advocates of science began to claim that it offered not only practical techniques, and thus material plenty, but also the cultural and political benefits that flowed from mainline Protestantism, without the divisive theological claims and metaphysical baggage. The expectation that science would provide civic resources as well as technical knowledge shaped not only descriptions of scientific inquiry, but also the direction and results of research programs in a wide range of fields. Much intellectual work of the late nineteenth and early twentieth centuries took its shape from a desire to demonstrate that a scientifically grounded democracy could work – that human beings were constituted such that they could bring their institutions into line with their needs, and thus sustain self-governance, without converging on a shared Protestant worldview.

As a number of historians have shown, the resulting understandings of science had a great deal in common with the liberalized forms of "low church" Protestantism they claimed to transcend.[1] Less frequently noticed is the fact that many of science's leading advocates and practitioners consciously sought to take over from Protestant leaders not only the interpretation of physical and biological nature, but also the core social function of cultural reproduction – the very formation of individuals. In the wake of the Civil War, a new generation of educational reformers began to argue that science could make ethical citizens. It is unclear how many ordinary Americans agreed that science carried with it an ethical orientation suited to democratic citizenship. But an ethical and ultimately political reading of science strongly conditioned the rise of the scientific disciplines and the modern research universities.

This understanding of science had a particularly profound effect on the growth of the human sciences.[2] But it is easy to miss the ethical and political impulses that animated so many figures in those disciplines. In fact, a burgeoning literature on the American human sciences after 1870 suggests that their practitioners systematically disengaged from public culture. According to the usual story, human scientists retreated institutionally from the public sphere and spoke only to other specialists, while building a high intellectual wall around their disciplines by sharply differentiating science and values. This line of argument holds that professionalism and "scientism" – an epistemological and methodological approach in which investigators aim at value-neutrality by rigidly suppressing their emotions and normative commitments – held sway in the human sciences by the 1920s.

Many interpreters argue further that the ostensible disengagement of science-minded scholars from public concerns actually had powerful political effects. The advocates of science, this account asserts, "naturalized" the beliefs and values of an emerging managerial elite by reading them into reality itself and selling the resulting forms of knowledge back to the mushrooming groups of professional administrators who ran increasingly bureaucratic organizations in the private sector, the philanthropic world, and the state apparatus. In short, the advocates of value-neutral professionalism enlisted on the side of the new managerial class against its radical, populist, egalitarian, and democratic challengers. Thus, the human sciences, despite – or rather, because of – their professed neutrality, became the ideological bulwark of a powerful new structure of social hierarchy.[3]

---

[1] See especially George M. Marsden, *The Soul of the American University: From Protestant Establishment to Established Nonbelief* (New York: Oxford University Press, 1994) and Julie A. Reuben, *The Making of the Modern University: Intellectual Transformation and the Marginalization of Morality* (Chicago: University of Chicago Press, 1996).

[2] I use this term – more familiar to European scholars than their American counterparts – to encompass the social sciences, philosophy, and closely related fields of natural science and the humanities.

[3] These interpreters often assert that the social sciences, and the universities centered on them, arose precisely to fill the knowledge needs of industrial capitalism and the administrative state. This

Such an approach to the historical study of science and American politics – what we might call the "disengagement thesis" – reflects a strong tendency among critical scholars since the 1960s to doubt that science can be
a progressive force in society. Organization, administration, rationalization,
bureaucracy, materialism: these are the social phenomena with which many
commentators today habitually associate science. This critique, which echoes
Weber's more radical interpreters, treats science as synonymous with an instrumental rationality that buttresses the rule of the dominant elite by claiming to
offer only technical means to externally determined ends. The recent flourishing of critical theory, interpretivism, and poststructuralism has fueled an out-
pouring of critical histories of the human sciences, as dissident practitioners
have joined with professional historians to rewrite each discipline's twentieth-
century career as a story of defeat and alienation at the hands of professional-
izers and value-neutralists.

But the story of value-neutrality's ascent, however well told, is not a substitute for a full-fledged political history of scientific thought in the United States.
The value-neutrality narrative does not include the whole range of claims
about science's political meaning. This book challenges the preoccupation with
professionalism and scientism that characterizes so many recent studies, finding instead that the ideal of engagement with public discourse and normative
questions remained central in the human sciences until the 1940s. To be sure,
many scholars at the time honored that ideal only in the breach or, more typically, pursued it by taking up interpretive questions whose relevance to public
concerns is difficult to grasp without a keen understanding of the specific disciplinary contexts. Still, even those figures who embraced normative engagement more clearly in theory than in practice believed that science found its true

analysis treats science and "corporate liberalism" – an expertly managed form of capitalism – as
two sides of the same coin: e.g., Stanley Aronowitz, *Science as Power: Discourse as Ideology
in Modern Society* (Minneapolis: University of Minnesota Press, 1988); Clyde W. Barrow,
*Universities and the Capitalist State: Corporate Liberalism and the Reconstruction of American
Higher Education, 1894–1928* (Madison: University of Wisconsin Press, 1990). See also the
essays in George Steinmetz, ed., *The Politics of Method in the Human Sciences: Positivism and Its
Epistemological Others* (Durham: Duke University Press, 2005). A slightly softer version of this
analysis appears in John M. Jordan, *Machine-Age Ideology: Social Engineering and American
Liberalism, 1911–1939* (Chapel Hill: University of North Carolina Press, 1994). A related body
of work identifies modern science's naturalism – its elimination of theological commitments –
as the source of its value-neutrality: Edward A. Purcell Jr., *The Crisis of Democratic Theory:
Scientific Naturalism & the Problem of Value* (Lexington: University Press of Kentucky, 1973);
Marsden, *The Soul of the American University*; Reuben, *The Making of the Modern University*;
Jon H. Roberts and James Turner, *The Sacred and the Secular University* (Princeton: Princeton
University Press, 2000); Christian Smith, ed., *The Secular Revolution: Power, Interests, and
Conflict in the Secularization of American Public Life* (Berkeley: University of California Press,
2003). In many of the latter group of texts, the desired alternative to modern, naturalistic science
is a religiously committed science that takes God's existence as its starting point. By contrast,
most critics of value-neutrality look instead to moral commitments drawn from nontranscendent
sources: the arts and literature, for example, or shared cultural or subcultural identities, or
established traditions of moral reasoning or hermeneutic interpretation.

raison d'être in reshaping the public mind, not in providing the knowledge base for rationalized state administration or industrial production.[4]

This revised account of science's desired political effects suggests new interpretations of its actual political impact in the twentieth century. Political historians now stress the contingent, contested, and internally fractured character of the "New Deal order" that appeared so robust to the critics of the 1960s and 1970s, even as it began to crumble under their feet.[5] Intellectual historians, however, have not yet interrogated the longstanding assumption that scientism and a technocratic, managerial liberalism were hegemonic in the mid-twentieth century. In fact, of the leading cultural elements that mobilized and divided Americans in the twentieth century, divergent beliefs about the character of the natural and human worlds have been by far the least well integrated into scholarly understandings of American politics. Historians know a great deal about racial divisions and class identities, and even more about changing views of the relationship between the state and the economy – views that largely structure the American party system today. But the other key element of today's party system – namely, widespread disagreement on foundational scientific claims about natural and social phenomena – continues to be poorly understood. Anti-statism and evangelical Protestantism appear everywhere in the new histories of twentieth-century America, but debates over the personal qualities required of democratic citizens and the relative capacities of science and religion to produce those qualities have been largely ignored.[6]

---

[4] A number of important books have paved the way by adopting a broadly political approach, although their interpretations differ from mine: James T. Kloppenberg, *Uncertain Victory: Social Democracy and Progressivism in European and American Thought, 1870–1920* (New York: Oxford University Press, 1986); Dorothy Ross, *The Origins of American Social Science* (New York: Cambridge University Press, 1991); Jeffrey P. Sklansky, *The Soul's Economy: Market Society and Selfhood in American Thought, 1820–1920* (Chapel Hill: University of North Carolina Press, 2002); and Howard Brick, *Transcending Capitalism: Visions of a New Society in Modern American Thought* (Ithaca: Cornell University Press, 2006). Several essays in David A. Hollinger's *In the American Province: Studies in the History and Historiography of Ideas* (Baltimore: Johns Hopkins University Press, 1985) and *Science, Jews, and Secular Culture: Studies in Mid-Twentieth-Century American Intellectual History* (Princeton: Princeton University Press, 1996) highlight the cultural and political ambitions of scientific thinkers.

[5] A good summary is Jefferson Cowie and Nick Salvatore, "The Long Exception: Rethinking the Place of the New Deal in American History," *International Labor and Working-Class History* 74, no. 1 (2008): 3–32.

[6] Theoretically, my account is indebted to work in the overlapping fields of history of science and science and technology studies (STS), work that has opened the possibility of understanding science's intersection with political commitments very differently. Scholars in these fields have come to view science in a thoroughly historicist manner, recognizing that "science" is a linguistic category, not a preexisting natural object. As such, its meaning is essentially fluid, contingent, and contested across all of its domains of application. To be sure, historians of science, like their counterparts elsewhere, have generally assumed that the story of objectivity claims and their impact is *the* story of science and politics in the twentieth century. Revealingly, the leading long-range histories of objectivity and quantification ignore American developments before World War II, whereas Cold War America often appears as the culmination of the political transformations associated with modern science: e.g., Robert N. Proctor, *Value-Free Science? Purity and Power in*

Although the visage of the philosopher John Dewey graces the cover of this book, the attentive reader will note that Dewey's own writings play a relatively small role in the narrative. But as the leading theorist of the push to make America scientific and a universally recognized symbol of that cultural project, Dewey casts a powerful shadow over the narrative. Historians' treatment of Dewey neatly encapsulates recent interpretive tendencies in the history of the human sciences. In 1968, when the United States Postal Service issued the stamp from which the cover image is taken, heterodox philosophers had begun to rehabilitate Dewey's reputation in their field as part of a broader attack on the highly specialized, technical approaches dominating it. So, too, had critical social scientists frustrated with the political quiescence of their own disciplines. Meanwhile, Dewey's writings on school reform had inspired a new generation of progressive educators to focus on the whole child. And student activists had recaptured Dewey's political ideal of a democracy centered on vigorous political participation by ordinary citizens.

As they unfolded through the 1970s and 1980s, these overlapping "Dewey revivals" rescued Dewey from the charges of critics such as Clarence Karier and R. Jeffrey Lustig who deemed him a consummate representative of "corporate liberalism." Like his fellow Progressives, Karier and Lustig contended, Dewey sought to build up a strong administrative state and an accompanying network of bureaucratic "parastate" organizations that would stave off radical challenges to capitalism by using social-scientific expertise to mitigate the most disruptive effects of the boom-and-bust cycle.[7] By contrast, Dewey's new champions recognized that he was a lifelong critic of corporate liberalism, a radical democrat who sought to put power back into the hands of the people rather than simply transferring it from business tycoons to social scientists, managers, and other middle-class experts. Turning afresh to Dewey's epistemological and political claims in the light of their own era's challenges to the liberal mainstream, these interpreters portrayed Dewey as a forceful but increasingly isolated advocate of a mode of Progressive thought that called for participation by citizens and normative public engagement by intellectuals, rather than administration by scientific experts.[8]

Modern Knowledge (Cambridge: Harvard University Press, 1991); Theodore M. Porter, *Trust in Numbers: The Pursuit of Objectivity in Science and Public Life* (Princeton: Princeton University Press, 1995); Lorraine Daston and Peter Galison, *Objectivity* (New York: Zone Books, 2007). The same focus on objectivity characterizes a newer body of literature centered on the emotions, scientific selfhood, and modes of personal discipline: e.g., Rebecca M. Herzig, *Suffering for Science: Reason and Sacrifice in Modern America* (New Brunswick: Rutgers University Press, 2005).

[7] Clarence Karier, "Making the World Safe for Democracy: An Historical Critique of John Dewey's Pragmatic Liberal Philosophy in the Warfare State," *Educational Theory* 27 (1977): 12–47; R. Jeffrey Lustig, *Corporate Liberalism: The Origins of Modern American Political Theory, 1890–1920* (Berkeley: University of California Press, 1982).

[8] Robert B. Westbrook's authoritative intellectual biography represented the culmination of the new approach to the study of Dewey: *John Dewey and American Democracy* (Ithaca: Cornell University Press, 1991).

But crucial insights were lost in this otherwise salutary interpretive shift, because it sheared away two central features of Dewey's life and work. First, his new champions tended to ignore or downplay one of the central organizing principles of Dewey's thought, namely his deep confidence that modern science contained within itself the seeds of an egalitarian, democratic culture, if only its reach could be extended into the realm of human behavior. Second, these treatments of Dewey have left the impression that he was virtually alone in carrying his brand of Progressivism forward into the 1920s and 1930s. After World War I, this literature suggests, Dewey was a kind of democratic lone wolf, holding out bravely against the tidal surge of corporate liberalism.[9]

Thus, the new body of historical writing on Dewey has profoundly altered our understanding of his thought, along with that of a few other Progressives rediscovered as democratic heroes, most notably William James, Jane Addams, and W. E. B. Du Bois. But Dewey's new champions have left the background picture of his interwar milieu largely untouched, essentially as the theorists of corporate liberalism rendered it. Having plucked Dewey and a few others from the political swamp, these interpreters write off the rest of the interwar intelligentsia as "managerial liberals" or "administrative progressives" who viewed science as value-neutral and sought to make it the centerpiece of a technocratic polity.[10]

In truth, however, a great many of Dewey's contemporaries shared his broad view of science's cultural promise and social role, even though few could match his firm commitment to deliberative democracy. This book shines a powerful light on both Dewey and his time by reconstructing a forgotten

[9] Westbrook codified this image of a heroic, embattled figure at odds with interwar American thought. On the one hand, according to Westbrook, Dewey was the saving remnant of the golden age of normatively grounded public participation in the late nineteenth and early twentieth centuries. On the other, he was a forefather to the rebirth of participatory democracy and associated modes of teaching and philosophizing in the 1960s. Westbrook attributes to most twentieth-century liberals the theory that ordinary citizens cannot and must not play a significant role in public decision-making. Thus, he writes, Dewey's participatory, democratic vision made him a "deviant" from the "liberal-realist" stance that dominated American thought from World War I onward. To be sure, Westbrook, unlike many of Dewey's recent champions, takes stock of Dewey's democratically tinged view of science. "The literature on Dewey's social thought," he notes, "is plagued by a failure to give such key terms as 'scientific intelligence,' 'social control,' and 'adaptation' the meanings he intended." In a pivotal section of the biography, Westbrook demolishes the assumption that Dewey embraced a narrow, value-neutral understanding of scientific inquiry. Yet he does not apply the same interpretive lens to Dewey's equally science-minded counterparts in the interwar period. *John Dewey and American Democracy*, xiii–xvi, 120–147. Earlier in his career, Westbrook had followed the corporate-liberal line on Dewey (558).

[10] Philosophers interested in recovering American naturalism as a full-fledged movement have connected Dewey to his closest philosophical interlocutors, but have provided little sense of his engagements across disciplinary lines. John Ryder, ed., *American Philosophic Naturalism in the Twentieth Century* (Amherst, NY: Prometheus, 1994); Victorino Tejera, *American Modern, The Path Not Taken: Aesthetics, Metaphysics, and Intellectual History in Classic American Philosophy* (Lanham, MD: Rowman & Littlefield, 1996); John P. Anton, *American Naturalism and Greek Philosophy* (Amherst, NY: Prometheus, 2005).

tradition of thinking about the democratic possibilities of modern science that
dated back to the Civil War and extended forward to the Cold War. It restores
science to its proper place at the center of Dewey's thought, and at the same
time identifies him as merely one among the innumerable American thinkers
who sought to weave a modern culture out of scientific materials. The narrative situates Dewey in a rich academic milieu whose leading figures shared
both his commitment to a scientific public culture and his expectation of its
eventual emergence as a result of scholarly efforts. Although Dewey shaped
that discourse more powerfully than any other single participant, he was
hardly the sole contributor.

This book thus offers a new and relatively comprehensive – though hardly
exhaustive – account of the relationship between the growth of scientific
authority and changes in American political culture during the late nineteenth
and early twentieth centuries. It draws on a wide range of literatures, pulling
together an institutionally fractured and decentralized discourse on science and
American politics that has taken what coherence it musters from the widespread assumption that objectivity is the main story, rather than from sustained
cross-disciplinary dialogue. At the same time, the book turns away from the
focus on objectivity, proposing a new center of coherence by exploring the
arguments of a heterogeneous group of American thinkers who shared a sense
that science could meet the cultural needs of democratic citizens.

To tell a new story about science and politics from the Civil War to the Cold
War is also to rewrite much of the wider narrative of American intellectual life
during that era. The question of science's political meaning engaged some of the
most influential and original thinkers of the day and ran through many well-
known intellectual episodes: the founding of the modern American universities
by reformers such as Andrew Dickson White, Daniel Coit Gilman, and Charles
W. Eliot; the innovations of "ethical economists" such as Richard T. Ely and
John Bates Clark; the early development of pragmatism by Charles S. Peirce,
William James, and Dewey; the pioneering sociological accounts of Edward
A. Ross and Charles Horton Cooley; the formulation of new modes of liberalism by the *New Republic* theorists Herbert Croly, Walter Weyl, and Walter
Lippmann; the articulation of cultural relativism by the anthropologists Franz
Boas, Margaret Mead, and Ruth Benedict; the creation of the "New History"
by James Harvey Robinson and Charles Beard; the debate between Dewey
and Lippmann on the prospects for democracy in an age of propaganda; the
struggle between William F. Ogburn and his sociological critics over the possibility of value-neutrality; the rise of the "culture and personality" school of
anthropologists, psychologists, and psychiatrists; and, of course, the emergence
of the national security state and associated forms of postwar liberalism. This
book casts each of these oft-studied phenomena in a new light, showing the
remarkable power of a belief that science found its highest purpose in changing
the normative commitments of the American people.

Moreover, it also shows that these episodes were simply the most visible
outcroppings of a much more extensive series of debates over science and

politics that gave American academic thought much of its distinctive energy and flavor for nearly a century. The question of science's place in the modern democratic project preoccupied vast numbers of American scholars, including many who figured prominently in the cultural discourses of their time but have since been forgotten. The book restores to their proper place in the story a series of important intellectual phenomena – including the partial professionalization of philosophy, emergent evolutionism, philosophical naturalism, contextual histories of science, and many others – that figured centrally at the time but are now remembered only by a few scattered specialists in precincts far beyond intellectual history.

As readers listen to the many hundreds of voices presented in the book, they will repeatedly encounter unfamiliar uses of the charged terms "science" and "scientific." Although it is impossible to fully avoid projecting today's concerns onto historical actors, the process of coming to grips with past arguments requires not only learning new information but also forgetting, or at least bracketing, the contemporary meanings of key terms of art. To grapple constructively with such texts, one must look closely at the conceptual vocabularies of those who created them and keep in mind that the linguistic categories now used to carve up the world of experience are contingent and fluid rather than fixed and given. As the later chapters show, the current equation of the term science with a strictly value-neutral conception of knowledge, along with the narrowing of its boundaries to include only the natural sciences and related technological pursuits, stem from mid-twentieth-century intellectual transformations. But that is not the science that most earlier thinkers had in mind when they set out to make America scientific.

Throughout this book, I apply the label "scientific democrats" to the large and varied group of American thinkers who contended that science, as they understood it, offered the basis for a cohesive and fulfilling modern culture. By using that label, I do not mean to suggest that the figures in question concentrated on expanding the circle of suffrage to include women, African-Americans, and other excluded groups. Nor do I mean "democratic" in the even stricter sense of those contemporary theorists who call for direct participation by citizens in decision-making.[11] In fact, scientific democrats thought surprisingly little about the formal mechanisms of governance. Simply assuming that public opinion mattered centrally in American governance, they focused on making an impact on the minds of citizens. Thus, my invocation of democracy is relatively colloquial, echoing the vernacular connotation of a polity defined by popular sovereignty – a polity in which the will of the people reigns supreme, in general if

[11] Today's democratic theorists also attend carefully to how public opinion is formed, counseling extreme caution about claims to authority in the spheres of knowledge and culture. By that measure, hardly anyone in the late nineteenth and early twentieth centuries qualified as a democrat. Most scientific democrats, like most religious leaders and cultural critics at the time – or, indeed, today – hoped to see a particular set of ideas and their spokespersons granted considerable authority by the public.

not in every detail.[12] Like so many other Americans, then and since, scientific democrats assumed that the nation's policies reflected the beliefs, opinions, values, and virtues of the people. They sought to change those policies by changing the underlying cultural substrate.

In other words, when I speak of scientific democrats my emphasis is on the "scientific" side of the phrase. The label is not designed to mark off a group of thinkers committed to maximizing political participation from others we would now consider technocrats. Instead, it serves to differentiate those who saw science's social effects largely in terms of its capacity to improve the process of formulating the public will from another group of far better studied figures who focused on science's capacity to generate instrumentally useful knowledge. Scientific democrats claimed that science could dramatically improve democratic practice not only by fostering technological growth, improving administrative techniques (both inside and outside of government), and giving citizens the technical information needed to participate constructively in policy debates, but also, and more importantly, by shaping their moral character, normative commitments, and discursive practices.[13]

Science, in short, promised thoroughgoing cultural change, rather than simply the augmentation of the nation's knowledge base. In this understanding, science denoted a personal orientation, not just a body of knowledge or a set of institutions. Being scientific meant much more than simply using empirical methods; it meant behaving in accordance with specific ethical tenets or exhibiting particular ethical virtues. It entailed a mode of speaking, a form of interpersonal relations, even a comprehensive way of life. Scientific democrats portrayed the scientific enterprise – the whole complex of practices, institutions, knowledge claims, and persons – as a concrete manifestation of an underlying ethical orientation that was perfectly suited to the needs of a modern democracy.[14]

Dewey and thousands of other leading scholars thus sought to expand science's authority because they believed it offered moral as well as cognitive

[12] I have otherwise largely avoided using common terms whose meaning is implicitly or explicitly contested – Enlightenment, technocratic, scientism, and the like – as shorthand descriptors for the figures and ideas treated herein. Likewise, I have eschewed the adjectival references to landmark theorists (Weberian, Habermasian) that dot many works of intellectual history. Using such labels saves innumerable words, but at the steep cost of eliding crucial meanings.

[13] Of course, virtually all of science's champions in the late nineteenth and early twentieth centuries employed multiple strategies for justifying and validating it in public debate. But those I call scientific democrats consistently emphasized the cultural influence of science above all.

[14] On its technocratic end, scientific democracy shaded off into the view that the needed cultural orientation among citizens entailed simply deference to the authority of scientific experts. Beyond that line, a few genuine technocrats sought to use state power to enforce scientific findings from the top down. These rare calls for an efficient, functionally oriented polity run directly by experts resolved tensions between scientific authority and democratic participation by simply eliminating the latter. See William Akin, *Technocracy and the American Dream: The Technocrat Movement, 1900–1941* (Berkeley: University of California Press, 1977) and Jordan, *Machine-Age Ideology.*

resources to the citizens and officials of a democracy. They argued that only the creation of a thoroughly scientific public culture could sustain democratic governance in the face of potent threats – prominently including science's other, more disruptive social effects. This belief powerfully shaped the course of American intellectual and political life from the Civil War era, when science still figured in the popular imagination primarily as a source of economically relevant "useful knowledge," to the early Cold War years, when the authority of scientific experts and the scale of national research spending reached staggering heights. Although the resulting social formations rarely met scientific democrats' expectations, their belief in science's political promise operated as a central driving force in the development of the American scientific enterprise.

This was particularly true in the human sciences, where the strongest impulse in the early twentieth century actually pointed toward engagement with American public culture rather than disengagement. On the question of science's relation to the public, Dewey spoke for the majority of interwar human scientists, including those who ignored his deliberative understanding of democratic practice. Historians who view American intellectual life between the wars through the lens of the "elite theory of democracy" articulated in the 1950s – the view that public opinion is politically irrelevant until it becomes so virulent and irrational that it interferes with the process of governing through expert administration and negotiation between elites – build their narratives on a small number of figures who stood well outside the main body of interwar thought in the human sciences.

Interpreters often account for Dewey's surprising influence among even his epistemological opponents by pointing to the legendary awkwardness of his prose and suggesting that his contemporaries misunderstood his relentless campaign to reform American education and democracy. But there is a simpler explanation: most of his contemporaries understood and agreed with him, at least on the question of science's historic vocation. Dewey mattered so much to the scientific thinkers of his age because he became the public voice of their democratic conscience, calling them back to what they considered their public responsibilities through his writings and personal example. Of course, he not only articulated clearly and forcefully the claim that democracy was a form of culture, but also spelled out, in myriad domains of intellectual life, what he took to be the philosophical underpinnings and expressions of that culture. But it was less these specific claims than Dewey's broader cultural project that inspired so many of his contemporaries. Dewey recognized that a polity armed with technical knowledge could not operate successfully without a citizenry capable of carefully assessing specific scientific claims and assertions of scientific authority, rather than categorically accepting or rejecting them. His influence in the second quarter of the twentieth century rested on his powerful articulation of a widespread hope that a fully normative – yet also fully scientific – culture could emerge from the modern disciplines and take its place as the foundation for a newly inclusive and egalitarian American polity. In short,

Dewey stood for the expectation that science and morality would fuse in a unified, post-Christian, and intrinsically democratic public culture.

The term scientific democracy highlights the contrast with other groups of historical actors who favored very different cultural projects, at a time when industrialization, urbanization, and immigration cut deeply into established patterns and altered both political institutions and the conditions of everyday life. Some aspiring cultural leaders lodged their hopes in the churches and synagogues, believing that the fuller internalization of a religiously grounded morality would place American democracy on firm footing. Others found a potent antidote to the ills of industrial society in the arts and literature. Still others looked to specific occupational groups, crediting modern businessmen, industrial workers, or professional administrators with the vision, skills, and virtues needed to lead the nation forward. In placing their hopes on a public versed in scientific modes of thought and behavior, the scientific democrats rejected each of these alternative plans for cultural organization and leadership.

Scientific democracy also stood in tension with several other intellectual and political projects carried out in the name of science, although few at the time recognized conflicts that seem glaringly obvious today. These alternative programs, too, aimed at facilitating the public acceptance of science by stressing its contributions to the collective good. But they emphasized other features of science. In the rapidly expanding "helping professions" and associated areas of disciplinary knowledge production – educational research, medical research, psychiatry, and psychology – champions hailed science's contributions to individual health and development. Another argument centered on science's promotion of industrial efficiency through the development of chemistry, engineering, many fields of physics, and later industrial psychology. A third and particularly well-studied culture of science identified the expansion and application of the social sciences as the key to reforming the administrative practices of the American state. Each of these modes of advocacy touted science's practical utility to a specific group of potential clients or patrons. Their proponents emphasized science's stability and reliability rather than joining Dewey in stressing the flexible and revisable character of scientific knowledge. Such arguments thus clashed with scientific democracy, insofar as most scientific democrats highlighted science's human origins and endemic fallibility.

At the time, few scientific thinkers worried about the tension between science's critical and positive sides – between its denaturalizing assault on established ways of knowing and the naturalizing tendency associated with its own production of specific knowledge claims. In the long run, however, these clusters of arguments about science carried very different implications for American public life. Even when scientific democracy clearly served to rationalize professional goals, it did so in a different manner than the other cultures of advocacy, and with different consequences. Neither science-based techniques nor a direct client relationship anchored scientific democracy and its associated conceptions of social and political order. It did not rely on a presumption of scientific neutrality or an image of inevitable progress, but instead presented science as a mode of public service in the interest of the community at large. Its proponents sought to create a new set of shared understandings of self and society, of human behavior and its sources, and with them a new avenue of approach to problems of belief and action. In fact, scientific democracy resembled the missionary enterprise more than an engineering project because it focused so heavily on transforming beliefs and values and viewed science as a species of ethical practice rather than an expression of instrumental reason. Offering even more expansive claims than the era's technological utopians or avatars of administration, scientific democrats declared that science could not only remake the American economy and state, but also provide a comprehensive intellectual-cultural foundation for a morally vigorous public culture.

Institutionally, scientific democracy's center of gravity lay in the universities, although scientific democrats also appeared at times in journalism, K-12 education, and even the federal bureaucracies.[15] By the interwar years, scientific democracy was particularly well established in disciplines with little direct relevance to the other modes of scientific advocacy. Thus, for example, it was central to the development of anthropology, history, and philosophy. Major outcroppings also appeared in sociology, psychology, and the emerging field of linguistics. Some political scientists and economists leaned toward scientific democracy as well, although their fields were also tied to the project of administrative reform. Scientific democracy even appeared in the natural sciences between the wars, in areas of biology where there were few obvious points of social application unless one supported eugenics. In the 1920s, much work in American biology focused quite explicitly on answering social, cultural, and moral questions.

By the 1930s, scientific democracy had also become particularly well represented in certain regions and types of universities. The land-grant schools and other public universities featured an ethos of service that could authorize attempts at cultural change – framed as the extension of both the material and the spiritual rewards of modern society to previously excluded social groups – as well as administrative reforms and improvements in industrial and agricultural production. The Midwest and upper South proved particularly fertile ground for scientific democracy, as these regions featured a high density of public institutions whose faculties sought to civilize the surrounding areas by importing cultural materials from the Northeast. Scientific democracy also flourished in private metropolitan universities such as Chicago and especially Columbia. These institutions became key sites for the articulation of scientific democracy, because their twin commitments to public service and to liberal rather than practical education led members of their faculties to see themselves as cultural leaders in their communities.

---

[15] For an example of the latter phenomenon, see Andrew Jewett, "Philosophy, the Social Sciences, and the Cultural Turn in the 1930s USDA," *Journal of the History of the Behavioral Sciences* (forthcoming).

Within the academic sphere, scientific democrats achieved many substantial successes. They worked with other reform-minded figures to profoundly alter American higher education. In the late nineteenth century, scientific democrats oversaw the creation of the modern American research university, an institution wherein highly educated professors simultaneously undertake original investigations and impart their findings to a new generation of future leaders and scholars. After World War I, they helped to create academic majors and general education requirements, believing that the free election of courses by undergraduates had failed to promote the needed scientific culture. Finally, from the 1930s onward, economic and military crises fostered powerful ties between the academic disciplines and the federal government.

These institutional relationships joined with generational and demographic shifts and other national and international developments to produce new versions of scientific democracy suited to the straitened political environment of the early Cold War years. Significantly muted, though hardly silenced, the voice of scientific democracy now spoke in a more chastened, less critical register. The perceived meaning of science itself changed, as political circumstances seemed to render the formation of citizens and public opinion less relevant to the task of promoting the public good.

But many older scientific democrats remained profoundly dissatisfied with American culture, even as World War II gave way to what is often described as the golden age of American science. Their dogged efforts seemed to them radically unfulfilled, despite the fact that the universities had moved toward the center of the nation's political economy and governance structure. In fact, such institutional advances merely heightened these scientific democrats' sense of the yawning chasm separating ideals from reality. After all, their ultimate aim had not been to produce high-tech consumer goods or even new tools of public administration. Instead, they had sought a transformation of the ways in which ordinary Americans thought about themselves, one another, and their social and political institutions.

My goal in the following chapters is not primarily to demonstrate the existence and persistence of scientific democracy but rather to explore its many varieties and myriad consequences for academic thought in the late nineteenth and early twentieth centuries. As a widely shared and highly mutable sensibility, rather than a monolithic school of thought or movement, scientific democracy cut across a vast range of local contexts. Because it seemed uncontroversial and axiomatic to its proponents, moreover, scientific democracy often took form only in subsidiary debates over technical issues related to its main terms and claims. The tasks of the book, then, are to explain what the scientific democrats meant by "science" and to identify some of the most important intellectual and institutional projects that emerged from their view of science as a democratic resource. Instead of simply showing the bare presence of scientific democracy or tabulating lists of its supporters and critics, the book takes stock of where and how scientific democracy mattered in the course of American thought, tracing its varying contours and illuminating its nuances and effects over time.[16]

Part I centers on the hopes that leading scholars invested in the new universities during the decades immediately following the Civil War. The ethical orientation characteristic of scientific democracy emerged in liberal Protestant circles during the second half of the nineteenth century. The early scientific democrats drew particularly heavily on the legacy of Whig political thought, which combined technological enthusiasm with humanitarianism and Protestant moralism. Yet these reformers lamented what they saw as the failure of the Protestant churches to advance the cultural values necessary for self-government. They hoped that new, science-centered universities could take over much of the cultural work of the churches, placing a rapidly industrializing nation on firm footing.

As the nineteenth century moved toward its close, a younger group of scientific democrats turned against the identification of their cultural program with economic *laissez-faire*. Other scientific democrats of that era struggled to accommodate the "New Psychology," with its emphasis on the profoundly irrational character of human action. Many embraced new philosophies of science that replaced a discredited common-sense psychology but retained a longstanding emphasis on the ethical and communicative dimensions of scientific practice. Scientific democrats drew on these views of knowledge as they worked to build the modern disciplines in the decades around the turn of the century. It quickly became clear, however, that the mere proliferation of scientific knowledge would not produce the cultural change they sought.

Part II details the concerted efforts of early-twentieth-century scientific democrats to connect academic knowledge to public discourse and to explore the cultural realm empirically. These figures worked to popularize existing knowledge, to reshape that knowledge in keeping with its allotted cultural tasks, and to make science's underlying ethical orientation widely available. Here, as always, professional self-interest intermingled with specific political values, a broad commitment to popular sovereignty, and an unshakeable belief in the applicability of scientific methods to human affairs.

---

[16] In terms of genre, this book combines synthetic ambitions with a monographic methodology. Like a synthesis, it surveys a broad terrain and cites an equally wide array of specialized secondary literatures. But though scholars have occasionally looked at the cultural ambitions of particular scientific intellectuals, none has systematically surveyed the kinds of politics associated with those ambitions. As in a monograph, then, my narrative is grounded in a close reading of original sources, undertaken after surveying the existing literature. Attempting to bracket what historians had written about well-known primary texts, I pored over them myself to develop my own interpretations. I also went far beyond familiar sources, finding troves of material on relatively unstudied subjects. Only after reading thousands of printed sources, honing my views on each, determining the overall argument and structure of the book, and boiling down lengthy readings of sources to far shorter analyses did I go back to the existing secondary literature to see where I could credit other historians with sharing the same insights, or at least addressing the same topics.

Unlike Part I, which spans the modern university as a whole, Part II centers on what figures from the time occasionally called the "cultural sciences." This trans-disciplinary formation ranged from philosophy to anthropology, history, psychology, psychiatry, linguistics, political science, sociology, economics, and even biology. In good empiricist fashion, scientific democrats in these fields subjected the entity they hoped to influence – culture – to intense scrutiny. They analyzed the role of cultural phenomena in social processes and explored the contours of American culture in particular. Many of the characteristic intellectual tendencies of the early Cold War years drew on the interwar flowering of the cultural sciences.[17] Although practitioners disagreed vociferously on key points of theory and strategy, they shared a common commitment to spreading a scientific form of ethics throughout the population.

Part III chronicles shifts in scientific democracy between the 1930s and the 1950s. Early in that period, scientific democrats developed sophisticated new theories of knowledge and culture as they applied empirical methods to language and other cultural phenomena. The ideological battles of the 1930s also highlighted the cultural and linguistic embeddedness of all thought, including even scientific thought. At the same time, however, the looming conflict between democracy and totalitarianism provoked most scientific democrats to defend American ideals and institutions in no uncertain terms. Facing stiff challenges from the political right, many sought to preserve the institutional gains of the Roosevelt years – and to protect their own hides – by claiming that they were simply implementing a consensual framework of liberal values that underlay mainstream American opinion.

In that process, the cultural sciences gave way to the more narrowly defined and politically restrained "behavioral sciences." By the McCarthy era of the early 1950s, few science-minded scholars openly sought to transform American public culture. The immense and practically unchallengeable national projects that dominated American politics in the middle decades of the twentieth century – first saving the nation from economic ruin, then defeating fascism, and lastly warding off the Soviet threat – gradually integrated science into the state apparatus and blunted the critical edge of scientific democracy. This development produced the irony – many would say the tragedy – of Cold War science: the massive enlistment of ostensibly neutral expertise for clearly normative purposes that so riled a new generation of radicals in the 1960s.

Within the discursive world of scientific democracy, one thus finds considerable diversity and change over time. Scientific democrats shared a deep frustration with extant political practices and a desire to remake those practices by altering American public culture, but they endorsed a variety of means to that common end and harbored competing visions of the final result. They described the practice of science in many different methodological and epistemological

---

[17] For more on these intellectual continuities across World War II, see Brick, *Transcending Capitalism*, esp. 9–10.

terms, and they relied on divergent conceptions of the self, society, and the state. A particularly deep and stable division, especially after 1920, separated those who thought science could access relatively permanent facts or even laws from those who, like Dewey, stressed its denaturalizing function and assumed that scientific theories would change constantly, like the social realities they indexed. Less obviously, though perhaps more consequentially, scientific democrats also differed in their specific political goals. By the early twentieth century, most occupied the broad ground between socialism and classical liberalism that historians variously designate Progressivism, social liberalism, or social democracy. Where some scientific democrats wanted to fundamentally restructure working conditions and labor relations, however, others focused primarily on widening the distribution of consumer goods. Likewise, some called for extensive participation by ordinary citizens, but others foresaw a greater role for experts in public life. For a variety of strategic and psychological reasons, scientific democrats often bracketed hard questions about their specific political goals.[18]

To bring some order to this picture, readers can think about the characters in this book as having occupied four concentric circles, with Dewey near the center. The first, outer ring included all of the scientific democrats – all of those who thought science provided the cultural resources needed to restore democracy in the wake of industrialization. Part I of the book focuses largely on a second, narrower circle of figures: those scientific democrats of the late nineteenth century who toiled in the universities rather than other spheres of practice. Most of these figures thought about engaging – and remaking – the public in very different terms than did their nonacademic peers. The third concentric circle included those who argued, beginning in the 1880s and accelerating after the turn of the century, that the development and popularization of the human sciences would turn Americans away from competitive capitalism rather than toward *laissez-faire* ideals. I term the figures in this third circle "Progressive theorists" and then "post-Progressive theorists," with World War I representing the dividing line between them. Virtually all of the university-based scientific democrats I discuss in Parts II and III occupied this domain. Finally, the innermost circle contained those twentieth-century scientific democrats who sought to drive cultural change by developing the human or cultural sciences. Most practitioners in those fields concerned themselves with the workings of culture precisely because

---

[18] Concerned as it is with public representations of science for expressly American audiences, this book operates within a national rather than transnational frame and is based on printed sources rather than archival collections. Frank M. Turner has described such arguments as "public science": *Contesting Cultural Authority: Essays in Victorian Intellectual Life* (New York: Cambridge University Press, 1993), 201–228. However, that term can imply that scholars' arguments were always mere rhetorical devices used to win support from patrons, rather than deeply rooted cultural assumptions that shaped careers, personae, and theoretical contributions. In any case, I focus on scientific democracy and refer only in passing to other public claims about science's social value.

they saw something profoundly amiss with American economic and political institutions and sought to remake them at their cultural roots.

Instead of artificially constructing a single, national debate on science and democracy, the book's chapters range across a series of more or less local settings. A persistent and widely shared preoccupation with the contributions of science to the formation of democratic citizens took many different forms and structured many different areas of intellectual and institutional work, suggesting that it fulfilled deep needs. The book explores not only explicit contributions to political theory, such as the 1920s writings of Dewey and Lippmann, but also the many vernacular traditions of political thought that were embedded in specific academic cultures and subcultures. Texts from these vernacular traditions need to be read carefully, with special attention to the nuances of their intellectual contexts, in order to understand how they spoke to broader social questions. Yet that interpretive effort pays great dividends. As the chapters move from one partial clearing in the academic wilderness to another, they demonstrate the remarkable frequency with which scientific democracy appeared and reappeared in a variety of intellectual contexts. By reading scholarly texts closely as contributions to moral, social, and political theory, the book recovers a forgotten culture of scientific advocacy, a lost world of political argumentation that is alien to contemporary ears and largely untouched by historians.

Not that it is any great feat to find arguments about the relation between science and democracy in sources from the time. Rare is the scholarly publication from that era that failed to speak to the question in one manner or another. Yet thinkers often engaged the matter of science and democracy in the concrete terms of a specific discipline or subdiscipline rather than the general terms of political theory. To see scientific democracy in the round, then, one must look beyond the texts that explicitly defended one or another version of it. One must learn to sense its subterranean presence in a host of specific domains and seemingly unrelated arguments, to pull out and recombine threads that formed a whole fabric in the minds of historical actors but rarely took coherent form in a single text. In short, one must track arguments about the political meaning of science through the thickening academic underbrush and into the dark forests of specific disciplinary discourses.

Much additional research is needed to establish more precisely where and how scientific democracy operated, to identify parallels between its American expressions and related impulses elsewhere, and to trace connections to the periods before the Civil War and after the 1950s. It is a sign of scientific democracy's enormous range and influence that its manifold expressions burst the bounds of a single book, even one this inclusive. *Science, Democracy, and the American University* certainly does not say the final word about the subject it treats. It does, however, propose a starting point, an angle of approach for further study. And it suggests that looking at the events of the period through a synoptic lens, at this level of generality, is crucial for understanding where American scholars and citizens are today, and where they can hope to be tomorrow.

As the account unfolds, it will become clear to readers that I sympathize more with the scientific democrats than with their opponents, that I prefer those in the early-twentieth-century human sciences to their Gilded Age predecessors, and that I lean toward the Deweyan strand of interwar scientific democracy rather than the value-neutralist model. But those looking for neatly demarcated sets of heroes and villains will not find them here. All of the scientific democrats exhibited profound flaws. The power of their paired visions of an ethical science and a scientific culture led them to ignore science's other social and political effects, including its highly ambiguous contributions to the development of corporate capitalism and the bureaucratic state. Already short on political savvy and badly overestimating the historical power of ideas, most scientific democrats also assumed that their favored ethical tenets were intrinsic to the scientific enterprise, whose every advance thus appeared as a sign of moral progress, especially when it augmented science's cultural authority. Meanwhile, scientific democracy also led proponents to assume that human beings were more alike and more transparent to external observation – in short, more prone to a generalizing form of empirical analysis – than they really are.

Still, for all of its shortcomings, scientific democracy also inspired some of the same individuals to work to strengthen the public sphere of democratic discourse, at a time when many other cultural reformers sought to limit it. Scientific democrats believed that the United States featured a tempered form of deliberative democracy in which the state of public opinion dictated the overall shape of national policy. This meant that those seeking to change the nation's direction could do so legitimately and effectively only by influencing the public mind. An ambition to engage the public did not always reflect a principled commitment to deliberative democracy so much as a practical need to make change in what seemed like the most effective manner. Yet we cannot assume that all scientific thinkers of the late nineteenth and early twentieth centuries sought to replace citizens with experts. In fact, many agreed with Dewey on the need to ensure the democratic legitimacy of policy decisions, even when those decisions also enjoyed the imprimatur of science. The following chapters trace these intertwined political effects of scientific democracy through a complex series of intellectual episodes and across a range of scholarly contexts. In doing so, they reveal many of the insights and blind spots that can result from a view of science as a cultural foundation for democratic politics.

PART I

THE SCIENTIFIC SPIRIT

"How could society escape destruction if, when political ties are relaxed, moral ties are not tightened? And what can be done with a people master of itself if it is not subject to God?"

– Alexis de Tocqueville

In the late nineteenth century, the American scientific enterprise expanded rapidly and gained solid institutional footing as a series of reforms in higher education placed science at the center of an emerging system of modern universities. This was not, however, merely an intellectual shift with potent political ramifications. Instead, it was a shift at once intellectual and political in its import. Much of science's attraction came from its apparent potential to solve festering problems of governance, and not just by creating a body of neutral, technical knowledge that the American state could use to better understand the population. More importantly, according to the early scientific democrats, science could shape new kinds of citizens. The rise of the scientific disciplines and "nonsectarian" universities in the United States was powerfully structured by a sense that these institutions could perform a political function that had been widely attributed to the pan-Protestant establishment in the preceding decades. No less than their counterparts in the churches, Gilded Age advocates of science thought the contours of the nation's culture made all the difference, politically. In fact, much of the energy of antebellum liberal Protestantism flowed directly into scientific channels after the Civil War and continued to course there until the early Cold War period.[1]

---

[1] David Hollinger frames the continuities slightly differently in "Justification by Verification: The Scientific Challenge to the Moral Authority of Christianity in Modern America," in *Religion and Twentieth-Century American Intellectual Life*, ed. Michael J. Lacey (New York: Cambridge University Press, 1989): 116–135. Other interpreters, such as the sociologist Christian Smith, see a "secular revolution" in the late nineteenth century. Smith contends that scientific professionals, led by social scientists, displaced Protestant leaders from the helm of American public life and installed themselves as, in effect, the high priests of a thoroughly deracinated culture; Smith, ed., *The Secular Revolution: Power, Interests, and Conflict in the Secularization of American Public*

Part I offers a new account of the epistemological, methodological, and cultural understandings of science that emerged in tandem with the modern American university system. It demonstrates how strongly these understandings were shaped by the claim that scientific work aimed primarily at changing the culture in support of democracy, and only secondarily at generating knowledge for its own sake. As in many subsequent instances, traumatic events in American politics – in this case, the twin shocks of civil war and pell-mell industrial expansion – led many scholars to reexamine their views about the nature and the cultural effects of their knowledge practices. Out of this political-intellectual cauldron came scientific democracy.

Of course, some scientific thinkers continued to espouse a thoroughly Christian culture during the Gilded Age. This approach has always been fairly common among physical scientists and engineers, who are more able to sustain a "separate spheres" division between science and religion than are their counterparts in many other provinces of science. In the late nineteenth century, however, even the most resolutely Christian advocates in the so-called hard sciences still argued that the scientific enterprise carried, and took its shape from, a set of ethical commitments. The question was simply whether those commitments could ground a public culture. How could people continue to be good, and to govern themselves, if large swaths of intellectual terrain were now the province of science rather than religion? Did institutional Christianity simply need to be brought to bear more forcefully on the American people, or could a scientific culture sustain democracy?

Scientific democrats held that a scientific culture could perform that task admirably. They sought results far beyond what we now call "scientific literacy" or the "public understanding of science." Their ultimate goal was not simply improved knowledge and informed citizens. Instead, scientific democrats reasoned that, because science carried with it a set of ethical resources, it could ground a democratic culture in the absence of a central religious authority, and thereby take over the core political functions of the pan-Protestant establishment. Different groups of scientific democrats disagreed on the precise content of the ethical orientation embodied in science. But in these circles, it remained commonplace well into the twentieth century that the term science referred to a full-fledged way of life, not simply the body of knowledge or the set of investigative techniques that way of life had produced.[2]

Life (Berkeley: University of California Press, 2003). But Gilded Age advocacy of a scientifically grounded culture is not well captured by the concept of "secularization," even if that term is used to refer to a social process rather than a mere intellectual shift. The vast majority of scientific democrats hoped that religion would play a significant role – not the role Smith and like-minded critics advocate, but a role nonetheless – in a scientific culture. Very few saw science and religion as incompatible except in certain specialized domains of knowledge.

[2] In the late nineteenth century, in fact, commentators often implied that science only produced reliable knowledge because investigators adopted this ethical orientation. The ideology of absolute disinterestedness that spread widely in the twentieth century is a descendant of this view, as it relies on an ethical capacity – that of self-denial or "self-elimination" – of individual scientists. For a striking example, see Theodore M. Porter, *Karl Pearson: The Scientific Life in*

When the first generation of scientific democrats took up their work in the era of the Civil War, an amalgam of evangelical Protestantism and republicanism, sustained by the cultural apparatus of the pan-Protestant establishment, dominated American political thought. At their point of overlap, these two idioms stressed self-sacrifice, mutual service, and an active, personal commitment to the common good. Republicans believed that citizens needed to possess certain civic virtues in order to preserve self-government against the depredations of would-be demagogues. Fortunately, according to most American theorists, evangelical Protestantism offered the needed virtues as well as the moral restraint necessary to forestall crimes against person and property. These qualities seemed especially crucial by the 1840s and 1850s, as the early stages of industrialization began to threaten the agrarian communities and individual freeholders that were central to earlier strands of American political thought. Leading antebellum thinkers believed that the commitment to the public good fostered by the Christian faith would effectively counterbalance the moral dangers of market competition and widespread prosperity. To render self-rule safe, reformers merely needed to spread the gospel.

The early scientific democrats attacked this assumption, holding that institutional Christianity could not sustain republican virtue. They charged that the churches, rather than fostering public-mindedness, actually fueled denominational strife by demanding belief in specific, unchanging doctrines. But these thinkers found an alternative at hand: science. They insisted that scientific practices embodied and transmitted the virtues cherished by republicans. In the rough-and-tumble frontier days, American scientists had justified scientific development primarily in terms of its contributions to national expansion and the accompanying economic gains. They had described the contribution of their work to self-government primarily in terms of its capacity to sustain economic autonomy, which the republican framework deemed necessary for civic virtue. Now, the scientific democrats of the Gilded Age proposed a much more direct connection between science and politics. They argued that learning and practicing science turned individuals away from narrow partisanship and versed them in forms of communication conducive to the common good, making them ideal citizens of a democratic republic.

In keeping with their emphasis on science's role in shoring up republican virtue, the early scientific democrats viewed science as the embodiment of a form of *virtue ethics* – an ethical system that identifies the good with a specific set of personal characteristics rather than particular motives or consequences. They often described the virtuous person as one who possessed the "scientific spirit." Such a person, these scientific democrats argued, exhibited a specific orientation toward the arguments put forth by others. The truly scientific individual engaged with competing arguments, listened patiently to the evidence presented on their behalf, accepted or rejected the arguments based solely on the evidence, and then, if the argument had been rejected, proceeded to respond to the interlocutor by patiently and non-coercively offering evidence for an alternative argument. In short, those who possessed the scientific spirit would actively engage in processes of mutual persuasion, employing evidence and reasoning as the sole means of changing minds and treating their own arguments as no more valid than those of others in the court of reason.

Although the scientific spirit was typically framed as a set of personal virtues, it also represented an ethic of communication: a description of the good in terms of a particular mode of communication. This form of communication was widely viewed as the foundation for democratic politics – as the source of a truly enlightened public opinion and of harmony between citizens. Many scientific democrats argued that the kinds of communication undertaken by those possessing scientific virtues could simultaneously ensure social progress by producing new knowledge and guarantee social order by generating intellectual agreement. These figures worked to spread the virtues and the communicative style they associated with science throughout the citizenry.

The early scientific democrats did not break in every regard with the pan-Protestant establishment, however. The initial formulations of scientific democracy emerged within liberal Protestant circles and drew heavily on a preexisting body of Christian political thought. Antebellum Whigs, especially, had combined humanitarian sentiments with a fervent hope in the promise of technological advancement, while arguing that a common Christian culture could compensate for the fragmenting effects of market capitalism. Many had also leaned toward an essentially ethical reading of Protestantism as a general mode of interpersonal behavior rather than a series of specific doctrinal tenets. These figures had infused American politics with a fiery millennial piety from the 1830s onward, seeking to build a thoroughly progressive, Christian civilization based on free markets, industrial innovation, and universal faith.

The early scientific democrats carried forward key elements of this Christian political project but rejected its explicitly theological grounding. As the moral clarity of the antislavery campaign gave way to a brutal war and then to shocking instances of corruption in the party of Lincoln, it seemed to many scientifically minded thinkers that the market had burst its cultural bonds. They concluded that the personal virtues inculcated by the pan-Protestant establishment were either too frail or too exclusive to guarantee social cohesion in an industrializing nation. The traumatic conflicts of the Civil War era and the rampant self-seeking of the early Gilded Age fueled the emergence of scientific democracy by convincing many scientific thinkers that the Protestant churches had failed to advance the cultural values necessary for self-government. Seeking solutions close at hand, they concluded that science could do what institutional Christianity could not – namely, serve as a cultural adhesive and a counterweight to capitalist excess, offering the ethical resources needed to knit together a highly centrifugal nation and forestall a looming crisis of democratic governance. A scientific culture could prevent the grand experiment in self-government from collapsing under the weight of economic self-interest. To be sure, the practical contributions of science to economic production and state administration also played significant roles in Gilded Age arguments for science, but for many, science's ethical orientation loomed largest. Scientific democrats traced both sides of social progress to science: the industrial innovations that were revolutionizing the economy and the dawning cultural transformation that would harmonize industrial growth with human needs and democratic values.

The early scientific democrats drew much of the content of their ethics from theologically liberal forms of the pan-Protestant establishment's "low-church" evangelicalism. Although the emerging rhetorics of science were sometimes framed as rejections of Christianity, many scientific democrats instead described their outlook as a purification of Christianity and often claimed that working scientists were more Christian than were ministers and theologians. They also carried forward key assumptions about the practical application of their ethical principles.[3] Most importantly, these figures believed that the scientific spirit called for strict limits on state power. Of course, a basic foundation of civil order and national defense promoted the interest of all, in their view. Outside a liberal polity there was only anarchy, and no one could prefer violence to civil peace. But these scientific democrats insisted that the state could not go much beyond its minimal functions without abridging the freedom of thought they deemed central to both science and Christianity.

Still, Gilded Age interpretations of the scientific spirit, and the corresponding defenses of *laissez-faire* policies, varied. The scientific democrats in the new universities differed in important ways from those employed in business and the professions. Standing far closer to ministers and theologians than to the captains of industry or the architects of the "organizational revolution," they shared Protestant leaders' fraught relationship with industrial capitalism. They welcomed industrialization for the material prosperity it made possible, but they found the rampant hedonism of the Gilded Age profoundly disturbing. Viewing the scientific spirit as a potent antidote to the selfish, profit-driven orientation of the market, scientific democrats in the universities worked to retrofit a rickety Christian public philosophy with new foundations. Unlike their

[3] The adoption of formal theories, in ethics as elsewhere, tends to accompany a search for practical results. The advocates of abstract ethical systems generally expect specific real-world outcomes to follow from their widespread adoption. Moreover, unless these advocates are especially rigorous and self-limiting, they tend to use the outcomes in question as a proxy for the ethical theory itself, assuming that those who agree with them on practical matters have employed the proper mode of ethical reasoning, whereas those who disagree with their conclusions have not employed it. This tendency among scientific democrats to conflate broad ethical principles with specific political conclusions remained essentially unchecked in the nineteenth century, when those on all sides viewed knowledge as a unitary system, each part relating intimately to the others. Few in those days could have imagined an ethical system that stood apart from politics.

counterparts outside the universities – and like their Whig-Protestant predecessors – they embraced industrial capitalism only to the extent that it appeared to be a mode of humanitarian aid, flowing from benevolent intentions toward others.

This delicate balancing act would not last. Two younger groups of university-based scientific democrats, coming to intellectual maturity in the 1870s and 1880s, pushed the first generation's ethical-political program in opposite directions, setting the stage for a bitter clash. In the 1870s, a small band of sociologists and economists, with William Graham Sumner in the lead, advanced the urban professionals' view of the market, not as a tool for welfare policy, but rather as a sorting mechanism that enabled self-reliance by faithfully translating divergences of natural endowment and character into differential social results. By the 1880s, a third set of scientific democrats, led by the economist Richard T. Ely, had reinterpreted the humanitarian side of the first generation's program in the light of the Protestant Social Gospel and begun to challenge *laissez-faire* itself in an explicitly moral idiom. Although the fractious "economists' debate" of the late 1880s and early 1890s produced a partial *rapprochement* that blunted the edge of Ely's critique, his morally grounded attack on *laissez-faire* would shape the work of leading Progressive Era scholars and define the mainstream of American social science through the 1930s.

This Gilded Age dispute over science, ethics, and politics, along with the equally bitter debates over the religious implications of Darwinian evolution in the 1870s and 1880s, led philosophically minded scientific democrats to examine more closely the nature of scientific reasoning and its relationship to ethical commitments. The early scientific democrats carried forward from the antebellum period a "common-sense" theory of knowledge holding that properly constituted evidence and argumentation would produce consensus. Agreement remained elusive, however, on the relation between science and religion and utterly imploded when it came to the relation between the state and the economy. Coming to doubt the common-sense framework itself, many science-minded philosophers based new theories of knowledge on an emerging, physiologically grounded form of psychology that stressed the importance of nonrational motives in human behavior. The pragmatist philosophers William James and John Dewey combined resources from this New Psychology with Charles S. Peirce's definition of truth as a matter of consequences. With various positivists, they threw off the common-sense understanding of knowledge as a fit between a mental faculty of reason and the world. Unlike many positivists, however, these early pragmatists retained a strong emphasis on the ethical and communicative dimensions of scientific practice.

Philosophical idealists influenced by Kant and Hegel also joined the conversation about the nature of science. In fact, as the American Philosophical Association began operation at the start of the new century, professionalizing philosophers argued bitterly over how they could advance the cause of a scientific culture without simply ceding their already diminished role in the collegiate curriculum to scientists. Acutely aware of their field's failure to match the advances of the sciences, these philosophers raised probing questions about the balance of interpretive freedom and intellectual authority that underpinned science – and democracy. But these were merely new formulations of the same concerns that had driven scientific democrats since the 1860s: What did the unprecedented capacity of physical science to produce agreement among practitioners mean for the conduct of human affairs? Which of its methods could be applied to social and political disagreements, and how? As the nineteenth century gave way to the twentieth, scientific democrats in the universities found themselves more divided than ever on such fundamental questions.

I

Founding Hopes

The rise of the modern American university system in the wake of the Civil War is usually traced to one of three factors: the desire of natural scientists to create institutions and facilities for producing new knowledge rather than simply disseminating existing knowledge; the knowledge needs of a rapidly industrializing economy; or the secularizing impact of Darwinian evolution on American intellectual life. But another impulse operated powerfully in that era, integrating each of these elements and transcending them in importance, at least in the minds of the actors at the time. This was the urge to remake American culture by spreading the ethical virtues, character traits, and social practices that scholars increasingly associated with science – and increasingly failed to discern in the dominant forms of Protestantism.

Viewed from this angle, the new universities did not represent as sharp a break with the denominational colleges of the mid-nineteenth century as either their founders or most interpreters today would insist. Historians have recently stressed the persistence of theological tenets associated with liberal Protestantism in the new university system. But an equally consequential and more deeply rooted point of continuity was a long-standing view of how democracy worked: institutions of higher education shaped the nation's culture, which in turn shaped its political system.[1] By reimagining scientific practice as a form of interpersonal ethics, the early scientific democrats in the universities allowed themselves and others to imagine a largely secular institution that produced new knowledge in the sciences without abandoning its social role of buttressing democracy through the formation of individual character. Under the pressure of national political crises, they crafted this ethical conception of science out of the cultural materials of liberal Protestantism and Whiggish technological enthusiasm.

[1] It has long been recognized that Americans tend to view education – though I would say culture more broadly – as a solution for any and all political ills. Less well-known is the centrality of this dynamic to modern American intellectual life, where the education in question was collegiate education.

The widespread perception that a scientific university could perform its cultural and political tasks far more effectively than could a Christian college helped fuel the rapid transformation in higher education that was such a striking feature of late-nineteenth-century intellectual American life. In fact, the university reformers occasionally argued that a properly scientific university system could have forestalled both the battle over slavery and the pecuniary politics of the Gilded Age. At a time when Radical Republicans sought to remake the South and the Knights of Labor rose up against business interests, the first generation of scientific democrats looked to higher education to solve the nation's political woes. In place of the economically independent freeholder or the devout Christian, they held up the individual imbued with the "scientific spirit" as the model citizen for an industrialized American republic.

The transformation of American higher learning between the 1860s and the 1890s also took much of its shape from the political views and financial largesse of the wealthy industrialists who founded most of the key private universities. By the turn of the century, the influence of a business-centered culture could also be seen clearly in the ever-growing public demand for academic credentials in new white-collar fields and the networking opportunities provided by fraternities and football clubs. At the level of the faculty, however, the new institutions actually embodied a powerful reaction *against* a key element of modern capitalism, namely its competitive ethos, which the scientific democrats thought had bled into the nation's political life with appalling consequences.

AGAINST THE CLASSICAL MODEL

The perceived link between the character of higher education and the state of political life dated back to the 1636 founding of Harvard, and beyond that to the Old World. Conceived at a time when civil and religious authority overlapped substantially, Harvard and its successor colleges of the colonial era – William and Mary (1693), Yale (1701), Pennsylvania (1740), Princeton (1746), Columbia (1754), Brown (1764), Rutgers (1766), and Dartmouth (1769) – prepared students simultaneously for political offices and careers in the traditional professions of medicine, law, and ministry. Although religious tenets framed the colonial curriculum, students did not concentrate on theology. Instead, collegiate leaders embedded essentially secular subjects – the medieval fields of rhetoric, logic, metaphysics, and mathematics; the languages and literature of classical Greece and Rome, so important to Renaissance scholars; and the newer fields of natural and moral philosophy – in a broadly Christian interpretive framework.[2]

---

[2] Christopher J. Lucas, *American Higher Education* (New York: St. Martin's Griffin, 1994), 103–121. With the exception of Dartmouth and William and Mary, each of the colonial colleges originally operated under a different name.

This "classical" model persisted through the first two-thirds of the nine-teenth century. The intense denominational struggles associated with the Second Great Awakening fueled a massive proliferation of upstart colleges on the American frontier, as graduates from Yale and other evangelical bastions fanned out into the new territories. Yale's faculty also helped the defenders of the classical curriculum stave off challenges from reformers who thought the commercial expansion of the era called for a more practical, less gentlemanly form of higher education. The Rensselaer Polytechnic Institute, launched in 1824, offered technical, vocationally relevant courses to the merchants, arti-sans, and even farmers who now staked their claims to political offices and pro-fessional roles. Among established institutions, Thomas Jefferson's University of Virginia stood alone in allowing its students to set aside the classical subjects altogether, but Union College, Columbia, and Harvard also experimented with elective courses in the natural sciences, modern languages, and political econ-omy. But the "Yale Report" of 1828 offered a potent defense of the classical subjects and helped the Christian, classical college maintain its status as the paradigmatic institution of American higher education. By the start of the Civil War, roughly 250 of these fragile enterprises dotted the landscape. Hundreds more had come and gone, or appeared only on paper. The leaders of the denom-inational colleges, like their colonial forebears, argued that classical learning fostered godliness and political virtue while laying the vocational foundation for professional service to the community.[3]

In the 1860s, however, as successive political crises appeared to convict the colleges of inadequate cultural leadership, the critics of classical educa-tion – including the first generation of scientific democrats – gained the upper hand. The assumption of a tight link between politics and higher education led American scholars to experience the Civil War and the era of political corrup-tion that followed as potent signs that they needed to rethink their own contri-butions. Deeply concerned for the nation's future, many practitioners of modern subjects such as chemistry and history would come to portray their fields not as extensions of a set of Christian tenets that could ground a democratic culture, but rather as relatively freestanding pursuits embodying the ethical practices of a true, purified faith that could better fill the needed political role.

In seeking to build up a scientific university, early reformers such as Andrew Dickson White, Charles W. Eliot, and Daniel Coit Gilman could draw on a strong reservoir of public support for a more scientific form of higher educa-tion. The vocationally minded figures who spoke for this public played a cru-cial role in pushing through Senator Justin Morrill's Land-Grant College Act

[3] *Ibid.*, 131–134; Edward Danforth Eddy Jr., *Colleges for Our Land and Time: The Land-Grant Idea in American Education* (New York: Harper & Brothers, 1956), 10; Paul H. Mattingly, "The Political Culture of American Antebellum Colleges," *History of Higher Education Annual* 17 (1997), 85–87; James Findlay, "Agency, Denominations, and the Western Colleges, 1830–1860: Some Connections Between Evangelicalism and American Higher Education," in *The American College in the Nineteenth Century*, ed. Roger L. Geiger (Nashville: Vanderbilt University Press, 2000), 115–126.

of 1862, having steadily ratcheted up their challenges to the existing system's focus on "unknown tongues, abstract problems and theories, and metaphysical figments and quibbles" in the 1840s and 1850s. These powerful figures meant something very different by "science" than did the scientific democrats. Based largely outside the colleges, they wanted science to be taught as an adjunct of manufacturing and agriculture, emphasizing specific techniques and their economically relevant applications rather than general processes and principles. Some of the vocational reformers were merchants and industrialists who sought to promote widespread prosperity (and, for many, morality as well) by teaching Americans how to apply science for economic gain. Others were journalistic and legislative champions of the industrial classes who called for a form of education pegged to their vocational needs. The scientific democrats feared that these practically minded constituencies would elevate science's technical side over its ethical side, allowing self-interest to run riot. But the status quo seemed even worse. Support from vocational reformers played an important role in the scientific democrats' victories at Cornell, Harvard, and Johns Hopkins, and the two groups also collaborated at the land-grant schools as they took shape in the late 1860s and 1870s.[4]

The scientific democrats' first major breakthrough came at Cornell, where one of their own took the helm of a powerful new institution of higher learning. Andrew Dickson White, a Yale graduate, had tried in the early 1860s to turn his personal inheritance into a university that would combat the loose morality of the business class. When his funds fell short, he sought outside help. New York's acceptance of the Morrill grant in 1864 gave White, recently elected to the state legislature, the break he needed. White convinced the self-made telegraph magnate Ezra Cornell to add a substantial contribution to the land-grant funds. Cornell, born a Quaker but forced to leave the fold when he married a Methodist in 1832, believed that technology rather than institutional religion embodied the true force of divinity in the world. He firmly supported White's plan for a science-centered institution but was far more attuned to vocational training than was White. Just as it combined land-grant and private funds, Cornell University reflected, in the persons of its twin founders, the divergent claims of the two reform contingents about the purpose and value of scientific study.[5]

A system of parallel tracks embodied Cornell's hybrid public-private status and its accommodation of both curricular philosophies of science. White's

[4] *American Higher Education*, 147–148; Eddy, *Colleges for Our Land and Time*, 15, 21–27, 36. On the land-grant colleges, also see the special issue "The Land-Grant Act and American Higher Education: Context and Consequences," *History of Higher Education Annual* 18 (1998): 5–129; and Geiger, ed., *The American College in the Nineteenth Century*.

[5] Glenn C. Altschuler, *Andrew D. White, Educator, Historian, Diplomat* (Ithaca: Cornell University Press, 1979), 42; George M. Marsden, *The Soul of the American University: From Protestant Establishment to Established Nonbelief* (New York: Oxford University Press, 1994), 114–115. For the tension between White and Cornell, see Frederick Rudolph, *Curriculum: A History of the American Undergraduate Course of Study Since 1636* (San Francisco: Jossey-Bass, 1977), 129–130.

initial report to the trustees in 1866 proposed separate faculties for each of the two prevailing forms of scientific education. A division of practical sciences and arts would offer technical material in agricultural and mechanical fields, while a second division of science, literature, and the arts would provide a broader course of cultural education. When Cornell opened with more than 400 students in 1868, it instead offered a choice of several courses of study, ranging from agriculture, engineering, and mining to exclusively liberal tracks that combined classical and scientific material. This new institution immediately set a powerful example for the far smaller land-grant schools elsewhere, which likewise struggled to find an educational identity by reconciling the practical and cultural arguments for scientific study.[6]

The university reformers gained a second major foothold the following year. When the Reverend Thomas Hill stepped down from Harvard's presidency in 1869, the trustees turned to the young MIT chemist Charles W. Eliot. Already the author of a pair of influential articles on the "new education," Eliot immediately became the national mouthpiece for university reform. He would retain that position through his forty-year tenure at Harvard. To the chagrin of White and many other scientific democrats, Eliot's vigorous speeches, writings, and leadership of the National Education Association's "Committee of Ten" on the standardization of the high school curriculum kept Harvard's model of university reform at the center of the national discussion on higher education.[7]

Despite their embrace of science, neither Cornell nor Harvard offered much institutional support for the generation of experimental knowledge during the 1870s. Johns Hopkins, which opened in 1876, pointed the way toward today's massive research enterprise. Yet its president, Daniel Coit Gilman, attended closely to science's ethical impact. Traveling through Europe with White in 1854, Gilman had written his parents, "I long for an opportunity to influence New England minds." When the railroad investor Johns Hopkins – another Quaker – endowed a new university in Baltimore, Gilman jumped at the chance. Returning east after an ill-fated stint as president of the University of California, where he was hounded by Grangers, the radical land reformer Henry George, and other advocates of vocational higher education, Gilman pledged Hopkins to "intellectual freedom in pursuit of the truth" and "the broadest charity toward those from who we differ in opinion."[8]

[6] Andrew Dickson White, *Report of the Committee on Organization* (Albany: C. Van Benthuysen, 1867), 4–5, 8; Eddy, *Colleges for Our Land and Time*.

[7] Richard Hofstadter and Walter P. Metzger, *The Development of Academic Freedom in the United States*, 360; Hugh Hawkins, *Between Harvard and America: The Educational Leadership of Charles W. Eliot* (New York: Oxford University Press, 1972). On White's dislike of Eliot, see Altschuler, *Andrew D. White*, 148.

[8] Quoted in Simon Flexner, *Daniel Coit Gilman: Creator of the American Type of University* (New York: Harcourt, Brace, 1946), 7; Lucas, *American Higher Education*, 144; Marsden, *The Soul of the American University*, 141, 150–153 (quote on 151). See also Hugh Hawkins, *Pioneer: A History of the Johns Hopkins University, 1874–1899* (Ithaca: Cornell University Press, 1960). Julie A. Reuben addresses character training in *The Making of the Modern University: Intellectual Transformation and the Marginalization of Morality* (Chicago: University of Chicago Press, 1996), 74–77, 90–92.

Although Johns Hopkins generated far more research – and researchers – than any other university of the era, Gilman defined its primary goals as character training and cultural change. He frequently complained that the phrase "promotion of research" failed to reflect Hopkins's aim, suggesting that "the discovery and advancement of truth and righteousness" or "zeal for the advancement of learning" better captured its dedication to "ideality" and "spiritualism." Like other university reformers, Gilman wanted his institution to imbue students, and by extension the wider community, with both specific knowledge and upright moral and intellectual habits. This held true even at the graduate level, where Hopkins became the first school to peg the doctorate to original research and inaugurated the long-standing American practice of relying heavily on graduate students for undergraduate teaching.[9] Hopkins put the university reform project over the top, as its highly trained, science-minded PhDs fanned out across the nation. Although few institutions copied its graduate-heavy organization, Hopkins pushed many others, most notably Harvard, toward the new model of graduate education.[10]

In the land-grant schools, meanwhile, pedagogical difficulties and weak public demand for the technical tracks combined to push the struggling new institutions toward the liberal model. Finding few qualified teachers and even fewer students for the practical courses in farming techniques and pursuits such as cabinetmaking, blacksmithing, telegraphy, and photography, the new land-grant schools usually followed Cornell's mixture of technical and liberal courses, creating an odd curricular amalgam symbolized by a Florida State professorship in "Agriculture, Horticulture, and Greek." Much of the public clearly shared the steel baron Andrew Carnegie's dim view of collegiate study's practical payoff. The land-grant universities appealed primarily to aspiring but poor students who could now get a liberal education for free. As a result, the scientific democrats found fertile ground in these institutions for their combination of useful knowledge and ethical lessons.[11]

At the same time, however, their political concerns multiplied. Everywhere they looked in those years, the early scientific democrats saw their fellow citizens utterly ignoring the dictates of morality and reason. Although they had typically supported the Northern cause in the Civil War, they were deeply troubled that it had taken such tremendous violence to erase the evil of slavery. If antebellum Americans had focused their colleges on natural science and

---

[9] Daniel Coit Gilman, "The Johns Hopkins University in its Beginning," in *University Problems in the United States* (New York: Century, 1898), 33; Gilman, "Thoughts on Universities," *Science* 8, no. 179 (July 9, 1886), 39, 41. The latter policy reflected both Gilman's commitment to interpersonal exchange and sheer necessity – he could entice only a handful of established scholars away from their positions.

[10] Francesco Cordasco, *The Shaping of American Graduate Education: Daniel Coit Gilman and the Protean Ph.D.* (Totowa, NJ: Rowman and Littlefield, 1973).

[11] John S. Brubacher and Willis Rudy, *Higher Education in Transition: An American History, 1636–1956* (New York: Harper & Brothers, 1958), 161; Eddy, *Colleges for Our Land and Time*, 68–69, 57, 59, 67, 75; Lucas, *American Higher Education*, 145, 149–151 (quote on 150–151).

political science rather than the classics, MIT's William P. Atkinson insisted, it would not have taken "a bloody war to open the eyes of the nation to its plainest duties." The scientific democrats applied the same moral lens to the politics of the early Gilded Age. Southern Reconstruction struck them as having brought out rapacious instincts on all sides. They saw the same dynamic in the emerging industrial economy of the North, where both workers and business owners increasingly looked to government to promote their interests. All these views seemed to indicate a systematic failure of American leadership, and thus to prove that the classical colleges threatened the nation and Christianity itself. Facing "no enemy except its own inherent weakness," explained Michigan's Charles Kendall Adams, the denominational college had failed to produce virtuous leaders.[12]

ANTECEDENTS

How would science help in this regard? In addition to political troubles, the period since the Yale Report had also seen cultural and intellectual shifts that paved the way for the emergence of scientific democracy. The spread of theological liberalism, evangelical humanitarianism, and faith in the moral impact of technological development had provided the cultural materials for an image of science that harmonized it with Christianity in a manner quite distinct from the "natural theology" of the antebellum years. Natural theology placed the "book of nature" – the natural world, accessible through sensory evidence – alongside the Bible (the "book of Scripture") as a second source of evidence of God. "If God is," explained one Harvard professor, "he must have put his signature on his whole creation," not merely on the Scriptures. This move underlay the widely used "argument from design" for God's existence: that natural structures were so complex that they could not have arisen except through conscious design by an omnicompetent power. Within the framework of natural theology, then, science supported Christianity by providing empirical proof that the natural world revealed the hand of an intelligent creator.[13] By contrast, the first generation of scientific democrats in the universities identified ethics, rather than God's handiwork in nature, as the point of contact between science and Christianity. They portrayed science as the highest expression of a mode of interpersonal behavior prescribed by God.

[12] Quoted in Thomas L. Haskell, *The Emergence of Professional Social Science: The American Social Science Association and the Nineteenth-Century Crisis of Authority* (Urbana: University of Illinois Press, 1977), 111; John G. Sproat, *"The Best Men": Liberal Reformers in the Gilded Age* (New York: Oxford University Press, 1968), esp. 72–75, 303; Nancy Cohen, *The Reconstruction of American Liberalism, 1865–1914* (Chapel Hill: University of North Carolina Press, 2002), 89–93; Charles Kendall Adams, "Ought the State to Provide for Higher Education?" *New Englander and Yale Review* 37, no. 3 (1878), 366. On the distinctive moral contributions of science to politics, see Andrew Dickson White, "Science and Public Affairs," *Popular Science Monthly* 2, no. 12 (1873), 737–738.
[13] Quoted in Reuben, *The Making of the Modern University*, 21.

This vision of science as an ethical practice emerged at the intersection of three preexisting impulses. The early university reformers, like antebellum theological liberals, viewed cognitive change (originally, conversion) as dependent on social trust and physical and emotional well-being. Meanwhile, they joined the Protestant activists associated with the Whig Party in viewing private enterprise and technological innovation as potent instruments of moral progress. Finally, these scientific democrats agreed with the moral philosophers, whose teachings anchored the antebellum collegiate curriculum, that morality needed to be justified empirically rather than biblically. Under the pressure of alarming political events, reformers combined these tenets into a new understanding of science and concluded from it that only a thoroughly scientific university could fulfill the cultural-political functions hitherto assigned to the classical college. They portrayed science as a purified expression of Protestant insights – a practice perfectly suited to producing genuine intellectual conviction.[14]

The reformers' understanding of how individuals changed their minds reflected new Protestant views of the conversion process. After 1800, American Protestants recoiled sharply from their flirtations with rational Christianity and deism, embracing a religion of the heart. The powerful revivals of the Second Great Awakening infused American public culture with both spiritual energies and new forms of personal discipline. The popular religiosity of the era often moved in the realm of unfettered emotionality, but many Protestant leaders instead presented ethical virtues as the proper alternative to rational principles. By mid-century, in fact, several theologians had come close to redefining Christianity as a system of personal ethics. Henry Ward Beecher insisted that Christianity referred to "the state of man's soul" and found its issue in "disposition and conduct." Christianity, he insisted, should never be confused with "the Church, the Bible, [or] the Creed." Horace Bushnell downplayed the doctrinal side of Christianity even further. He brought down heresy charges in 1849 by arguing, on the basis of a sophisticated philosophy of language, that the teachings of Scripture should be read as symbolic rather than literal truths. Although few took these liberal tenets as far as Beecher or Bushnell,

[14] Jeffrey P. Sklansky notes the use of conversion as a model for psychologists: *The Soul's Economy: Market Society and Selfhood in American Thought, 1820–1920* (Chapel Hill: University of North Carolina Press, 2002), 163, 169–170. On the continuities between Whig thought and its modern successors, see Louise L. Stevenson, *Scholarly Means to Evangelical Ends: The New Haven Scholars and the Transformation of Higher Learning in America, 1830–1890* (Baltimore: Johns Hopkins University Press, 1986), esp. 5–6. Alternative interpretations of the links between mid-nineteenth-century Protestantism and subsequent scientific thought include James Turner, *Without God, Without Creed: The Origins of Unbelief in America* (Baltimore: Johns Hopkins University Press, 1985); Bruce Kuklick, *Churchmen and Philosophers: From Jonathan Edwards to John Dewey* (New Haven: Yale University Press, 1985); Marsden, *The Soul of the American University*; Reuben, *The Making of the Modern University*; D. G. Hart, "The Protestant Enlightenment Revisited: Daniel Coit Gilman and the Academic Reforms of the Modern American University," *Journal of Ecclesiastical History* 47, no. 4 (1996): 683–703; and Jon H. Roberts and James Turner, *The Sacred and the Secular University* (Princeton: Princeton University Press, 2000).

they enjoyed a wide following nonetheless. Theological liberals relegated doctrinal fidelity to the margins of their thought and viewed personal character and commitment as the touchstone of Christian faith.[15]

The new emphasis on ethics often accompanied a belief that the whole-hearted cognitive commitment characteristic of full conversion depended on emotional, and thus material, preconditions. "It is the heart which governs the intellect," Henry Ward Beecher's father Lyman insisted. Other Protestant liberals likewise argued that rational conviction could have little purchase if the "lower" elements of human nature provoked resistance. Only an emotionally receptive listener would heed rational appeals. But how were they to increase the American public's emotional receptivity to divine truth? Revivals offered one avenue. For many Protestant liberals, however, humanitarian aid seemed an even more powerful tool for spreading the gospel, as well as a moral imperative in its own right. One could build trust, and therefore receptivity to the Christian message, by treating those in need kindly and offering them material comfort and security. Many liberals argued that enjoying substantial financial resources without using them to promote Christian aims represented a blatant violation of God's law.[16]

On the other hand, the focus on humanitarian aid suggested that possessing wealth was perfectly compatible with Christian faith, provided one used it appropriately. For socially prominent liberal Protestants associated with the Whig Party, the humanitarian project neatly harmonized industrial enterprise, and the resulting profits, with Christian ethics. These figures defined the market economy itself as the most important and effective field of humanitarian endeavor, given its apparent promise of widespread prosperity. Recasting the principles of classical political economy as guidelines for the implementation of Christian benevolence, they portrayed market competition, checked by personal virtue, as an expression of mutual service rather than self-interest.

Technological innovation, which made new goods and services widely available by lowering their prices, played a key role in this Whig-Protestant mindset. On this view, applying technology to factory processes to reduce production costs and create new consumer goods exemplified Christian benevolence in action. *Contra* Thomas Malthus, Whig-Protestant thinkers believed that industrial progress would ensure an ever-expanding economic pie for competitors to

[15] Quoted in Reuben, *The Making of the Modern University*, 57; E. Brooks Holifield, *Theology in America: Christian Thought from the Age of the Puritans to the Civil War* (New Haven: Yale University Press, 2003), 460.

[16] Quoted in Turner, *Without God, Without Creed*, 105. On evangelical activism in general, see Daniel Walker Howe, "The Evangelical Movement and Political Culture in the North during the Second Party System," *Journal of American History* 77, no. 4 (March 1991): 1216–1239; Howe, *The Political Culture of the American Whigs* (Chicago: University of Chicago Press, 1979); Ronald G. Walters, *American Reformers: 1815–1860* (New York: Hill and Wang, 1978); and Richard J. Carwardine, *Evangelicals and Politics in Antebellum America* (New Haven: Yale University Press, 1993).

divide and thus prevent the development of antagonistic classes. By fostering individual comfort and social harmony, industrialization would prepare the ground for the universal acceptance of the divine word.[17]

At the same time, however, the spread of market relations also presented an unprecedented threat to the Christian project. Liberal Protestants worried about selfishness, the dark shadow of Christian benevolence. In the murky realm of personal motives, only the thinnest of margins separated healthy competition from rampant self-seeking. Markets offered innumerable temptations to cross the line. Still, Whig-Protestant activists believed that they could teach Americans to recognize and police the boundary between selflessness and hedonism by infusing public culture with Christian virtues. By encasing the engine of material progress – the competitive industrial economy – in a culture of benevolence, they held, a Christian civilization could keep itself on the path of virtue. Antebellum liberal Protestants assumed that cultural institutions, including churches, schools, and colleges, could check the corrosive ethical effects of market competition and enable Americans to harness industrial power to moral ends. The citizens of a Christian republic would retain their precarious position atop the industrial dynamo by holding one another to high moral standards in the civil sphere.[18]

A final building block of the new image of science was a mode of Christian apologetics resting on empirical and rational evidence, rather than Scriptural texts. This approach characterized the moral philosophy course that capped off the classical curriculum in the antebellum colleges. Taught in the senior year, frequently by the president himself, moral philosophy embodied the goal of conserving and promoting the moral standards of Christian civilization. Yet its proponents, leaning heavily on natural theology, sought to appeal to a modern, skeptical audience by mobilizing scientific data wherever possible.[19]

In addition to offering the powerful argument from design, which was also stressed in many antebellum science courses, moral philosophers such as Harvard's Francis Bowen taught that empirical study revealed a system of moral laws, or "moral government," embedded in the universe. Bowen called his subject "a general science of Human Nature," comprising "the special sciences of Ethics, Psychology, Aesthetics, Politics, and Political Economy." Such courses derived specific rules of behavior from an ethic of obligation and service, describing moral philosophy as a "science of duty" that steeled students for constant vigilance against the "stain of selfishness."

[17] Howe, *The Political Culture of The American Whigs*, esp. 96–122; Donald H. Meyer, *The Instructed Conscience: The Shaping of the American National Ethic* (Philadelphia: University of Pennsylvania Press, 1972), 99–107.

[18] Meyer, *The Instructed Conscience*, 77–86.

[19] Holifield, *Theology in America*, esp. 173–196; Reuben, *The Making of the Modern University*, 17–23; Allen C. Guelzo, "'The Science of Duty': Moral Philosophy and the Epistemology of Science in Nineteenth-Century America," in *Evangelicals and Science in Historical Perspective*, ed. David N. Livingstone, D. G. Hart, and Mark A. Noll (New York: Oxford University Press, 1999); and Marsden, *The Soul of the American University*, 48–64, 90–93.

Moral philosophers sought to equip each graduate with an internal gyroscope, to use David Riesman's analogy, and thereby to create a well-ordered society in the absence of either a formal religious establishment or a highly centralized state.[20] It was only a small step for the scientific democrats to eliminate biblical evidence entirely from this process, justifying both the moral tenets of Christianity and the teachings of political economy on purely empirical grounds.

Synthesizing the elements of ethical Christianity and Whig technological enthusiasm, the first scientific democrats often moved imperceptibly away from the moral philosophers' correspondence theory of truth – the view that scientific theories amounted to more or less faithful pictures of a reality "out there," rather than merely useful devices for navigating a reality whose essential character could not be known. Systematic, self-conscious alternatives to the correspondence theory would take shape only later, beginning with Charles S. Peirce's formulation of the pragmatic definition of truth as that which works. Long before Peirce's pioneering articles of the late 1870s, however, leading scientific democrats had concluded from their liberal Protestant heritage that the essence of truth lay in its consequences. A purely ethical interpretation of Christianity defined doctrinal tenets in terms of their contributions to action, which suggested that beliefs should be judged in terms of the results attained by acting on them. This orientation elevated character or behavior over belief. It also suggested that believing was itself a form of behavior, valuable in terms of its contribution to the Christian project rather than its conformity to any set of a priori principles. For those willing to follow the lead, ethical Christianity implied that the truth could be known only by its results.[21]

A consequentialist reading of ethical Christianity offered new justifications for patient, detailed research in both the natural and human sciences. If one sought ultimately to create a Christian society, it became more important to determine which acts – including acts of belief – actually furthered that end than to stick fast to a particular interpretation of Scripture. The early scientific democrats reinterpreted natural science itself as a process of using lessons learned from past experience to guide future behavior. Describing science in this manner also directed new attention to the scientific analysis of human experience – the empirical study of human acts and their real-world consequences. The early scientific democrats extrapolated broadly from the success of the physical sciences, concluding that setting aside the Bible and systematically attending to experiential evidence would always prove the best way to discern God's will. The book of history, they reasoned, rather than the book of Scripture or the book of nature, offered the most reliable access to God's ethical teachings. They sought to extend science's empirical approach – which

---

[20] Quoted in Reuben, *The Making of the Modern University*, 20; Guelzo, "The Science of Duty"; quoted in Meyer, *The Instructed Conscience*, 80.
[21] Cf. Turner, *Without God, Without Creed*, esp. 227–246.

produced a kind of natural ethics, a set of guidelines for proper behavior in the natural world – into the field of social ethics.[22]

Meanwhile, the Whig-Protestant portrait of technological industry as a mode of humanitarian aid provided an additional justification for work in the natural sciences. To the extent that scientific research facilitated the spread of technological industry, this view held, it performed an essentially Christian task, if only an indirect one. By undertaking investigations that spun off new technologies, the scientist served God and fellow man alike. Focusing on natural science's contribution to material welfare freed it from its close identification with the argument from design. Like ethical Christianity, an expansive view of industrialization's moral impact harmonized with an image of science as generating behavioral lessons – prescriptions regarding what human beings should do – rather than revealing a divinely created order. The Whigs' sacralization of economic progress provided a distinctively Christian motive for redefining natural science in ethical or technological terms, rather than purely cognitive ones.

THE SCIENTIFIC SPIRIT

The early scientific democrats brought together these latent implications of existing modes of thought in the 1860s, when a series of political crises led them and other educational reformers to challenge the entire system of denominational colleges. Steeped in the tenets of liberal Protestantism, technological enthusiasm, and moral philosophy, leading scientific democrats concluded from the catastrophe of the Civil War and the corruption of the Gilded Age that the denominational colleges, and behind them institutional Christianity, had failed to create a polity immune to selfishness. But they believed that a more effective source of cultural restraint stood at hand, in science. Or, more accurately, they thought the scientific enterprise reflected the true spirit of Christianity at work – the spiritual fount to which the churches, like all human institutions, needed to return in order to moralize an industrial society. Institutional and intellectual change went hand in hand as the early scientific democrats worked to build a new form of higher education that comported with their emerging understanding of science.

Looking upon the apparent failure of the colleges to maintain the political rectitude of the public and its leaders, these reformers identified three shortcomings in particular: first, the relatively small role played by science and other nonclassical subjects in the curriculum; second, the adoption of strict doctrinal tests – actually honored more in the breach than in practice – in faculty hiring decisions; and third, the underlying identification of Christianity with correct belief rather than benevolent intention.

---

[22] This entailed a subtle move from the free interpretation of the Bible, a foundational Protestant principle, to the free interpretation of God's lessons wherever they could be found.

Scientific democrats sharply contrasted the "scientific spirit" of their ideal university to what the popular evolutionist John Fiske called the "persecuting spirit" of "denominational" or "traditional" Christianity. According to the reformers, the existing churches and colleges rested on a belief that failure to adhere to a specific religious body's doctrinal creed would bring eternal damnation. This, in turn, reflected an assumption of infallibility for the creed in question. Reformers deemed this tenet deeply selfish and directly opposed to God's system of moral government. No denomination could possibly know ahead of time that its creed offered an exclusive road to salvation. The proper view, which Fiske labeled "the relative truth of opinions," held that because every individual has part of God's truth, each needed to learn from the others, aided by empirical evidence. In this framework, God did not judge every belief held by individuals. Instead, God judged how individuals acted, including how – not what – they believed. God demanded of humanity a type of character, a set of behavioral virtues, rather than adherence to a list of intellectual tenets. University reformers identified science with a form of Christianity that, as Cornell president Andrew Dickson White put it, "finds its realization in righteous men and not in gush."[23]

When these early scientific democrats spoke of the "scientific method," they referred as much to a particular style of free and open dialogue as to a set of empirical techniques. They portrayed scientific communication as a form of humanitarian aid, carried out in the spirit of benevolence and involving no imposition of coercive authority. The scientist, rather than invoking the specter of damnation or demanding obedience, simply offered evidence to suggest that the hearer would benefit from believing or acting in a particular way. Here, as elsewhere, the university reformers invoked the core Protestant principles of free interpretation and voluntary conviction. According to them, the scientist's humanitarian approach to communication outdid the alternatives both ethically and practically, because it acknowledged that the listeners were free beings in control of their own destinies.

This ideal of scientific communication underwrote the entire university reform project and provided the primary axis on which reformers distinguished the scientific university from the denominational college. They insisted that the denominational colleges failed to achieve their cultural ends because doctrinal religion violated the deepest modern insights about the nature of conviction. Most importantly, it played to the lower motives, especially fear. Traditional religion coerced intellectual compliance by threatening public ridicule and expulsion in the short term and eternal punishment in the long term.

---

[23] John Fiske, "The Philosophy of Persecution," *North American Review* 132, no. 290 (1881), 1, 13; quoted in Altschuler, *Andrew D. White*, 177. E. L. Youmans drew a similar contrast between methods of attaining truth in "The Science of Biblical Criticism," *Popular Science Monthly* 19, no. 3 (1882), 408. For alternative accounts of the reformers' critique of traditional Christianity, see Marsden, *The Soul of the American University*, esp. 113–121; and Reuben, *The Making of the Modern University*, esp. 4, 29–30, 56–58, 95–101.

Critics charged that these base appeals were unable to produce real conversion, leading to reluctant obedience at best and hostile resistance at worst. "Not what claims our respect, but what gains it, is our true faith, and the basis of our religion," pointed out the philosopher Chauncey Wright. Against the doctrinal strategy, they argued that God's central injunction to benevolence extended to communicative behavior. The true Christian guided others to the truth by allowing them full latitude of interpretation, eschewing coercive appeals and sharing interpretive resources freely.[24]

In the contemporary era, university reformers insisted, doctrinal strife had caused the core truths of Christianity to fall out of favor. They reasoned that Americans, repelled by the evangelicals' communicative style, had refused to hear the evidence offered on behalf of their ethical teachings. The problem was not the churches' emphasis on persuasion; university reformers, too, sought to change the minds of the American population. They inherited a vision of civil society as the realm, not of voluntary association *per se*, but rather of voluntary association for the purpose of moral suasion in the name of Christianity. For the early scientific democrats, as for their evangelical forebears, persuasive appeals provided the basic currency of civic action.[25] But their version of Christianity led them to emphasize intellectual freedom at all costs. Science, they believed, offered the only truly Christian mode of persuasion, the only one that jealously guarded the listener's integrity.

To guide every form of interpersonal relations, the scientific democrats espoused an ethic of communication: a set of normative guidelines dictating how individuals should interact discursively. This ethic can be difficult to discern, because the scientific democrats framed their communicative norms as personal virtues rather than formal tenets. Yet enumerations of the scientist's discursive virtues recur constantly in the literature of the period, especially in laudatory biographical sketches and in disputes over the public presentation of controversial theories. Taken together, they amounted to a fairly coherent set of ground rules for proper argumentation, violations of which incensed and dismayed the scientific thinkers of the era.

Humility topped the list of scientific virtues. One biographer paid the physicist Joseph Henry the highest possible compliment by calling him "as modest as he was pure." But humility did not lead the scientist to shy away from critical engagement. Instead, an editorial explained, science "must covet the scrutiny of every eye, and must be generous ever in the acknowledgement of its shortcomings." Remaining always "catholic and liberal toward others," the scientist refused to "magnify differences" or "impute motives." The Johns Hopkins physicist Henry A. Rowland warned against two opposite extremes: passively conforming to the prevailing opinion or developing one's ideas in total isolation. The truly responsible individual, wrote Rowland, possessed both "full respect for the opinions of those around him" and "such discrimination that

[24] Quoted in Turner, *Without God, Without Creed*, 221.
[25] Walters outlines this understanding of civil society in *American Reformers*, 32–34.

he sees a chance of error in all, and most of all in himself." When faced with criticism, the scientific communicator listened respectfully and then offered an equally respectful response. The philosopher George Sylvester Morris, who gave John Dewey his first university post at the University of Michigan, explained that those holding a questionable view deserved, not "a vigorous *assertion* that they are in error," but rather "a dispassionate and objective demonstration that they are so." Daniel Coit Gilman likewise ascribed to the scientist "that sort of courage which neither dreads nor provokes controversy," a trait that Michigan president James Burrill Angell called "essentially democratic in the best sense of the term." The dialectic of interpretive humility and discursive engagement anchored reformers' reading of scientific communication.[26]

Indeed, reformers portrayed scientific observation itself as a form of communication, a discursive transaction with God via experience. Many described it as a form of reverential listening, rather than employing the metaphors of clear sight that we habitually associate with science. A Zen-like adage of the time captured this communicative reading of the empirical method: "Ask the grasshopper how many legs he has. Don't go to the book. The book may lie. Grasshopper will tell you the truth." According to the university founders, nature's array of guileless voices offered the clearest expression of God's word, whereas the multiple layers of human interpretation weighing down religious and classical texts rendered them useless as guides to action.[27]

But the existing colleges, reformers charged, taught students both by example and by prescription to obey authority rather than judge freely. The colleges' heavy reliance on textbooks topped the list of pedagogical complaints; reformers deemed book-learning the primary obstacle to ethical training. At that time, colleges relied heavily on the "recitation method," wherein students repeated written texts from memory. According to reformers, this method allowed students to succeed without ever internalizing the material. Worse yet, it squelched student interest in that material by demanding deference to teachers and texts, rather than granting free interpretive rein. Lecturing from textbooks produced similar results, they believed. A student "shot at" by "the ablest lecturers for months," declared Wisconsin's P. A. Chadbourne, learned far less than one entering into "real intellectual conflict with a great man." John Fiske averred that even science, when taught from a book, fostered "the pernicious habit of accepting statements upon authority." Fiske and many other reformers argued

---

[26] A. M. Mayer, "Joseph Henry," *Science* 1, no. 11 (September 11, 1880), 130; [Samuel H. Scudder], "The Future of American Science," *Science* n.s. 1, no. 1 (February 9, 1883), 3; Henry A. Rowland, "The Physical Laboratory in Modern Education," *Science* n.s. 7, no. 177 (June 25, 1886), 573; George S. Morris, "Correspondence," *Science* 3, no. 80 (January 14, 1882), 7; quoted in Reuben, *The Making of the Modern University*, 91; James B. Angell, "The Higher Education: A Plea for Making It Accessible to All," in *Selected Addresses* (New York: Longmans, Green, 1912), 60. See also the endorsements of intellectual diversity by G. Stanley Hall, in Hawkins, *Pioneer*, 109; and Henry Adams, in Seymour Martin Lipset and David Riesman, *Education and Politics at Harvard* (New York: McGraw-Hill, 1975), 95.

[27] Quoted in Eddy, *Colleges for Our Land and Time*, 75–76.

that the denominational colleges hampered the spread of Christian ethics by employing recitations and lectures rather than experiential, interactive methods. By contrast, the books of nature and history lay open to all. Reading them required no training in any textual tradition, whether biblical or classical, and could not fail to provide ethical lessons.[28]

Science, too, had its texts. But the reformers believed that their communicative ideals could be implemented in writing as well. They identified the authorial voice as the centerpiece of a writer's orientation toward readers. William Jay Youmans, son of the famed journalist E. L. Youmans, compared a text that would "ask a man or a child to assume what can not be proved" to another "which lays no claim to any kind of privilege, and which, therefore, can not force the belief of any one." According to the reformers, any claim whatsoever could be presented as a matter for discussion rather than worship or deference. Reformers lamented that scientists' critics routinely failed to note the tentative tone in which they presented their findings. In this view, writers could be just as scientific as speakers, by humbly presenting their ideas for consideration rather than expecting or demanding obedience.[29]

Like the moral philosophers before them, the early scientific democrats held that scientific communication offered the only way to reach the public, because modern Americans would listen only to empirical arguments. In the true spirit of Protestantism, the argument ran, Americans had become deeply suspicious of human claims to authority. University of Wisconsin president John Bascom wrote that, although truth remained "just as capable of eliciting enthusiasm among young men to-day as ever before," it could only be "simple" truth, "offered in its own light."[30] Bascom and like-minded figures found empirical evidence that Americans had adopted a proto-scientific attitude in such contradictory phenomena as the flowering of evangelism and the flouting of Christian ethical teachings. They claimed that centuries of physical science and decades of democratic progress had prepared the public to respond solely to non-coercive appeals backed by empirical evidence. But the colleges lagged behind.

The university founders hoped to change the churches as well as higher education. They sometimes spoke directly to religious leaders, urging the latter to embrace a scientific style of communication and thus to ensure a continuing public role for the churches. The public had adopted a "new standard of intellectual sincerity," Eliot wrote, having learned "that discussion often elicits

[28] P. A. Chadbourne, "Colleges and College Education," *Putnam's Monthly* 14, no. 21 (1869), 341; John Fiske, "Essays on a Liberal Education," *North American Review* 107, no. 220 (1868), 128. In reality, the heavy teaching loads necessitated by endemic financial difficulties largely determined the pedagogical priorities of the denominational colleges. Reformers blamed curricular theory instead.

[29] [W. J. Youmans,] "A Minority But Not a Sect," *Popular Science Monthly* 36, no. 1 (1889), 122; "Address of Professor Asaph Hall," *Science* 1, no. 11 (September 11, 1880), 127.

[30] John Bascom, "Atheism in Colleges," *North American Review* 132, no. 290 (1881), 39. Cf. White's emphasis on the need for "facts cogently presented," quoted in Altschuler, *Andrew D. White*, 74.

truth, that controversy is useful on many difficult subjects, and that in some circumstances many heads are better than one." Henceforth, the minister would shape public culture only through "the purity and strength of his character," "the vigor of his intelligence and the depth of his learning," and "the power of his speech." Eliot traced the new scientific outlook to democracy. He explained that the lesson that self-government could produce a high standard of living had undermined both the doctrine of human depravity and "monarchical and military" conceptions of God. Now that democracy had stripped away all doctrinal encrustations from the pure ethical core of true Christianity, Eliot continued, ministers could retain a leading place in society only by adopting a scientific mode of persuasion.[31]

Reformers demanded a complete break from the denominational model in both the colleges and the churches. They believed that the current system could not be changed from within, because its institutional structures reinforced its intellectual flaws. Denominational institutions prevented their employees from adopting new views by linking job security to their obedience to rigid creeds. Eliot wrote that ministers lacked "room enough to grow" in the churches; they were constantly threatened with punishment or even banishment for changing their minds on matters entirely irrelevant to "Christian character or right living."[32] Professors could not serve as effective leaders of public opinion either, as they too were constrained by their denominations. The colleges stamped out mental independence by demanding deference and intellectual rigidity from the faculty, who then modeled these character defects for their students. A truly Christian education, by contrast, would allow professors and students alike to follow God's truth wherever it led, even if that meant abandoning doctrinal orthodoxy altogether.

The political impetus of the reformers' critique was clear. Their ideal scientist – at once cooperative and independent, engaged and humble – provided a perfect model for citizens charged with reconciling their differences peacefully. According to the university reformers, Americans stood on the cusp of a new phase of human history, in which they would complete the scientific transformation of society by extending the scientist's ethical orientation to the entire population. William Jay Youmans explained the dual character of this process. He admitted that science "tends to produce social ferment by continually introducing new ideas and continually unsettling commercial arrangements." On the other hand, Youmans averred, if science's ethical lessons were widely disseminated, it would "do vastly more" to "knit, than it possibly can in any

---

[31] Charles W. Eliot, "On the Education of Ministers," in *Educational Reform: Essays and Addresses* (New York: Century, 1898), 66–70. Cf. F. W. Clarke, "Scientific Dabblers," *Popular Science Monthly* 1, no. 5 (1872), 598–599; and Reuben, *The Making of the Modern University*, 33–34, 76–83. "The standard biography of Eliot is Hawkins, *Between Harvard and America*.

[32] Eliot, "On the Education of Ministers," 85. It should be noted that the dominance of the classical curriculum also exacted financial penalties against those with a scientific bent. With few opportunities for professional employment, choosing a career in science over the ministry or a classical field usually brought a lifetime of material disadvantage.

other way to loosen, the bonds of society." Beginning with the transformation of American higher education, the early scientific democrats argued, the next stage of human history would culminate in the transformation of American politics, wherein the public, having adopted the scientific spirit itself, demanded that spirit of its political leaders.[33]

Yet, the elder Youmans warned, the entire experiment would fail if either science's critics or its less perceptive supporters managed to "keep science in its old physical grooves." Elaborating scientific techniques without expanding the circle of scientific communication to include all topics and all individuals would surely doom the democratic experiment. The Youmanses and other reformers discerned the germ of a Christian polity in the emerging republic of science and hoped to extend its egalitarian, non-coercive form of discourse into all areas of inquiry and of public life, producing an ever-wider sphere of social agreement.[34]

Strategically, however, the task of the moment was to build a system of scientific universities, both by carving out space for scientific courses in the undergraduate curriculum and by eliminating all traces of coercion from teacher-student relations. Reformers, both inside and outside the colleges, feared that the formidable communicative barriers in American higher education threatened to snuff out the scientific republic in its infancy. Their humanitarian analysis of the conversion process suggested that the ethical orientation they hoped to transmit to students would spread through personal inspiration and act as a contagious force. Eliot argued that students undertook ethical growth only through sustained personal contact with deeply Christian individuals, by which he meant scientists. White likewise called for extensive student-teacher interaction in his planning report for Cornell, writing that "hearty, manly sympathy" between "a young man and a man of thought, learning, character and experience, is worth more than all educational programmes and machinery." If White had gotten his way at Cornell, he would have docked the pay of professors who failed to make themselves available to students and fired those who argued publicly with one another, given the disastrous impact of such friction on popular perceptions of science. "Better to have science taught less brilliantly, than to have it rendered contemptible," he declared in his report. University reformers also rejected the denominational colleges' strict control of student behavior. They believed, with White, that student discipline could be sustained through "pleasant, extra-official intercourse," in which professors would relate to students as "a body of friends," not as "police." Armed through such means with "good general culture," reformers believed, students would go out and breathe the scientific spirit into other citizens, eventually bringing the entire polity into the circle of scientific communication.[35]

[33] [W. J. Youmans,] "Science and Civilization," *Popular Science Monthly* 38, no. 5 (1891), 704.
[34] [E. L. Youmans,] "Science and Social Reform," *Popular Science Monthly* 6, no. 4 (1875), 504.
[35] Charles W. Eliot, "Inaugural Address as President of Harvard College," in *Educational Reform*, 17; White, *Report of the Committee on Organization*, 21–22, 37, 20.

Thus, the reformers' charge that denominational colleges obstructed "academic freedom" or "intellectual freedom" represented much more than a call for basic research in the natural sciences. Deeming existing modes of communication unscientific, unproductive, and immoral, they sought to break down the communicative barriers between individuals that they thought plagued denominational higher education. "We should be jealous of the integrity of the mental processes of the student," noted Wisconsin's Bascom, "for we wish him to show this integrity in the actual encounters of the world." An assumption that science embodied and fostered open communication, and could thereby provide a reliable solvent for political conflicts, shaped the reformers' pedagogical and curricular initiatives at every turn.[36]

DEFENDING A MODERN CURRICULUM

A scientific university could hardly hope to take over key cultural functions from the denominational colleges and Protestant churches if it could not attract students and stave off public criticism. The early scientific democrats crafted a complex and potentially self-contradictory set of arguments for the new university's legitimacy and value. Overall, they contended, a remade curriculum would solve the nation's political problems by offering a balance of material rewards and ethical training. Science's vocational applications would draw students into the university, where they would imbibe science's ethical lessons. By spinning off new technologies, finally, the university would also drive industrial progress, even as it moralized the individuals engaged in it. In sum, a scientific university would represent the greatest imaginable expression of humanitarianism – the perfect device for spreading ethical Christianity.

It is crucial to recognize that very few scientific democrats sought a purely scientific curriculum in the modern sense of that word. They did not intend to restrict the universities to the natural sciences. Both the British discourse of science and the German concept of *Wissenschaft*, on which Americans drew so heavily in building up their own disciplines in the late nineteenth century, included various forms of social analysis and even pursuits such as geography and philology – indeed, any field in which causal relations could be isolated and explored without the aid of supernatural explanations. Scientific democrats sometimes applied labels such as "ethical studies" or "ethical sciences" to these fields, previously subsumed under the heading of moral philosophy. But their curricular proposals went still further. The desired "modern subjects" included not only the natural sciences, history, and political economy, but also the English, French, and German languages and literatures. Reformers portrayed this set of fields as the modern equivalent to the classical curriculum, arguing that it would give students the general knowledge and the ethical training needed for any of life's pursuits.[37]

[36] Bascom, "Atheism in Colleges," 40.
[37] Eliot defined this course of study most systematically: "What Is a Liberal Education?" in *Educational Reform*, 89–122. See also Shirley W. Smith, *James Burrill Angell: An American*

Substantial disagreement over institutional strategies persisted. Most of the scientific democrats would give professors greater freedom to choose their course offerings. Many would give students some degree of choice regarding their courses, whether this was a choice between tracks or the fully decentralized "elective system" that became associated with Eliot's administration at Harvard. Others feared, with Youmans, that students would choose the "traditional and fashionable studies" over the more valuable scientific fields, or, more commonly, that they would specialize exclusively in the natural sciences, ignoring the university's wider moral and political lessons.[38] Most reformers would give professors a chance to engage in original research. Some would establish graduate programs. All, however, agreed on the need to bring students into direct contact with the book of history, especially the most recent chapters. At each of the new universities, whether public or private, scientific democrats sought to introduce students to the ethical and cognitive resources of modern civilization.

Unlike their vocationally minded counterparts, scientific democrats in the new universities did not need to build a new set of teaching fields from scratch. Science had always fit neatly, if peripherally, into the existing colleges, which steadily added individual courses in botany, chemistry, zoology, and other fields through the antebellum period. In theory, these courses tied science to natural theology; professors were officially expected to emphasize the argument from design. Vocational reformers and scientific democrats alike deplored the assumption that science performed its educational task by revealing the divine order rather than providing technical resources or ethical lessons. In practice, however, the argument from design figured only vestigially in collegiate science courses by the 1860s, and it took little effort to incorporate these courses into a modern curriculum.[39]

In the other modern subjects, scientific democrats recast existing courses more thoroughly by bringing the scientific spirit to bear on traditional fields. Literary study offers a good example. Whereas classicists often deemed the ancient civilizations morally superior to their modern counterparts, Eliot turned the claim around, arguing that "Greek literature compares with English" as did "infantile with adult civilization." Meanwhile, Eliot pointed out, one could not "get at the experience of the world upon any modern industrial, social, or financial question" without facility in French and German.[40] He argued that studying modern languages and literature would expose students to the best ethical thought and facilitate the communicative exchange that defined science. Most reformers gave more credit to the ancients than did Eliot, yet echoed his

Influence (Ann Arbor: University of Michigan Press, 1954), 95; and Reuben, *The Making of the Modern University*, 212–213.

[38] [E. L. Youmans,] "Sir Josiah Mason's Science College," *Popular Science Monthly* 18, no. 2 (1880), 266.

[39] Lucas, *American Higher Education*, 134; Stanley M. Guralnick, *Science and the Ante-Bellum American College* (Philadelphia: American Philosophical Society, 1975), ix.

[40] Eliot, "What Is a Liberal Education?" 98, 103.

unswerving commitment to the superiority of the present age.[41] Reformers' belief in the moral worth of modern civilization fueled their embrace of modern literatures and languages.

In history, too, university leaders built on existing foundations, expanding and updating the old moral philosophy and classics courses. Many prominent reformers, including White and his student Charles Kendall Adams, worked in the emerging field of modern history.[42] A few of them wanted students to focus exclusively on modern nations, but most recast classical study as one element in a kind of comparative anthropology of ancient and contemporary institutions. John Fiske explained that studying the "life, manners, moral ideas, and superstitions" of ancient societies alongside their modern equivalents would make students more "cosmopolitan and hospitable." Ancient history, he continued, when presented in a comparative framework, offered invaluable "lessons in political conduct" by showing "the people of bygone times as men like ourselves, alike yet different, actuated by like passions, but guided by different opinions and different conceptions." The editor of *Science* agreed, writing that science students, even more than their peers, needed a "sense of perspective in the intellectual history of mankind," a feeling for "the oneness of human history."[43]

Political economy represented the heart of the new curriculum, however. In fact, these reformers believed that courses in modern history served primarily to illustrate the general principles of political economy, which had also figured prominently in the antebellum course of study, both as an independent field and as the backbone of moral philosophy. Dubbed the "science of wealth" by Eliot, political economy amounted to a set of historically derived imperatives relating to "social theories and the moral effects of economic conditions."[44] Like the moral philosophers, the early scientific democrats thought that the lessons of modern history favored the Whig-Protestant combination of *laissez-faire* governance, Christian charity, and technological industry. Many made an exception to the *laissez-faire* principle in the case of public education.[45] But the early scientific democrats felt strongly that the state should stay out of economic affairs, and that the student could learn no more valuable lesson from collegiate study.

The political preoccupations of these science-minded reformers can be seen most clearly in their repeated attempts to establish graduate programs for future public leaders, the curricula of which would center on political economy. As early as 1862, White imagined an advanced course of study for future civil servants, teaching "moral Philosophy, History and Political

[41] E.g., William P. Atkinson, "Liberal Education of the Nineteenth Century," *Popular Science Monthly* 4, no. 19 (1873), 8–14.

[42] Dorothy Ross notes this connection in *The Origins of American Social Science* (New York: Cambridge University Press, 1991), 67–68.

[43] Fiske, "Essays on a Liberal Education," 126; [Samuel H. Scudder,] "Sir Charles Lyell. II," *Science* (2nd series) 1, no. 3 (February 16, 1883), 69–70.

[44] Eliot, "What Is a Liberal Education?" 107.

[45] E.g., Altschuler, *Andrew D. White*, 51.

Economy unwarped to suit present abuses in Politics and Religion." Later, at Cornell, White bracketed his *laissez-faire* sentiments and sought federal funding for the training of administrators. He also proposed parallel, publicly funded programs for those who would exercise public influence "through the newspaper or in the forum" after graduating. For his part, Gilman not only thought that an ideal undergraduate education would consist of equal parts "Physical and Natural Science" and "Political and Moral Science," but also hoped to install the latter subject as the centerpiece of "an institute for the education of publicists." Two universities actually created such programs toward the end of the founding period. Under John W. Burgess, Columbia's School of Political Science opened in 1880 to prepare students for public careers, and Michigan's School of Political Science, led by Charles Kendall Adams, followed the next year. During the decade that followed, as civil service reformers struggled against their legislative opponents on the national political stage, university leaders worked behind the scenes to turn out virtuous public leaders committed to restricting the scope of state power.[46]

At the same time, the scientific democrats faced an array of vocal critics. Local religious leaders, for example, blasted the university reformers for their theological heterodoxy, charging that the new institutions peddled a spiritually subversive materialism. The reformers called for a "nonsectarian" university – a form of higher education that was essentially Protestant, but only in the broadest sense, standing apart from denominational institutions and creeds. Although a few of the critics charged that holding nondenominational chapel services in public universities violated the constitutional principle of church-state separation, many more of the commentators found the new universities practically atheistic. "To be without a creed is to be without a belief," declared the editor of one Baptist journal, who warned that nonsectarian education would fuel "indifferentism" or even "positive infidelity." Michigan and Cornell, with their large student bodies and outspoken presidents, drew particularly heavy fire from church leaders in their communities.[47]

Another potent challenge in the name of Christianity emanated from Princeton and Yale, the most notable holdouts from the new model of higher education during the Gilded Age. Their respective presidents, James McCosh and Noah Porter, both taught moral philosophy. Although they were hardly orthodox defenders of the antebellum curriculum, McCosh and Porter worried that re-centering the curriculum on the sciences threatened to leave

---

[46] Quoted in Haskell, *The Emergence of Professional Social Science*, 193; Brubacher and Rudy, *Higher Education in Transition*, 160, 162, 181; Daniel Coit Gilman, "Present Aspects of College Training," *North American Review* 136, no. 319 (1883), 534; quoted in Hawkins, *Pioneer*, 55; Frederick A. P. Barnard, "Columbia College as a University," *American Journal of Education* 31, no. 2 (1881), 253; Reuben, *The Making of the Modern University*, 160–161. On these ambitions in general, see Ross, *The Origins of American Social Science*, 67–70.

[47] Marsden, *The Soul of the American University*, 84–93, 168–170; Reuben, *The Making of the Modern University*, 76–87 (quote from 82); Altschuler, *Andrew D. White*, 94. On the nonsectarian ideal, see also Hart, "The Protestant Enlightenment Revisited."

undergraduates without moral standards to guide them through the "noisy tumult" of modern industrial life. At Princeton, McCosh identified the classical world as the source of these standards and its study as an integral component of any viable curriculum. He added that the Christian tenor of the prevailing curriculum was necessary to prevent students from descending into "idleness or dissipation." Porter further drew out this equation of religious belief with ethical virtue, arguing that the latter required acceptance of a spare but irreplaceable set of Christian axioms: "that there is a personal and self-existent Creator … ; that man is morally responsible to himself, and therefore to God, and needs guidance and help from God; [and] that he is destined to, and capable of, another life." Porter defined Christianity in nondenominational but essentially intellectual terms, rather than ethical ones.[48]

To parry the charge that nonsectarian higher education subverted Christian civilization, reformers relied on two lines of argument, one more than a century old and the other quite new. The first argument extended the traditional framework of natural theology to include the book of history as well as the book of nature, insisting that the results of empirical investigation in these fields would inevitably square with the essentials of Christianity. Because "one truth is never in conflict with another truth," Gilman declared, "the ethics of the New Testament will be accepted by the scientific as well as the religious faculties of man: to the former, as law; to the latter, as gospel." A second argument, arriving on American shores in earnest during the 1870s, rested on the work of German scholars who had painstakingly traced the historical provenance of various biblical narratives. Their "higher criticism" embedded the Bible in secular history, describing it as a collection of stories written by fallible human beings seeking to understand the ways of God. Led by White, university reformers embraced the higher criticism's suggestion that biblical texts should be interpreted in the light of modern empirical inquiry, rather than the other way around. The Bible, they insisted, offered symbolic or poetic truths that buttressed, rather than displaced, the empirical truths of science. These scientific democrats used both natural theology and the higher criticism to deflect attacks on their nonsectarian institutions.[49]

[48] Quoted in Lucas, *American Higher Education*, 168; quoted in Marsden, *Soul of the Modern University*, 200; Noah Porter, *Science and Sentiment* (New York: Scribner, 1882), 79. Porter emphasized that Christianity offered the philosophical basis for science itself (366). See also J. David Hoeveler, *James McCosh and the Scottish Intellectual Tradition: From Glasgow to Princeton* (Princeton: Princeton University Press, 1981); Stevenson, *Scholarly Means to Evangelical Ends*, esp. 30–37; Marsden, *The Soul of the Modern University*, 123–133; Caroline Winterer, *The Culture of Classicism: Ancient Greece and Rome in American Intellectual Life, 1780–1910* (Baltimore: Johns Hopkins University Press, 2002), esp. 111–117; and George Levesque, "Noah Porter Revisited," *Perspectives on the History of Higher Education* 26 (2007): 29–66.
[49] Daniel Coit Gilman, "The Utility of Universities," in *University Problems in the United States*, 61; Andrew Dickson White, *The Warfare of Science* (New York: Appleton, 1876), esp. 115; Bruce Kuklick, *A History of Philosophy in America, 1720–2000* (New York: Oxford University Press, 2001), 102–104.

It took additional work to counter the charge, levied by defenders of the classics, that centering the curriculum on science rendered the colleges incapable of fulfilling the goal of liberal education: preparing students for ethical leadership by lifting their eyes from material to spiritual ends.[50] Critics in this vein ignored the ethically charged image of science put forth by the university reformers and portrayed science narrowly as the study of physical entities with an eye to practical rewards. Defined in this manner, science stood doubly condemned by its material object of study and its intimate connection to commerce.

Like their successors today, the science-minded educators of the late nineteenth century struggled to overcome a widespread equation of science with metaphysical and ethical "materialism." They brought science under the umbrella of liberal education by defining the latter in terms of motive rather than content and describing science's educational function as character training rather than knowledge acquisition. According to Eliot, the "humanity" of a given course stemmed from "its power to enlarge the intellectual and moral interests of the student, quicken his sympathies, impel him to the side of truth and virtue, and make him loathe falsehood and vice." Science reliably turned students away from thoughts of worldly gain toward spiritual ends, Eliot contended. In fact, he went so far as to suggest that the true university would teach science in the "literary spirit." MIT's Atkinson likewise argued that both chemistry and Greek could be pursued "with a selfish eye to the loaves and fishes, or with an aim at the higher rewards of true culture, and the higher advancement of man's estate."[51]

But what about the obvious link between science and industry? The scientific democrats could hardly disavow this connection, given that they sought to interest workers, employers, and legislators in a seemingly impractical, if nonclassical, form of higher education. They reconciled the anti-utilitarian and utilitarian cases for science by arguing that the true scientist followed a broadly humanitarian desire to advance the cause of mankind as a whole but eschewed all specific worldly interests, whether economic or political. And while scientists chose their careers, and perhaps even their research problems, with an eye to the benefit of humanity rather than their own personal or group interests, the argument continued, they set aside even this humanitarian bias during the research process itself. Yet the resulting knowledge had inestimable value, the liberal-practical case for science concluded: the disinterested investigation of nature produced technological innovations "as a by-thing, but constantly." This argument allowed the scientific democrats to claim, to a practically minded public, that teaching science in a liberal fashion would still produce technological

[50] E.g., E. R. Sill, "Should a College Educate?" *Atlantic Monthly* 56, no. 334 (1885), 209.
[51] Eliot, "What Is a Liberal Education?" 105; Eliot, "Inaugural Address as President of Harvard College," 1–2; Eliot, "The New Education. Its Organization," *Atlantic Monthly* 23, no. 136 (1869), 215; Atkinson, "Liberal Education of the Nineteenth Century," 25. Leaders at Johns Hopkins took particular care to emphasize that the school, despite its unprecedented attention to research, prioritized character development: e.g., Gilman, "The Johns Hopkins University," 19; Rowland, "The Physical Laboratory in Modern Education," 574.

plenty, while reassuring traditionalists – and themselves – that the scientific scholar cared nothing for such material outcomes, at least while working in the laboratory or the field.[52]

Michigan's Angell drew out most clearly the institutional implications of reformers' hybrid conception of methodology, in the form of a proposed deal with the state legislature: if political leaders stayed out of the university's affairs, the university would serve an invaluable public purpose by pursuing scientific inquiry in the proper spirit. To harness scientists' public-minded concern, Angell and other scientific democrats insisted, Americans needed to free their institutions of higher education from religious and political control.[53] Shorn of its Christian overtones, this argument would reverberate down through the decades and lodge at the heart of the Cold War-era science establishment, as Chapter 10 will show. In that context, it would serve as a potent resource for scientists who sought federal funding but worried about federal control.

For the time being, though, the promise of technological spinoffs from the professoriate's research would hardly suffice to fill the classrooms with students or win financial support from skeptical industrialists and legislators. To this end, scientific democrats took a page from their counterparts in the vocational reform movement and emphasized the immediate, individual benefits of the modern curriculum. They insisted that the universities would equip students from all class backgrounds to compete effectively in the industrial economy. Such arguments took on a stronger political valence in the 1870s, when a sharp economic downturn revealed a deep ideological rift between capital and labor. To workers, scientific democrats stressed that, because knowledge brought power, universal access to a scientific form of higher education would arm laborers against the potential domination of an aristocracy that held superior wealth and superior knowledge. They promised the industrial classes that the new universities would welcome their children with open arms, moving them into the vanguard of national progress. Meanwhile, scientific democrats urged wealthy employers to "weigh the cost of the mob and the tramp against the expense of universal and sufficient education." White told one well-off audience that to keep the poor uneducated was to "place a powder magazine beneath your habitations." The reformers insisted that the new curriculum could stave off class conflict by ensuring the steady, orderly advancement of the industrial classes through its direct vocational benefits.[54]

[52] H. Newell Martin, "The Study and Teaching of Biology," *Popular Science Monthly* 10, no. 3 (1877), 299. In many circles, this conception of science accompanied a sharp differentiation between two realms that roughly corresponded to the contemporary division between science and technology. Ronald Kline explores this phenomenon in "Construing 'Technology' as 'Applied Science': Public Rhetoric of Scientists and Engineers in the United States, 1880–1945," *Isis* 86, no. 2 (1995): 194–221.

[53] James B. Angell, "Inaugural Address, University of Michigan," in *Selected Addresses*, 30, 28. Cf. Andrew Dickson White, *The Warfare of Science* (New York: Appleton, 1876), 146.

[54] Rossiter W. Raymond, "The Requirements of a Scientific Education," *Popular Science Monthly* 4, no. 2 (1873), 211; [E. L. Youmans,] "The Recent Strikes," *Popular Science Monthly* 1, no.

University reformers worked to make this meritocratic ideal a reality by seeking to lower tuition costs, even at elite private colleges. But the universities would never enroll every citizen, no matter how inclusive and affordable they became. Still, the reformers insisted, the universities would serve the whole public. Those who did not attend would nevertheless reap the benefits of the new curriculum. Many reformers emphasized the economic impact of technology, arguing that the universities would bring low-cost goods to all. Some added the old Whig argument that technology fueled moral progress across all classes. To buttress the latter claim, scientific democrats extracted powerful lessons regarding science's moral impact from the course of modern history. For example, the physicist Joseph Henry credited science with abolishing slavery. A second writer suggested that new modes of transportation and communication had spread moral virtues and other aspects of refined living across the continent. Others argued that the leisure produced by labor-saving devices promoted morality. William P. Atkinson wrote that scientific education represented the masses' "emancipation" into "the privileges of intellectual life," hitherto monopolized by the wealthy. Science, he continued, would "spiritualize material things by enabling us to put them to higher uses." Through its technological impact, these reformers argued, the scientific university would promote the moral and material interests of all Americans.[55]

Many scientific democrats also contended that the university would wield its moral influence through interpersonal communication between college graduates and their fellow citizens. The university would perfect democracy indirectly, as its graduates circulated knowledge and virtues ever more widely among the populace. Some versions of this trickle-down theory took the general public to be essentially passive. *Science* editor Samuel H. Scudder wrote forthrightly that "the improvement of any age or any people, depends upon great men; great men are nurtured by great ideas; [and] great ideas are developed by higher education." Most commentators, by contrast, granted the public a more active role in the process, assuming that citizens would recognize the value of higher learning when the influence of graduates made it personally available to them in their communities. Angell thus assured populist critics that his students could never monopolize the rewards of higher education. Even if their explicit intentions were selfish, he wrote, Michigan's graduates could not

5 (1872), 624; John Eaton, "Report of the Commissioner of Education for 1877," *American Journal of Education* 29, no. 1 (1878), viii; White quoted in Altschuler, *Andrew D. White*, 52. These educators insisted that workers' children would feel welcome in even the most prestigious private universities: Denison Olmsted, "On the Democratic Tendencies of Science," *American Journal of Education* 1, no. 2 (1855), 165; Angell, "The Higher Education," 48–49; Eliot, "Inaugural Address," 16.

55 Abram S. Hewitt, "Iron and Civilization," *Popular Science Monthly* 1, no. 3 (1872), 343, 342; Joseph Henry, "On the Importance of the Cultivation of Science," *Popular Science Monthly* 2, no. 6 (1872), 644; Olmsted, "On the Democratic Tendencies of Science," 167; Atkinson, "Liberal Education of the Nineteenth Century," 23–24.

help but shed "life-giving beams" on others. "Intellectual activity is necessarily luminous, outgoing, diffusive, reproductive," Angell explained. These scientific democrats insisted that higher education for the few would benefit the many, especially if that few represented a broad economic cross-section of society.[56]

All the while, the question of science's – or rather, the scientist's – authority in the wider culture remained muddy. Scientific democrats engaged in reforming American higher education occasionally worried that, in altering the surrounding public culture, they would install scientists as a kind of priestly class. Perhaps Americans would miss their historic opportunity for mental liberation, simply shifting their allegiance from the classics, or ministers and the Bible, to scientists. In the 1860s and 1870s, however, such an outcome seemed remote. Technology had long impressed the public, but science did not possess the widespread authority it would enjoy after 1900. Moreover, the reformers' Protestant backgrounds reassured them that the unfolding of God's plans required no imposition of human authority; cooperation and consensus would emerge through purely voluntary means. The real danger, in their view, came from the narrow-minded graduates of the classical colleges. Furthermore, the question of scientific authority was moot as long as religious leaders and classicists kept scientists from giving the public the cultural tools it needed to harness the industrial machine. For the time being, scientific democrats concentrated on disseminating science's ethic of humility, in order to combat what they saw as the denominational colleges' authoritarian tendencies.

Their efforts changed the nation. In the three decades after Cornell opened its doors in 1868, the scientific democrats and their allies fundamentally altered the American educational landscape. Today's constellation of leading universities had largely taken shape by 1900. New private institutions such as Stanford and Chicago; public universities in Michigan, California, and elsewhere; and eventually even Princeton and Yale, the last holdouts among the elite colleges, joined Cornell, Harvard, and Johns Hopkins on the university path. In their new incarnations, each of these schools bore the distinguishing mark of the modern American university: an institutional philosophy grounded in an expansive conception of science's ethical promise as well as its material applications. Yet the promised regeneration of American public culture remained elusive, forcing scholars to think anew about the nature of science and its implications for social ethics and political practice. The bitter political disputes of the late 1880s and early 1890s would reveal deep divisions within the republic of science itself, leading to feverish attempts to purify American science by shoring up its research base, clarifying its methodologies, and strengthening the disciplinary structures that facilitated communication.

[56] [Samuel H. Scudder,] "Higher Education and the Masses," *Science* (2nd series) 4, no. 76 (July 18, 1884), 53; Angell, "The Higher Education," 50–51; Angell, "State Universities," in *Selected Addresses*, 117.

2

Internal Divisions

The first generation of scientific democrats promised their fellow citizens – and fervently believed themselves – that adopting the scientific spirit in all realms of human activity would produce an ever-widening sphere of social agreement. Yet as the disciplines became more specialized and generated a growing body of empirical knowledge in the late nineteenth century, divisions within the ranks challenged this easy assumption of latent mental and social harmony.

The founders of the new universities disagreed among themselves on a number of strategic points. For example, Eliot's elective system generated tensions: Given the benefits of a scientific curriculum, should students enjoy the freedom to choose their courses or should they be required to take modern subjects? A more fractious question concerned public funding. Eliot and other leaders of established private universities insisted that it would infuse academia with the small-mindedness of politics, while leaders of land-grant schools and fledgling institutions such as Cornell, Johns Hopkins, and MIT portrayed the university as the cornerstone of the national infrastructure, serving a public purpose and deserving public support.[1]

Indecorous though they were, these strategic disagreements hardly challenged the underlying assumption that the spread of the scientific spirit would fuel social harmony. By contrast, the Darwinian debates of the 1870s and 1880s

---

[1] Charles W. Eliot, "The Exemption from Taxation," in *American Contributions to Civilization and Other Essays and Addresses* (New York: Century, 1897), 322–324; Glenn C. Altschuler, *Andrew D. White, Educator, Historian, Diplomat* (Ithaca: Cornell University Press, 1979), 147, 181; Henry J. Anderson, "Physical Science," *American Journal of Education* 1, no. 4 (1855): 516–531; [E. L. Youmans,] "State Education," *Popular Science Monthly* 19, no. 5 (1881), 703; James B. Angell, "Inaugural Address, University of Michigan," in *Selected Addresses* (New York: Longmans, Green, 1912), 3–4; Charles Kendall Adams, review of "American State Universities," reprinted in Richard Hofstadter and Wilson Smith, eds., *American Higher Education: A Documentary History, Volume II* (Chicago: University of Chicago Press, 1961), 669; William P. Atkinson, "Liberal Education of the Nineteenth Century," *Popular Science Monthly* 4, no. 19 (1873), 1, 16; N. H. Winchell, "The State and Higher Education," *Science* 2, no. 62 (September 3, 1881), 414–415; W. S. Barnard, "Zoological Education," *Popular Science Monthly* 17, no. 5 (1880), 668.

and a highly public dispute over national economic policy that stretched into the 1890s cut more deeply into the scientific democrats' project. These bitter conflicts seemed to indicate either that many of the thinkers claiming the mantle of science had failed to internalize its distinctive spirit or that the empirical basis from which they reasoned was faulty. Both analyses suggested that the scientific community would need to harmonize its own members' beliefs before it could bring the public into the circle of unforced consensus.

STEPPING BACK

Though daunting in practice, this task seemed simple in theory to the early scientific democrats. Since the earliest days of the Scientific Revolution, when Francis Bacon challenged the medieval scholastics' reliance on deductive reasoning from theological premises in the name of a thoroughgoing empiricism, science's public advocates had insisted on the need to scrap all claims not grounded in agreement among working scientists. Yet they remained confident that a body of reliable knowledge would soon emerge. Both the Protestant critique of authority and the scholastics' portrayal of reasoning as reliable and intrinsically persuasive led practitioners and theorists of science to expect intellectual and political harmony as a result of their work.

By the nineteenth century, a well-worn conception of the human mind underpinned this assumption for American scientists. The faculty psychology framed by Thomas Reid and other philosophers of the Scottish common-sense school divided the mind into a series of discrete organs or faculties. These portions of the mind were attuned to various elements of the divine order. Thus, the human mind naturally fixed on instances of goodness, beauty, and truth in the universe, just as one magnet was drawn to another. Freely willing human beings knew or felt when they were on the path of righteousness – when their actions comported with the divine plan for humanity. The common-sense theory of science framed knowledge as a snug fit of the mind with the world, a sympathetic identification of the faculty of reason – the divine element in the individual – with the divine element in experience.[2]

This approach virtually eliminated the need for the active interpretation of facts. Instead, it assumed that individuals passively registered the meaning inherent in evidential data. Moreover, common-sense epistemology described all human minds as essentially alike, and thus insisted that every mind would reach the same conclusion from a given body of experiential evidence. This suggested that, in science, as elsewhere, one could guarantee consensus on a problem or theory simply by gathering the relevant evidence and disseminating it to a receptive audience. The same held true of political action: every situation evinced a single path of moral duty, and every individual would follow

[2] Julie A. Reuben, *The Making of the Modern University: Intellectual Transformation and the Marginalization of Morality* (Chicago: University of Chicago Press, 1996), 22, 25–26; George M. Marsden, *The Soul of the American University: From Protestant Establishment to Established Nonbelief* (New York: Oxford University Press, 1994), 82.

that path once it became clear. The moral philosophers, who relied heavily on common-sense views, had expected Americans to converge on a shared political framework, once freed from the grip of the passions and given the needed facts. So, too, did their successors in the emerging modern universities. The common-sense theory of knowledge embraced by the university founders and other scientific democrats in the early Gilded Age held that the knowing process took the form of an intuitively perceived fit between the mental faculty of reason and an external phenomenon in the world.[3]

From the standpoint of the broad campaign to make America scientific, this conception of knowledge had a great deal going for it. Its emphasis on the universality of the mind's basic structure militated against overt scholarly partisanship, and it extended the reach of empirical investigation – and thus the prospect of impending agreement – beyond physical questions to moral and social ones. Common-sense thought suggested that political consensus could flow from intellectual consensus, and, thus, that scholars could contribute to national comity as well as progress.

The common-sense framework also authorized disciplinary organization and the attendant modes of communication. Many different motives fueled professionalization and specialization in the late nineteenth century. At the level of legitimation, however, the common-sense framework offered a simple justification: because all investigators would ultimately find the same thing, the work of each increased the storehouse of knowledge for all. Each scientist served as a kind of reporter for humanity, bringing back universally valid observations from the field or laboratory. Professional organizations merely increased the efficiency of the process by facilitating the clarification of definitions and the diffusion of essentially interchangeable, impersonal findings.[4]

Few, in the halcyon days of the early Gilded Age, could imagine a scientific community characterized by permanent interpretive divisions. After all, when disputes arose, the common-sense philosophy offered a simple recipe for restoring harmony: establish a common starting point by stepping back to basic principles and reliable facts, then make these available to all discussants, while clearing away the emotional and material obstacles to their interpretive freedom. The common-sense approach held that agreement on the principles and facts of a particular case would invariably produce agreement on the proper interpretation and the resulting course of action, at least among right-thinking

---

[3] E. Brooks Holifield, *Theology in America: Christian Thought from the Age of the Puritans to the Civil War* (New Haven: Yale University Press, 2003), 174–180; Marsden, *The Soul of the American University*, 90–93; Reuben, *The Making of the Modern University*, 19–22, 36–39; D. H. Meyer, *The Instructed Conscience: The Shaping of the American National Ethic* (Philadelphia: University of Pennsylvania Press, 1972), 5, 35–42, 138.

[4] E.g., [Samuel H. Scudder,] "The Import Duty on Scientific Journals," *Science* n.s. 1, no. 21 (June 29, 1883), 590. On the importance of public audiences to late-nineteenth-century professors, see Richard F. Teichgraeber, "The Academic Public Sphere: The University Movement in American Culture, 1870–1901," in *Building Culture: Studies in the Intellectual History of Industrializing America, 1867–1910* (Columbia: University of South Carolina Press, 2010): 79–105.

individuals.[5] Thus, disagreement in the latter realm signaled either that the parties to the dispute had begun from different starting points or that one or both of them had formed their opinions in an illegitimate manner, either because they had been coerced or because they had let their emotions prevent them from hearing all sides of the argument.[6]

The early scientific democrats were confident that dealing with these sources of misinterpretation would generate universal consensus, even in highly disputed areas of religious and political doctrine. After all, the common-sense framework seemed to embody an insight shared by both democracy and post-Enlightenment Protestantism: give the people their intellectual freedom, and they will find their way to God's truth. What many nineteenth-century Americans meant by "science" was this process of stepping back, of clearing away empirical and emotional obstacles to the unforced interpretation of disputed questions. Advances in the physical sciences had proved the fruitfulness of this method, and it now needed to be implemented in all areas of human life. As the chemist John W. Draper explained, such a procedure offered the only reliable way to change another person's beliefs without exercising human authority.[7]

Indeed, attempts to get back to common ground and then reason forward again characterized many areas of American intellectual life in the mid-nineteenth century. Liberal Protestantism itself represented a version of the stepping-back maneuver. Its practitioners sought to bring about wider acceptance of their doctrinal commitments and their political visions for the nation – their specific versions of "moral government" – by returning to the core ethical principles of Christianity and eliminating the distortions produced by material want and doctrinal encrustations. "Begin by looking at everything from the moral point of view," the British educator Thomas Arnold famously wrote, "and you will end by believing in God." Similarly, moral philosophers sought to prove detailed conclusions by establishing a consensual foundation of first principles, then reasoning deductively from them. A deep faith in the universality of human minds and reasoning processes informed both of these dynamics.[8]

[5] Marsden, *The Soul of the American University*, 91.

[6] Commentators at the time frequently implied, and occasionally stated, that the success of scientific inquiry depended on certain personal virtues, not the recipe-like practices that twentieth-century thinkers would describe as "scientific methods": e.g., Charles W. Eliot, *Educational Reform* (New York: Century, 1898), 110–111; Lewis H. Morgan, "American Association for the Advancement of Science," *Science* 1, no. 10 (September 4, 1880), 110; [John Michels,] untitled editorial in *Science* 2, no. 59 (August 13, 1881), 377.

[7] John W. Draper, "Science in America," *Popular Science Monthly* 10, no. 3 (January 1877), 326. Cf. Reuben, *The Making of the Modern University*, 49, 73–74, 186–187.

[8] Quoted in Allen C. Guelzo, "'The Science of Duty': Moral Philosophy and the Epistemology of Science in Nineteenth-Century America," in *Evangelicals and Science in Historical Perspective*, ed. David N. Livingstone, D. G. Hart, and Mark A. Noll (New York: Oxford University Press, 1999), 267; Meyer, *The Instructed Conscience*, 140; Marsden, *The Soul of the American University*, 90–93, 211–215.

Likewise, such a faith informed the university founders' attempts to step back even farther, from the claims of Christianity to the truths of science. The early scientific democrats believed that their Protestant predecessors, by demanding of students and readers a commitment to certain theistic doctrines, had stopped short of truly common ground – they had not left enough room for uncoerced interpretation. White, Gilman, and other members of the founding generation believed that a scientific education, stripped down to the absolute core of human truth, would generate social consensus in short order.

But it did not. Rather than issuing in concord, the advent of a scientific form of higher education led to the sorry sight of earnest, highly trained scholars fighting over the most basic principles of natural history and metaphysics in the 1870s. What had gone wrong? Perhaps the divergence signaled hidden weaknesses in the empirical foundation. Scholars rushed to undertake new investigations, presuming that better data would surely produce a universally accepted conclusion. In fact, a widespread belief that interpretive divisions revealed empirical inadequacies fed into the intense burst of institution-building that characterized the last years of the nineteenth century, when scholars launched the national organizations and academic departments that now define our disciplinary matrix. In these emerging disciplinary bodies, researchers sought to get back to the basics, to establish foundational facts and definitions on which all investigators could agree.[9]

SCIENCE AND SPECULATION

Academic struggles over Darwin's theory of natural selection and its religious implications presented the early scientific democrats with the first major obstacle to their project of building intellectual and political harmony. In the 1870s and 1880s, American science journals featured a running argument over whether scientists could legitimately speculate in areas beyond their immediate expertise – in particular, the realm of philosophy, where Darwin's theory of evolution fueled controversy between materialists, atheists, agnostics, and various kinds of theists. The speculation debate revealed important uncertainties about how to balance the virtues comprising the scientific spirit. Should scientists emphasize caution or engagement? How should they approach colleagues and fellow citizens advocating different views than their own? Epistemological questions typically appeared in these implicitly political forms during the early Gilded Age.[10]

[9] Roger L. Geiger summarizes the era's disciplinary growth: *To Advance Knowledge: The Growth of American Research Universities, 1900–1940* (New York: Oxford University Press, 1986), 20–39.

[10] [E. L. Youmans,] "The Charges Against 'The Popular Science Monthly,'" *Popular Science Monthly* 20, no. 3 (January 1882), 404; Reuben, *The Making of the Modern University*, 134. On the importance of Darwinism for late-nineteenth-century debates over the nature of science, see Reuben, *The Making of the Modern University*, 36–50 and Paul Jerome Croce, *Science and Religion in the Era of William James: The Eclipse of Certainty, 1820–1880* (Chapel Hill: University of North Carolina Press, 1995), 87–148.

Participants in the speculation debate feared that science's public standing lay in the balance. In the hands of the influential British naturalist Herbert Spencer, evolutionary theory had become wrapped up with agnosticism – the view that human beings could never attain the kind of knowledge needed to validate either theism or atheism. But Spencer's critics charged that his agnosticism amounted to atheism in disguise. To profess to leave unresolved the question of God's existence, they argued, was in fact to answer that question in the negative. By and large, the American reading public agreed with the charge. Appleton's publishing house, which had dropped the *North American Review* after it printed pieces by the notorious freethinker Robert Ingersoll, faced pressure to do the same when the *Popular Science Monthly* published Spencer's works. Meanwhile, Harvard's Board of Overseers scotched the appointment of the historian and outspoken agnostic John Fiske to the faculty.

Many working scientists, possibly guided by strategic considerations as well as personal convictions, recoiled against what they saw as the agnostics' confusion of science with philosophy. Johns Hopkins' Henry A. Rowland ruled out all trespasses on philosophical terrain, demanding that the scientist "abstain from having opinions on subjects of which he knows nothing." Rowland's abstemious stricture appealed to those who feared Darwinism's impact on Christianity. The noted geologist Joseph LeConte, a sharp critic of materialist readings of Darwinism, argued that science, having taken theology's place as the "seat of power and fashion" in modern societies, now served as home base for dogmatism, which always accompanied cultural authority. A third writer discerned a widespread search for "a sort of scientific Nicene Creed" that researchers would be forced to believe and espouse even without supporting evidence. Such hypotheses as natural selection and the atomic theory had their uses, he allowed, provided they were "confined within a very small compass, and employed only to stimulate rather than satisfy inquiry." But these heuristic devices said nothing about "the actual constitution of Nature" and could not be installed as doctrinal tests.[11]

The novel view that imaginative hypotheses guided scientific investigation could cut both ways in the speculation debate, however. Under the influence of British naturalists such as William Whewell, as well as the Darwinian controversy, a few Americans began to abandon a strictly empiricist philosophy of science in the 1870s. Orthodox empiricists held that scientists began with indubitable facts of observation and reasoned from them in a purely inductive fashion, building upward from specific phenomena to general laws of causation. Whewell countered that scientists began not with concrete facts, but rather with the proposed laws or theories themselves. Empirical investigation, in his portrayal,

---

[11] Henry A. Rowland, "The Physical Laboratory in Modern Education," *Science* 7, no. 177 (June 25, 1886), 57; Joseph LeConte, "Science and Mental Improvement," *Popular Science Monthly* 13, no. 1 (May 1878), 101; "J. C. D.," "The Dangers and Securities of Science," *Popular Science Monthly* 3, no. 3 (July 1873), 241. Cf. J. B. Stallo, "Speculative Science," *Popular Science Monthly* 21, no. 2 (June 1882), 164.

served to confirm or refute the initial hypotheses, not to generate them out of whole cloth. In the context of the controversy over agnosticism, the American chemist and science journalist Edward L. Youmans took Whewell's emphasis even further, prefiguring Karl Popper's mid-twentieth-century view that empirical data could disconfirm theories but never absolutely confirm them. On these grounds, Youmans insisted that science offered the only means of attaining knowledge in any and all fields, including philosophy and theology.[12]

Both Spencer and the American critics mentioned previously offered versions of what historians of science and religion call a "separate spheres" model: one that confines science and religion to distinct and unrelated realms of inquiry rather than expecting their findings to harmonize, after the fashion of natural theology. Thus, Spencer cordoned off a set of theological and metaphysical questions – God's existence, the ultimate nature of reality, and so forth – to which the empirical data examined by scientists could never speak, no matter how much of this data they gathered. Rowland and the others simply denied that Spencer had drawn the boundary properly, insisting that he had strayed into philosophical territory.

By contrast, Youmans interpreted the relationship between science and its intellectual neighbors quite differently. Although he affirmed Spencer's agnostic conclusions and devoted a great deal of his considerable energy to making Spencer a household name among Americans, Youmans adopted an even more imperialistic attitude toward philosophers and theologians than did his British idol.[13] Rather than identifying two distinct realms of subject matter and assigning to these separate methods of inquiry and different groups of inquirers, Youmans instead echoed the French theorist Auguste Comte's suggestion that scientific methods offered the only possible source of knowledge and should replace theological and philosophical methods in all fields. Like Comte, Youmans expected scientists to take a strong leadership role in society,

[12] Additional research is needed on how American scientific democrats drew on the mid-nineteenth-century British "men of science" – influential naturalists such as Whewell, Spencer, John Stuart Mill, Thomas Henry Huxley, John Tyndall, and William Kingdon Clifford – who offered influential accounts of the scientist's role in a democracy. The present account looks past the similarities to the unique political and religious configurations that led many Americans to substantially modify the arguments of the British theorists.

[13] For a time in the early 1870s, Spencer enjoyed a larger audience in the United States than in England, thanks to Youmans' publication of *The Study of Sociology* and serialization of several chapters from it in *Popular Science Monthly*. For biographical information, see "Prof. Youmans Dead," *New York Times* (January 19, 1887): 8; and Charles M. Haar, "E. L. Youmans: A Chapter in the Diffusion of Science in America," *Journal of the History of Ideas* 9 (1948): 193–213. The fullest accounts of Youmans' thought remain William E. Leverette Jr.'s "Science and Values: A Study of Edward L. Youmans' *Popular Science Monthly*, 1872–1887" (PhD dissertation, Vanderbilt University, 1963) and "E. L. Youmans' Crusade for Scientific Autonomy and Respectability," *American Quarterly* 17, no. 1 (Spring 1965): 12–32. On Youmans' dealings with Spencer, see also Mark Francis, *Herbert Spencer and the Invention of Modern Life* (Ithaca: Cornell University Press, 2007) and Barry Werth, *Banquet at Delmonico's: Great Minds, the Gilded Age, and the Triumph of Evolution in America* (New York: Random House, 2009).

speaking honestly and fearlessly for what they took to be the truth. He did not believe that scientists could simply halt at the border to philosophy, or even religion.

Indeed, Youmans denied the very existence of such borders, or at least of fixed borders. He held that the scientific method applied to all realms. To the extent that one could identify a boundary between spheres of thought, this boundary merely divided the truths of science from the vague gropings of other thinkers in the residual territory not yet explored scientifically. He identified the scientific method as the source of reliable knowledge in any field, including morality and even theology. To his mind, a truth claim either characterized the world – in which case empirical inquiry could never contradict it – or it did not. "Intellect, feeling, human action, language, education, history, morals, religion, law, commerce," and much else counted as part of nature, in Youmans' book. All of these areas of human endeavor, he explained, involved "accessible and observable phenomena" in causal relations, and were thus amenable to scientific study. In short, science revealed all of God's truths – everything that human beings needed to know to behave correctly in the world. Youmans assumed that the domains of science and of other forms of inquiry, rather than being fixed by the character of their questions, were separated by a dividing line which moved over time and which scientists themselves drew in the course of their ongoing researches. In the end, this border was nothing more than the line between knowledge and ignorance or fraud, set by the historical state of science rather than the intrinsic character of reality.[14]

Youmans' conception of science, however, illustrated an emerging emphasis on the fallibility of all knowledge, whereas Comte's work of the 1830s and 1840s had borne the mark of the age-old quest for interpretive certainty. Not just initial hypotheses, Youmans insisted, but also the settled theories that Whewell deemed confirmed, served merely heuristic purposes. A theory, he explained, gained adherents because of its "superiority to the views it seeks to supersede," not its "everlastingness": "Does it involve fewer assumptions? Does it account for more facts? Does it harmonize conflicting opinion? Does it open new inquiries and incite to fresh research?"[15] Even as Youmans declared science sovereign over all realms of thought, he cautioned that it could never achieve absolute certainty. Citizens needed to heed its teachings at all times, not losing faith even when these changed dramatically.

[14] [E. L. Youmans,] "Purpose and Plan of Our Enterprise," *Popular Science Monthly* 1, no. 1 (May 1872), 113; [Youmans,] "Mr. Godwin's Letter," *Popular Science Monthly* 3, no.1 (May 1873), 116–117. Cf. Reuben, *The Making of the Modern University*, 51–53; George Levine, "Scientific Discourse as an Alternative to Faith," in *Victorian Faith in Crisis: Essays on Continuity and Change in Nineteenth-Century Religious Belief*, ed. Richard J. Helmstadter and Bernard V. Lightman (Stanford: Stanford University Press, 1990); and Frank M. Turner, *Contesting Cultural Authority: Essays in Victorian Intellectual Life* (New York: Cambridge University Press, 1993), esp. 150.

[15] [Youmans,] "Mr. Godwin's Letter," 116. On Youmans' exchange with Parke Godwin and late-nineteenth-century "progressivist" theories of science more generally, see Reuben, *The Making of the Modern University*, 41–50.

Youmans applied this position to the all-important case of scientific education. He attacked those, including Whewell, who held that the schools should teach "only those subjects, the truths of which are demonstrated and settled forever." Even the foundations of mathematics were disputed, Youmans noted, and physics had just undergone a fundamental theoretical shift. Meanwhile, students could not be kept from the best available knowledge. In the case of evolution and other controversial theories, Youmans urged ordinary citizens to heed the early efforts in a scientific field just as they would the later, more refined theories. He wanted the schools to teach the latest scientific theories, including Darwinian evolution, whether or not they conflicted with the deeply held beliefs of the public.[16]

Historians frequently use Youmans' writings and his *Popular Science Monthly* as a window into the thought of scientific intellectuals during the founding decades of the modern American universities.[17] But these texts illustrate only one of several positions on the nature and relations of science then prevailing in the United States – and one that found relatively few adherents within the emerging universities. Major epistemological and political fault lines separated Youmans from most of the other figures discussed in the previous chapters. To be sure, Youmans was a committed scientific democrat and educational reformer – in fact, he was one of the era's most vigorous supporters of scientific education as the cultural basis for self-government. Every issue of Youmans' *Popular Science Monthly* brimmed with arguments for educational change, and his 1867 edited volume, *The Culture Demanded by Modern Life*, offered even more arguments. "Deeper than all questions of Reconstruction, Suffrage, and Finance," he insisted, "is the question, 'What kind of culture shall the growing mind of the nation have?'"[18] But Youmans departed from most other early scientific democrats in his views on the character of scientific knowledge and its relation to social problems.

Even in the world of scientific journalism, Youmans' endorsement of Spencerian agnosticism and imperialistic attitude toward other intellectual pursuits raised numerous hackles. Over at *Science*, editor John Michels insisted that the scientific spirit dictated caution, not wild theological speculation. Youmans and Michels sparred repeatedly over the question of agnosticism. On the question of religion, Youmans followed Spencer in viewing the world in frankly materialistic terms, while positing a vague "Unknowable" as the basis for a modern faith. Michels deemed this a "sickening hypocrisy"

---

[16] E. L. Youmans, "Introduction – On Mental Discipline in Education," in Youmans, ed., *The Culture Demanded by Modern Life: A Series of Addresses and Arguments on the Claims of Scientific Education* (New York: Appleton, 1867), 35–36; [Youmans,] "Mr. Godwin's Letter," 117.

[17] To take an early example, Richard Hofstadter called Youmans "the self-appointed salesman of the scientific world-outlook": *Social Darwinism in American Thought: 1860–1915* (Philadelphia: University of Pennsylvania Press, 1944), 14.

[18] E. L. Youmans, preface to *The Culture Demanded by Modern Life*, v. Cf. James B. Angell, "Inaugural Address, University of Michigan," in *Selected Addresses* (New York: Longmans, Green, 1912), 19 and John Eaton, "Report of the Commissioner of Education for 1880," *American Journal of Education* 30, no. 1 (1881), v.

and thought Youmans' reading of evolutionary theory equated "the perfect man" with "the perfect hog": "the one whose nervous organization is perfectly adapted to surrounding physical conditions." Michels believed that associating science with Spencer's views could only bring it harm, given that Americans already confused science with atheism. Moreover, he insisted that science had proven empirically the Christian truth that "an intelligent Creator has designed and pre-arranged the order of both matter and mind." Any other view, Michels continued, served to "undermine society itself by denying the intrinsic value of morality." Adopting essentially the same position as Noah Porter on the relation between science and Christianity, Michels declared atheists and agnostics far more dangerous than even religious traditionalists.[19]

Despite his clear-cut theological stance, Michels insisted that a scientific journal should not take sides in philosophical or theological disputes. In *Science*, he wrote proudly, "no editorial bias has been given to any particular set of views," and nothing whatsoever was said on matters of theology. Like many theorists of liberal Protestantism, Michels portrayed a basic foundation of Christian theism as the neutral, nonpartisan conclusion of all right-thinking individuals; to be neutral or disinterested entailed actively endorsing this view rather than opposing it. This assumption enabled Michels to aver that genuine scientists "do not care to interfere with their neighbor's [sic] religious opinions," much less to "force atheistical views upon them," especially in a nation characterized by religious freedom. Blaming "too rapid generalization" for the materialistic viewpoint of many Darwinians, Michels recommended to scientists a "conservative spirit": "[D]o not attempt to run before you are sure you can walk." Michels' successor at the journal, Samuel H. Scudder, agreed that science should always exhibit "a due consideration" in "its relations with other departments of knowledge."[20] Michels and Scudder adopted the broad version of the separate spheres argument endorsed by many of Spencer's other American critics.

The same line of division that set Youmans against Michels also separated what are typically said to be the inaugural texts of the modern "conflict thesis" regarding science and religion: John W. Draper's *History of the Conflict Between Religion and Science* (1874) and Andrew Dickson White's *The Warfare of Science* (1876), a precursor to his better-known 1896 text *A History of the Warfare of Science with Theology in Christendom*. The pugnacious Draper, whose

[19] Reuben, *The Making of the Modern University*, 54; [John Michels,] untitled editorial in *Science* 3, no. 80 (January 14, 1882), 1–2. On the challenges posed to Christian thinkers by views of science such as Spencer's, see Charles D. Cashdollar, *The Transformation of Theology, 1830–1890: Positivism and Protestant Thought in Britain and America* (Princeton: Princeton University Press, 1989). More broadly relevant is James Turner, *Without God, Without Creed: The Origins of Unbelief in America* (Baltimore: Johns Hopkins University Press, 1985).

[20] [John Michels,] untitled editorial in *Science* 2, no. 79 (December 31, 1881), 607; [Michels,] editorial of January 14, 1882, 2; [Michels,] untitled note in *Science* 1, no. 12 (September 18, 1880), 141; [Samuel H. Scudder,] "The Future of American Science," *Science* (2nd series) 1, no. 1 (February 9, 1883), 3.

book Youmans solicited for his International Science Series, took his cues from the likes of Comte and Spencer. He identified two opposing methods for seeking truth in any sphere: the static, intolerant, and authoritarian method of revelation and the progressive, empirical approach of the scientist. White, by contrast, differentiated science from "Ecclesiasticism" or "sectarianism," not from genuine faith. In fact, he insisted that "God's truths must agree," though they would do so only if each category of truth were pursued according to its own distinctive method. Sounding very little like Draper or Youmans, and much like the liberal Protestant theologians of his day, White closed his 1876 book by calling for scientific and religious leaders alike to champion "the living kernel of religion" against "the dead and dried husks of sect and dogma."[21]

SMALL-STATE SCIENCE

Following hard on the heels of the controversy over Darwinism and agnosticism, a bitter divide between scientific thinkers over the legitimacy of economic regulation thrust them into the turmoil of national politics and further threatened the image of science as a reliable source of consensus. In this case, empirical data seemed not to have produced agreement even within the domain of science itself. The result was an all-out, highly visible struggle between the nation's leading economic experts over the question of science's relationship to morality. To understand the import of this controversy, we must delve more deeply into the *laissez-faire* consensus that dominated the early universities and grasp the continuities between the versions of *laissez-faire* endorsed by most scientific democrats and the views of the ethical economists who urged the application of Christian principles to economic relations in the 1880s and 1890s.

When rapid industrialization led to a sharp downturn in the mid-1870s, labor strife of a type never before seen on American soil erupted around the nation. In response, scientific democrats based in the new universities urged restraint by all parties, including politicians. As detailed in the previous chapter, these figures believed that adopting the scientific spirit in the study of history revealed the legitimacy of *laissez-faire* principles of political economy. They defended free markets and free trade on the grounds of Christian charity, insisting that in the long term such policies provided the most effective means of raising the standard of living for all.

However, theirs were not the only voices in the *laissez-faire* chorus. A second group of science-minded thinkers, which included some academic scholars but was dominated by nonacademic professionals in commercial centers such as New York, offered a very different version of the argument against economic

[21] Donald Fleming, *John William Draper and the Religion of Science* (Philadelphia: University of Pennsylvania Press, 1950); John W. Draper, *History of the Conflict Between Religion and Science* (New York: Appleton, 1874), xi; Andrew Dickson White, *The Warfare of Science* (New York: Appleton, 1876), 145, 8, 151. Cf. [Michels,] editorial of January 14, 1882, 2. For a different take, see Reuben, *The Making of the Modern University*, 58–59.

regulation. Youmans, a key figure in this latter group, again serves as a useful foil to the university-based scientific democrats, highlighting key elements of their outlook. He and other hard-edged advocates of *laissez-faire* argued that science revealed absolute laws of social organization that could not be abridged. Their case for the irrelevance of Christian charity to economic causation pointed toward twentieth-century theories of value-neutral science.

As mentioned previously, Youmans believed that scientific findings were infallible nowhere, but authoritative everywhere. He also took them to be extraordinarily reliable in the case of political economy. Here, at least, Youmans felt sure that scientific laws approximated the contours of reality with sufficient faithfulness to render them essentially certain. Those laws, he believed, dictated severe limitations on state action. Like Spencer, Youmans called for a policy of strict *laissez-faire* in the realm of economic regulation.

However, Youmans bypassed key elements of Spencer's thought, most notably his vision of society as a biological organism, his preoccupation with social differentiation and functional integration, and his search for laws of cosmic evolution that subsumed social change within larger processes of biological change. Essentially ignoring the content of Darwinian biology, at least as a political resource, Youmans derived his *laissez-faire* theory from a very different starting point: individual freedom of action, properly constrained only by the stern, unchanging laws of nature. Unlike Spencer, Youmans seems to have felt no pressure to justify his emphasis on individual freedom in the terms of overall social functioning. As a result, his writings have a very different flavor than do Spencer's, which are easily mistaken at first glance for paeans to socialism. In place of the inexorable laws of social development on which Spencer hoped to base a master science of sociology, Youmans found in science detailed behavioral rules for heads of household charged with protecting their families under conditions of scarcity. At the core of his analysis stood a central image of classical economics: the male head of household locked in a struggle with the environment, delaying gratification in order to ensure survival and, eventually, attaining a degree of happiness for his family. Through Youmans' individualistic lens, institutions and cultural forms held meaning only insofar as they frustrated or supported the family in its struggle against nature. "Man's first and his life-long concern," he wrote, "is with his environment, the objective universe of God, the theatre of his activity, ownership, ambition, enjoyment, and the multifarious instrumentality of his experience and education."[22]

Youmans did, however, follow Spencer (and Jeremy Bentham before him) in flatly rejecting the "old ascetic misconception" that duty or altruism could ever guide human behavior. He saw only two motivating forces: pleasure and pain. Of the two, Youmans viewed pleasure as the more effective and permanent spur to action; the search for pleasure encompassed avoidance of pain but added its own appeal. He declared that all human institutions should be designed to make maximum use of this potent force, the "love of enjoyment."

[22] Youmans, "Introduction," 44.

Science loomed large here, because it taught individuals how to maximize pleasure and avoid pain in a world governed by strict, all-inclusive laws of behavior. Youmans described science as nothing less than "the revelation to reason
of the policy by which God administers the affairs of the world." The causal
relations it explored were God's means of punishing those who failed to obey
his rules.[23] Youmans departed from most scientific reformers of his era in finding nonscientific subjects, which did not offer reliable behavioral injunctions,
utterly valueless. "Greek," he wrote acerbically, "is not so ennobling a study as
that of sewerage." Despite its offensive subject matter, the latter field forced the
student to "trace out the obscure laws of our own and of surrounding nature,
so as to get command of natural agencies for beneficent ends."[24] Like all scientific studies, it revealed God's system of moral government.

No component of this system struck Youmans as more consequential than
the injunction to solve one's own problems without recourse to state power.
He equated state action with the manipulation of the inferior motive of fear of
pain and viewed private action as an expression of the more productive reliance on the search for pleasure. Whereas virtually all other scientific advocates
of *laissez-faire* made exceptions in the cases of science and education, Youmans
steadfastly insisted that these most potent sources of social rectitude must be
kept out of the hands of political operators. To his mind, education could only
function properly as a system of scientifically informed philanthropy. If schooling could not be pried away from the state, he feared, then there would be no
way to produce the kind of independently minded citizens who would ignore
the temptation to seek handouts and follow science's central lesson by attending to their affairs through private means. Youmans found Americans' near-
universal support for public education utterly exasperating, given the power
of the schools to shape citizens, and thus institutions. "Having affirmed the
voluntary principle of religion," he charged, Americans arbitrarily exempted
education from the universal truth that individuals always "know what is best
for themselves."[25]

Youmans thus endorsed a surprising combination of thoroughly scientized
politics and militant *laissez-faire*. On the one hand, he reduced the sphere of
politics to the mere application of science's findings, declaring that public officials could legitimately do no more than align public laws with the underlying

[23] *Ibid.*, 51, 48. Youmans lambasted religious thinkers who forgave men for their sins and, to his
mind, taught only license: e.g., [E. L. Youmans,] "Goldwin Smith on Scientific Morality," *Popular
Science Monthly* 20, no. 6 (April 1882), 847.

[24] [E. L. Youmans,] "The Study of Sewerage in London," *Popular Science Monthly* 18, no. 3
(January 1881), 414.

[25] [E. L. Youmans,] "Politics against Political Science," *Popular Science Monthly* 17, no. 4 (August
1880), 559–560; [Youmans,] "How New York Got a College," *Popular Science Monthly*
13, no. 1 (May 1878), 107. In Youmans' usage, "politics" and "science" represented directly
opposed principles for the guidance of public institutions – indeed, he employed these terms as
synonyms for "falsehood" and "truth." Thus, he called even the classical colleges "workshops for
the manufacture of politicians." "Politics against Political Science," 559.

"constitution of society," as revealed by scientists. On the other hand, he refused to abridge the *laissez-faire* principle even for the purposes of changing the mind of a public that sought something entirely different from its political representatives. If the vast majority of citizens failed to understand God's natural laws, or even to recognize the need of harmonizing their individual and collective behavior with such laws – if they were, as Youmans put it, in "Pilate's state of mind in regard to truth" – then the only scientifically valid solution was for more upright citizens to step in and educate their fellows by private means. Youmans viewed the state as an illegitimate expression of human power, its very existence a violation of God's policy. The state represented the use of pain, rather than pleasure, as a motivating force. Thus, all political proposals, no matter how strongly rooted in scientific findings, would need to work their way through the people themselves. In the end, Youmans believed, slow, evolutionary changes in public opinion, assisted by philanthropy "wisely conformed to facts," would eventually create a citizenry that looked to science for evidence of God's will and demanded that the state conform itself to science's teachings. God's commands would come to rule over society without any human actors having abridged the first and most important of these commands: appeal to pleasure, not to pain.[26]

Youmans gave voice to a rigorous, market-friendly vision of science and politics that resonated powerfully among the older commercial elite and the emerging professional-managerial class, especially in Youmans' home base of New York City. It seems fair to say that such a vision helped to broker a merger of these two elites in the 1870s and 1880s.[27] *Popular Science Monthly* thus filled an important niche in American intellectual life. The elder Oliver Wendell Holmes, a famed skeptic and medical man as well as poet, gushed to Youmans that each issue "comes to me like the air they send down to people in a diving bell."[28] Combining British versions of naturalism with a distinctively American form of economic conservatism, Youmans popularized a vision of science ready-made for the modernizing leaders of the nation's bustling commercial cities.

Within the ivied walls of academia, this unsentimental formulation of *laissez-faire* found its leading champion in Yale sociologist William Graham Sumner. Like Youmans, Sumner viewed science as the expression of a version of Christianity rooted in empirical facts and centered on rigorous adherence to God's laws. And he, too, outlined a competitive model of social relations

---

[26] [E. L. Youmans,] "Sociology and Theology at Yale College," *Popular Science Monthly* 17, no. 2 (June 1880), 265; [Youmans,] "Professor Martin on Scientific Education," *Popular Science Monthly* 10, no. 3 (January 1877), 369; [Youmans,] "Science and Social Reform," *Popular Science Monthly* 6, no. 34 (February 1875), 505.

[27] Thomas L. Haskell, *The Emergence of Professional Social Science: The American Social Science Association and the Nineteenth-Century Crisis of Authority* (Urbana: University of Illinois Press, 1977); Sven Beckert, *The Monied Metropolis: New York City and the Consolidation of the American Bourgeoisie, 1850–1896* (New York: Cambridge University Press, 2001).

[28] Quoted in Haar, "E. L. Youmans," 201.

that owed little to evolutionary metaphors, despite his retrospective identification as a "social Darwinist."[29] For Sumner, as for Youmans, "nature" meant the world surrounding the individual head of household, not a series of grand evolutionary laws.

However, Sumner was less enamored with utilitarian psychology than was Youmans. In fact, he defined his commitment to *laissez-faire* in normative terms. Sumner equated both social justice and social freedom with the achievement of a perfect correspondence between material rewards and individual merit, understood as a combination of effort, character, and natural gifts. Thus, a just society would produce a pattern of "unequal results ... proportioned to the merits of individuals." For Sumner, advancing the cause of justice meant eliminating all institutions that interfered with the translation of meritorious behavior into concrete outcomes. He believed that market capitalism performed this task flawlessly, and he deemed the growing inequality of the Gilded Age a sure sign of progress. "A drunkard in the gutter," wrote Sumner, "is just where he ought to be, according to the fitness and tendency of things."[30]

An independent science of sociology emerged as a central component of Sumner's project, in a way that it had not for Spencer and Youmans. Spencer essentially reduced social science to evolutionary biology, while Youmans reduced it to Bentham's pleasure-pain psychology and what he took to be the corresponding imperative to radically curtail state power. By contrast, Sumner sought to distinguish socially derived inequalities from natural differences in merit, in order to eliminate the former and leave each individual "no troubles but what belong to Nature." Sociology, as the science that revealed the dividing line between nature and society – between those forms of human suffering traceable to "the struggle with Nature for existence" and those "due to the malice of men, and to the imperfections or errors of civil institutions" – figured centrally for Sumner, even if its lessons amounted to little more than the *laissez-faire* ideal of classical political economy.[31]

Sumner's analysis added a new role for science, one not found in Youmans' model and centered on technology rather than political economy. In the manner of the Whigs before him, he described investment in industrial production as a potent form of private charity, because it mitigated material discomfort by maximizing the returns on a given quantity of human effort. Scientific and

---

[29] On the applications of this label, see Geoffrey M. Hodgson, "Social Darwinism in Anglophone Academic Journals: A Contribution to the History of the Term," *Journal of Historical Sociology* 17, no. 4 (December 2004): 428–463.

[30] William Graham Sumner, *What Social Classes Owe to Each Other* (New York: Harper, 1883), 164, 131. A short biography is Bruce Curtis, *William Graham Sumner* (Boston: Twayne, 1981). See also Robert C. Bannister, *Sociology and Scientism: The American Quest for Objectivity, 1880–1940* (Chapel Hill: University of North Carolina Press, 1987), 87–110; Dorothy Ross, *The Origins of American Social Science* (New York: Cambridge University Press, 1991), 85–88; and Jeffrey Sklansky, *The Soul's Economy: Market Society and Selfhood in American Thought, 1820–1920* (Chapel Hill: University of North Carolina Press, 2002), 105–136.

[31] Sumner, *What Social Classes Owe to Each Other*, 121, 17–18; Sumner, *Collected Essays in Political and Social Science* (New York: Holt, 1885), 78.

technological research stood even higher on Sumner's moral scale, as these improved, rather than merely perpetuating and spreading, industrial processes. These forms of humanitarian aid fostered prosperity without redistribution by "opening the chances" for merit to produce rewards. Opening chances, Sumner explained, added to the total value available to the community, enabling an increase of satisfaction for one or more persons without an accompanying decrease for others. By contrast, the redistribution of existing resources by political means left the sum of available value untouched, such that to "lift one man up" meant to "push another down." Worse still, Sumner added, when this process involved redistribution down the economic scale, it decreased the overall efficiency of the community by channeling material resources to those who did not value such resources enough to attain them through hard work. Such political meddling served only to inoculate the vicious against the invaluable lesson of nature's penalty.[32]

By focusing on the struggle between the individual and nature, Sumner and Youmans located economic activities beyond the government's sphere of jurisdiction. Each presented the same stark choice between two paths in political economy: the path of "individual liberty" and that of "paternalism, discipline, and authority." Further, their assumption of a natural harmony of economic interests meant that, as Sumner explained, "the duty of making the best of one's self individually is not a separate thing from the duty of filling one's place in society ... the two are one, and the latter is accomplished when the former is done." Indeed, Sumner declared that even voluntarily checking the pursuit of one's own interests inevitably eroded the freedom of others. The model polity portrayed by Sumner and Youmans would feature uncoerced interaction among private individuals who possessed knowledge of their own interests and desires, the power to pursue their ends, and the wisdom and self-restraint not to be tempted down mischievous paths.[33]

Many Gilded Age scientists worried about pinning science's hopes solely on the vagaries of private initiative, even when they looked to the latter for virtually all else. Whereas Youmans sought to keep science, like the churches, fully separate from the state, other scientific thinkers believed that the state should actively promote a kind of established faith, using the schools and universities to spread a scientific mode of thinking – including the free-market gospel – among leading citizens. Johns Hopkins' Simon Newcomb emerged as the leading advocate of this position in the 1870s. A polymathic astronomer who also published seminal works in statistical theory and economics, Newcomb was both America's best-known academic scientist and one of the most vigorous advocates of a science-friendly version of *laissez-faire*.[34]

Newcomb conceived of the economic matrix for science in very different terms than did Youmans and Sumner. Neither of them acknowledged the

[32] Sumner, *What Social Classes Owe to Each Other*, 164–166, 128.
[33] *Ibid.*, 98, 113.
[34] Albert E. Moyer, *A Scientist's Voice in American Culture: Simon Newcomb and the Rhetoric of Scientific Method* (Berkeley: University of California Press, 1992).

large-scale, bureaucratic organizations emerging all around them. By contrast, Newcomb viewed mass production and bureaucratic administration as the characteristic tendencies of the age. He deemed precise logical thought, as exemplified in the sciences, the appropriate mode of mental discipline for a modern, corporate society. The corresponding form of education should be neither classical nor merely technical, in Newcomb's view. Only training in the logic of the exact sciences, aimed at inculcating the underlying mode of thought rather than specific techniques and results, could do the trick. "No want from which our nation suffers," he declared during the 1876 centennial, "is more urgent than that of a wider diffusion of the ideas and modes of thought of the exact sciences."[35]

Newcomb endorsed what Nancy Cohen calls an "administrative state": one that adopted the paradoxical task of directly shaping the public mind to create a base of cultural support for limited government. He expected the state to fund the popular education in *laissez-faire* that Youmans assigned to private philanthropy. Newcomb also sought a munificent program of government funding for what would come to be called "basic science" – research into fundamental principles, undertaken with no thought of its potential application but transformative in the aggregate and in the long run. As an unqualified boon to all, yet directly profitable to no one, basic science required an exception to the *laissez-faire* rule, Newcomb argued. Precisely because he believed that scientific education offered a deeper and more permanent alternative to legislative enactments, Newcomb pushed for publicly funded research and education in the modern sciences.[36]

Newcomb's vision of a science-friendly administrative state meshed fairly well with the goals of the prominent Northeastern "mugwumps" who tried to stem rampant corruption and bridge incipient economic chasms by declaring independence from both parties and seeking to depoliticize national governance through various proposals for civil service reform. Like Newcomb, men such as *The Nation*'s founding editor E. L. Godkin stood one step away from strict *laissez-faire*, favoring an administrative state that would create public support for free-market policies. The vagaries of public opinion could not be left to chance, Godkin and other civil service reformers believed. Outside

[35] Simon Newcomb, "Abstract Science in America, 1776–1876," *North American Review* 122, no. 250 (January 1876), 91, 122–123.

[36] Nancy Cohen, *The Reconstruction of American Liberalism, 1865–1914* (Chapel Hill: University of North Carolina Press, 2002); Newcomb, "Abstract Science in America," 88, 118, 122. Despite his emphasis on the universal need for logical thinking, Newcomb did not envisage a program of scientific education for the masses. Instead, he outlined a kind of trickle-down theory, calling on his fellow citizens to generously fund "a wide and liberal training in the scientific spirit and the scientific method" for a small group of leaders. This handful of men, he explained, would "direct the society of the future," as they had during industrialization, even though their social influence would operate in a fashion "occult to the ordinary mind." Private initiative, Newcomb insisted, simply could not generate adequate financial support for research into the abstract, fundamental principles of nature, which promised immediate profit to no single individual but immense long-term gains for everyone. "What is a Liberal Education?" *Science* n.s. 3, no. 62 (April 11, 1884), 435; "Abstract Science," 121, 88.

the realm of economic policy, these figures worried less about the size of the state as such than about its administrative uprightness and its ability to sustain widespread agreement on the principles of classical political economy.[37]

Civil service reform, however, intersected only partially with the project of Newcomb and other scientific democrats in the universities. As we saw in Chapter 1, many of the latter initially hoped to create schools of political science that would produce a morally vigorous, administratively competent elite to run the kind of state the civil service reformers sought. But even those scientific democrats who actively supported civil service reform typically viewed the universities, rather than the state itself, as the leading instruments of national change. In their view, no merely legislative change could provide the needed regeneration of public virtue. Conversely, Newcomb averred, a public that properly valued logical thought would inevitably see to it that the "scientific offices" of public administration went to scientists and the "political offices" to politicians, no matter what official rules were in place.[38]

This difference had important ramifications for how the two groups thought about democratic participation. As the Gilded Age progressed, frustrated civil service reformers sometimes sought suffrage restrictions when the principles of popular sovereignty and social-scientific rectitude seemed to conflict. But scientific democrats in the universities expected to change government indirectly – and more effectively – by transforming the underlying body of public sentiment.[39] The prevalence of this expectation explains why university professors played a relatively minor role in the story of civil service reform. For the early scientific democrats, the state of the culture represented a far more immediate and important problem than the culture of the state.[40]

[37] In fact, Cohen sees in this theoretical move, designed to promote economic *laissez-faire*, the roots of Progressivism and New Deal liberalism. Liberal reformers endorsed other limited forms of state action to ensure good governance, provided these were carried out under the guise of neutral administration rather than in the spirit of partisan politics: *The Reconstruction of American Liberalism*. See also Mary O. Furner, "Social Scientists and the State: Constructing the Knowledge Base for Public Policy, 1880–1920," in *Intellectuals and Public Life: Between Radicalism and Reform*, ed. Leon Fink, Stephen T. Leonard, and Donald Reid (Ithaca: Cornell University Press, 1996).

[38] Simon Newcomb, "Exact Science in America," *North American Review* 119, no. 250 (October 1874), 302.

[39] Beckert, *The Monied Metropolis*, esp. 218–224. Cohen reminds us that all of the Gilded Age liberal reformers saw their primary role as teaching the public, even if they occasionally augmented this task with formal political engagement. *The Reconstruction of American Liberalism*, 13. See also Leslie Butler, *Critical Americans: Victorian Intellectuals and Transatlantic Liberal Reform* (Chapel Hill: University of North Carolina Press, 2007).

[40] Among those Cohen identifies as leaders of the civil service reform movement, only Sumner, Andrew Dickson White, and the classical economist Francis Amasa Walker held academic posts. Moreover, many of its academic supporters clustered in the Brahmin stronghold of Harvard, where President Eliot vigorously supported civil service reform from his bully pulpit. And even Eliot ultimately pinned his hopes for reform on the spread of academic knowledge among the public, not the direct influence of college graduates in politics. Cohen, *The Reconstruction of American Liberalism*, 12; Butler, *Critical Americans*, 195.

Like Youmans' utilitarian version of *laissez-faire*, then, civil service reform found its most vocal advocate – Godkin – and its base of support outside the universities, in an emerging class of white-collar professionals loosely centered in New York City. Scientific democrats in the universities did not constitute the core of *Popular Science Monthly*'s audience, or even the leadership cadre of the American Social Science Association (ASSA), formed in 1865 to promote the nascent field of inquiry. The ideas of Youmans and Godkin appealed primarily to professional men dependent on market competition and technical knowledge for their success. As labor strife increased in the 1870s and 1880s, many lawyers, doctors, and other professionals joined businessmen in supporting a strict version of *laissez-faire*, while increasingly defining this policy as a firm conclusion from scientific facts rather than a normative principle rooted in Christianity. Employed as experts and staking their livelihood on the belief of clients that they wielded reliable scientific knowledge, these professionals also pushed toward the familiar twentieth-century conception of science as a rigorously neutral source of immediately applicable knowledge. In fact, they made something of a fetish of their distaste for expressions of benevolence. In the intellectual realm, these non-academic professionals held that benevolence, if it had any meaning at all, could refer only to the practice of holding interlocutors to rigorous standards of argumentation. Politically, meanwhile, they demanded the strictest constraints on alms-giving and other deviations from a system of resource distribution regulated by market exchange.[41]

By contrast, most academic scholars, like most ministers, identified *laissez-faire* as an expression of Christian benevolence, after the manner of the antebellum Whigs. Whereas Youmans and Godkin declared that ethical action entailed strict adherence to the specific social laws revealed by science, most university-based scientific democrats stressed harmonious interpersonal relations and identified science with what Gilman called the "Yale spirit": "intelligence, industry, order, obedience, community, living for others, not for one's self, the greatest happiness in the utmost service."[42] In their view, it was not a series of detailed social laws, but rather a set of personal virtues rooted in the call to mutual service, that mediated God's relationship to man and pointed out the road to salvation.[43] The version of virtue ethics

---

[41] Such views were particularly well represented at meetings of the ASSA. See, for example, Haskell, *The Emergence of Professional Social Science*, 105, 147.

[42] Daniel Coit Gilman, *The Relations of Yale to Letters and Science* (Baltimore: [n.p.,] 1901), 41–42. The ethic of benevolence could also generate a lofty evolutionary idealism that urged students, teachers, and investigators alike to "commune with the Spirit of Life and Truth and Love Eternal": Jacob Gould Schurman, "Reply to the Addresses," in *Proceedings and Addresses at the Inauguration of Jacob Gould Schurman* (Ithaca: Cornell University, 1892), 21; cf. Angell, *Selected Addresses*, 3, 29, 43. This was particularly true in the state universities of the Midwest and California, which, as Marsden explains, were strongly shaped by Whig values: *The Soul of the American University*, 85.

[43] This is easy to miss, given that figures such as White discerned plenty of incontrovertible "facts" and "laws," particularly regarding the limits of state action. But they viewed agreement on these facts and laws as the product of the scientific mode of communication.

that these early scientific democrats tagged with the phrase "scientific spirit" often issued in the policy prescriptions held by Youmans or, more commonly, Newcomb, but the underlying reasoning differed substantially. Echoing the moral philosophers and Whig-Protestant activists, most scientific democrats in the new universities sought to imbue students with certain character traits, not a set of hard and fast beliefs.[44] Youmans and his like-minded counterparts swam against a powerful academic mainstream in declaring Christian ethics irrelevant to public policy, if not in trumpeting broadly *laissez-faire* sentiments.

THE ETHICAL ECONOMISTS

But what if Christian benevolence actually conflicted with *laissez-faire*? By the mid-1880s, a third group of scientific political economists had entered the fray, declaring that the scientific spirit called for active intervention by the state in economic affairs. After the bitter strike year of 1877 gave way to another economic bust in the following decade, the "labor question" – how Americans should respond to the endemic inequality and periodic depressions caused by industrial capitalism – bitterly divided the nation and its new universities. A new generation of scientific democrats, led by men such as Wisconsin's Richard T. Ely, concluded that the new conditions created by industrialization meant that the dictates of classical political economy no longer applied. The replacement of small proprietorships by massive corporations had fundamentally changed the economic game, they argued. Now, the proper expression of humanitarian ideals was an activist state that would intervene in economic relations to preserve the freedom and promote the flourishing of all. Revealingly, these new scientific democrats saw potential allies in older counterparts such as Andrew Dickson White – who, like their younger counterparts, insisted on the relevance of Christian ethics to public policy – even as they fought tooth and nail against the likes of Youmans, Sumner, and Newcomb. Yet as Ely and company ran into stiff opposition from the latter group, they gradually abandoned strong normative claims and presented their principles as purely empirical conclusions, a position that protected them from criticism by business-minded university trustees and seemed to promise greater rhetorical purchase on a public culture increasingly susceptible to claims of objectivity. Theoretically, meanwhile, the emergence of marginalism, with its orientation toward consumption rather than production, provided a broad umbrella under which all of the scientific democrats could find room. But as Part II shows, the core impulses of ethical economics resonated powerfully throughout the human sciences in the twentieth century.

[44] Even Eliot, whose distaste for public funding extended to the Morrill Act and infrastructural projects such as canals, rooted his hardheaded emphasis on the educative value of competition in an ethic of Christian benevolence. Altschuler, *Andrew D. White*, 181; Charles W. Eliot, *The Happy Life* (New York: Thomas Y. Crowell, 1896), 25–26; Eliot, *Educational Reform*, 37, 68; Eliot quoted in Rush Welter, *Popular Education and Democratic Thought in America*, 195. On this aspect of moral philosophy, see Marsden, *The Soul of the American University*, 51.

Ethical economics was in many respects an outgrowth of the Social Gospel movement among liberal Protestants.[45] Appalled by the inequality, misery, and violence of the Gilded Age, Social Gospelers such as Ely insisted that industrial capitalism violated the fundamental tenets of Christianity, and thus of truth itself.[46] Despite the threat of materialism posed by Darwin's theory of natural selection, on which many social scientists drew, many leaders of the American Protestant denominations embraced the social sciences as a potent new tool in the Christian arsenal, and sometimes essayed into the field themselves. Although they struggled to accommodate the harsher implications of Darwin's emphasis on competition, these Protestants harbored little doubt that science would prove compatible with – indeed, indispensable for – the evangelization of a polyglot urban population and the creation of a fully Christian society.[47]

The scientific turn in American Protestant thought found its apotheosis in W. D. P. Bliss' massive *Encyclopedia of Social Reforms* (1897), packed with tables and charts, and the even larger 1908 revision.[48] Even more than a penchant for statistics, however, a personal and theoretical commitment to the collective good united the Progressive Era thinkers we anachronistically sort into "religious" and "scientific" contingents. Both ministers and economists emphasized the priority of the common good over that of the individual. "Sociology is a science of sacrifice; of redemption; of atonement," noted the Social Gospel firebrand George D. Herron, because "[s]ociety is sacrifice. It is the manifestation of the eternal sacrifice of God, as it was unveiled on the cross, in human relations." The literary scholar Susan L. Mizruchi has shown that images of sacrifice pervaded Western scientific discourses in the late nineteenth and early twentieth centuries. As we will see in subsequent chapters, the Christian ideal of mutual service meshed well with either a language of rigid self-negation or a softer, more characteristically American ethic of collectively aided self-realization. Advocates of both approaches identified the "immanence of the social" as the very "modern note" itself.[49]

---

[45] Ely employed this term in "Ethics and Economics," *Science* n.s. 7, no. 175 (June 11, 1886), 530, and several recent historians have adopted it.

[46] Paul T. Phillips, *A Kingdom on Earth: Anglo-American Social Christianity, 1880–1940* (University Park: Pennsylvania State University Press, 1996); Gary Scott Smith, *The Search for Social Salvation: Social Christianity and America, 1880–1925* (Lanham, MD: Lexington Books, 2000); Gary J. Dorrien, *The Making of American Liberal Theology: Imagining Progressive Religion, 1805–1900* (Louisville: Westminster John Knox, 2001).

[47] Cashdollar, *The Transformation of Theology*, 419–422; Gary J. Dorrien, *Social Ethics in the Making: Interpreting an American Tradition* (Malden, MA: Wiley-Blackwell, 2009), esp. 6–59.

[48] W. D. P. Bliss, ed., *The Encyclopedia of Social Reforms* (New York: Funk & Wagnalls, 1897); Bliss, ed., *The New Encyclopedia of Social Reform* (New York: Funk & Wagnalls, 1908).

[49] George D. Herron, "The Christian Society" (1894), in Eldon J. Eisenach, ed., *The Social and Political Thought of American Progressivism* (Indianapolis: Hackett, 2006), 184; Susan L. Mizruchi, *The Science of Sacrifice: American Literature and Modern Social Theory* (Princeton: Princeton University Press, 1998); Albion W. Small, review of Edward A. Ross, *Social Control*,

From a theological standpoint, the concept of "social salvation" anchored the Social Gospel outlook. "It is not a matter of saving human atoms, but of saving the social organism," summarized Walter Rauschenbusch in his influential *Christianity and the Social Crisis* (1907). "It is not a matter of getting individuals to heaven, but of transforming the life on earth into the harmony of heaven." Protestant progressives asserted that God worked his transformation of humanity through cultural and social processes, including the growth of natural and social knowledge and the emergence of humanitarian institutions. Walking side by side with the ethical economists, sociologically minded ministers and theologians hoped to use public institutions to implement the new progressive spirit.[50]

Back in the early 1880s, Lester Frank Ward had framed a similar political project in more secular terms. The central figure, along with Sumner, in the early development of American sociology, Ward was a sometime botanist and long-serving federal bureaucrat rather than a university professor. Unlike New York's industrialists and professionals, the naturalists of Gilded Age Washington saw government as an unalloyed good – indeed, a necessary catalyst of progress. Ward embraced a form of evolutionary idealism directly opposed to Spencer's version, echoing Spencer's focus on society as a single unit but linking social progress directly to the growth of governmental initiative.[51]

Ascribing to the human mind an active role in relation to nature, Ward described conscious thought as a means of turning nature's causal relations to human advantage. He further reasoned that mental activity stood in the same relation to human desires – the building blocks of social reality – as it did to natural resources. Desires provided the motive force in the social world, while the intellect steered and channeled their concrete expressions, and thus their outcomes. Ward thus defined all social institutions, including the state, as products of human intelligence that promoted individual flourishing far more effectively than market competition. Rather than struggling in solitude, Ward believed, Americans should wage a collective war on nature, aided by a

---

American Journal of Sociology 9 (1904), 580. See also David A. Hollinger, "Inquiry and Uplift: Late Nineteenth-Century American Academics and the Moral Efficacy of Scientific Practice," in *The Authority of Experts*, ed. Thomas L. Haskell (Bloomington: Indiana University Press, 1984); and George Levine, *Dying to Know: Scientific Epistemology and Narrative in Victorian England* (Chicago: University of Chicago Press, 2002).

[50] Quoted in William R. Hutchison, *The Modernist Impulse in American Protestantism* (Cambridge: Harvard University Press, 1976), 173.

[51] Michael J. Lacey and Mary O. Furner, eds., *The State and Social Investigation in Britain and the United States* (Washington: Woodrow Wilson Center Press, 1993); Philip J. Pauly, *Biologists and the Promise of American Life* (Princeton: Princeton University Press, 2000), 44–70. On Ward, see also Bannister, *Sociology and Scientism*, 13–31; Wilfred M. McClay, *The Masterless: Self and Society in Modern America* (Chapel Hill: University of North Carolina Press, 1994), 120–133; Gillis J. Harp, *Positivist Republic: Auguste Comte and the Reconstruction of American Liberalism, 1865–1920* (University Park: Pennsylvania State University Press, 1995), 109–153; Sklansky, *The Soul's Economy*, 196–201; and Edward C. Rafferty, *Apostle of Human Progress: Lester Frank Ward and American Political Thought, 1841–1913* (Lanham, MD: Rowman & Littlefield, 2003).

scientifically designed state. Freedom and prosperity would flow from good government, not small government.

Ward soundly rejected all attempts to justify the dog-eat-dog competition of a modern capitalist economy by identifying it as an extension or analog of the biological process of natural selection. In place of natural selection's "destruction of the weak," he argued, an advanced humanitarian society practiced "artificial selection," in the form of "the *protection* of the weak." Ely augmented this benevolent impulse with a robust ethic of self-realization that demanded far more than material aid to one's fellows. "There are powers in every human being capable of cultivation," he wrote, and it was the purpose of the individual to cultivate these powers to the fullest. This was also the purpose of social institutions, and above all economic institutions, given that "the economic life is the basis of this growth of all higher faculties – faculties of love, of knowledge, of aesthetic perception, and the like, as exhibited in religion, art, language, literature, science, social and political life." According to Ely, economists, as citizens, had an ethical obligation to seek "such a production and such a distribution of economic goods as must in the highest practicable degree subserve the end and purpose of human existence for all members of society."[52]

In place of Ward's Comtean and Spencerian influences, Ely and most other ethical economists took their cues from the Social Gospel. With the exception of the Jewish E. R. A. Seligman, these figures had close personal and professional ties to the Social Gospel search for a cooperative commonwealth. Even after his controversial economic views forced him to leave Cornell for Michigan, Henry Carter Adams, the son of a Congregationalist missionary to Iowa, urged his fellow economists to help realize "the old Christian conception of a just price, and the modern Christian conception of equal opportunities for all." Adams and his counterparts argued that these ethical imperatives required social scientists to side with labor and build up an activist state to check corporate power.[53]

Lessons learned overseas reinforced the ethical economists' Social Gospel commitments. Unlike the substantially self-taught Youmans, Newcomb, and Ward, these figures cut their scholarly teeth in Germany, where American scholars flocked for doctoral work under the world's leading scientists after the Civil War. The ethical economists brought home three leading tenets of German "historical economists" such as Karl Knies and Gustav von Schmoller, who were friendly to Chancellor Bismarck's new welfare state. First, as Ward had also taught, human beings did not simply discover economic institutions, but created them. They could thus re-create those institutions to channel economic impulses toward desired ends. Second, society should be seen as a

[52] Lester Frank Ward, "Mind as a Social Factor," *Mind* 9, no. 36 (October 1884), 570; Ely, "Ethics and Economics," 531–532.
[53] Ross, *The Origins of American Social Science*, 102–122; quoted in Marsden, *The Soul of the American University*, 176. Ross calls these figures "evangelical economists" and points out that the inheritance of liberal Christianity (or Judaism, in Seligman's case) distinguished them from their free-market rivals (103–104).

growing, changing organism, not a static, mechanical structure. This meant that its economic activities could not be studied in isolation from its other social functions, including patterns of moral regulation. It also implied that generalizations about the social organism's past or current structures could not be presumed to hold true for all stages of its life. To the extent that scholars could discern economic "laws," these were contingent, relational, contextual, and statistical, operating very differently than the absolute, transcendent prescriptions of classical political economy.[54]

Finally, the ethical economists learned from their German teachers to see a fundamental break in the recent course of history: the social organism had entered a new industrial "environment" that presented a radically altered set of conditions for survival. At the heart of the ethical economists' departure from classical principles stood a claim of historical discontinuity that undermined any conception of scientific truths as timeless universals. The only "permanent and universal fact" about the social organism was the "law of its own development," Adams explained; "all other facts are relative truths; and those systems of thought based upon them, temporary systems." The ethical economists claimed that, in the new industrial environment, market competition no longer offered the most effective means of ensuring human flourishing. Only its corrosive effects on personal virtue remained.[55]

Between 1884 and 1886, the younger economists and their orthodox elders aired their differences in a vigorous series of print exchanges. The ethical economists called for scientific activism, drawing on their historicist, organicist principles. To the small-state thinkers, however, any historical change that had occurred in economics amounted to nothing more than the progressive revelation of universal principles of economic organization, which were as true for industrial societies as for agricultural and commercial ones. Either there were solid cause-effect relationships that held true for all times and places, Newcomb insisted, or science and collective public action both failed as tools for creating social order. If economic laws could not be relied on to stand still, then science could not perform the political tasks charged to it by society. All economists could agree, claimed Newcomb, "that the state ought to interfere where it is really necessary to the public welfare," namely in "education, the public morals, and the public health." Yet *laissez-faire* remained the best economic arrangement. That system, said Newcomb, promoted the good of all "by giving the adult individual the widest liberty within the limits prescribed by considerations of public health and morality."[56]

[54] Nicholas Balabkins, *Not by Theory Alone...: The Economics of Gustav Von Schmoller and Its Legacy to America* (Berlin: Duncker & Humblot, 1988), 86–110. A recent study of the school is Yūichi Shionoya, *The Soul of the German Historical School: Methodological Essays on Schmoller, Weber, and Schumpeter* (New York: Springer, 2005). See also Daniel T. Rodgers, *Atlantic Crossings: Social Politics in a Progressive Age* (Cambridge: Belknap, 1998), 76–111.

[55] Henry Carter Adams, "Economics and Jurisprudence," *Science* n.s. 8, no. 178 (July 2, 1886), 16; Ross, *The Origins of American Social Science*, 106–111.

[56] Richard T. Ely, "The Economic Discussion in *Science*," *Science* n.s. 8, no. 178 (July 2, 1886), 5; Adams, "Economics and Jurisprudence," 16; quoted in Mary O. Furner, *Advocacy & Objectivity:*

In this debate, questions about the character of scientific laws and their relation to moral principles loomed large. Newcomb deemed it "a contradiction in terms to call a talk about what ought to be, science." One could certainly attempt to promote ethical ends by engaging in empirical investigation, he allowed. But one could never confuse science with ethics. To Newcomb's mind, the ability to distinguish between facts and wishes represented both the heart of science and the most pressing cultural need in modern society. Arthur T. Hadley, a younger *laissez-faire* theorist, compared the economist to a mechanical engineer charged with ensuring that a bridge would not fail. Political economy, he argued, told legislators, "[b]eyond certain limits, all legislation fails," not "[s]uch and such legislation will produce the best results." This, stated Hadley, "is the natural relation of a science to the art."[57] For the ethical economists, by contrast, economic "laws" were statistical generalizations that tracked patterns of behavior shaped by human institutions rather than structural characteristics of a pre-social world. And economic laws could change over time, if they were not etched in the universe itself.

The ethical economists' stance opened room for normative commitments: What patterns of behavior should economic institutions seek to produce? At the same time, it also authorized an extensive program of empirical inquiry. To understand an ever-changing social organism, after all, one needed a flexible and continual program of detailed investigation, especially at moments of rapid transition such as the post-Civil War era. Both the normative and the empirical elements of the ethical economists' program shone through in the 1885 founding statement of the American Economic Association (AEA). Ely framed the document to exclude "the Sumner crowd," while including all of those who viewed economics in ethical terms, from the Social Gospel minister Washington Gladden to Andrew Dickson White and other advocates of an administrative state. Yet the statement also featured indicators of the less value-laden rhetoric that most AEA members would adopt by the mid-1890s. Ely dismissed classical political economy as mere "speculation," deduced from the false premise of a purely self-interested individual. By contrast, he wrote, his group would offer "an impartial study of actual conditions," wherein ethical commitments served as an important motive in economic action. "From a purely scientific standpoint," Ely explained, "we do not live for ourselves alone, but for one another as well as ourselves."[58]

A Crisis in the Professionalization of American Social Science, 1865–1905 (Lexington: University Press of Kentucky, 1975), 97; Simon Newcomb, "Aspects of the Economic Discussion," *Science* n.s. 7, no. 176 (June 18, 1886), 538, 541.

[57] Newcomb, "Aspects of the Economic Discussion," 539–540; Arthur T. Hadley, "Economic Laws and Methods," *Science* n.s. 8, no. 180 (July 16, 1886), 48.

[58] Quoted in Furner, *Advocacy and Objectivity*, 69, 73; Bradley W. Bateman, "Race, Intellectual History, and American Economics: A Prolegomenon to the Past," *History of Political Economy* 35, no. 4 (2003), 715; quoted in Ely, "Report of the Organization," *Publications of the American Economic Association* 1, no. 1 (March 1886), 17, 7, 12. Newcomb fumed that Ely's organization amounted to "a sort of church, requiring for admission to its full communion a renunciation of ancient errors, and an adhesion to the supposed new creed." Quoted in Furner, *Advocacy and Objectivity*, 73.

Political pressures subsequently led the ethical economists to stress the empirical side of their program and downplay its normative orientation. By the late 1880s, most middle-class Americans had come to see political subversion rather than Christian ethics or scientific realism in the ethical economists' activities.[59] Ely's ambiguous protest that no one in his circle supported "socialism, pure and simple," hardly stemmed a wave of public criticism that swelled with the Haymarket bombing in 1886 and reached new heights during the bitter labor struggles accompanying the downturn of 1893.[60] Meanwhile, *laissez-faire* theorists in the universities mobilized against Ely and his compatriots, joining hands with business-minded trustees to shut down the political activism of their colleagues. Even the other leading ethical economists worried about Ely's public activities, which threatened to undermine their theoretical project. Seligman, temperamentally more cautious than Ely and largely uninterested in the movement's activist side, tried to minimize Ely's influence by inviting the traditionalists into the AEA, whereupon the group dropped its platform denouncing *laissez-faire*. Seligman's campaign to rein in Ely enjoyed tacit support from the other two leading ethical economists, Adams and John Bates Clark. The latter had steadily soured on labor, and had long since turned his attention to theoretical questions. For his part, Adams read organic analogies more conservatively than did Ely, emphasizing the stability of the social organism over its dynamism.[61] With all of these forces arrayed against Ely, something would have to give.

Under fire from all sides, Ely stepped down as AEA secretary in 1892. Over the next few years, he and other ethical economists who continued to engage in political activism – most notably John R. Commons and Edward Bemis – faced the prospect of losing their jobs unless they tempered their public statements. Bemis persisted, at the cost of his career, but Ely stepped back from the brink. In October 1894, he declared himself "a conservative rather than a radical," and even "in the strict sense of the term an aristocrat rather than a democrat." For the time being, Ely largely confined himself to technical studies in rural economics and banished explicit policy statements from his writings and speeches. Cutting his ties with Christian socialists, he instead followed the path blazed by Adams, who had concluded several years earlier that economists should play only a technical or managerial role in the formation of public policy.[62]

[59] Bradley W. Bateman and Ethan B. Kapstein, "Between God and the Market: The Religious Roots of the American Economic Association," *Journal of Economic Perspectives* 13, no. 4 (Autumn 1999), 254.

[60] Quoted in Ross, *The Origins of American Social Science*, 113; Bateman and Kapstein, "Between God and the Market," 254.

[61] Furner, *Advocacy and Objectivity*, 115–123, 185; Ross, *The Origins of American Social Science*, 115. In fact, Adams denied that moral exhortation of the type favored by Ely could do any good at all. Taking a utilitarian line, he argued that only the fear of punishment associated with legal constraints could push Americans toward a more just economic system. "Economics and Jurisprudence," 17.

[62] Furner, *Advocacy and Objectivity*, 123, 200–203, 158 (quoted), 162; Adams, "Economics and Jurisprudence," 7. On the controversy in general, see also Ross, *The Origins of American Social Science*, 115–118.

Thus, under the pressure of political and professional resistance, something like today's value-neutral conception of science began to emerge from within the fiery furnace of Christian radicalism, just as it had taken shape among small-state thinkers. Ely's adoption of a more technical approach did not mean that he had converted to *laissez-faire*, of course. Instead, making a virtue out of political necessity, he gambled on the prospect that the American people would choose to put technical knowledge to progressive moral and political uses. The cultural activities of the Protestant denominations, inspired by the Social Gospel, provided the missing link in this conception of reform: Ely assumed that a Christianized citizenry would naturally look to the technical products of social science in seeking to realize its deepest commitments. By providing an accurate description of the emerging social mechanism and its multiple ethical potentialities, social scientists would empower Americans to choose something other than the status quo. At the level of emotions and motives, however, the churches and newspapers would go it alone in stoking the fires of reform sentiment. Political action, in this conception, still involved both moral and practical dimensions. But Christianity provided the tenets of political morality, while scientists concentrated on practical matters. Viewing the beloved community of the Social Gospel as an immanent – and imminent – prospect in history, Ely and his counterparts presumed that the public, guided by progressive church leaders, would use scientific knowledge to rein in big business.

The development of marginal theory in economics rendered this accommodation more palatable to the young critics of *laissez-faire* in the 1890s. This approach embodied a new way of thinking about the key economic concept of "value." Shifting the focus from production to consumption, the marginalist framework defined the value of any given good in terms of the market demand for it, rather than the labor that had gone into producing it. In this definition, value stemmed from largely ineffable individual preferences and could be determined empirically by looking at prices. This move had several important implications. For one, it disengaged the concept of value from the finitude of labor and tied it to the virtually infinite plane of consumer desire. At the same time, marginalism ensured that not only physical items, but also services and even experiences, could be assigned discrete values.[63]

To be sure, many of the early marginal theorists upheld *laissez-faire*, arguing that markets, rather than charitable institutions or the state, could best mediate the conflicts arising between individuals in pursuit of their own interests.

---

[63] Ross, *The Origins of American Social Science*, 118–122, 172–195; Michael A. Bernstein, "American Economists and the 'Marginalist Revolution': Notes on the Intellectual and Social Contexts of Professionalization," *Journal of Historical Sociology* 16, no. 1 (March 2003): 135–180. Although the line between historical economics and marginalism seems sharp in retrospect, the boundary was quite fuzzy at the time: Bateman and Kapstein, "Between God and the Market," 255. On the political and cultural resonances of marginalism, see also James Livingston, *Pragmatism and the Political Economy of Cultural Revolution, 1850–1940* (Chapel Hill: University of North Carolina Press, 1994), 49–83. Cf. Richard Wightman Fox, "The Culture of Liberal Protestant Progressivism, 1875–1925," *Journal of Interdisciplinary History* 23 (1993): 639–660.

Even within a market framework, however, the new theory made room for a potentially infinite range and diversity of human values – values that could vary between individuals and groups. Many ways of life, including that recommended by Christ, now seemed compatible with property rights and market relations. At the theoretical level, most of the ethical economists sought first and foremost to displace "the assumption of self-interest as the sole regulator of economic action." Thus Clark, a key architect of marginal theory, saw in it the culmination of his long struggle to "find a place in the system for the better motives of human nature." By historicizing values but leaving market structures largely untouched, marginalism provided a meeting ground for the backtracking ethical economists and their *laissez-faire* critics.[64]

The figures described in this chapter were hardly philosophers of science in the deep sense of the phrase. They developed their views on the nature of scientific laws, evidence, and reasoning in the context of intellectual projects that overlapped substantially with contemporary political controversies. The same, of course, was true of those who committed more of their time and resources to the systematic study of science. But the latter began to depart, as the disputants over political economy did not, from the common-sense theory of knowledge. The "foundationalist" postulates that human minds and truths were universal and that knowledge involved a fit between the mind and the external world comported fairly well with both Youmans' emphasis on the ultimate uncertainty of all such knowledge and the ethical economists' insistence that the social organism – the object of empirical study for social scientists – had changed its character fundamentally in recent years.

As we have seen, however, this view of human knowledge had descriptive shortcomings in a world where the establishment of scientific universities and the accumulation of empirical evidence seemed only to have deepened the divides between working scientists on questions of religion and political economy. Additionally, accelerating immigration and the proliferation of goods and lifestyles presented many theorists with the impression of an essential pluralism in the human condition that stood in significant tension with the common-sense theory. And the apparent irrationality of Gilded Age politics also called into question the dominance of reason over human behavior. Addressing all of these concerns at once, the New Psychology that emerged in the 1880s and 1890s fed into powerful new theories of knowledge that fundamentally reconceived science's relationship to democracy.

---

[64] Quoted in Ross, *The Origins of American Social Science*, 118; E. R. A. Seligman, "Change in the Tenets of Political Economy With Time," *Science* [2nd series] 7, no. 168 (April 23, 1886), 381.

3

Science and Philosophy

In the last decades of the nineteenth century, debates over Darwinism and the labor problem helped to fuel a broader set of disagreements regarding the nature of scientific inquiry. According to the inherited framework of faculty psychology and the associated common-sense theory of knowledge, all unbiased investigators should see the truths of nature in the same light. The presence of bitter interpretive divisions among highly trained and presumably honest scholars challenged this presumption. Common-sense principles offered a ready explanation for such fractiousness: some of the investigators had not fully imbibed the scientific spirit and labored under pervasive personal or social biases. But for those standing between the poles in these religious and political-economic debates, as well as those disputants who hoped to convert rather than outmaneuver their enemies, disagreements over the state's economic role called into question the common-sense presumption that scientific truth is universal and easy to discern.

As a result, American thinkers began to develop more sophisticated, historically grounded conceptions of the knowing process. The common-sense framework slowly fractured as scientific democrats parted ways on fundamental questions of scientific methodology, the cognitive impact of investigators' personal commitments, and the very nature of the mind. Although many still flew the flag of common-sense philosophy as the new century dawned, the intellectual ferment of the Gilded Age generated new conceptions of science that prefigured much twentieth-century thought on the matter. Like the arguments discussed in the previous chapter, these proliferating philosophies of knowledge highlight the breadth of scientific democracy – the number of different ways that American thinkers committed to the general proposition that science could sustain a democratic culture filled out this diffuse claim with specific intellectual and institutional content.

The political stakes became even clearer after 1900, when philosophers struggled to specify the relationship between their field and the sciences. Drawing direct parallels between the success of science in generating reliable, consensual knowledge and the hoped-for capacity of a modern democracy to achieve cultural coherence, many of these figures portrayed their own discipline as the needed link between the two. But they disagreed on how sciences and polities worked. Should philosophy become a one-party system in which all endorsed a common "platform" of consensually held truths? Should it remain a fully open polity of discussion in which individual disagreements would continue unabated? Or would something in between – perhaps a two-party or three-party system – best approximate the conditions that had produced such spectacular successes in the sciences?

POSITIVISM

Two broad families of theories, under the loose headings of positivism and pragmatism, initially contended to replace the common-sense framework in the last decades of the nineteenth century, although the full flowering of each awaited the twentieth century. Adherents of both approaches reflected on the historical and contemporary practices of physical scientists, whom they understood to have largely resolved the problem of attaining reliable, consensual knowledge. They then sought to generalize the lessons of the physical sciences into systematic theories regarding the nature and scope of science.

But the two groups reached quite different conclusions from their explorations. Positivists declared that failures to reach consensus stemmed from the improper application of empirical methods to questions or realms to which they did not apply. Positivists believed it was possible to determine ahead of time where empirical inquiry would produce interpretive consensus and where it could never do so because of the intrinsic nature of the subject matter. Pragmatists, on the other hand, concluded just the opposite from the past success of the sciences: one could never presume to know in advance where consensus would or would not emerge. The only possible response to this fact, they believed, was to reserve the term "truth" for those areas of working consensus that had already appeared, while promoting sustained empirical investigation in all other spheres of inquiry.

To state the point in different terms, positivists and pragmatists adopted divergent formulations of the scientific spirit embraced by the scientific democrats of the previous generation. Positivists applied the core virtue of humility to specific realms of thought. Whether skeptical or respectful of their counterparts in fields outside the empirical sciences, they worked to distinguish science sharply from these other areas, in which, they presumed, intellectual consensus could never emerge. In an important sense, positivist theories of science echoed the stepping-back move at the heart of common-sense thought, even when their advocates abandoned the conception of the mind as a collection of discrete faculties tuned to various phases of the world. They sought to define, once and for all, the terrain to which investigators needed to step back in order to find common ground. They did so by parceling up the sphere of intellectual activity and declaring only certain questions or forms of evidence amenable to scientific treatment.

Pragmatists, by contrast, echoed part of the strategy adopted by E. L. Youmans in the Darwinian conflict. They sought to render science's epistemological claims more modest, yet at the same time extend its empirical method and mode of social organization into all fields – especially those ruled off limits by positivists. To their minds, the virtue of humility required investigators to refrain from making claims in advance about where their distinctive modes of inquiry could produce a scientific consensus, let alone truly universal truths that would enlist the unforced assent of all humanity.

Versions of the positivist tendency to delimit the sphere of empirical inquiry appeared repeatedly in the controversy over evolutionary theory's philosophical and theological implications. Yet the term "positivism" had initially referred to something quite different. It was the coinage of the early-nineteenth-century French philosopher Auguste Comte, who also gave the world "sociology" and "altruism." Although Comte sharply distinguished science from theology and metaphysics, he portrayed these as divergent methods, not discrete subject areas. Indeed, Comte expected scientists, proceeding inductively from incontrovertible facts, to provide a firm empirical foundation of "positive knowledge" on which to rebuild the higher forms of human thought from the ground up. Facts loomed as large in Comte's positivism as in the common-sense framework. These hard, incontrovertible atoms of knowledge, from which no sane mind could deviate, represented the building blocks of all valid human thought. Yet for Comte and his followers, positivism – a commitment to the creation of true, positive knowledge in every field of human endeavor – remained a *philosophy*. It would play the same role as traditional philosophical systems, answering not only questions about the physical world, but also moral and political questions. Many Anglo-American theorists, including Herbert Spencer, John W. Draper, and Youmans, followed Comte in defining science as a universally applicable method of inquiry and assuming that it would fall disastrously short of its historical destiny if it did not provide normative guidance for all aspects of life. Comte even proposed a "religion of humanity," rooted in scientific facts.[1]

Positivism in its more familiar sense took shape in the second half of the nineteenth century, as scientific materialists and agnostics clashed with theologians and theistic philosophers. Many in the former group tried to wall science off from theological questions, hoping thereby to protect science's public image as a fount of consensus and reliable knowledge. This version of positivism formalized John Michels' rather loose conception of separate intellectual spheres by limiting science to questions regarding the natural world. Scientists, in this view, should restrict themselves to inquiries in which the existence of a solid, external object – nature – ensured eventual agreement. Some positivists in this vein argued that part of social reality fell within the domain of science – that

[1] Charles D. Cashdollar, *The Transformation of Theology, 1830–1890: Positivism and Protestant Thought in Britain and America* (Princeton: Princeton University Press, 1989), 7–12; Gillis J. Harp, *Positivist Republic: Auguste Comte and the Reconstruction of American Liberalism, 1865–1920* (University Park: Pennsylvania State University Press, 1995), 10–22.

something like "social nature" existed. (This was the view toward which Ely and most other ethical economists leaned by the mid-1890s.) Others denied that claim, designating an entirely different method for the study of the social realm. But all agreed that scientific methods extended only so far, and could not, by definition, provide answers to all of life's dilemmas.[2]

Meanwhile, science-minded thinkers committed to metaphysical idealism – the view that reality was not, in fact, material in composition but essentially mental or spiritual – sometimes adopted a different version of positivism in the hopes of conceptually disentangling science from materialism and agnosticism. This version of positivism differentiated science from metaphysics – questions about the ultimate constitution of reality – rather than theology. In the early 1880s, for example, the German-born jurist and philosopher J. B. Stallo, a fervent opponent of the atomic theory, insisted that scientific evidence said nothing whatsoever about the character of the underlying substance or substances composing the universe. Whether the world was constituted by mind or by matter, by particles or by fields – these were questions for philosophers, not scientists. In fact, Stallo argued that the habit of fixating on empirical results utterly disqualified scientists for philosophical speculation on the constitution of reality. The true scientist, he said, roundly rejected "the delusion that every momentary device for sorting and grouping facts is to be hailed as a new scientific revelation." Toward the end of the nineteenth century, the German physicist Ernst Mach and the British statistician Karl Pearson became the leading advocates of this kind of positivism. Deploring the frequent identification of science with materialism, Mach and Pearson protested that genuine scientists made no claims whatsoever about the nature of reality. Science, they argued, described how the world worked – how its various parts functioned in practice – rather than how the world was constituted in some ultimate sense.[3]

Pearson's *The Grammar of Science* (1892) proved particularly influential among American scientific thinkers for its dismissal of metaphysics and emphasis on the scientific investigator's rigorous suppression of personal feelings and values in the process of inquiry. Yet Pearson, like earlier theorists such as Comte and John Stuart Mill, nevertheless believed that the positive knowledge thus created could serve as the basis for a comprehensive approach to the world, a "social morality" providing guidance in all spheres of life. Hardly a cautious

[2] Adherents to this view rarely considered themselves positivists, a term that philosophers of science – as opposed to critical social scientists – restrict to the view detailed in the following paragraph. In practice, however, many of the professionalizing scholars of the late nineteenth century hewed to this position. Ceding the spheres of political and cultural struggle to other groups of leaders, they sought to weed out amateurs in their fields, especially those who persisted in speculating on matters beyond the reach of the human sensory apparatus. Still, they hardly believed, like their twentieth-century successors, that science offered no normative conclusions regarding human behavior.

[3] J. B. Stallo, "Speculative Science," *Popular Science Monthly* 21, no. 2 (June 1882), 164; Gerald Holton, "Ernst Mach and the Fortunes of Positivism in America," *Isis* 83, no. 1 (March 1992): 27–60; Theodore M. Porter, *Karl Pearson: The Scientific Life in a Statistical Age* (Princeton: Princeton University Press, 2004).

thinker, Pearson declared that science dictated a total overhaul of modern politics, in keeping with socialist, feminist, and eugenicist principles.[4]

A final group of positivists, less inclined to spell out their views in systematic terms, lopped off both of the problematic domains and several others as well. Their kitchen-sink positivism differentiated science not just from theology and metaphysics, but also from morality and politics – indeed, from all realms of normative commitment and subjective experience, including aesthetic and emotional pursuits as well as ethical ones. Positivists of this variety assumed that human subjectivity was a realm of deep, irremediable differences, such that the incontrovertible facts needed to build up generalized knowledge claims would prove elusive in all areas of life – virtually the whole of it – where human subjectivity played a part. This left to scientists only those limited areas in which observers could discern unchanging causal connections on the basis of unimpeachable sensory evidence. And even there, this version of positivism held, scientists could learn nothing about the ultimate character of reality, as Mach and Pearson had argued. On this view, scientific knowledge was true in a technological rather than normative sense. It spoke to means, not ends, telling its users how to do what they sought, not which ends they should seek. Science generated only conditional, if-then statements; it offered no direct imperatives, no absolute commands. Viewed from a political standpoint, this view held that science set strict boundaries to action. Yet within those boundaries it said nothing about the choice between various actions.[5]

This form of positivism is largely in keeping with the contemporary American usage of "science," which refers to a specific, clearly delimited domain of nonnormative questions rather than a universally applicable method of inquiry. It appeared only sporadically in the late nineteenth century. But some *laissez-faire* theorists, seeking to rule out redistributive policies, began to gravitate toward this position. William Graham Sumner's debate with Lester Frank Ward is revealing on this score. Ward adopted a nonnormative, essentially technological understanding of physical science. But in the social sciences, he was a positivist of the Comtean variety. Ward identified broad laws of progress covering every domain of social experience, whether moral or practical. By contrast, Sumner argued that social science offered only neutral knowledge of natural forces, and that these forces possessed no moral weight. "Science is colorless and impersonal," he declared. "The moral deductions as to what one ought to do are to be drawn by the reason and conscience of the individual man," after that man had assimilated science's lessons about the effects of various courses of action. "There is no injunction, no 'ought' in political economy at all," Sumner repeatedly emphasized, even as he declared that "the

---

[4] Quoted in Porter, *Karl Pearson*, 292. Cf. George Levine, "Science and Citizenship: Karl Pearson and the Ethics of Epistemology," *Modernism/Modernity* 3 (1996): 137–143.
[5] Robert N. Proctor, *Value-Free Science? Purity and Power in Modern Knowledge* (Cambridge, MA: Harvard University Press, 1991), 65–208. A good example is Arthur T. Hadley, "Economic Laws and Methods," *Science* n.s. 8, no. 180 (July 16, 1886), 47–48.

natural laws of the social order are in their entire character like the laws of physics."[6]

PRAGMATISM

The intellectual pruning operations undertaken by positivists were not the only available response to the endemic disagreement among scientists, however. The new experimental psychology that emerged in the 1880s pointed toward an alternative understanding of the knowing process, even as it generated a body of empirical evidence that challenged the postulate of cognitive universality, and indeed the entire common-sense framework. The leading New Psychologists – German experimentalists such as Wilhelm Wundt and their American students, led by William James and G. Stanley Hall – rejected the idea of sharply delineated mental faculties and even blurred the line between mind and body. Hall, for example, wrote in 1879 that "the active part of our nature is the essential element in cognition and all possible truth is practical." This suggested that noncognitive motives played a central role in human behavior, distorting and often even obscuring the operations of reason. By positing an intimate connection between the mind and the nervous system, the New Psychology undermined both the common-sense picture of the mind as a collection of discrete faculties and the associated definition of truth as a fit between mind and world. The emerging framework suggested that human minds differed substantially, and that rational behavior – and rational agreement – was difficult to come by in human affairs.[7]

What could science mean, if this were the case? How did largely nonrational minds come to rational agreement? Steeped in the tenets of Protestant liberalism, the pragmatists drew out the side of the scientific spirit that directed attention to the *how* of believing rather than the *what* of belief. On the level of systematic theory, at least, hardly any scientific scholar of the Gilded Age would follow Sumner in stepping back to a place prior to the core ethical

[6] Sumner, *What Social Classes Owe to Each Other* (New York: Harper, 1883), 154, 159–160, 156; Sumner, *Collected Essays*, 96. Sumner equated social-scientific neutrality with open advocacy for "the Forgotten Man": "an honest, sober, industrious citizen, unknown outside his little circle, paying his debts and his taxes, supporting the church and the school, reading his party newspaper, and cheering for his pet politician." *What Social Classes Owe to Each Other*, 145.

[7] G. Stanley Hall, "Philosophy in the United States," *Mind* 4, no. 13 (January 1879), 105. On the other hand, the system's emphasis on process over content and active rather than passive learning harmonized with the belief of scientific democrats that science represented a personal orientation rather than a body of knowledge: e.g., John Fiske, "Essays on a Liberal Education," *North American Review* 107, no. 220 (July 1868), 128, 131; William P. Atkinson, "Liberal Education of the Nineteenth Century," *Popular Science Monthly* 4, no. 1 (November 1873), 4; Hall, "Philosophy in the United States," 105. On the New Psychology, see Jeffrey Sklansky, *The Soul's Economy: Market Society and Selfhood in American Thought, 1820–1920* (Chapel Hill: University of North Carolina Press, 2002), esp. 162–170; Ernest Keen, *A History of Ideas in American Psychology* (Westport, CT: Praeger, 2001), 17–74; and Christopher G. White, *Unsettled Minds: Psychology and the American Search for Spiritual Assurance, 1830–1940* (Berkeley: University of California Press, 2009).

commitments of liberal Christianity. The pragmatists, like most scientific democrats, were not seeking mere agreement; they were seeking a specifically ethical agreement – agreement on normative goals for human behavior. They hoped to create a community of ethical sentiment, a beloved community. In the late nineteenth century, leading American scholars assumed that Christian ethical truths were universal, even if human beings differed in their political views – or in their views of the natural world. Even John Dewey, who left the church entirely in the 1890s, never doubted the moral framework he inherited from liberal Protestantism. Like the other scientific democrats of the era, the pragmatists believed that their fellow citizens needed a moral framework above all, rather than technical knowledge *per se*. They viewed the turn away from common-sense assumptions as a means of re-engaging with ethics, not of disengaging from it.

The pragmatists drew out an equation that lay implicit in the Gilded Age discourse on the scientific spirit, linking science with a specific social process of belief formation. That discourse's emphasis on the social side of inquiry enabled James and Dewey to accommodate the alarming results of the New Psychology – in whose articulation James played a central role – without abandoning the claim of science's cognitive superiority to other ways of knowing or the prospect of a scientifically grounded system of social ethics. Like Youmans, though to a far greater degree, the pragmatists ratcheted down science's presumptions to certainty in order to preserve its foothold in normative fields such as ethics and politics. They traced scientific disputes to an inescapably noncognitive element in human thought, while seeking to keep these interpretive disputes from obstructing collective, ethically grounded action. Charles S. Peirce, James, and Dewey described all scientific truths as human creations – more or less reliable tools for navigating an ineffable external world – while locating core ethical commitments deeper in the universe or the human person. The pragmatists defined scientific inquiry as both the result and the instrument of ethical behavior.

At the center of the philosophical tradition inaugurated by Peirce is the "pragmatic criterion of truth," the claim that we can legitimately assign the label "true" only to those ideas that prove useful in our relations to the world. Truth, James famously wrote, "*happens* to an idea."[8] The pragmatists rejected the "correspondence theory of truth" central to the common-sense approach. Instead, they argued that truth is not, or at least is not merely, a function of the fit between an idea and the world. We simply cannot know whether a given idea fits the world, because there is no way to step outside the process of interpretation and perceive this fit directly. Instead, we decide that an idea is true when, and only so long as, it seems to work. This, the pragmatists reasoned, was what the physical scientists had done, producing their remarkable interpretive success.

[8] William James, *Pragmatism: A New Name for Some Old Ways of Thinking* (New York: Longmans Green, 1907), 201.

But the pragmatic concept of "working" is ambiguous, comporting with various visions of what human beings seek to achieve by interpreting the world. The three early pragmatists interpreted the term in very different ways. To Peirce, "working" referred to the process of building consensus within a community of investigators seeking to explicate the natural world. James instead emphasized the psychological empowerment that came from believing certain things and disbelieving others. Dewey, finally, read the term "working" with reference to social goals – specifically, the building of the ethical economists' cooperative commonwealth, or what he came to call simply "democracy."[9]

Peirce, the son of the famed Harvard mathematician Benjamin Peirce, launched pragmatism with a series of path-breaking articles in the late 1870s. He reframed the scientific spirit in collective rather than individual terms, while emphasizing its competitive side. Indeed, Peirce portrayed science as the intellectual equivalent of market competition: a morally dangerous practice in which individuals nonetheless bore a moral obligation to participate, because it promised unparalleled benefits to all. He described a scientific conclusion as the collective product of fierce contests between individuals sharing a common intellectual problem and disciplined by a commitment to employ only empirical evidence and logical reasoning in staking their claims.[10]

The product of this contentious social process was the closest approximation of God's truth attainable by flawed human beings, Peirce held. Only when a group of thinkers committed to rigorous mutual criticism had attained consensus could one safely say that the truth had emerged. The community's conclusion could then be said to have worked: it had produced uncoerced agreement among all participants. Empirically and logically grounded competition between the participants would have eliminated all partial human perspectives – all exercises of merely human authority – and produced a universal viewpoint shaped solely by the contours of the objective world. Famously contentious himself, Peirce assumed that competition led to intellectual purification, as each contestant sought to neutralize potentially debilitating criticism from the others. By trading the niceties of civilized conversation for a sterner form of communicative exchange, scientists provided their fellow citizens with increasingly reliable guidelines for engaging with the world. To Peirce, this represented the maximum of selflessness that could be expected of human beings. He contrasted the disciplined argumentation of scientists to the utterly unconstrained clashes currently wracking the cultural sphere, wherein contestants challenged their opponents on the basis of *ad hominem* attacks, outright falsehoods, and self-interested flights of fancy. Peirce described scientific argumentation as the

---

[9] Bruce Kuklick, *A History of Philosophy in America, 1720–2000* (New York: Oxford University Press, 2001), 156, 181–182.

[10] This communicative dimension is crucial to Peirce's theory of science. It is no coincidence that one of the first systematic philosophers of science in the United States was also one of its most important – if unappreciated at the time – theorists of rhetoric and symbolism: e.g., Mats Bergman, *Peirce's Philosophy of Communication: The Rhetorical Underpinnings of the Theory of Signs* (London: Continuum, 2009).

most important form of communication, because it was the only effective way to change another person's beliefs without coercion.[11]

Still, Peirce recognized that human beings could never fully transcend self-interest, even when acting collectively as communities of inquiry. In keeping with Protestant tenets, he held that all human thought betrayed the mark of partiality. Peirce addressed this dilemma by locating the scientific community's imagined point of consensus in the infinitely distant future. Today, tomorrow, and ten years from now, he reasoned, its conclusions would remain limited and fallible. But at any given time, these were the only results on which one could confidently rely, because they had been subjected to organized empirical criticism and thus approximated the truth more closely than the musings of a single individual. In practice, then, Peirce assigned the term "truth" to the hypothesis currently prevailing among the trained experts comprising the relevant community of inquiry.

Unlike many positivists of his era, Peirce believed that science's communicative techniques applied to the domain of moral reasoning. He expected empirical inquiry to prove the centrality of love and mutual service in the natural process. In his metaphysical ruminations on "evolutionary love," the apparently hard-boiled logician revealed himself to be a starry-eyed romantic. Peirce explained that the universe's "movement of love" was "circular, at one and the same impulse projecting creations into independency and drawing them into harmony." He envisioned a time "when there shall be no more poetry, for that which was poetically divined shall be scientifically known." Like many theorists of market behavior, Peirce expected vigorous competition – in this case, intellectual competition – to create a godly community, albeit through a long, slow process of collective moral growth.[12]

Peirce also thought mutual criticism would undermine philosophical materialism and produce a more satisfying metaphysical foundation for science.

---

[11] This and the following paragraph draw on Thomas L. Haskell, "Professionalism versus Capitalism: R. H. Tawney, Emile Durkheim, and C. S. Peirce on the Disinterestedness of Professional Communities," in Haskell, ed., *The Authority of Experts* (Bloomington: Indiana University Press, 1984), 202–216; Daniel J. Wilson, *Science, Community, and the Transformation of American Philosophy, 1860–1930* (Chicago: University of Chicago Press, 1990), 12–39; W. Christopher Stewart, "Peirce on the Role of Authority in Science," *Transactions of the Charles S. Peirce Society* 30, no. 2 (1994): 297–326; Paul Jerome Croce, *Science and Religion in the Era of William James: The Eclipse of Certainty, 1820–1880* (Chapel Hill: University of North Carolina Press, 1995), 177–224; Kuklick, *A History of Philosophy in America*, 132–149; and Christopher Hookway, "Truth, Reality, and Convergence," in Cheryl J. Misak, ed., *The Cambridge Companion to Peirce* (New York: Cambridge University Press, 2004).

[12] Charles S. Peirce, *Chance, Love, and Logic* (New York: Harcourt, Brace, 1923), 269; quoted in R. Jackson Wilson, *In Quest of Community: Social Philosophy in the United States, 1860–1920* (New York: Wiley, 1968), 34. On Peirce's politics, see Douglas R. Anderson, "A Political Dimension of Fixing Belief," in Jacqueline Brunning and Paul Forster, eds., *The Rule of Reason: The Philosophy of Charles Sanders Peirce* (Toronto: University of Toronto Press, 1997), 223–240; James Hoopes, *Community Denied: The Wrong Turn of Pragmatic Liberalism* (Ithaca: Cornell University Press, 1998), 29–51; Robert B. Westbrook, *Democratic Hope: Pragmatism and the Politics of Truth* (Ithaca: Cornell University Press, 2005), 25–40.

James and Dewey, on the other hand, took a page from Mach and Pearson on this score. Like those positivists, they sought to bracket the vexed question of what kind of stuff made up the universe, although they did so to clear the way for consensus on ethical and political values, not claims about the natural world. In other words, whereas Mach and Pearson excluded empty metaphysical disputes from the domain of natural knowledge, James and Dewey instead sought to isolate ethics from their corrupting influence. The pragmatic criterion of truth helped them do so. The two men reasoned that, if the truth is what works, then a distinction that does not do any work – that makes no difference for how we act in the world – is not worth arguing about. James and Dewey declared, to Peirce's acute dismay, that metaphysical claims did no work in the world because they did not dictate specific forms of behavior. Thus, they were neither true nor false in pragmatic terms. All that mattered, according to these pragmatists, was how the entities comprising the universe behaved, not whether they were constituted by spirit or by matter. Developing their mature views in the bitterly contentious 1890s, James and Dewey believed that scholars, charged with healing the nation at a time of immense social strain, shirked their duty when they wasted time and resources dabbling in pointless metaphysical speculations.

Beyond their shared rejection of metaphysics, the two differed on many points of social ethics and politics, though less so than the usual contrast between an individualistic James and a collectivistic Dewey would suggest. Acutely concerned to realize the ethical potential of modern science, James and Dewey each hoped that pragmatism would bring thinkers of different philosophical bents into a kind of intellectual coalition, based on a shared ethical framework for social action. But James saw the major threat to this alliance coming from the side of scientific advocates of materialism, whose narrow reading of sensory evidence seemed to banish the possibility of a scientifically informed ethics. By contrast, Dewey, who was confident that the ethical values he associated with science held an intrinsic appeal to the citizenry, worried more about the cultural influence of doctrinal religion.

James, though trained in medicine, a pioneer in the application of laboratory methods to psychology, and the author of the leading early textbook in that field, nevertheless spent much of his time attacking what he saw as the pernicious philosophical consequences of the rise of the sciences. He charged that scientific materialists imposed on the world an *a priori* metaphysical framework that kept them from acknowledging much of what was empirically true about it. James instead urged a radically pluralistic understanding of the experiential world, arguing that individuals viewed it as a welter of discrete entities enmeshed in a complex web of interrelations. Each individual, he wrote, experienced the "great blooming, buzzing confusion" of the world as a complex, continuous "stream of consciousness." James further explained that at every moment, the observer faced an ethically charged and ultimately voluntary (if strongly habitual and culturally patterned) choice to attend to one element in the stream of consciousness rather than another. Scientific materialists, he argued, operated under the influence of the narrowly practical tone of the wider culture and chose to attend only to those elements of experience that upheld this view of reality. James thus levied a potent ethical and political critique against metaphysical materialism.[13]

James had obviously traveled a long way from the common-sense view of the mind as a universal set of faculties that received truths from the surrounding world. His version of the New Psychology described individual minds as contingent, diverse, and constructed entities, each of which took shape through a slow but steady process of accretion in a cultural context that provided ready-made lenses – such as materialism – through which to view experience. According to James, a "mind" was not a discrete entity at all, but rather a mental function, comprising a system of related items into which each new fact or relation had to be fitted through a process of mutual adjustment. Habits of perception or interpretation – established perceptual frames that operated below the threshold of willful effort – thus served as sources of mental inertia. Yet individuals could slowly change these frames by directing their attention to other elements of experience, and indeed were ethically obligated to do so when their thought patterns proved deleterious to others.

In keeping with his pluralistic worldview, James embraced a multitude of interpretive methods, each tuned to a different mode of experience. The world, he argued, contained "many interpenetrating spheres of reality, which we can thus approach in alternation by using different conceptions and assuming different attitudes." Common sense, James held, "is *better* for one sphere of life, science for another, philosophic criticism for a third; but whether either can be truer absolutely, Heaven only knows." In his view, the narrow, exclusively sensory definition of "evidence" employed by many scientists ruled out most of what was interesting in man's experience, especially the pervasive sense of the presence of a higher power or purpose in the world. James believed that theism of a sort was true, despite the absence of concrete proof, because it clearly made a difference in the world: those who embraced the "religious hypothesis" acted differently than those who did not. To be sure, James phrased this hypothesis in very thin terms, ranging from the existence of a "wider self through which saving experiences come" to an even more attenuated "possibility that things may be *better*." Yet he believed that ethical action depended on acceptance of the religious hypothesis, which, by revealing the possibility of change in the world, inspired believers to make that change.[14]

[13] William James, *The Principles of Psychology* (Cambridge: Harvard University Press, 1981 [1890]), 462. Helpful accounts of James' psychology include Sklansky, *The Soul's Economy,* 141–151 and Francesca Bordogna, "The Psychology and Physiology of Temperament: Pragmatism in Context," *Journal of the History of the Behavioral Sciences* 37 (2001): 3–25; see also Wilson, *Science, Community, and the Transformation of American Philosophy,* 12–39; Charlene Haddock Seigfried, *William James's Radical Reconstruction of Philosophy* (Albany: State University of New York Press, 1990); and Kuklick, *A History of Philosophy in America,* 150–178. James drew on Mach's critique of materialism: Holton, "Ernst Mach and the Fortunes of Positivism in America," 33–36.

[14] Quoted in Wilson, *Science, Community, and the Transformation of American Philosophy,* 131; quoted in Trygve Throntveit, "William James's Ethical Republic," *Journal of the History of Ideas* 72, no. 2 (April 2011), 264–265. On James' thin conception of theism, see also David

Dewey, by contrast, believed that claims about the existence or nonexistence of God did no more work in the world than did metaphysical tenets. He sought primarily to keep the empirical power of science trained on ethical and political questions, contending simultaneously against two parties to ethical debates: those theorists, whether Christian or Kantian, who located ethical commitments outside of experience in some nonempirical, transcendent realm; and the kitchen-sink positivists, who deemed these realms utterly inaccessible to empirical inquiry. Dewey insisted that ethical imperatives represented human conclusions from experience, and that social and political institutions should take their shape from such empirically informed commitments.

James had come of age as a Boston Brahmin in the mold of Charles W. Eliot, and he believed that large-scale organizations of all kinds threatened individual freedom by tending to standardize the minds of their inhabitants. The younger Dewey, however, shared with Richard T. Ely a belief that such organizations could facilitate as well as obstruct the process of individual self-realization. His political consciousness took shape around the labor problem in the late 1880s and early 1890s. Always concerned with the public functions of academic work, Dewey found particularly galling the tendency of both religious and scientific thinkers to legitimate the worst excesses of the industrial order by either "naturalizing" its core presumptions – declaring them proven, once and for all, by empirical evidence – or protecting them from scientific criticism by relegating them to a sphere beyond experience. He blamed this phenomenon for the middle-class public's support of violence against striking workers in the years after Haymarket.[15]

Dewey shared the Social Gospel radicalism of the ethical economists, and he likewise followed them in viewing economic institutions as human constructs, conceiving of society as an organism, and seeing in recent history a tremendous rupture that amounted to the emergence of a new technological and economic

A. Hollinger, "'Damned for God's Glory': William James and the Scientific Vindication of Protestant Culture," in *William James and a Science of Religions: Reexperiencing The Varieties of Religious Experience*, ed. Wayne Proudfoot (New York: Columbia University Press, 2004), 28–29. Contributions to the vigorous debate over James' politics include James T. Kloppenberg, *Uncertain Victory: Social Democracy and Progressivism in European and American Thought, 1870–1920* (New York: Oxford University Press, 1986); James Livingston, *Pragmatism and the Political Economy of Cultural Revolution, 1850–1940* (Chapel Hill: University of North Carolina Press, 1994), 158–172; Deborah J. Coon, "'One Moment in the World's Salvation': Anarchism and the Radicalization of William James," *Journal of American History* 83 (1996): 70–99; Hoopes, *Community Denied*, 52–74; Daniel S. Malachuk, "'Loyal to a Dream Country': Republicanism and the Pragmatism of William James and Richard Rorty," *Journal of American Studies* 34 (2000), 98; Westbrook, *Democratic Hope*, 52–73; and Francesca Bordogna, *William James at the Boundaries: Philosophy, Science, and the Geography of Knowledge* (Chicago: University of Chicago Press, 2008), 189–217.

15 The standard biography is Robert B. Westbrook, *John Dewey and American Democracy* (Ithaca: Cornell University Press, 1991). See also Stephen C. Rockefeller, *John Dewey: Religious Faith and Democratic Humanism* (New York: Columbia University Press, 1991); Alan Ryan, *John Dewey and the High Tide of American Liberalism* (New York: W. W. Norton, 1995); and Westbrook, *Democratic Hope*, esp. 74–98.

environment for the social organism. Throughout his life, Dewey implicitly took ethical economics and its successor movements to represent the promise of the scientific method – an empirically informed answer to the era's most pressing question of social ethics. Yet Dewey was a philosopher, not an economist. Attuned to developments in psychology and epistemology, he spied in the pragmatism of Peirce and James a new way to attack *laissez-faire* by undermining what he took to be its cultural foundation: a widespread belief that the empirical, critical method characteristic of modern science applied only to physical nature. Rather than presenting the basic conclusions of ethical economics as transparent representations of an emerging social object, as Ely and company tended to do by the 1890s, Dewey presented ethical knowledge, and indeed all knowledge, as a construction, valid only to the extent that it enabled the manipulation of reality in the interests of human beings.[16]

In short, Dewey refused to naturalize the cooperative commonwealth by cloaking it in the authority of infallible truth. Instead, he redefined science, including modern economics, as a set of tools that human beings happened to find useful. The political events of the Gilded Age left Dewey with a lifelong belief that granting absolute authority to any statement regarding social ethics inevitably aided the powerful by either sacralizing or naturalizing – in his mind, the two were the same – the existing social order. Indeed, Dewey found it just as abhorrent when scientists endorsed absolutes as when religious leaders did. He identified the psychological desire for absolute truth with a kind of political-economic Platonism that placed institutions beyond the reach of criticism, imposing a particular vision of the future by forcibly walling it off from public discussion. To his mind, making transcendent claims always meant fixing, and thus imposing on others, a self-interested picture of a desired future. Dewey argued that the social crisis occasioned by industrialization obligated scholars to help the public see that the principles of *laissez-faire* capitalism, like all other supposedly absolute truths, were human constructs, not natural laws.[17]

Dewey extended this denaturalizing analysis to claims about the existence (or nonexistence) of God, leading him to transfer his loyalty from the church to the democratic community. In 1892, two years before he took up a post at John D. Rockefeller's new University of Chicago, Dewey announced to the Students' Christian Association at Michigan that democracy, Christianity, and science were simply different names for the same historical process. Democracy, he argued, represented nothing less than "the freeing of truth" from all boundaries between persons and groups. At the same time, its success also served as a form of empirical proof, validating Christianity's ethical claims by providing an affirmative answer to the question, "Does a life loyal to the truth bring freedom?" Even after Dewey left the church in 1894, he viewed his thought as an

---

[16] Sklansky details Dewey's psychological engagements in *The Soul's Economy*, 151–162.
[17] On Dewey's politics, see especially Westbrook, *Democratic Hope*, 74–98. By contrast, John Bates Clark adopted the naturalizing approach: "The Scholar's Political Opportunity," *Political Science Quarterly* 12, no. 4 (December 1897), 600.

extension and expansion of liberal Protestantism. He believed that the struggle against authority went on simultaneously in the realms of science and religion, as part of a united campaign against the immoral claim to possess absolute truth. Throughout his life, Dewey embraced the Christian ethical canon and cooperated politically with liberal Christians, despite his inability to accept the theological premises of Christian ethical theories.[18]

Thus, while James worried about the ethical impact of a too narrowly defined empiricism, Dewey identified empiricism as a highly endangered mode of thought in a culture that was deeply religious, but in all the wrong ways. He focused on defending empirical knowledge against those who deemed it contrary to Christian principles. Although Dewey rejected scientific materialism himself, he believed so strongly that the scientific method embodied a broad and fundamentally democratic ethic that he saw only promise in its extension to any and all domains. Meanwhile, he saw behind organized religion a socially and politically dangerous attempt to place certain values or beliefs beyond the reach of human criticism, in an imagined "supernatural" realm. Dewey thus focused primarily on the dangers of traditional religiosity, rather than those of the rising empiricist outlook.

Dewey's remarkably sanguine take on the extension of science's cultural power stemmed in large part from his faith that science embodied the rejection of illegitimate authority. He excused the errors of existing scholars, believing that science's democratic essence would shine through in the end. His unwavering commitment to intellectual transformation also rested on his belief that one group of scientists – theorists of *laissez-faire* in the vein of Sumner and Youmans – already enjoyed a hegemonic role in American public culture, holding sway over the middle-class public's understanding of science and its relation (or lack thereof) to social ethics. "I think professional people are probably worse than the capitalists themselves," Dewey wrote his wife during the Pullman strike of 1894. Among the emerging professional classes, he saw a prevailing belief that social and moral truths were of a different type than were physical truths – that they were transcendent, not subject to empirical criticism. Dewey devoted his life to challenging that presumption and thereby clearing the way for the expansion of science's empirical, antiauthoritarian method into the domain of social ethics. He argued that genuine science, understood as a form of communication and a tool for ethical change, offered reliability without certainty and persuasion without coercion.[19]

---

[18] Westbrook, *John Dewey and American Democracy*, 35–36, 79–80; John Dewey, "Christianity and Democracy," in Jo Ann Boydston, ed., *The Early Works of John Dewey, 1882–1898, Volume 3* (Carbondale: Southern Illinois University Press, 1967), 5, 8; Westbrook, *Democratic Hope*, 78–79. As many have noted, Dewey's conception of God was already so immanentist that his abandonment of the term represented little more than a semantic difference.

[19] Westbrook, *John Dewey and American Democracy*, 87 (quoted), 142–147. Dewey's commitment to naturalism has gained considerable interest of late: e.g., Terry Hoy, *Toward a Naturalistic Political Theory: Aristotle, Hume, Dewey, Evolutionary Biology, and Deep Ecology* (Westport, CT: Praeger, 2000); Thomas C. Dalton, *Becoming John Dewey: Dilemmas of a Philosopher and*

Yet Dewey still thought that Americans needed a philosophy to guide them in their daily lives and to shape their character. As it developed, Dewey's "instrumentalism" offered a spare, flexible conception of human behavior as a compound of variable environmental factors with four underlying "native impulses," namely the impulses "to communicate, to construct, to inquire, and to express in finer form." Dewey added the postulate that self-realization and collective realization harmonized, declaring, "IN THE REALIZATION OF INDIVIDUALITY THERE IS FOUND ALSO THE NEEDED REALIZATION OF SOME COMMUNITY OF PERSONS OF WHICH THE INDIVIDUAL IS A MEMBER; AND, CONVERSELY, THE AGENT WHO DULY SATISFIES THE COMMUNITY IN WHICH HE SHARES, BY THAT SAME CONDUCT SATISFIES HIMSELF." Human beings made social change as they sought to realize themselves as collectively constituted actors – to express their native impulses to the fullest under the extant cultural and institutional constraints.[20]

They made knowledge, too. Dewey's commitment to a secularized version of Christian ethics powerfully shaped his conception of scientific inquiry. Whereas Peirce held that doubt was represented by a clash of opinions between individuals, Dewey argued that the most consequential social conflicts registered not between but *within* individuals, where the conflicts appeared as struggles between competing ends or ideals. He thought individuals operated within a linguistically constructed "social sensorium" that functioned to harmonize these potentially conflicting desires. Science, for Dewey, was coeval with this process of cultural harmonization; it was the name for both the process by which individuals reconciled social conflicts through intellectual work and the knowledge produced therein – knowledge that could by definition inspire social agreement. He soon added Darwinian evolution to the picture, developing a biologically rooted view that treated the Christian ethical canon as uniquely aiding the survival of the species and the self-realization of individuals. Although knowledge was indispensable, it mattered only indirectly in Dewey's system. His individual was always an "agent-patient, doer, sufferer, and enjoyer," but only occasionally a knower. To Dewey's mind, both factual and evaluative judgments were fallible products of human beings engaged in solving specific problems, in specific situations, in the pursuit of specific future results. From his philosophical perspective, Dewey noted, any "so-called scientific judgment appears definitely as a moral judgment."[21]

Dewey harbored grand hopes for a scientific transformation of society, despite his personal modesty. He chafed at the cloistered life of the academic,

Naturalist (Bloomington: Indiana University Press, 2002); Jerome A. Popp, *Evolution's First Philosopher: John Dewey and the Continuity of Nature* (Albany: State University of New York Press, 2007); and Peter T. Manicas, *Rescuing Dewey: Essays in Pragmatic Naturalism* (Lanham, MD: Lexington Books, 2008).

[20] Quoted in Westbrook, *John Dewey and American Democracy*, 98, 62.

[21] Quoted in *ibid.*, 126; John Dewey, "Logical Conditions of a Scientific Treatment of Morality" (1903), in Jo Ann Boydston, ed., *The Middle Works of John Dewey, 1899–1924, Volume 3* (Carbondale: Southern Illinois University Press, 1976), 19.

calling the typical professor a "new Scholastic" who merely "criticises the criticisms with which some other Scholastic has criticised other criticisms," such that "the writing upon writings goes on till the substructure of reality is long obscured." Dewey sought in various ways to change that substructure. Arguing in 1892 that the nation's communication networks would prove to be the most effective tools for constructing the Great Community, Dewey joined an abortive attempt to launch an intellectual newspaper, under his editorship, titled "Thought News." Following the failure of that media venture, he tried his hand at education, co-founding Chicago's Laboratory School to turn out citizens equipped with the knowledge and virtues needed to sustain self-government under industrial conditions. Drawing in part on his experiences at Jane Addams' Hull House, Dewey offered a theoretical explication of his democratic, science-centered form of education in *The School and Society* (1899). Finally, after moving to Columbia in 1904, Dewey worked with his student James H. Tufts on a more academic expression of pragmatic ethics. Their influential 1908 textbook codified Dewey's view of science's relation to values and applied that theory directly to the industrial crisis. Throughout his career, as we will see in subsequent chapters, Dewey worked to bring his democratic conception of science to bear on American public culture.[22]

SCIENCE AND DISCIPLINARITY

Peirce, James, and Dewey were not the only American philosophers to develop innovative new portraits of scientific inquiry. By the early decades of the twentieth century, fierce debates over science's character and lessons for philosophy echoed through the halls at meetings of the American Philosophical Association (APA), launched in 1902. There were clear professional stakes, given that philosophy's curricular status had declined precipitously with the rise of the sciences. Having ruled the roost in the antebellum colleges, American philosophers had struggled since the 1860s to craft a new guide for daily life by accommodating Darwinism with Christian morality. Yet the capstone moral philosophy course had been ousted from its central curricular place. Meanwhile, the social sciences had broken away from moral philosophy, forming their own professional organizations. From these experiences, philosophers drew a conclusion that few other scientific democrats would reach until after World War I: most Americans actively rejected the prospect of replacing the "pan-Protestant establishment" with a new public culture centered on scientific knowledge and virtues.[23]

[22] John Dewey, "The Scholastic and the Speculator" (1891–1892), in *The Early Works, Volume 3*, 151; Westbrook, *John Dewey and American Democracy*, 93–113, 151–166; Katherine Camp Mayhew and Anna Camp Edwards, *The Dewey School: The Laboratory School of the University of Chicago, 1896–1903* (New Brunswick: Transaction, 2007). See also Jennifer Welchman, *Dewey's Ethical Thought* (Ithaca: Cornell University Press, 1995); and Abraham Edel, *Ethical Theory & Social Change: The Evolution of John Dewey's Ethics, 1908–1932* (New Brunswick, NJ: Transaction, 2001).

[23] On the debates analyzed here, see also Daniel J. Wilson, "Professionalization and Organized Discussion in the American Philosophical Association, 1900–1922," *Journal of the History of*

Few of these philosophers thought pragmatism solved the problems of knowledge and cultural authority facing an industrial democracy. As the twentieth century dawned, idealism, a broad metaphysical view holding that the constitutive elements of reality are mental rather than material in character, dominated American philosophy. From the writings of the late-eighteenth-century German thinker Immanuel Kant and his early-nineteenth-century successor G. W. F. Hegel, idealism had spread across the Western intellectual world by the late nineteenth century. Many proponents thought it could replace Christianity as a source of intellectual guidance in the modern world. Thus, philosophical idealism stood in a complex relationship to science, whose advocates had their own cultural ambitions.[24]

On American soil, idealism and scientific democracy both sprang up in the wake of the Civil War. William Torrey Harris and the "St. Louis Hegelians" founded the *Journal of Speculative Philosophy* in 1867, and Dewey began as a Hegelian as well. Meanwhile, some of the most fervent American advocates of a scientific philosophy built on Kant. The two traditions mingled freely in Paul Carus' Open Court Publishing venture and his journal *The Monist*, founded in 1887 and 1890, and in the more professional *Philosophical Review*, launched at Cornell in 1892. By that time, problems about the intelligibility of the material world had become paramount in Anglo-American circles and Hegelian theories – especially those influenced by the vigorous tradition of British idealism – centered on the metaphysical postulate that the stuff of the universe was mental. Idealists of this variety argued that the knowing mind could gain a purchase on the universe only if a single substance constituted both entities. They viewed history as a process through which a rational ideal of interpersonal harmony and clear understanding gradually took on concrete form, progressively displacing irrational forms of organization, behavior, and belief.[25]

Kant's argument that pre-given mental categories structured all knowledge offered American theorists additional weapons against materialism, while also pointing toward a new defense of disciplinary specialization. Kant famously held that the fundamental categories of reality allegedly discovered by scientists – causality, time, space, number – were actually the mind's own contributions to human experience, representing logical preconditions for the possibility of that experience. He also insisted that both scientific determinism and the common-sense understanding of freely willed behavior

Philosophy 17, no. 1 (1979): 53–69; Wilson, *Science, Community, and the Transformation of American Philosophy*, 131–149; Cornelis de Waal, "Introduction," in de Waal, ed., *American New Realism, 1910–1920, Volume I* (Bristol: Thoemmes, 2001); and James Campbell, *A Thoughtful Profession: The Early Years of the American Philosophical Association* (Chicago: Open Court, 2006).

[24] Kuklick, *A History of Philosophy in America*, 111–128.

[25] James A. Good, *A Search for Unity in Diversity: The "Permanent Hegelian Deposit" in the Philosophy of John Dewey* (Lanham, MD: Lexington Books, 2006); Kuklick, *A History of Philosophy in America*, esp. 111–122, 190; Harold Henderson, *Catalyst for Controversy: Paul Carus of Open Court* (Carbondale: Southern Illinois University Press, 1993).

were correct, the apparent contradiction merely reflecting the adoption of different vantage points on the same complex reality.[26] As the American disciplines took shape around 1900, arguments in this Kantian vein provided one potent defense of intellectual specialization. In this understanding, each discipline reflected one of many perceptual filters through which human beings could choose to view the world. Each filter was unique and useful, and none offered an unmediated view that gave its users a monopoly on truth.

In defining and promoting the discipline of philosophy, the APA's first president, the Canadian-born Kantian J. E. Creighton, combined this perspectivalist argument with a community of inquiry model similar to Peirce's. Creighton had earned his degree at Cornell in 1892 and would teach there until his death in 1924. He thus knew a thing or two about the importance of practical results as the source of a university's public reputation. Although philosophy was no panacea, he argued, it still needed to "bake *some* bread" in order to enjoy the cultural authority that modern citizens afforded the suppliers of reliable, useful knowledge. Creighton saw the APA as the germ of a truly productive discipline: a "social group of cooperating minds" characterized by vigorous mutual criticism. Philosophers, he argued, could not regain their lost prestige by clinging to the outmoded image of the scholar as a heroic pioneer going it alone. Science's success demonstrated the need to band together, focus collectively on the era's "main drift of problems," and "fight it out with an open mind."[27]

Despite this common mode of social organization, however, Creighton argued that "[p]hilosophical science" was neither an auxiliary of one of the natural sciences nor a synthesis of all of them. Instead, it represented an equally valid – and ultimately more important – reading of "the facts of experience," produced by the adoption of a different perceptual filter. Natural scientists, said Creighton, acknowledged only the objective, material dimension of experienced reality, while psychologists adopted an opposing and equally reductive focus on subjective purposes unmoored from the material world. Philosophers put these partial perspectives together, viewing experience in its fullness as the locus of conscious human action. Here, said Creighton, lay a practical contribution unparalleled in the academic world: philosophy could "humanize" the results of the natural sciences, fostering human brotherhood and creating the foundation for a modern social order.[28]

Creighton's perspectivalist approach rested on a conception of experience similar to that of James. Hardly "a clear and transparent medium that presents to us facts in unambiguous and unmistakable form," he wrote, experience was instead "so many-sided and complex, in some relations so shifting and

---

[26] Murray G. Murphey, "Kant's Children," *Transactions of the Charles S. Peirce Society* 4 (1968): 3–33.

[27] J. E. Creighton, "The Purposes of a Philosophical Association," *Philosophical Review* 11, no. 3 (May 1902), 224, 228, 234.

[28] *Ibid.*, 226, 237; J. E. Creighton, "The Standpoint of Experience," *Philosophical Review* 12, no. 6 (November 1903), 594.

unstable, as to be capable of yielding various and even contradictory readings." The very facts of experience changed according to observers' "conceptions and presuppositions." Thus, a loaf of bread could be viewed with equal validity "as an object in space possessing certain physical and chemical properties, as a complex of sensations, or as an object of desire and will." Creighton contended that each of the resulting forms of inquiry qualified as a science, even if their observations proved utterly irreconcilable.[29]

Creighton thus portrayed philosophy as one of many parallel sciences, each proceeding from its own presuppositions and each equally correct within its given domain and for its particular purpose. A science, in this definition, entailed the social process of organized mutual criticism on the basis of experiential data, not the forced adoption of the natural scientists' partial, materialistic filter. According to Creighton, the various interpretive perspectives of the sciences were orthogonal and mutually correcting, not antagonistic and mutually exclusive. Each deserved a niche in the academic ecology, because the full interpretation of experience required a coordinated array of specialists. Like the theory of emergent evolution that emerged in the 1920s, Creighton's post-Kantian formulation authorized disciplinary pluralism: the peaceful coexistence of interpretive pursuits approaching the complex world of experience from divergent but equally valid perspectives, for different purposes.[30]

Yet disputes over the nature of knowledge, and the intimately related question of philosophy's role in the academic polity, churned on through the Progressive Era. Most American philosophers shared Creighton's distaste for materialism, agreeing that "the so-called empirical philosophy" of David Hume and his British successors illegitimately elevated a partial perspective – the natural scientist's view of experience as an array of discrete, immediate sensations – into a description of reality as a whole. But they generally found other ways of integrating minds, purposes, and other non-materialistic phenomena into their accounts of reality.[31]

Like Creighton's perspectivalism, the pragmatism of James and Dewey represented an embattled minority position at the time, despite its prominence in our historical accounts. On one side of the pragmatists stood Josiah Royce and other powerful defenders of idealism, who saw a latent materialism in their equation of truth with practical success. On the other stood an emerging group of "New Realists," led by Royce's young Harvard colleague Ralph Barton Perry, who

[29] Creighton, "The Standpoint of Experience," 593–594; Creighton, "The Purposes of a Philosophical Association," 235–237.

[30] Institutionally, Creighton hoped the American Association for the Advancement of Science would become a broad "federation" of scientific researchers that included philosophers: "The Purposes of a Philosophical Association," 235–237.

[31] Creighton, "The Standpoint of Experience," 594. All kinds of philosophers could play the "more scientific than thou" game. At the 1905 APA meeting, Brother Chrysostom (John Conlon of Manhattan College) argued that neo-Thomism harmonized perfectly with modern science: "The Fifth Meeting of the American Philosophical Association," *Journal of Philosophy, Psychology and Scientific Methods* 3, no. 3 (February 1, 1906), 76.

sought to ground philosophy in the natural sciences. They charged that the pragmatists smuggled in metaphysics and wrongly made the mind, rather than the external world, the arbiter of knowledge. Although James and Dewey hoped to bracket pedantic disputes and foster ethical harmony, they spent most of their energy in the Progressive years crafting technical defenses of their systems.[32]

Nor was science the only reference point for American philosophers. The broad division of knowledge known as "the humanities" had begun to take shape, offering another possible allegiance for philosophers. Centered on the textual interpretation of artistic and literary productions, the humanities found a professional identity largely by portraying themselves as the antisciences. Scholars such as Harvard's Charles Eliot Norton described literature and the arts as the needed source of moral guidance in the modern world, a potent corrective to the intrinsically amoral and deterministic perspective of science. Yet philosophy, like philology and history, straddled the still indistinct boundary between the sciences and the humanities. Rather than claiming scientific status, philosophers could also enlist with the humanists by identifying their work as an antidote to science's morally enervating effects.[33]

Many went at least partway down this road, insisting that philosophy should be primarily a teaching field that aimed at inspiring self-understanding and moral insight, rather than a research field modeled on the sciences. Creighton immediately raised this group's hackles, declaring pedagogy a matter of individual rather than collective concern and proposing that the APA exclude teaching from its purview. But a steady stream of articles placed the emphasis squarely on philosophy's pedagogical functions, decrying its shrunken curricular role. Even James, in his famous attack on the "Ph.D. octopus," rejected Creighton's emphasis on research. He expressed deep skepticism about the

[32] Ralph Barton Perry, "Realism as a Polemic and a Program of Reform. II," *Journal of Philosophy, Psychology and Scientific Methods* 7, no. 14 (July 7, 1910), 370; Westbrook, *John Dewey and American Democracy*, 120–137; John R. Shook, "John Dewey's Struggle with American Realism, 1904–1910," *Transactions of the Charles S. Peirce Society* 31, no. 3 (Summer 1995): 542–566; Eugene I. Taylor and Robert H. Wozniak, eds., *Pure Experience: The Response to William James* (Bristol: Thoemmes Press, 1996). Dewey's colleague Frederick J. E. Woodbridge later lamented the pragmatists' submersion in specialized debates: Campbell, *A Thoughtful Profession*, 109.

[33] Julie A. Reuben, *The Making of the Modern University: Intellectual Transformation and the Marginalization of Morality* (Chicago: University of Chicago Press, 1996), 211–229; James Turner, *The Liberal Education of Charles Eliot Norton* (Baltimore: Johns Hopkins University Press, 1999); Linda Dowling, *Charles Eliot Norton: The Art of Reform in Nineteenth-Century America* (Lebanon: University of New Hampshire Press, 2007). On "the humanities" as a category, see also Laurence R. Veysey, "The Plural Organized Worlds of the Humanities," in *The Organization of Knowledge in Modern America, 1860–1920*, ed. Alexandra Oleson and John Voss (Baltimore: Johns Hopkins University Press, 1979); Bruce Kuklick, "The Emergence of the Humanities," *South Atlantic Quarterly* 89, no. 1 (1990): 195–206; Jon H. Roberts and James Turner, *The Sacred and the Secular University* (Princeton: Princeton University Press, 2000), 75–122; Caroline Winterer, *The Culture of Classicism: Ancient Greece and Rome in American Intellectual Life, 1780–1910* (Baltimore: Johns Hopkins University Press, 2002), 99–183; and Steven Marcus, "Humanities from Classics to Cultural Studies: Notes Toward the History of an Idea," *Daedalus* 135, no. 2 (2006): 15–21.

APA in its early years. James thought professors should be hired on the basis of moral character, not degrees, given that their role was to augment "the class of highly educated men" by serving as "the jealous custodians of personal and spiritual spontaneity." Similarly, Princeton's Norman Kemp Smith rejected the standards of research productivity and direct applicability while promising a more diffuse form of social relevance by seeking to equip students with a mind-set appropriate to the cultural challenges of a modern democracy.[34]

Ironically, one of the leading advocates of a more humanistic, pedagogical self-understanding was an ardent enthusiast of science, namely Morris R. Cohen of the City College of New York. The Jewish, Russian-born Cohen sought a modern substitute for the moral philosophy course that offered a coherent social viewpoint, synthesized the sciences, and instilled in students a relentless drive to harmonize the parts with the whole, thereby equipping them to mediate between their specialized areas of endeavor and the broader life of the nation. Yet Cohen denied that a single synthesis could ever win over philosophers as a group, because "temperamental preference" mattered in their field. "Let philosophy resolutely aim to be as scientific as possible," he declared, but "let her not forget her strong kinship with literature." Philosophy, like the arts and social sciences, produced many true pictures. Going beyond Creighton's system of equally valid disciplines, Cohen championed intellectual diversity within the discipline itself.[35]

Meanwhile, the New Realists and other young philosophers inspired by science's capacity to generate interpretive consensus took aim at this presumption of intrinsic pluralism. At the 1908 APA meeting, two mathematically trained thinkers proposed that philosophers adopt a common intellectual platform to guide collective explorations of the type Creighton proposed. Christine Ladd-Franklin looked to a body of scientific conclusions that would set the content of philosophy courses and define the research frontier in the other sciences, whereas Karl Schmidt favored a set of shared problems for those purposes. But both sought to ensure intellectual order by elevating the collective judgment of the profession over those of individual investigators. In their view, philosophers could perform their social duties only if they adopted the scientists' recipe for achieving consensus without coercion, replacing "the petty criticizing spirit" with "the mutual understanding of cooperation" and substituting a firm structure of

[34] William James, "The Ph.D. Octopus," *Harvard Monthly* 36, no. 1 (March 1903); Norman Kemp Smith, "The Problem of Knowledge," *Journal of Philosophy, Psychology and Scientific Methods* 9, no. 5 (February 29, 1912), 117. Cf. Arthur Ernest Davies, "Education and Philosophy," *Journal of Philosophy, Psychology and Scientific Methods* 6, no. 14 (July 8, 1909): 365–372.

[35] Morris R. Cohen, "The Conception of Philosophy in Recent Discussion," *Journal of Philosophy, Psychology and Scientific Methods* 7 (July 21, 1910), 406, 409; Campbell, *A Thoughtful Profession*, 133–136; Campbell, "The Ambivalence toward Teaching in the Early Years of the American Philosophical Association," *Teaching Philosophy* 25, no. 1 (March 2002): 53–68. Cf. George R. Dodson, "The Function of Philosophy as an Academic Discipline," *Journal of Philosophy, Psychology and Scientific Methods* 5, no. 17 (August 13, 1908): 454–458. Cohen's biography is David A. Hollinger, *Morris R. Cohen and the Scientific Ideal* (Cambridge, MA: MIT Press, 1975).

authority for the prevailing "anarchy" in the field. Ladd-Franklin and Schmidt sought to end the age-old individualism and subjectivism of philosophy.[36]

Schmidt couched his proposal in strikingly political terms. Unlike Ladd-Franklin, who foresaw a single paradigm for the field, Schmidt advocated a two-party system. He called for a "recognized doctrine and school" to act as a "ruling party" by establishing canons of truth – agreed-upon standards of argumentation. Lacking these, he wrote, "we can not know whether we are really advancing, *or even what an advance means*." With such canons, the field would enjoy the steady "momentum of the school," as a judicious blending of new and old forestalled "erratic and premature changes." But the system could not work without an opposition party to provide a "safety-valve" against "sudden revolutions" by loyally applying "systematic and persistent criticism." This dynamic, said Schmidt, explained how "absolute freedom of research and opinion" generated "order and uniformity" in the sciences. Science, in Schmidt's portrait, did not involve the absence of authority, but rather the exercise of the same mode of authority – relatively stable, yet always revisable and liberating – found in a democracy. By submitting to the authority of a revocable platform, he promised, philosophers would attain "independence and true originality."[37]

But the platform proposal did not become a focus of disciplinary debate until 1910, when Perry and five other New Realists – Edwin B. Holt of Harvard, W. P. Montague and Walter B. Pitkin of Columbia, Walter T. Marvin of Rutgers, and E. G. Spaulding of Princeton – issued a collective statement intended to give American philosophers a concrete basis for unity.[38] The New Realists inclined toward James' radical empiricism and pluralist metaphysics. Perry, their unofficial leader, had studied under James and now taught alongside him at Harvard. But the group disavowed the pragmatic criterion of truth. Inspired by Bertrand Russell and other critics of British idealism, they contended that minds enjoyed unmediated access to objects in the external world. Given that science's conclusions proved reliable, the New Realists reasoned, they must represent accurate pictures of a mind-independent reality.[39]

---

[36] Karl Schmidt, "Concerning a Philosophical Platform," *Journal of Philosophy, Psychology and Scientific Methods* 6, no. 25 (December 9, 1909): 680–685; Frank Thilly, "Proceedings of the American Philosophical Association: The Eighth Annual Meeting," *Philosophical Review* 18, no. 2 (March 1909), 180–182; Campbell, *A Thoughtful Profession*, 145–148. The 1908 meeting also revealed considerable interest in other scientific topics. For example, more than half the presenters addressed logic, while only one each spoke to theology and aesthetics. Harold Chapman Brown, "The Eighth Annual Meeting of the American Philosophical Association," *Journal of Philosophy, Psychology and Scientific Methods* 6, no. 2 (January 21, 1909), 45, 49–51.

[37] Schmidt, "Concerning a Philosophical Platform," 679–681. Schmidt drew on European work in mathematical logic: "Studies in the Structure of Systems," *Journal of Philosophy, Psychology and Scientific Methods* 10, no. 3 (January 30, 1913): 64–75.

[38] Edwin B. Holt et al., "The Program and Platform of Six Realists," *Journal of Philosophy, Psychology and Scientific Methods* 7, no. 15 (July 21, 1910): 393–401.

[39] Bruce Kuklick, *The Rise of American Philosophy, Cambridge, Massachusetts, 1860–1930* (New Haven: Yale University Press, 1977), 338–350, 417–434; Kuklick, *A History of Philosophy in America*, 202–207; de Waal, "Introduction"; Campbell, *A Thoughtful Profession*, 110–114.

The New Realists also proposed a thoroughly collectivist model of philosophy, centered on rigorous analytical methods. In a pair of programmatic articles, Perry announced the group's intention to salvage a long-abandoned founding aim of modern philosophy, namely the reduction of "philosophical discourse to logical form." Perry expected this formalization of the discipline to produce a set of shared propositions, the fundamental building blocks of "an impersonal system of philosophical knowledge." Only when philosophies were stated in systematic form, like "Euclidean geometry or Newtonian mechanics," could they be subjected to logical criticism, producing a steadily expanding space of agreement. Perry did not accept a two-party system. He predicted that even the New Realism would eventually disappear "as propaganda," giving way to a "science of philosophy in which all investigators should work together without party or school designation." This would bring the discipline the public recognition it deserved by undermining its prevailing image as "unscientific, subjective, or temperamental."[40]

The writings of the New Realists generated intense controversy. To be sure, even more rigorous platform proposals had been floated. Earlier in 1910, Dewey's student Savilla Alice Elkus had proposed that a hierarchical structure of philosophical "laboratories" could gradually eliminate the differences between competing systems. Dewey himself welcomed a broad platform of "anti-idealism," though he disagreed with the New Realists on a central epistemological tenet, the "doctrine of external relations." But most American philosophers sharply rejected the platform idea, often in explicitly political terms. Stressing the free play of ideas, they portrayed freedom and authority – no matter how revocable or flexible – as opposites. Creighton denied that even the natural sciences exhibited the pre-made frameworks discerned by Schmidt and insisted that there, too, intellectual harmony emerged only "in and through the process of emphasizing differences." He and many others adopted the view of belief formation associated with liberal Protestantism and its secular successors, holding that the combination of absolute interpretive freedom with a spirit of mutual aid and fair play would generate consensus.[41]

Others thought the same combination would produce a pluralistic system of coexisting schools. In a nearly daylong discussion of "Agreement in Philosophy" at the 1912 APA meeting, Norman Kemp Smith joined Bryn Mawr's Theodore De Laguna and Cornell's Frank Thilly in calling the platform idea profoundly authoritarian. Against the "tyranny" of the New Realists, Kemp Smith cited

[40] Ralph Barton Perry, "Realism as a Polemic and Program of Reform. I," *Journal of Philosophy, Psychology and Scientific Methods* 7, no. 13 (June 23, 1910), 337–338; Perry, "Realism as a Polemic and Program of Reform. II," 371, 377; Holt et al., "The Program and Platform of Six Realists," 393.

[41] Savilla Alice Elkus, "A Philosophical Platform from Another Standpoint," *Journal of Philosophy, Psychology and Scientific Methods* 7, no. 1 (January 6, 1910), 19–20; John Dewey, "The Short-Cut to Realism Examined," *Journal of Philosophy, Psychology and Scientific Methods* 7, no. 20 (September 29, 1910), 553–554; Campbell, *A Thoughtful Profession*, 148–149, 152–156; J. E. Creighton, "The Idea of a Philosophical Platform," *Journal of Philosophy, Psychology and Scientific Methods* 6, no. 6 (March 18, 1909), 142–143.

Woodrow Wilson's description of a genuine collective opinion as "a living thing" comprising "the vital substance of many minds, many personalities, many experiences," joined through "face to face debate ... the direct clash of mind with mind." Yet even if philosophers attained the needed "liberality of thought," Kemp Smith argued, full consensus would prove elusive. Instead, the outcome would be a stable pluralism in which "the three fundamental types of philosophical thinking," namely "naturalism, scepticism, and idealism," would each gain from mutual correction by the others. Deeming agreement unforced and unplanned by definition, Kemp Smith decried claims of absolute certainty and called for resistance to the philosophical incursions of the "scientific philistine."[42]

Schmidt and Perry denied advocating absolutism or tyrannous unity. They reasoned that considerable agreement on both specific truths and general canons of truth already existed, given the fact that philosophers could engage in rational discussion. They simply wanted to facilitate progress by stating more systematically the points of agreement and disagreement.[43] These protestations did little good; discussion of the platform idea faded after the 1912 APA meeting, and the New Realism broke apart as a coherent movement in 1914. But in his December 1916 presidential address, Arthur O. Lovejoy of Johns Hopkins made one last attempt to revitalize the platform model. Rigidly distinguishing public rationality from private emotions, Lovejoy insisted that interpretive freedom and intellectual diversity were not sufficiently important goods to justify leaving many in "error about some of the greatest matters of human concernment." On such matters, he contended, a "drab unanimity" was the lesser evil.[44]

Lovejoy drew a firm line between two types of cultural leaders: the artist, who aimed at exhilaration, appreciation, or at best edification; and the "man

[42] Norman Kemp Smith, "How Far is Agreement Possible in Philosophy?" *Journal of Philosophy, Psychology and Scientific Methods* 9, no. 26 (December 19, 1912), 701, 706, 708–710; Campbell, *A Thoughtful Profession*, 152–156.

[43] Karl Schmidt, "Concerning a Philosophical Platform: A Reply to Professor Creighton," *Journal of Philosophy, Psychology and Scientific Methods* 6, no. 9 (April 29, 1909): 240–242; Schmidt, "Agreement," *Journal of Philosophy, Psychology and Scientific Methods* 9, no. 26 (December 19, 1912), 715–717; Ralph Barton Perry, "If the Blind Lead the Blind: A Reply to Dr. Brown," *Journal of Philosophy, Psychology and Scientific Methods* 7, no. 23 (November 10, 1910), 627–628. Marvin's description of the platform idea as "a return to dogmatism" could not have helped the cause, however: *A First Book in Metaphysics* (New York: Macmillan, 1912). For yet another variant of the platform idea, framed in equally political terms, see Walter B. Pitkin, "Is Agreement Desirable?" *Journal of Philosophy, Psychology and Scientific Methods* 9, no. 26 (December 19, 1912): 711–715.

[44] Ralph Barton Perry, "William Pepperell Montague and the New Realists," *Journal of Philosophy* 51, no. 21 (October 14, 1954), 606; Kuklick, *A History of Philosophy in America*, 207; Arthur O. Lovejoy, "On Some Conditions of Progress in Philosophical Inquiry," *Philosophical Review* 26, no. 2 (March 1917), 133; Campbell, *A Thoughtful Profession*, 165–180. Cf. Perry, "If the Blind Lead the Blind," 628. On Lovejoy, see Daniel J. Wilson, *Arthur O. Lovejoy and the Quest for Intelligibility* (Chapel Hill: University of North Carolina Press, 1980) and James Campbell, "Arthur Lovejoy and the Progress of Philosophy," *Transactions of the Charles S. Peirce Society* 39, no. 4 (Fall 2003): 617–643.

of science," who sought verification through "rigorous, objective and conceptually communicable reasoning." He allowed that concrete philosophies would always be tinged by personal factors. But this merely redoubled the need for the "methodized precautions against error" that characterized every "depersonalized science." Philosophy, Lovejoy insisted, could not justify its existence at all, were it to remain "a disclosure of unstandardized private reactions upon the universe." He imagined a "metaphysical testing laboratory, where the materials to be used in the construction of philosophical engines are first systematically subjected to all conceivable strains and stresses" to determine their possibilities for "dialectical locomotion."[45]

Turning to political theory, Lovejoy invoked the British banker-philosopher Walter Bagehot's ideal of "government by discussion." Like Bagehot, Lovejoy found "the desire to conclude promptly" the greatest of all communicative dangers and hoped to foster stability by forestalling collective action until every possible viewpoint had been carefully considered and tested. In his ideal philosophical polity, every claim would take the form of a set of hypothetical postulates and a set of conclusions believed to follow from these postulates. The claims would be withheld from the wider public until each of their components had been thoroughly and collectively vetted by the discipline, using the best available methods of error correction. In exchange for this exertion of self-discipline, Lovejoy promised, philosophers would reach "cleared areas within which there need be no controversy."[46]

However, Lovejoy's proposal had a far more authoritarian ring than most invocations of government by discussion, and it alarmed his fellow philosophers. Even Perry, after all, had felt obliged to identify logical reasoning as an auxiliary to free interpretation rather than an instrument of coercion. By contrast, Lovejoy stated bluntly that philosophers should hang up their hats if their claims could not serve as "instruments for coercing the judgment of stubborn dissenters." Echoing the common-sense thinkers, he contended that if a philosophical dispute centered on a genuine question, "an identical logical reaction to it will be obtainable from all rational minds capable of understanding it." But this sounded much too authoritarian for most American philosophers of that era. The idealist William Ernest Hocking countered with a characteristically organicist sentiment, writing that philosophical systems, "like political constitutions," emerged most effectively through "a mixture of a little making with a good deal of growth." The other responses to Lovejoy's address were even less sympathetic.[47]

[45] Lovejoy, "On Some Conditions of Progress in Philosophical Inquiry," 131, 133, 143, 147; Lovejoy, "Progress in Philosophical Inquiry," *Philosophical Review* 26, no. 5 (September 1917), 543.

[46] Lovejoy, "On Some Conditions of Progress in Philosophical Inquiry," 152–153, 157–158. Bagehot's book was *Physics and Politics* (New York: D. Appleton, 1906 [1873]), published in E. L. Youmans' series.

[47] Lovejoy, "On Some Conditions of Progress in Philosophical Inquiry," 131, 159; William Ernest Hocking, in "Progress in Philosophical Inquiry and Mr. Lovejoy's Presidential Address," *Philosophical Review* 26, no. 3 (May 1917), 330; Campbell, *A Thoughtful Profession*, 175–176.

The Progressive Era debate over science's meaning and relevance for the discipline of philosophy paralleled larger debates about the nature and possibilities of democracy. Like the pragmatists, other philosophers wrestled with an issue facing all scientific democrats seeking to bring disciplinary expertise to bear on American public culture: How could they reconcile freedom of opinion with the findings of scientific inquiry, when these diverged? They resolved this tension in various ways, drawing on different understandings of freedom and authority. James and especially Dewey demoted scientific "truth" to a species of consensual agreement. The platform advocates claimed that empirical evidence and logical reasoning exerted no authority at all; the term "authority" applied only to the exercise of partial, human interests. In this view, true freedom rested on conformity to the structure of existence, as revealed by the sciences and mathematics. Indeed, it was both necessary and salutary to exert some human authority to keep individuals in line with this natural structure. Still other philosophers preferred slow, unforced, "organic" modes of intellectual integration to the "mechanical" procedure of the platform-builders. In their view, preserving interpretive freedom would inexorably lead all individuals to robust, representational truths. Strongly echoing Progressive Era arguments about the scope and functions of the American state, the platform debate highlighted the diverse conceptions of science, cultural authority, and democracy at work in the push to make America scientific.

It also revealed important commonalities among the participants. All of the philosophers discussed here identified the achievement of civil relations among themselves as a prerequisite for the achievement of civil relations among citizens. These philosophers traced social cohesion, not to market competition or majority rule, but rather to a discursively conditioned, yet ultimately voluntary, orientation toward the common good on the part of each individual. Disliking exercises of self-interest by individuals or groups, they undertook to create democratic citizens, either by drawing out ethical lessons implicit in science or by providing ethical lessons that were unavailable from science but had been rendered all the more necessary by its alteration of economic processes and social forms. In short, these philosophers hoped to sustain democracy in an industrial age by reconciling science with values. But their explorations into the precise nature of scientific forms of argumentation and social organization would find few parallels until the 1930s, when science and language became subjects of intensive analysis. During the Progressive Era and the 1920s, most scientific democrats would explore more practical questions, asking how best to put scientific knowledge and virtues into the hands of the people.

PART II

THE SCIENTIFIC ATTITUDE

"Under the old monarchy the French took it as a maxim that the king could do no wrong, and when he did do wrong, they thought the fault lay with his advisers. This made obedience wonderfully much easier. One could grumble against the law without ceasing to love and respect the lawgiver. The Americans take the same view of the majority."

– Alexis de Tocqueville

A well-worn account of American thought holds that by the 1920s, human scientists had decisively chosen the path of professionalization and normative disengagement. That decade, the story goes, witnessed the consolidation of an epistemological "objectivism" or "scientism" that declared all value judgments utterly extraneous to scientific inquiry. On such a view, scientists could speak only to descriptive questions of what existed, not to normative questions about what people ought to do. In fields such as psychology, sociology, and political science, scholars struggled to purge their work of all subjective elements. As the sociologist William F. Ogburn famously put it, the true scientist could "crush out" all emotions and values, leaving only the "pure gold" of objective knowledge.[1] Ogburn and other "neo-positivists" adopted new methodological techniques, especially quantification, to eliminate their biases more effectively. Experiments – artificial forms of experience – also made their way into social-scientific practice in the 1910s and 1920s. The spread of experimental methods both reflected and reinforced a belief that social scientists could, like natural scientists, create and analyze relatively pure forms of experience, leading to highly reliable conclusions.[2] Both of these conceptions of social science, the

---

[1] William Fielding Ogburn, "The Folk-Ways of a Scientific Sociology," *Scientific Monthly* 30 (April 1930), 306.

[2] There is an extensive literature on experimental methods in the American social sciences, especially psychology: e.g., Jill G. Morawski, ed., *The Rise of Experimentation in American Psychology* (New Haven: Yale University Press, 1988); Kurt Danziger, "Making Social Psychology Experimental: A Conceptual History, 1920–1970," *Journal of the History of the Behavioral Sciences* 36 (2000): 329–347.

story continues, generated powerful political effects, as advocates of the new modes of social inquiry came to believe that they could guide society more effectively than the public or professional politicians.

In truth, however, the turn toward scientism has been magnified by the prevailing interpretive habit of simply tracking objectivity claims rather than assessing the full range of scientific thinking. The usual portrait of a few democratic islands in a sea of technocrats dramatically exaggerates the dominance of the technocratic mindset. The 1920s brought a largely subterranean epistemological divergence, rather than the open triumph of one epistemology over another. The push to bring science to bear on American public culture hardly required "objectivity talk," and in important ways it militated against such an epistemological stance. John Dewey and many others denied the possibility of an interpretive end run around normative commitments – and around ordinary citizens. These scientific democrats increasingly stressed the constructed, contextual, and dialogic character of scientific knowledge, declaring value-neutrality impossible in their fields and calling the search for it politically dangerous. In their view, science produced neither incontestable facts nor infallible laws, but only what the economist John M. Clark called "generalizations that are significantly true," in a phrase where "significant" rang as loudly as "true."[3] These scholars worked to build up a body of social knowledge infused with the tenets of the thoroughly progressive culture whose emerging outlines they thought they discerned. Rather than a "view from nowhere," they sought a view from the perspective of the deepest needs of the American people.

These scientific democrats had inherited from their Enlightenment predecessors an unshakable faith that the scientific method was a universally applicable means of understanding the world. For them, it was literally unthinkable that social life could offer anything that was beyond the reach of the scientific method. If the social world proved to be less regular or less predictable than the natural world, then the techniques of science would simply have to be altered to take account of those differences. A science that could not explain social behavior was no science at all, for it could never perform its central social function: ensuring solidarity by fostering cultural consensus.

More broadly, most American scholars in the late nineteenth century, and many in the early twentieth century, agreed that the human sciences were part and parcel of a unified, if heterogeneous, scientific enterprise that included the likes of Franz Boas's ethnographic investigations, Charles Beard's historical analyses, and perhaps even I. A. Richards' literary criticism. A smaller group, including Dewey, also held that science, understood as a process of exchanging experience and reasoning collectively about its meaning, would generate some consensus in the moral realm, though certainly not the absolute, timeless moral laws of the nineteenth century. Another group, again including Dewey,

[3] Quoted in Yuval P. Yonay, *The Struggle Over the Soul of Economics: Institutionalist and Neoclassical Economists in America Between the Wars* (Princeton: Princeton University Press, 1998), 93.

challenged the idea of absolute physical laws in the same manner, identifying the natural sciences as irreducibly interpretive and projective, just like the human sciences.

Until the 1930s, however, scholarly controversy was not organized along the line of epistemological disagreement. Part II highlights the intellectual and political bridges across what is usually portrayed as a yawning epistemological chasm between advocates and critics of value-neutrality in the human sciences. It shows that in the Progressive Era and the 1920s, two shared sets of commitments – first, scientific democracy, with its imperative to build up the scientific disciplines and make their methods and findings available to the public, and second, a cluster of more specific assumptions centered on the image of society as a relational, psychological phenomenon – kept scholars with increasingly polarized epistemological views from falling into open conflict. Epistemology did not become a matter of vigorous contention among human scientists until the 1930s, when the economic crisis and then the New Deal rendered questions about science's relationship to governance more pressing and concrete.

Covering the first three decades of the twentieth century, Part II describes how new groups of scientific democrats sought to lay the theoretical and institutional groundwork for a culture that the human sciences would anchor and constantly refresh. During those years, scientific democracy became closely associated with the human sciences. Physical scientists found applications for their work in the industrial sector and largely backed away from claims about the cultural meaning of science. By contrast, scientific democracy figured centrally in the rapidly expanding human sciences, shaping the content of those disciplines as well as the descriptions of science their practitioners offered. Much knowledge work in these "cultural sciences" – including biology – reflected either a desire to prove that government through popular sovereignty could work or a search to identify and neutralize cultural threats to its operation. The cultural science approach privileged culture and language as social forces and held up science as a powerful tool for discursive engagement in public affairs. Taking shape during the Progressive Era and expanding in numerous directions between the wars, it shared much in common with Richard T. Ely's ethical economics. However, the new Progressive outlook centered on psychology, sociology, anthropology, linguistics, philosophy, and history rather than the older disciplines of political science and economics.

Political shifts helped to frame the era's reformulations of scientific democracy. Between 1901 and 1920, the nation engaged in a series of unprecedented experiments designed to bring industrial capitalism under the yoke of communal values and harmonize the social conflicts it had produced. Americans, it seemed, had come to grasp the inner meaning of modern science. But the wartime upsurge of nativism and antiradicalism, followed by the debacle at Versailles, the implosion of Progressive reform, and the rise of new propaganda techniques, led scientific democrats to conclude that the scientific reform of American culture had hardly begun. The push to form scientific citizens began anew in the 1920s.

Some scientific democrats worked to make their ethical lessons more widely available and attractive through popular writings and curricular reforms in the schools and universities. Others took a self-reflexive approach, turning their critical gaze back on themselves. Fearing that the nation's cultural problems originated in their own disciplines, these scholars worked to eliminate the taint of nineteenth-century individualism so that their knowledge might serve as a more effective tool for cultural criticism and reconstruction. Neo-positivism and the democratically committed alternative took shape as alternative ways of purging the capitalist residue from the human sciences. The neo-positivists presented a thoroughly non-ideological science as the needed successor to the nineteenth century's false, bourgeois science, whereas Dewey and his allies, deeming neutrality impossible, instead sought a science infused with the proper cultural values – the values of the people, not the values of big business.

These views also reflected divergent paths away from the Christian framework of social ethics that had guided the university reformers of the late nineteenth century. Those Gilded Age figures had joined their more thoroughly theistic counterparts in defining ethics in terms of a set of personal virtues: honesty, benevolence, and the like. Why, though, were those virtues good? In truth, there was no need to ask; one could simply rest content with a system of virtue ethics that equated the good with the exercise of the virtues held by a moral exemplar, namely Jesus. But it was common, as we saw in Chapter 2, for Gilded Age scientific democrats to incline toward either motives or consequences in delineating the ethical orientation of science. In the twentieth century, these emphases became more clearly separate within the dominant rhetorics of science.

Those scientific democrats who read the scientific spirit through the lens of *motives* emphasized the need to eliminate all selfish impulses and interests. It took just a short step to convert this ethical injunction into the epistemological claim that only research undertaken in a thoroughly disinterested fashion could reveal the truth. That view also echoed the republican virtue of autonomy, foundational to both American political thought and Western scientific thought.[4]

By the end of the nineteenth century, however, American theorists had become increasingly likely to deny that absolute disinterestedness was possible, even for the most virtuous scientist. Such a view of human cognition comported badly with the New Psychology, and the scholarly conflicts of the late nineteenth century had thrown it into disrepute as well. Nor did total self-denial offer a very good model for the formation of citizens, although Karl Pearson gave it a go in Britain. Many American theorists nevertheless retained disinterestedness as the ethical centerpiece of science by methodizing and collectivizing its attainment. They introduced a series of mechanical and social technologies into the research process, seeking to cancel out individual biases and ensure the disinterestedness of the discipline's collective judgments. This

---

[4] Steven Shapin, *A Social History of Truth: Civility and Science in Seventeenth-Century England* (Chicago: University of Chicago Press, 1994).

approach had the side effect of disaggregating the ethics of scientific practice from the ethics of opinion formation in the public sphere, because citizens clearly could not apply experimental methods in their daily lives. The new emphasis on what has been called mechanical objectivity tended to accompany highly attenuated versions of scientific democracy that defined the scientific ethic as a matter of deference to evidence – meaning, in the case of the public, deference to experts.[5]

But there was a way to retain a more robust and culturally applicable understanding of science's ethical meaning while acknowledging the ubiquity of human subjectivity. This approach involved thinking about the ethics of science in terms of *consequences*, rather than virtues or motives. A consequentialist theory of ethics is one that defines the good in terms of actions that produce certain consequences for others. In this light, following the example of Jesus did not mean rigorously suppressing the self. Instead, it entailed discovering how one's actions affected others and building those consequences into one's behavioral habits with an eye toward maximizing the flourishing of all. The emphasis of such a view on cause-effect relations offered a different point of connection to science, which obviously had something to say about patterns of causation. In fact, a consequentialist approach meant that a properly constituted social science could teach individuals how to behave ethically – especially if it also, in its guise as psychology, explained the contours and conditions of human flourishing. Philosophies of knowledge compatible with this consequentialist ethic – most pertinently pragmatism, which embedded consequentialism in its very definition of truth – also lay close at hand for those scientific democrats who emphasized consequences rather than motives.[6]

Best of all, a consequentialist orientation seemed perfectly suited to both the realm of moral and political discourse and that of the sciences. Whereas it was difficult to understand what it would mean for citizens to eliminate their subjectivities through the political equivalents of microscopes or regression analyses, a consequentialist ethic simply said that individuals should seek to promote the well-being of others with the aid of reliable causal knowledge. Balancing their interests against those of others seemed far more feasible for citizens than the kind of absolute self-effacement that Pearson proposed. Increasingly invoking a "scientific attitude" rather than a "scientific spirit," many scientific democrats in the early twentieth century began using the term "scientific" to signal a particular orientation toward the *needs* of others, rather than toward their *arguments*.[7]

[5] On the deeper, European roots of this view, see Lorraine Daston and Peter Galison, *Objectivity* (New York: Zone Books, 2007).
[6] Positivism, too, could serve that function, although as we have seen, it proved more difficult to square with prevailing religious and political commitments.
[7] This linguistic change reflected the growing centrality of psychology – whose practitioners developed the "attitude" concept as a theoretical tool in the 1920s – to the human sciences and scientific democracy. By the early twentieth century, to be a modern, scientific thinker meant to embrace the New Psychology's emphasis on the largely noncognitive roots of belief and action.

Few took this consequentialist orientation to mean that human beings possessed material needs alone. Instead, echoing John Stuart Mill's mid-nineteenth-century reformulation of utilitarianism, they portrayed human beings as richly endowed with emotional, aesthetic, and social needs – even spiritual needs, although not necessarily of the kind addressed by traditional theism. Many scientific democrats in the early twentieth century, especially in the universities, identified a consequentialist, humanitarian ethic as the central lesson of science, the kernel of any valid religion, and the essential meaning of democracy. They defined a true democracy as a society that designed its public institutions – and its public culture – so as to maximize the welfare of all. These scientific democrats imagined the ideal citizen as knowledgeable, but more importantly, as committed to seeking collective agreement on mutually beneficial plans of action, the political equivalent of scientific theories. Such citizens would incorporate the needs of others into their motives, and would possess the skills and knowledge to competently balance those needs against their own, in light of the prevailing conditions.[8]

This consequentialist understanding of science and politics shaped a great deal of work in the cultural sciences up to the era of World War II and in many cases well beyond. The ambition to make American culture more scientific by instilling in citizens a humanitarian ethical framework extended well beyond the borders of today's social sciences, bringing together into a single discursive network a surprisingly wide range of intellectual domains in which scholars applied, or advocated applying, scientific tools of analysis to human phenomena. By the end of the 1920s, the post-Progressive version of scientific democracy had become a dominant presence in sociology, political science, anthropology, history, and linguistics, as well as psychology itself and, to a lesser extent, philosophy. It had also made considerable inroads into economics and biology and had even attracted a few literary theorists. In short, it stretched across the entirety of the cultural sciences, giving them a coherence as a transdisciplinary formation that they would lose by mid-century.

Despite the implicit disagreement between those scientific democrats who defined science in terms of self-effacement and value-neutrality and those who placed at its center a concern for consequences, that epistemological split did not produce a sustained scholarly debate until the 1930s, when new economic and political conditions heightened the stakes. Prior to that, several shared commitments fostered unity across lines of epistemological division that seem glaring, even unbridgeable, in hindsight. First and foremost was the felt imperative to build up the human sciences. Whatever their ethical and epistemological commitments, all scientific democrats in the human sciences could agree

[8] The University of Chicago philosopher T. V. Smith identified the connection to Mill's work in "Philosophical Ethics and the Social Sciences," *Social Forces* 7, no. 1 (September 1928), 18–19. Many other scientific democrats, including Ralph Barton Perry, simply took for granted the "preeminent good of the general happiness" as "a fact ... independent of judgement or sentiment." "Realism in Retrospect," in *Contemporary American Philosophy*, ed. George P. Adams and William P. Montague (New York: Macmillan, 1930), 206–208.

on the need to develop their disciplines and make their outlooks more widely available. The human sciences expanded rapidly and became the leading edge of scientific democracy by the 1920s. Following economics, which had coalesced in the 1890s, the disciplines of sociology, anthropology, political science, psychology, and history took on their recognizably modern forms. But if those fields resembled their counterparts today, they hardly acted like them. Much of the era's discipline-building energy flowed from a sense that the human sciences could remake American culture and thereby restore the conditions for self-government. As physical scientists increasingly ceded ethical authority to the churches and equated the social value of their research with its industrial applications, human scientists became the most vocal advocates of scientific democracy.

Most of these early-twentieth-century scientific democrats, whether they stressed self-elimination or control by consequences, also followed the ethical economists in believing that a truly scientific citizenry would respond to the ills of industrial society by placing strong collective controls on economic behavior. Like their *laissez-faire* predecessors, these scientific democrats assumed they had grasped *the* scientific view of modern social organization – the view that would result from a properly inclusive and empirically grounded inquiry into the best way to adjust social institutions to human needs under industrial conditions. In fact, they tended to treat support for a thoroughly cooperative social order as a kind of proxy for the scientific outlook. To be sure, alternative political readings of scientific democracy still found homes in the human sciences, most notably in economics, where *laissez-faire* retained some support. But many young scholars were drawn to the burgeoning human sciences precisely because of their close association with the Progressive and post-Progressive versions of scientific democracy. Leading theorists in the human sciences believed that society represented a meaningful whole and that the good of all trumped the absolute freedom of the individual. They drew out the Social Gospel emphasis on reorienting cultural practices and remaking social institutions to enable the self-realization – material, intellectual, and spiritual – of all individuals. Their broadly humanitarian ethic, embedded in new conceptions of the psyche and of social relations, powerfully shaped the universities and disciplines in the early twentieth century.

Still, the commitment of these scientific democrats to the priority of society over individual freedom should not be equated, retroactively, with support for something like the New Deal state. Recent studies have shown that, as late as World War II, natural scientists profoundly distrusted the federal government, even as it moved toward lavishly funding their research. Less frequently recognized is the fact that many human scientists thought a thoroughly cooperative culture would be far more effective, and more ethical, than formal, top-down political control. Until the start of the Depression, and often thereafter, many scientific democrats believed that establishing a more scientific culture would obviate the need for strict governmental controls in crucial areas of social practice. Moreover, even those who embraced state expansion as a solution still saw a need for thoroughgoing cultural change, not only to gin up public support for the new governing apparatus but also to create a citizenry that could use it responsibly. Thus, while economic *laissez-faire* found few champions among early-twentieth-century human scientists, these same figures typically resisted the prospect of state-employed experts defining and promoting the public good, without the active and informed consent of the public. They expected a mobilized citizenry, equipped with a scientific understanding of the social order and the accompanying set of personal virtues, to steer a middle path that combined the political freedom of *laissez-faire* with the material benefits of state socialism. Science, they believed, could make that reconciliation possible.

4

Scientific Citizenship

During the Progressive Era, the idea of a society knit together by voluntary, scientifically informed cooperation, rather than by individual self-seeking or a strong central state, became a central theme in the human sciences. A group of pioneering theorists developed new understandings of social behavior that centered on the intersubjective realm of communicative exchange. They traced social integration to a realm of shared beliefs and values wherein the prevailing "culture," as the anthropologist Franz Boas influentially dubbed it, shaped adults and children alike. But modern societies avoided utter stasis, these theorists argued, because the creative work of scientific experts infused the culture with new knowledge and gradually altered it over time. At the same time, the human sciences made the citizenry less resistant to proposals for cultural change by imbuing individuals with the scientist's habitual tendency to identify existing forms and practices as historical and functional in character rather than absolute and transcendent.[1]

In the emerging understanding, citizens needed sufficient scientific knowledge to predict the effects of their actions and sufficient sympathy to orient those actions toward the needs of others as well as their own. Where did that sympathy come from, however? A hope of discovering absolute, empirically provable moral propositions had lingered in the early days of scientific democracy. Yet scientific democrats in the universities had always tended to follow the liberal Protestant conception of morality as a matter of behavior rather than belief. Pragmatist and positivist conceptions of knowledge joined with Darwinism and the postulate of a fundamental break in economic history to undermine the idea of an empirically discoverable moral order in nature. In the 1890s, a few theorists looked to biology as a source of the needed sympathy, seeing in the human animal an instinct of "gregariousness" – an inborn tendency to cooperate with others and to conform to the collective beliefs that underpinned

---

[1] Cf. Christopher Shannon, *Conspicuous Criticism: Tradition, the Individual, and Culture in American Social Thought, from Veblen to Mills* (Baltimore: Johns Hopkins University Press, 1996).

social order. But numerous problems doomed this theory of action, including its affinities with racism, its tendency to reduce social inquiry to the mere application of biological principles, and its inability to account for cultural change. By the era of World War I, psychologists and other social scientists had begun to follow anthropologists in rejecting the idea of innate behavioral instincts and ascribing human behavior to culturally derived habits.

But modern psychology's demotion of reason and the subsequent rejection of innate instincts left open the possibility of a form of moral sympathy, and thus social order, rooted in the emotions. Echoing the "sentimental Enlightenment" of eighteenth-century Scotland, many Progressive Era scientific democrats found the source of the needed emotional commitments in what Boas labeled culture and others simply called "society": a realm of communicative interchange standing apart from, and prior to, both the state and the economy. These theorists identified social order as an essentially mental phenomenon, rooted in a combination of shared values and beliefs and active discursive engagement around points of difference. As they explored the sphere of society or culture, they calibrated their conceptions of democracy and even selfhood to preserve the possibility of a thoroughly scientific, democratic polity.[2]

It is hardly surprising that the theorists of that generation, who had witnessed a bitter conflict of classes and an unprecedented expansion of ethnic diversity, would be preoccupied with identifying the source of social integration; nor that they would find it outside the realms of economic competition, which seemed to have sown only discord, and formal politics, which seemed hopelessly corrupted. Like the scientific democrats discussed in Part I, those of the early twentieth century found existing structures of belief and cultural authority inadequate to the task of keeping a self-governing society on track. They proposed that widespread, uncoerced adherence to science's findings could serve as the needed cultural glue in a democracy by orienting social action toward ethically desirable consequences.[3]

---

[2] Jeffrey Sklansky, *The Soul's Economy: Market Society and Selfhood in American Thought, 1820–1920* (Chapel Hill: University of North Carolina Press, 2002); Michael L. Frazer, *The Enlightenment of Sympathy: Justice and the Moral Sentiments in the Eighteenth Century and Today* (New York: Oxford University Press, 2010). Cf. Howard Brick, "Society," in *Encyclopedia of the United States in the Twentieth Century*, ed. Stanley I. Kutler (New York: Charles Scribner's Sons, 1996), 918–924; and Brick, *Transcending Capitalism: Visions of a New Society in Modern American Thought* (Ithaca: Cornell University Press, 2006), 13–14. An early overview of this phenomenon is A. H. Lloyd, "The Organic Theory of Society: Passing of the Contract Theory," *American Journal of Sociology* 6, no. 5 (March 1901), 577.

[3] Many figures of the era shared an even thicker set of assumptions, in the form of a biologically informed vision of society as an organism. The image of a "social organism" found American precedents in the work of several important Gilded Age thinkers, including the ethical economists Richard T. Ely and Henry Carter Adams and their sociological counterparts Lester Frank Ward, Albion W. Small, and Franklin H. Giddings. It also had deeper roots in the British naturalist Herbert Spencer's influential writings of the 1850s and 1860s. The organic analogy validated a number of commitments shared by most Progressive Era social scientists, starting with the primacy of the community over both corporations and the state.

Although an emphasis on the priority of society over the individual ran through much of the social and political thought of the Progressive Era, it is important to resist the assumption that when the scientific democrats of that period said "society," they meant the state. True, we know the period best for its innovations in the realm of government: bureaucratic commissions, regulatory agencies, the referendum and recall, and the like. The scientific democrats of the early twentieth century, however, continued to believe that rendering American culture scientific would drive political change much more effectively than would any amount of institutional tinkering. Although most scientific democrats sought to place limits on individual behavior, and especially economic behavior, they typically identified the cultural authority of science rather than the coercive power of the state as the proper means for implementing such limits. The classic works of Progressive Era social theory amount to blueprints for the creation of a scientific culture, rather than early sketches of the New Deal state.

The preference for cultural rather than governmental checks on individual self-seeking flowed from multiple sources. Many Progressive theorists carried forward their Gilded Age predecessors' suspicion of state-driven modes of political change, a view grounded in the liberal Protestant distaste for coercion. These figures also tended to assume, like so many Western thinkers through the centuries, that a nation's political life reflects the contours of its intellectual life. More specifically, they believed that the United States was already a deliberative democracy, its social, political, and even economic institutions shaped directly, if perhaps only broadly, by public opinion. This meant that reformers could only change American institutions from below, by redirecting the deep cultural forces that produced them.

For these reasons, Progressive Era scientific democrats targeted the inner sphere of beliefs, values, and motives. Although some also participated in the search for new knowledge relevant to policy formation and administration, they all worked to create the cultural foundation for a new institutional order, in the form of a collective public commitment to fulfilling the needs of all through secular means. They expected science to exert its beneficent influence *through* the people, rather than coming down to them from on high or operating entirely above their heads.[4]

Thus, writings by figures such as Edward A. Ross, Charles Horton Cooley, George Herbert Mead, and John Dewey do not simply emphasize the priority of social needs over the economic freedom of the individual in the interest of raising the overall standard of living. Their portraits of social reality also incorporate an emphasis on distributing science among the public, along with

---

[4] In fact, one of their main arguments against both revolutionary socialism and business unionism was that the political system would never truly change until the minds of the people were transformed. Marc Stears highlights the Progressives' emphasis on public deliberation in *Demanding Democracy: American Radicals in Search of a New Politics* (Princeton: Princeton University Press, 2010).

a conception of science as the vector for a powerful ethical orientation. These texts argue that society's mental grip on the individual accounts for the preservation of social order, while the spread of scientific ideas among the population makes social change possible and sets its direction. Rather than working to establish a base of political legitimacy for a strong administrative state, the leading Progressive Era theorists of scientific democracy sought instead to build a new culture – to create citizens oriented toward the good of others rather than their own, narrowly defined interests. In fact, as the last part of this chapter shows, even the founding editors of the *New Republic*, often described as pioneering advocates of a managerial state, deemed the widespread internalization of scientific values a necessary complement to economic regulation.

Viewing these Progressive Era works as expressions of scientific democracy highlights their similarities to two other major contributions of the era that are rarely viewed in relation to Progressive social thought or to one another: the conservative sociologist William Graham Sumner's 1906 exploration of "folkways" and "mores" and Franz Boas' early elaboration of the culture concept and the standpoint of cultural relativism. Like their contemporaries, Sumner and Boas described institutions, including the state itself, as reflexes of a larger, pre-political entity, namely society. And they found society coextensive with the beliefs, values, and attitudes of the public. Sumner and Boas, no less than Ross, Cooley, Mead, and Dewey, contributed to an emerging understanding of democracy and a corresponding vision of the scientific scholar's political role by identifying public opinion as the ultimate basis of all social institutions – in Ross' words, "the original *plasm* out of which various organs of discipline have evolved."[5]

BIG-STATE SCIENCE?

One could easily be forgiven for thinking Edward A. Ross an advocate of unbridled state power. Although he had studied under Richard T. Ely at the University of Wisconsin, Ross adopted the hard-boiled language of realism employed by Ely's enemies, those Gilded Age *laissez-faire* theorists who advocated strict limits on state intervention in the economy. Moreover, Ross embraced a biological racism that was increasingly drawing opposition among academically inclined scientific democrats, though in his wildly influential *Social Control* (1901) he used this racial theory less to paint a picture of conflict between nations or classes than simply to identify the fractiousness of the individual Westerner as – frustratingly, but also comfortingly – an endemic condition, a permanent racial trait. According to Ross, the ferocious independence of the "restless, striving, doing Aryan" threatened to tear apart Western nations, which were riven by latent material and psychological conflicts. Society needed to act with crushing force, ruthlessly imposing its will on individuals to maximize their freedom

[5] Edward A. Ross, *Social Control: A Survey of the Foundations of Order* (New York: Macmillan, 1901), 99.

and material well-being. "In the taming of men," Ross declared, "there must be provided coil after coil to entangle the unruly one.... The truth by all means if it will promote obedience, but in any case, obedience!"[6]

Yet Ross' expansive vision of social authority was only weakly tied to the central state, which he viewed as a fading vestige of past authoritarian regimes. For all his emphasis on authority and conformity, Ross portrayed the political sphere in a highly decentralized manner. He insisted in no uncertain terms that society should control the state, rather than the other way around. Ross viewed the development of social science as a means of preserving this balance between society and the state, steering between the two extremes of anarchic disorder and repressive statism. In fact, Ross' celebrated phrase "social control" emphasized the term "social" rather than "control." Given that every society featured a system of control, Ross reasoned, the task was to discover that which was most thoroughly *social*: that which took maximal account of individuals' needs, including what Ross viewed as their powerful, biologically rooted need for personal freedom. Social control, which took the good of all as its end and operated through formally voluntary means, was the opposite of "class control," in which a small group used direct coercion to bend the mechanisms of order toward its own interests.[7]

How could social control be achieved? How could the members of a race devoid of social instincts live together in peace without an exertion of oppressive force from above? Surveying the vast array of devices that social groups used to scotch egoism, Ross identified the internal sphere of "beliefs" as the locus of a means of control at once unparalleled in its potency and fully compatible with democracy. He proposed to "bind from within" the fractious Aryan, using "sweet seduction rather than rude force" to placate each individual with "the illusion of self-direction even at the very moment he martyrizes himself for the ideal we have sedulously impressed upon him." Rarely has a social theorist so openly and vigorously embraced the mode of internal behavioral control that Michel Foucault dubbed "governmentality." "Men do not need to be sheep," Ross explained, "in order to develop the ethos of the herbivore." He called for a wise, disinterested "ethical elite" to join hands in the "most stupendous enterprise of all time," the "campaign against the unsocial self."[8]

Sweet seduction did not entail the exercise of pure reason, however. Ross joined his idol (and uncle by marriage) Lester Frank Ward in arguing that

---

[6] Ross, *Social Control*, 3, 304; Dorothy Ross, *The Origins of American Social Science* (New York: Cambridge University Press, 1991), 229–240, 251–252; Sklansky, *The Soul's Economy*, 193–196, 200–204. See also Sean H. McMahon, *Social Control and Public Intellect: The Legacy of Edward A. Ross* (New Brunswick: Transaction, 1999).

[7] Ross, *Social Control*, 278, 347. Although Ross considered himself a socialist, he feared the expansion of state power that would accompany such policies as progressive taxation. Society could take care of itself; it was the individual that needed protection, both from his own nature and from his fellows. McMahon, *Social Control and Public Intellect*, 15–16.

[8] Ross, *Social Control*, 215, 244, 337, 347. Cf. Nikolas Rose, *Inventing Ourselves: Psychology, Power, and Personhood* (New York: Cambridge University Press, 1996), 116–149.

individual desires operated as the central motor of history, and that reason could at best divert or channel these elemental forces. Ross proposed that the ethical elite should use scientific methods to identify the "frontier between the individual and society" – the means of calibrating systems of social discipline to provide "the most welfare for the least abridgement of liberty" under given conditions. However, heeding the lessons of modern psychology, the ethical elite would couch these findings in appropriately inspirational language, mobilizing emotionally laden associations and valuations rather than the sterile, uninspiring conclusions of science. This elite would disseminate a scientifically grounded "social religion" – shades of Auguste Comte, one of Ward's heroes – that would guide individual desires into the channels of social good, thereby giving Anglo-Saxon democracy the needed substitute for innate communal instincts or a coercive state. By dispensing the social religion through the schools and churches, the ethical elite would sacralize modern sociology, lending its dictates an aura of essential rightness rather than mere expediency and creating the foundation of a profoundly integrative public culture. Ross declared that the "state-craft" of old was giving way to a new "folk-craft" that used "[s]uggestion, education, and publicity" to give "the higher traditions and knowledge" of the scholars to the masses. The ethical elite would exert its influence through public opinion, that "paternal, benevolent despot."[9]

Although Ross' vision gave social scientists a central role in the polity – he predicted that "the mandarinate" of professors and researchers would gain considerable prestige as society sought to control its affairs peaceably – he worried little that the social religion these experts helped to create would slip out of alignment with the public good. Ross did not invoke objectivity here; he did not argue that social scientists could bracket their biases. Instead, he looked to public opinion as a check on the work of experts. Ross held that the network of interrelations comprising society produced an emergent mental phenomenon, "a kind of collective mind evincing itself in living ideals, conventions, dogmas, institutions, and religious sentiments" that carried out "the task of safeguarding the collective welfare from the ravages of egoism." This social mind could not, of itself, produce new ideas; it could only chew on those put into circulation by leaders. But it gradually weeded out "those suggestions and ideas which are felt to be unfavorable to the social welfare," while implementing those innovations that passed the test.[10]

Ross proposed another check on the influence of social scientists as well, arguing that the social religion should incorporate, in addition to personal habits comporting with sociological findings, "certain convenient illusions and fallacies which it is nobody's interest to denounce" even though they had no

---

[9] Ross, *Social Control*, 217, 427, 355–358, 432, 375, 293.

[10] *Ibid.*, 88, 293, 327–328. Ross declined to equip social scientists with political power and even cautioned them against open dialogue with the public and most of its self-declared leaders. Only the ethical elite could safely apply scientific social knowledge, he argued; the egoists would use it to buck social control instead, necessitating a quick return to coercive instruments of force (441).

foundation in fact. Ross specified that the social religion should uphold the idea of "abstract, indefeasible rights" and the ideals of "Liberty, Justice, and Humanity," presenting these as natural facts rather than social constructs. Fearing that social control might produce, as it had in Germany, a sense that the interests of either the state or "the social ego" trumped those of individuals, Ross sought to preserve the framework of modern rights as a kind of noble lie, of the type later advocated by the political theorist Leo Strauss.[11]

Ross's defense of "moral engineering," which offered an ethically satisfying role for members of the burgeoning professional classes, made *Social Control* a runaway success. The volume's many admirers included Roscoe Pound, Oliver Wendell Holmes Jr., and Theodore Roosevelt; it was also widely used as a college textbook. The basic contours of his analysis – the image of a society knit together by shared beliefs and practices and guided by a small group of natural leaders who inaugurated beneficial changes without exerting direct force – could be mobilized for many political purposes, given that emphasizing the primacy of the collective good over individual self-interest left each term almost entirely undefined. In Ross' case, this lack of specificity produced some surprising results: he found socialism entirely compatible with economic competition, private property, and the bourgeois values of sobriety, monogamy, privacy, and decency. But others who followed his argument could slot their own values into the "selfish" and "social" categories.[12]

Social Control exemplifies the striking combination of strong claims on behalf of "society" (and social science, its designated voice) and profound fears of the state that resonated through much academic thought in the Progressive Era. But whereas Ross argued that social science would produce its effects indirectly through the social religion articulated by an ethical elite, many other texts from the era expected scientific experts to wield their authority over public opinion directly, albeit outside the channels of state power.

A good example of this approach is *The Transition to an Objective Standard of Social Control*, the doctoral dissertation filed by the University of Chicago sociologist Luther Lee Bernard in 1911. Conceiving of freedom and self-governance in even narrower terms than Ross, Bernard declared that "subjectivism in any form" stemmed from a failure to recognize that "social survival" took precedence over all other behavioral goals. Moreover, he stated flatly that a small elite controlled every stable social order. In the modern era, according to Bernard, this needed to be a scientific elite – the only one capable of objectively determining the functional needs of the social organism.[13]

[11] *Ibid.*, 415, 420–422.
[12] *Ibid.*, 5–6, 106, 363, 393–394.
[13] Luther Lee Bernard, *The Transition to an Objective Standard of Social Control* (Chicago: University of Chicago Press, 1911), 69, 8. On Bernard, see Robert C. Bannister, *Sociology and Scientism: The American Quest for Objectivity, 1880–1940* (Chapel Hill: University of North Carolina Press, 1987), 111–127; on technocratic tendencies more generally, John M. Jordan, *Machine-Age Ideology: Social Engineering and American Liberalism, 1911–1939* (Chapel Hill: University of North Carolina Press, 1994), 1–109.

But Bernard saw no conflict between democracy and his frank call for "centralized administration by experts." He insisted that these experts would simply carry out the popular will, attending to the technical details of its implementation rather than imposing it from the top down. Bernard's experts, however, would already have shaped that popular will in accordance with the needs of the social organism by using the schools to condition citizens to find pleasure only in activities that promoted social order. Here, said Bernard, lay the path to true happiness, which he defined as the ability to realize one's desires without experiencing the subjective feeling of being constrained by the group. Much like Karl Pearson in England, Bernard called for non-coerced – indeed, unconscious – submission by citizens, within the framework of political democracy, to the needs of the social organism, as determined by scientists. He derided as merely "pleas for personal license in more attractive form" all claims that liberty or self-realization should come before the good of the whole. Bernard's ideal polity would use the conditioning mechanisms revealed by modern psychology to align the behavior of citizens with the findings of modern sociology. His writings stood at the very boundary of scientific democracy, where it began to shade over into technocracy.[14]

SOCIAL SELVES

Many of the era's scientific democrats saw little need for the severe constraints on individual behavior envisaged by Ross and especially Bernard. Not that they thought people should run roughshod over one another. Instead, these theorists of the "social self" – most notably the University of Michigan sociologist Charles Horton Cooley, the University of Chicago philosopher George Herbert Mead, and Columbia's John Dewey, who built the concept of a social self into his influential program of progressive education – believed that the very process of growing to adulthood in a social context naturally imbued each individual with a positive emotional commitment to the good of others. The arrows of self-realization and social order naturally lined up, according to Cooley, Mead, and Dewey; they did not point in opposite directions and necessitate the repression of individuals by society. The fact that individuals were literally constituted by networks of social relations instilled in them a powerful need for participatory belonging, for a sense of contributing to a collective project. Normal individuals gravitated toward mutually supportive relationships because their sense of self had emerged in and through such relationships in childhood. If equipped with realistic, functional explanations of social causation, such individuals would naturally promote the public welfare in their daily lives, without any need for Ross' mechanisms of social discipline or Bernard's program of conditioning by experts. Social scientists could thus ensure social health and progress by simply pursuing their research and getting their findings out to the people in appropriately accessible forms.

[14] Bernard, *The Transition to an Objective Standard of Social Control*, 93, 41, 96.

The concept of a social self solved key problems facing the scientific democrats of the Progressive Era. For one thing, it provided a foundation for the impulse toward moral good that seemed compatible with the New Psychology. Human beings did not need to be naturally altruistic or gregarious to sustain social order. Nor did the scientific democrats need to rely on the churches to propagate benevolent motives, as had Ely and Ross. The intimate, face-to-face settings in which children grew to maturity could do the trick instead. Cooley, Mead, and Dewey held that socialization, in the bare sense of the word, proved sufficient to socialize, in a richer sense, the emotions and the will. Politically, meanwhile, they argued that the socialization process enabled individuals to govern themselves in the relative absence of coercive authority, because it instilled in citizens a felt obligation to balance their own interests with those of others through the give-and-take of exploratory communication.[15]

If the socialization process imbued each individual with a sturdy group consciousness, why did the current crop of Americans direct their moral energies toward such counterproductive ends? And how could reformers redirect those energies? Cooley, Mead, and Dewey identified scientific knowledge as the missing link. They argued that the inhabitants of modern industrial society did not yet understand how it worked, because the prevailing academic theories and cultural values had been formed in a bygone economic and social environment. Given the proper input from scientific scholars, however, a society of social selves would naturally tend toward a mutually satisfying form of integration that balanced flexibility with stability, easily shifting its contours to accommodate changes in the environment. Once again, the theorists of the social self portrayed the expansion of science's cultural authority as far preferable to the direct regulation of behavior by the state.[16]

Cooley developed the most extensive and detailed theory of the social self in a pair of books, *Human Nature and the Social Order* (1902) and *Social Organization* (1909). His writings offered a particularly powerful illustration of the capacity of scientific democracy to frame a scholarly career as a contribution to the betterment of political practice. Coming to the social sciences by way of a mechanical engineering degree at Michigan, Cooley followed his jurist father's path to a position at the Interstate Commerce Commission in the early 1890s, where he wrote his early works on the political economy of transportation networks. He soon recognized in these networks the material

---

[15] Those Progressive sociologists who did not employ the concept of a social self often held it to be trivially true and insufficient as a means of discipline, rather than false: e.g., Ross, *Social Control*, 23–24.

[16] Outside the universities, prominent scientific democrats in this vein included the pioneering settlement-house worker Jane Addams and Ethical Culture founder Felix Adler. See Addams, *Democracy and Social Ethics* (Cambridge, MA: Harvard University Press, 1964 [1902]), esp. 7, 273; Addams, "The Subjective Necessity for Social Settlements" and "The Objective Value of a Social Settlement," in Henry Carter Adams, ed., *Philanthropy and Social Progress* (New York: Thomas Y. Crowell, 1893), 1–26, 27–56; and Adler, *Life and Destiny* (New York: McClure, Phillips & Co., 1908).

foundation for a socially integrative form of communication and concluded that the emergence of truly national communication would solve an age-old political problem by making self-government possible in a very large nation, even under industrial conditions.

What kind of people could sustain self-government in this manner, however? In formulating his conception of the social self, Cooley drew little on American sociological writings of the era. Something of a recluse, he instead cobbled together ideas from his wide reading in the belletristic tradition, including Scottish Enlightenment meditations on the power of the sentiments as well as the German Romantics' emphasis on the constitutive role of ideas in history. Cooley thus served as a direct link between American theorists and the sentimental Enlightenment of the late eighteenth century. In fact, his version of the social self resembles Adam Smith's "looking-glass self," developed in dialectical relation to the judgments of others. Engaging with the full range of Western thought on the nature of modern societies and their forms of politics, Cooley offered his works as contributions to democratic theory, not simply sociological treatises.[17]

Cooley defined modern society as an organism composed of functionally integrated parts and posited that this organism possessed a social mind, "an organic whole made up of co-operating individualities, in somewhat the same way that the music of an orchestra is made up of divergent but related sounds." Rather than being located in a spatially distinct "brain," this social mind took the form of a centerless network of interconnected nodes or cells. In fact, Cooley followed idealist philosophers in denying that the social mind and the individual mind were even separate objects. He argued that these putatively distinct entities were merely the products of different scales of analysis, alternative angles of vision on a mental unity properly called "Human Life."[18]

Combined with the concept of a social self, Cooley's organicist vision of the networked society allowed him to abandon the longstanding liberal assumption, adopted even by Ross and Bernard, that society and the individual stood in a relationship of antagonism. As parts of a single whole, Cooley argued, neither of these preceded the other, either historically or morally. The individual and society had arisen simultaneously and changed in tandem, as part of a broader process of human growth. Enmeshed in a communicative web extending outward from personal intimates to the whole of humanity, past and present, Cooley's individuals developed a powerful instinct of sociability and expressed it through further acts of communication.

[17] Glenn Jacobs, *Charles Horton Cooley: Imagining Social Reality* (Amherst: University of Massachusetts Press, 2006), 8, 36. See also Marshall J. Cohen, *Charles Horton Cooley and the Social Self in American Thought* (New York: Garland, 1982); Ross, *The Origins of American Social Science*, 240–247; Caroline Winterer, "Happy Medium: The Sociology of Charles Horton Cooley," *Journal of the History of the Behavioral Sciences* 30 (1994): 19–27; Sklansky, *The Soul's Economy*, 205–209, 215–223; and Michael D. Clark, *The American Discovery of Tradition, 1865–1942* (Baton Rouge: Louisiana State University Press, 2005), 182–215.

[18] Charles H. Cooley, "Social Consciousness," *American Journal of Sociology* 12 (1907), 675; Cooley, *Human Nature and the Social Order* (New York: Charles Scribner's Sons, 1902), 1.

Meanwhile, another instinct or sentiment – the desire for self-esteem stressed by Smith – ensured that their communicative acts produced cohesion rather than antagonism. "Always and everywhere," Cooley asserted, "men seek honor and dread ridicule," and thus heed public opinion. This drive for approval turned egoism into altruism. What Cooley called the "primary groups" – the family, the play-group, and the local community – modeled for each child a form of social order that mobilized the powerful instinct of sympathy to harmonize the "self-assertion and various appropriative passions" of the individuals involved, rather than simply repressing these when they clashed. In a kind of transfer of training, each primary group prepared the individual to create harmony at the level of society as a whole, producing a polity "wherein individual minds are merged and the higher capacities of the members find total and adequate expression." Indeed, Cooley believed that his analysis provided a new justification for "the theory of a natural freedom modified by contract," which he had earlier knocked off of its transcendental foundations: "Natural freedom would correspond roughly to the ideals generated and partly realized in primary association, the social contract to the limitations these ideals encounter in seeking a larger expression." In an ironic echo of *laissez-faire*, Cooley's theory held that properly socialized individuals would automatically build such a "moral whole" by simply pursuing their own freedom and growth.[19]

Cooley avoided the seemingly relativistic implications of his identification of morality with adherence to communal beliefs by suggesting, in a vaguely Hegelian fashion, that social evolution itself entailed moral growth. He saw a movement in history toward "the use of higher and more rational forms" of discipline. Institutions of all kinds, wrote Cooley, increasingly employed "reason and conscience" to shape behavior, targeting "self-respect" rather than the body. Society was headed away from the dominance of the state and the church, and toward the healthy, diverse expression of all human personalities, harmonized via horizontal, non-coercive communication. In short, it was moving toward what Ross called "social control."[20]

Yet the cooperative commonwealth had not yet taken shape. Why not? Cooley answered that men lacked knowledge of the proper outlets for their moral sentiments, not having fully grasped the contours of a modern industrial society. Like Ross, he urged cultural leaders to set new moral exemplars before the public, using scientific knowledge of social causation to convert vague sympathies into a concrete system of behavioral prescriptions. Scholarly organizations would play a central role in the process. According to Cooley – who helped to define the field of sociology while rarely leaving Ann Arbor – the new communications technologies powerfully facilitated both the dissemination of knowledge and the clustering of "more discriminating minds," as in the

[19] Charles H. Cooley, *Social Organization* (New York, Charles Scribner's Sons, 1909), 28, 23, 33, 47. Cooley also used his theory to justify legal strictures and self-government as a whole (42, 45). Ross notes that Cooley added the primary group concept very late in the writing process: *The Origins of American Social Science*, 245.
[20] Cooley, *Human Nature and the Social Order*, 396–397.

academic disciplines. With the further spread of printing, Cooley predicted, modern societies would eventually come under the sway of "a true aristocracy of intelligence and character."[21]

The pragmatist philosopher George Herbert Mead, who was far more concerned than Cooley to establish his scientific bona fides, produced a theory of the social self that relied on the intellect rather than the emotions. From biological research, he concluded that experienced reality was unitary, encompassing both environments and actors. What the actor perceived as an "environment," he wrote, was merely "the projection of himself in the conditions of conduct," the "statement of the conditions under which his different conflicting impulses may get their expression." Morality, Mead continued, emerged through the interaction of individual agents and social environments as individuals voluntarily pursued paths of action promoting "the fullest life of the species." Science thus represented "the intellectual phase of moral conduct," in Mead's view, because it offered reliable knowledge of the consequences of action. This view identified science as the means human beings employed in seeking to follow an ethical system based on producing the best possible consequences for all.[22]

Why would individuals armed with reliable knowledge naturally build up the "organism of personalities" that characterized the moral community? Like Cooley, Mead argued that childhood socialization led individuals to freely pursue the collective good. He held that social disharmony produced parallel conflicts within individual personalities, who sought at all times to integrate the welter of echoing voices in their heads. As a result of the socialization process, Mead explained, multiple personalities existed in the mind of each individual, occupying what he called the "objective social field": the sum of all social voices available to the individual. The appearance of new, discordant voices necessitated reflective thought, which took the form of an internal argument aimed at quieting the intellectual din. Mead held that the individual experienced an external social conflict as a painful fracture within the self, wherein "certain values find a spokesman in the old self or a dominant part of the old self" and other values "find other spokesmen to present their cases."[23]

[21] Cooley, *Social Organization*, 52–53, 28, 88, 75. The sociologist Charles A. Ellwood recognized (and shared) Cooley's professionalizing impulse: Review of Cooley, *Social Organization*, *International Journal of Ethics* 20 (1910), 229–230.
[22] George Herbert Mead, "The Philosophical Basis of Ethics," *International Journal of Ethics* 18 (1908), 311, 315–316, 319–320, 322. Much as Cooley's emphasis on communication as the constitutive element of selfhood coexisted with personal reclusiveness, Mead's accompanied an inability to publish. His ideas gained their full influence only in the 1930s with the posthumous publication of three edited volumes of lecture notes. Yet they were well known in the inner circles of American social thought much earlier. Recent studies include Gary A. Cook, *George Herbert Mead: The Making of a Social Pragmatist* (Urbana: University of Illinois Press, 1993); Filipe Carreira da Silva, *G. H. Mead: A Critical Introduction* (Malden, Mass.: Polity, 2007); and Carreira da Silva, *Mead and Modernity: Science, Selfhood, and Democratic Politics* (Lanham, Md.: Lexington Books, 2008).
[23] Mead, "The Philosophical Basis of Ethics," 317, 322; Mead, "The Social Self," *Journal of Philosophy, Psychology and Scientific Methods* 10 (1913), 379, 378.

In the ensuing process of self-reconstruction, Mead cautioned, favoring the old self over the new voices signified "that lack of objectivity which we criticize not only in the moral agent, but in the scientist." If individuals did adopt objective, fully symmetrical views, however, any moral conflict could be resolved, not through a zero-sum "struggle between selves," but rather through "such a reconstruction of the situation that different and enlarged and more adequate personalities may emerge." Mead's mental "forum of reflection" represented a kind of internalized public sphere in which actors reconciled social interests conceptually and proposed appropriate forms of moral action. In fact, his formulation of the social self seemed to suggest that intelligent individuals could provide the basis for political reconstruction by simply performing integrative acts of thought and behaving accordingly. Even such a thorny problem as industrial conflict could be solved easily if the "different social interests" were "given full expression" in the minds of community members, he asserted.[24]

Mead based his model of reflective thought directly on the practice of science, comparing the individual's composition of the objective social field to the researcher's development of a hypothesis to fit the available facts. "The process," he wrote, "is in its logic identical with the abandonment of the old theory with which the scientist has identified himself, his refusal to grant this old attitude any further weight than may be given to the other conflicting observations and hypotheses." By taking this approach, the researcher literally gained a new self that harmonized a conflict. Mead held that moral reflection followed exactly the same method, merely replacing the physical data of science with "concrete personal interests." In effect, he portrayed each individual's current self as nothing more than a hypothesis generated by the interaction of inner impulses and the external environment – *I must act in this way so as to guarantee further harmonization.* When conflicts arose, individuals revised their hypotheses to better fit the evidence. Mead thus portrayed science as both the source of reliable knowledge regarding the social consequences of various actions and as a model of equitable conversation, perfectly suited for adaptation to moral conflicts. Through the scientific revision of selves, driven by an innate tendency to reconcile competing internal voices, he expected discursive harmony to extend outward from individual minds into the social phase of experience.[25]

Mead's writings illustrate how the concept of the social self made it possible for pragmatists to believe that their rejection of ethical absolutes would not spin off into purely self-interested action, and in fact would ground a democratic culture featuring widespread commitment to the public good. The social self concept suggested that human beings could discipline themselves to promote the interests of others even without a strong central state or transcendent moral laws – a crucial conclusion for a world where science seemed to rule out such laws and democracy seemed to rule out such a state. In a 1915 essay, Mead clearly articulated the political implications of his version of the social self. He

---

[24] Mead, "The Social Self," 378–379; Mead, "The Philosophical Basis of Ethics," 318.
[25] Mead, "The Social Self," 379.

portrayed his theory of politics as a description of how democracy actually worked, not an ideal picture of how it should work. The very structure of the democratic state, Mead argued, changed in response to the shifting contours of public debate. There, in the cultural stratum underlying political institutions, intellectuals carried out their work, ensuring the smooth accommodation of new interests and arguments. Mead stressed that most social conflicts could be resolved within this realm, through citizens simply changing their habits and attitudes as a result of moral reflection. Formal bodies such as courts or legislatures played a role only when there was a need to get a clearer picture, especially by bringing physically or socially separated parties into "immediate sympathetic relation." Having clarified the problem, governmental agencies then referred it back to the citizens, who implemented a solution in keeping with the new "common goods" that formal debate had revealed. Mead thus portrayed state institutions as specialized adjuncts of the collective and internal phases of deliberation underpinning opinion formation in a democracy.[26] His articles on the social self present one of the clearest examples of the tendency among scientific democrats to assume that something like deliberative democracy already existed in the United States, and that scholars needed only to infuse the public culture with new beliefs and values to change the nation's institutions.

Few theorists promoted that largely unstated assumption as vigorously as did John Dewey. Dewey stood at the center of the Progressive Era discourse on the social self, interacting closely with both Cooley and Mead. But he took the concept of a social self for granted and explored its manifold applications and implications, rather than defending its theoretical validity. In *Democracy and Education* (1916), Dewey held out the possibility of using the school system to arm individuals with reliable knowledge about modern industrial society. Playing the strategist to Cooley's theorist – Cooley had ignored schools, because they did not appear in all societies and thus could not account for the universality of the social sentiments – Dewey identified the school system as a modern society's primary instrument of self-replication. He focused on scientific training as the key to the school's success. In fact, Dewey suggested that, in an era when primary groups had been powerfully eroded by the anonymity of modern social relations, scientific education could replicate much of their moral impact, while also providing the knowledge needed to effectively channel the resulting social sentiments. More than any other theorist of his era, Dewey emphasized the whole range of social benefits attributed to science by the founders of the American universities. He portrayed science as the fount of

[26] George Herbert Mead, "Natural Rights and the Theory of the Political Institution," *Journal of Philosophy, Psychology and Scientific Methods* 12 (1915), 141, 152–153. Like Cooley, Mead found a basis for political rights in society itself, rather than in an imagined state of nature (150, 155). Cf. Beth J. Singer, "Mead: The Nature of Rights," in *Classical American Pragmatism: Its Contemporary Vitality*, ed. Sandra B. Rosenthal, Carl R. Hausman, and Douglas R. Anderson (Urbana: University of Illinois Press, 1999); see also Mary Jo Deegan, *Self, War, & Society: George Herbert Mead's Macrosociology* (New Brunswick: Transaction, 2008).

the intellectual, moral, and technological phases of modern society. *Democracy and Education* provided by far the fullest statement to date of Dewey's influential version of scientific democracy, combining his ethical theory and conception of the social self with his vision of the progressive school.[27]

All of the major themes of Progressive Era scientific democracy figured prominently in Dewey's 1916 book: the image of society as a communicative network, the postulate that communication engendered sympathy and thereby a sense of equality, the portrait of social institutions as extensions of the cultural substrate, and the provision of a central role for scientific intellectuals in cultural change. Dewey also followed many contemporaries in viewing societies as akin to living organisms. He pointed to one important difference between societies and organisms, however. Societies, he noted, did not need to die. Instead, adults could pass knowledge and values down to the young, who were "born immature, helpless, without language, beliefs, ideas, or social standards." Dewey explained that modern families delegated this task to the schools, expecting them to offer both knowledge of the "aims and habits of the social group" and a positive commitment to preserving and improving them. Through education, he summarized, a society turned "uninitiated and seemingly alienated beings into robust trustees of its own resources and ideals."[28]

Like the other theorists of the social self, Dewey expected the relationships between individuals in a democracy to feature sympathetic engagement and mutual improvement, rather than mere coordination of action. By breaking through "barriers of social stratification," he wrote, a proper education would give each individual "a cultivated imagination for what men have in common and a rebellion at whatever unnecessarily divides them." This "benevolent interest in others" would lead individuals "to free them so that they may seek and find the good of their own choice," however divergent those goods might be. The outcome would be full self-realization for all, finally giving the nation as a whole a satisfying culture. With Cooley and Mead – and Ely – and against Ross, Dewey held that industrial production and the cultural resources provided by science made it possible to harmonize social progress with individual self-realization.[29]

Science played a crucial role in Dewey's educational and political vision, given that he grounded the possibility of democracy in individuals' capacity to understand their impact on others. He described science as a general method, "that which we think *with* rather than that which we think about." Its use differentiated "authorized conviction" from mere opinion, he averred, and science thus properly stood at the center of a modern curriculum. Yet science's

[27] John Dewey, *Democracy and Education* (New York: Macmillan, 1916); Cooley, *Social Organization*, 23; Robert B. Westbrook, *John Dewey and American Democracy* (Ithaca: Cornell University Press, 1991), 167–182; Sklansky, *The Soul's Economy*, 151–162.

[28] Dewey, *Democracy and Education*, 3, 12. At Michigan, Cooley took a course with Dewey before joining him and Mead on the faculty. Dewey was Mead's closest colleague there, and he took Mead with him to Chicago in 1894. Louis Menand, *The Metaphysical Club* (New York: Farrar, Straus, and Giroux, 2001).

[29] Dewey, *Democracy and Education*, 141–142.

authority also presented a unique danger, Dewey warned. In its textbook formulations, science appeared hopelessly abstract, cut off from life – a matter
of dogmatic tenets handed down from above, rather than a living process of
human inquiry. Dewey sought to overcome this obstacle by proposing that
teachers should help students connect their native interests to the various bodies of scientific knowledge in the context of practical projects. By employing
such means to introduce students to science, he believed, educators would give
them both the knowledge most relevant to social progress and science's distinctive ethic of "directness, open-mindedness, single-mindedness (or whole-
heartedness), and responsibility."[30]

Although an intellectual product of the late 1880s and early 1890s himself,
Dewey added the elements of the scientific attitude to earlier formulations of
the scientific spirit. Like his nineteenth-century predecessors (and teachers),
he insisted that students steeped in scientific knowledge would illustrate "an
active disposition to welcome points of view hitherto alien." But, he added,
they would also exhibit "an active desire to entertain considerations which
modify existing purposes" – by which Dewey clearly meant social purposes.
Such students, he elaborated, would direct their attention to "avowed, public, and socially responsible undertakings" rather than "private, ill-regulated,
and suppressed indulgences of thought." Most fundamentally, scientific education would produce a "disposition to consider in advance the probable consequences of any projected step" and to make such consequences the basis for
moral action. In sum, it would produce just that type of personal character
called for in Dewey's pragmatic ethics and said by Cooley to be produced by
the primary groups.[31]

TOWARD CULTURE

Many other scientific democrats of the Progressive Era likewise portrayed a
society rendered stable by webs of communicative interchange and flexible
by constant injections of empirical knowledge of causal relations. Looking
beyond individuals to the larger interpersonal formations surrounding them,
Yale's William Graham Sumner and Columbia's Franz Boas analyzed what
Sumner called the folkways and mores and Boas called culture. The roughly
parallel theories developed by this politically disparate pair helped to embed
in American sociology and anthropology, respectively, the vision of society as
a largely psychological construct. At the same time, however, their writings
highlighted the tendency for the public mind to resist change – for the culture
to subject individuals to its powerful inertial influence. Even as the new theory
of society as a psychological network indicated the most profitable path to

[30] *Ibid.*, 222–223, 214–215, 225, 204. Cf. John L. Rudolph, "Turning Science to Account: Chicago
and the General Science Movement in Secondary Education, 1905–1920," *Isis* 96, no. 3 (2005):
353–389.
[31] Dewey, *Democracy and Education*, 206, 209.

reform for scientific democrats, it also explained the stubborn fact of wide-spread popular resistance to their political proposals.

Sumner's 1906 book *Folkways* provides a striking illustration of the ideo-logical range of the emerging understanding of a psychologically integrated society, for he remained an unreconstructed champion of *laissez-faire*. But Sumner's conception of social relations had changed substantially from the days when he saw little other than individuals struggling against nature and occasionally offering helpful inventions to their contemporaries. He joined Ross, Bernard, Cooley, Mead, and Dewey in theorizing the operation of what we now call culture, in order to assess the possibilities and limitations of schol-arly efforts to make America scientific. Sumner shared with these Progressive Era counterparts a tendency to sharply distinguish between selfish and selfless behavior. Yet he mapped that potent dichotomy onto economic policies in pre-cisely the opposite manner. Sumner's book portrayed *laissez-faire* as the highest expression of other-mindedness and blamed calls for economic regulation on undisciplined self-interest.[32]

Sumner also agreed with the Progressives that the nineteenth century had witnessed a disturbing political turn, but he held a very different view of that phenomenon. Whereas most scientific democrats in the Progressive Era viewed atomistic individualism as the characteristic ideology of the nineteenth cen-tury, Sumner instead saw that century as an age of "humanitarianism": the misguided belief that the weaker party to any social conflict exhibited moral superiority and deserved special favors. He called on the bourgeoisie to hold fast to its policy of constitutional liberties, rather than giving in to the tide of altruistic sentiment and squandering the effective political mechanisms it had built up slowly over time – or, worse still, setting a bad example by looking to government for its own handouts. As before, Sumner insisted that the regime of property rights and economic competition, coupled with technological innova-tions, offered the best hope for rich and poor alike.

Finally, Sumner had come to share with the Progressive theorists a sense that the source of continuing opposition to the socially beneficial institutions he championed lay in a stubbornly inflexible body of public opinion that resisted the teachings of social scientists. He now identified a tightly integrated, "super-organic" network of "relations, conventions, and institutional arrangements" – the folkways – as the immediate cause of social behavior, mediating between individuals and the environment. Sumner explained that the folkways emerged as individuals sharing a common natural and technological environment grad-ually discovered the most effective ways, under those circumstances, of min-imizing pain and maximizing pleasure in "the competition of life." Folkways

---

[32] William Graham Sumner, *Folkways: A Study of the Sociological Importance of Usages, Manners, Customs, Mores, and Morals* (Boston: Ginn, 1906). On this book, see Bannister, *Sociology and Scientism*, 98–110; and Steve J. Shone, "Cultural Relativism and the Savage: The Alleged Inconsistency of William Graham Sumner," *American Journal of Economics & Sociology* 63, no. 3 (July 2004): 697–715.

turned into mores, he continued, when the community deemed them socially productive and enforced them through a subtle social pressure from all sides. In short, mores were folkways made normative.[33]

Sumner adopted a strong form of cultural determinism, arguing that the mores gave each individual "his character, conduct, and code of life." In fact, they produced the morality of an era, the systematic philosophical and ethical principles employed by scholars, and even the canons of truth itself. Sumner repeatedly stressed that all supposedly universal principles of right or justice represented *post hoc* justifications of the current mores. Even the democratic ideals of equality and self-government were as much a product of the mores as a belief in witches, Sumner wrote pointedly.[34]

But Sumner believed that the mores in turn rested on a material base, taking their shape from the prevailing economic environment and the particular ruling elite that it produced. In the modern era, he wrote, both the ascendance of the bourgeoisie and its misguided, humanitarian democracy stemmed from the relatively uncrowded global situation generated by technological advances and the opening of new land, which had combined to temporarily disable the Malthusian dynamic of cutthroat competition for scarce resources. Under these highly unusual conditions, Sumner argued, the bourgeoisie had managed to wrest power away from landed property. In the course of the struggle, that new ruling class had discovered a form of politics that would, uniquely, remain valid for all economic conditions and ruling elites: the system of constitutional liberties, which provided "safety of person and property" to all. But it had also invented the chimerical concepts of natural rights and natural equality, which the workers had quickly turned into weapons against constitutionalism and which the bourgeoisie itself, forgetting its constitutional scruples, used to plunder the less fortunate at home and abroad. Humanitarian democracy could not last forever, Sumner asserted hopefully, because it rested on a temporary state of affairs wherein "the earth is underpopulated and there is an economic demand for men." The danger was that constitutionalism, too, would go down with the democratic ship.[35]

Sumner examined the operations of the folkways and mores, looking for means by which social scientists might teach the public to distinguish a valid constitutionalism from an invalid, humanitarian democracy. The folkways and mores tracked deep, environmental shifts in what Sumner called the "life conditions," the combination of natural resources and technological forms that determined the pleasure/pain balance produced by various forms of behavior. He also noted that the folkways and mores changed slowly even within a stable environment, as creative individuals discovered new techniques for attaining more pleasure and less pain under extant circumstances. Finally, Sumner identified direct, cultural leadership by "the classes" – the masses, he insisted,

[33] Sumner, *Folkways*, iv, 16.
[34] *Ibid.*, iii, 108.
[35] *Ibid.*, 169, 194. Cf. E. R. A. Seligman, quoted in Ross, *The Origins of American Social Science*, 149.

had never contributed anything to the civilizing process – as a third source of change in the folkways and especially the mores.[36]

Thus, social scientists could potentially change the prevailing mores directly. They could not, however, do so by reasoning critically with their fellow citizens. The average citizen was a thoroughly irrational being, Sumner insisted, no more able to tolerate the fallibility and uncertainty of scientific knowledge than to join in building up civilization. But would-be leaders could use "the old methods of suggestion" to subtly redirect the mores, if they understood their prevailing force and direction. Sumner urged sociologists and their allies "to perceive and hold fast to the truth, but also to know the delusion which the mass are about to adopt." In particular, he hoped that they would guide the "mysticism of democracy" and "transcendentalism of political philosophy" that gripped the masses in the direction of a sound constitutionalism.[37]

Sumner cautioned that philosophical systems and ethical principles could not be of help in this task, because they merely reflected the prevailing mores. But the social sciences could cut through the mores themselves, laying bare their roots in the prevailing life conditions and revealing how they operated to shape thought and values. By using such knowledge to nudge the mores this way or that, social scientists could render their society capable of adapting relatively easily to dramatic changes in the life conditions. Near the end of the book, in fact, Sumner unexpectedly waxed rhapsodic about the prospects for a scientific system of education. "When the schools are not too rigidly stereotyped," he wrote, "they become seats of new thought, of criticism of what is traditional, and of new ideas which remold the mores," in keeping with the innately rebellious tendencies of youth. Targeting the malleable young rather than hidebound adults, an education in the habit of critical inquiry might produce a public capable of thinking skeptically and grasping the fallibility of all truths and ideals.[38]

Comparing Sumner's ideas with those of the pioneering anthropologist Franz Boas, whose analysis of culture gave us our name for the communicative networks, practices, values, and beliefs on which so many Progressive Era thinkers focused, highlights the fact that both men embraced scientific democracy, despite the political gulf between them. Boas shared Sumner's presumption that a few individuals of particularly strong character – namely, social scientists – could break the grip of the mores and attain a critical perspective on these immensely powerful, innately conservative forces. Additionally, he included under the heading of culture essentially the same range of beliefs and practices that Sumner divided into folkways and mores, while tracing them to individuals' desire to maximize pleasure and minimize pain in a given environment. Boas diverged from Sumner primarily in describing the folkways, like the mores, as socially normative rather than freely chosen.

[36] Sumner, *Folkways*, 5–6, 84, 58.
[37] *Ibid.*, 79, 220.
[38] *Ibid.*, 231–232, 36, 94–95, 634, 633.

When historians think about Boas' political commitments, they invariably look to the issue of race relations. A German-Jewish émigré, Boas waged a relentless battle against social prejudice from his base at Columbia University after 1896. By describing individual behavior as a cultural product rather than an expression of biological tendencies, Boas and his cosmopolitan circle of students sought to invalidate racial determinism. But the cultural approach to behavior provided an alternative not only to biological determinism but also to a view of American political institutions and values as either reflections of innate, natural structures or objects of purely rational allegiance. Boas was centrally concerned with the psychological, emotional foundations of culture, even describing anthropology as a form of comparative psychology that could cut through the easy presumption that Western subjects represented specimens of humanity in the raw. The questions about human nature, beliefs, and behavior that his culture concept addressed stood at the heart of Progressive Era debates about the meaning and prospects of democracy. In this regard, as in his resistance to biological racism, Boas set the tone for much interwar work in the human sciences.[39]

Part of this tone was epistemological. Boas articulated a conception of scientific knowledge that comported well with the pragmatism of James and Dewey and found many takers between the wars. The fact that Boas collapsed Sumner's folkways and mores – along with scientific beliefs – into a single category of culture pointed to the crucial epistemological difference: Boas viewed scientific claims and all other culture elements as more or less functional abstractions from a richly detailed body of concrete experience. Whereas Sumner treated the normativity of the mores as a product of their unique capacity to bring down social ostracism for nonconformity, Boas argued that cultures amounted to habitual filters on the very process of individual perception. An inveterate cataloguer and classifier himself, even when attempting to define his own field, Boas described each culture as a taxonomic grid for sorting experiences into

[39] On Boas and his students, see especially Regna Darnell, *Invisible Genealogies: A History of Americanist Anthropology* (Lincoln: University of Nebraska Press, 2001). The antiracist aspect of his work is discussed, among other places, in Elazar Barkan, *The Retreat of Scientific Racism: Changing Concepts of Race in Britain and the United States Between the World Wars* (New York: Cambridge University Press, 1992), 76–95; Lee D. Baker, *From Savage to Negro: Anthropology and the Construction of Race, 1896–1954* (Berkeley: University of California Press, 1998), 99–126 and 143–167; and Vernon J. Williams, *The Social Sciences and Theories of Race* (Urbana: University of Illinois Press, 2006), 16–47. Herbert S. Lewis identifies important affinities between Boas' thought and pragmatism: "Boas, Darwin, Science, and Anthropology," *Current Anthropology* 42, no. 3 (June 2001): 381. Studies of the background to the culture concept include Anthony Darcy, "Franz Boas and the Concept of Culture: A Genealogy," in *Creating Culture: Profiles in the Study of Culture*, ed. Diane J. Austin-Broos (Boston: Allen & Unwin, 1987), 3–17; and Matti Bunzl, "Franz Boas and the Humboldtian Tradition: From *Volksgeist* and *Nationalcharakter* to an Anthropological Concept of Culture," in *Volksgeist as Method and Ethic: Essays on Boasian Ethnography and the German Anthropological Tradition*, ed. George W. Stocking Jr. (Madison: University of Wisconsin Press, 1996), 17–78.

conceptual categories. Firmly installed in the unconscious mind of every adult member, a culture shaped the group's understanding of reality itself.[40]

It was not through acts of convention-defying will that scientists transcended these cultural filters, according to Boas. The fact that each culture shaped the very perceptions of scientific observers meant that merely braving the danger of social judgment could hardly enable observers to break through pre-given cultural forms. Scientists could explore the operations of culture only by exerting their will in a different direction and choosing a particular analytic tool: the comparative method. Boas called on psychologists and other social scientists to draw their material from the widest possible circumstances, rather than simply defining Western perceptual categories as human universals. Meanwhile, he believed that the general public could also use a healthy dose of comparative analysis, as it would prevent them from denigrating other cultures and perhaps even lead them to take a few lessons from primitive societies. For Boas, human scientists were not those who saw reality clearly and without bias, as for Sumner, but simply individuals who looked across numerous schemes for defining reality itself, without invidiously judging those that differed from their own.

Still, Boas joined Sumner and many other Progressive Era theorists in identifying social institutions as products of the underlying cultural substrate, and in assuming that scientific scholars could help the people change their institutions without relying on state power. He saw in the comparative method a powerful new means of teaching both ordinary citizens and working scientists to bring their deeply embedded cultural assumptions to consciousness and then to reshape them. The Boasian concept of culture powerfully reinforced the emerging view that a thick web of psychologically complex relationships bound together even a modern, industrial society – and indicated the point at which scholars needed to intervene if they hoped to make lasting political changes in such a society.

CULTURE AND THE STATE

As we have seen, scientific democrats in the universities gravitated toward a mode of politics centered on the circulation of new ideas in the realm of public culture, rather than the use of technical knowledge by a powerful administrative state. The *New Republic*'s founding editors, Walter Weyl, Herbert Croly, and Walter Lippmann, developed a more state-friendly approach during the Taft and Wilson years. The state they imagined, however, would not be a managerial one, implementing social-scientific theories said to be purely empirical and value-neutral. Instead, it would be a force for the improvement of

---

[40] Boas' taxonomic accounts of his field include "Anthropology," *Science* n.s. 9, no. 212 (January 20, 1899): 93–96; and "The History of Anthropology," *Science* n.s. 20, no. 512 (October 21, 1904): 513–524.

ordinary citizens. The *New Republic* editors joined their scholarly counterparts in viewing the state and other social institutions as potent means to the end of individual self-realization, even as they took greater account of the stubborn persistence of group interests in a modern democracy.

The influential books written by Weyl, Croly, and Lippmann between 1909 and 1914 highlight the fact that, although the late Progressive period fostered a political sensibility less moralistic and more attuned to engineering models than its predecessor, the cultural visibility of avatars of efficiency such as Frederick Winslow Taylor did not necessarily indicate that all scientific democrats declared their knowledge value-neutral and hitched their wagon to a bureaucratic state. For one thing, the 1910s discourse of efficiency was itself moral. That language usually took its power from a felt obligation to fulfill the widest range of human needs with the given means, not from a parsimonious, cost-saving mentality. More importantly, the idiom of efficiency was hardly the only one available to Progressives for thinking about the political implications of science. In fact, Weyl, Croly, and Lippmann focused primarily on science's relationship to human values, even as they embraced more robust conceptions of state power than did their academic counterparts. None of these three figures was a prototypical "managerial liberal," calling for the application of value-neutral science by bureaucratic agencies. Although they envisaged new state structures, they primarily targeted what Cold War-era theorists would call the "political culture": the orientation of ordinary citizens – especially the politically influential middle class – toward the system of governance. The writings of the *New Republic* trio, even more clearly than those of academic figures, demonstrate that Progressive social thought did not simply prefigure the welfare state built during the New Deal years.[41]

Of the three, Walter Weyl devoted the least attention to the theoretical question of science's role in a democracy. Instead, he developed a particular scientific approach – the economic theory of his mentor Simon Patten – into a political project focused on liberating the public from the cultural clutches of "the plutocracy" (a term also employed by Sumner). Patten combined elements of marginalism with the claim that the United States was on the verge of becoming the first nation to break free of the Malthusian cycle of poverty and overpopulation. He believed that the emerging economy of abundance would rest on the active use of tax policy to redistribute resources, in keeping with what Patten saw as the essentially social character of wealth creation. Weyl, who had taken a PhD under Patten at the University of Pennsylvania, reiterated in his 1912 book *The New Democracy* that the state, not individuals, possessed "a primordial, intrinsic, underlying right to all property." Seeking to translate that theoretical observation into a practical political movement, he called for the creation of a new consumers' party whose members, drawn from all walks of life, would exert their shared interest in a more egalitarian

[41] Jordan offers this reading in *Machine-Age Ideology*, 68–78.

distribution of the "social surplus" produced by Americans' collective assault on the continent.[42]

Weyl combined Patten's emphasis on the social character of wealth with lessons from political experience about the difficulty of making manifest the public good. He believed that orthodox *laissez-faire* was no longer the greatest obstacle to the realization of a genuine democracy, as many business leaders had jumped onto the reform bandwagon and embraced child labor laws, technical education, internal improvements, peace, public health, and other humanitarian efforts, with the single caveat that these initiatives could not "interfere with business arrangements." More committed to their own interests than to free-market ideology, Weyl explained, these plutocrats hoped to steer radical social movements toward what historians would later label "corporate liberalism": economic regulations designed to temper the destructive effects of open competition and stabilize profit-taking.[43]

The second salient fact about the present political situation, in Weyl's view, was that the plutocracy exercised hegemonic control over the beliefs and values of the democratic public. Deep down, he insisted, Americans actually rejected the program of the business progressives, with its emphasis on maximizing aggregate national output. They instead embraced the utilitarian vision of "such a production, distribution, and consumption of wealth as will give the highest excess of economic pleasure over economic pain to the largest number of people for the longest possible time." Weyl insisted that the nascent democratic movement pointed toward a society in which "the final arbiter of all relations, industrial, political, and social, is the people" and "the ultimate standard of values, the ultimate sanction, is not legal but moral." But, he continued, the plutocracy had subverted this democratic impulse through a "subtle, devious, and anonymous campaign of suppression, misrepresentation, and falsehood" that instilled the values of the wealthy in "millions of like-minded poor men, *penniless plutocrats, dream-millionaires.*" In Weyl's view, the plutocracy, representing a small minority of the population, maintained its economic and political ascendancy solely through its influence over the public mind.[44]

Weyl urged intellectuals and reformers to break the American public's cultural chains by undertaking "the democratization of government, the socialization of industry, and the civilization of the citizen." In his view, the latter process was crucial, because democracy meant not merely the absence of a monarch but also the public's possession of "the mind and the will to rule itself." Until such time as citizens attained "their full intellectual and moral stature," Weyl reasoned, the advance of democracy would be a task for slow,

---

[42] Walter E. Weyl, *The New Democracy* (New York: Macmillan, 1912), 295, 244. See also Daniel M. Fox, *The Discovery of Abundance: Simon N. Patten and the Transformation of Social Theory* (Ithaca: Cornell University Press, 1967).

[43] Weyl, *The New Democracy*, 143–144.

[44] *Ibid.*, 119, 121, 150, 154. Cf. Thorstein Veblen, *The Theory of the Leisure Class* (New York: Macmillan, 1899).

steady, knowledgeable leadership by those armed with modern science's insights regarding the social nature of wealth and the importance of non-material elements in the "full life" ultimately sought by all individuals. Weyl urged public-spirited thinkers to work diligently to improve "the life, health, intellect, character, and social qualities of the citizenry."[45]

Weyl identified two key areas of intervention for reformers, in addition to helping build up the consumers' party that would ultimately implement true democracy. On the economic front, he believed that little institutional change would be necessary; the plutocracy itself had ensured the "socialization of industry" through its embrace of coordinated action and corporate organization. But citizens would not win a proper distribution of the industrial machine's benefits until reformers helped them regain control of the media marketplace, organizing as subscribers against the profit-driven "adulteration" of the news and editorial pages. The other crucial arena for reform activity was education. Here, too, the plutocracy worked to ensure its own demise, in Weyl's view. He insisted that even the private universities endowed by the Rockefellers and Vanderbilts of the world inevitably hewed to "the general direction of the popular mind," while giving it "tone, character, and an ethical interpretation." At the lower educational levels, Weyl argued that public high schools for the masses would "create a revolutionary force in the community" by helping Americans extend to the realm of consumption "the discipline and coördination which they have learned in production." Instilling a fundamentally social orientation, democratic schooling would combat the nation's "elephantiasis of consumption" by minimizing the "unwise consumption of wealth" and the "inept use of leisure," while also supporting full "social capillarity" between classes by ensuring a meritocratic distribution of opportunities. Virtually every opportunity for education broke another link in the public's chains, Weyl believed.[46]

Whereas Weyl echoed key Enlightenment themes, Croly proposed a seemingly paradoxical combination of medieval economic forms and up-to-the-minute scientific standards. A man of deep, if unorthodox, religiosity – he was the first American child baptized into Auguste Comte's positivistic religion of humanity – Croly was drawn to "guild socialism," a system based on occupational groups operating under the loose umbrella of the state. Guild socialists believed that they could gain the benefits of collectivism without a strong central state by investing quasi-political power in these modern successors of the medieval guilds. Broad groups with directly conflicting interests – management and labor, say, or industrial and agricultural workers – would negotiate solutions *en bloc*, facilitated by government oversight and enforcement. Within these occupational groups, the integrated nature of the tasks undertaken by more specialized subgroups – dockworkers and shipbuilders, for example – would ensure harmony, as the interests of all would naturally point in the

[45] Weyl, *The New Democracy*, 278, 319–321, 326.
[46] *Ibid.*, 130, 134–136, 329–330, 333, 352.

same direction. Attractive to many American Progressives and their Fabian counterparts in Britain, guild socialism served as a halfway house between the more radical views of the late nineteenth century and the later conception of governance by a "broker state" whose policy proposals were shaped by competing interest groups.[47]

But Croly believed that, in a democracy, such a system would require a firm foundation in public sentiment. In *The Promise of American Life* (1909) and *Progressive Democracy* (1914), Croly called for a new national feeling, a powerful commitment to the common good. He targeted a culture focused exclusively on the pursuit of private ends, believing that this orientation would render any and all institutional changes moot, while its replacement with a more social orientation would naturally generate a mode of governance in which the structure of the state aligned perfectly with that of the industrial economy.

Croly might have simply invoked the authority of social science on behalf of this vision, arguing that empirical research demonstrated its necessity under industrial conditions. However, he brought in science in a very different manner. In Croly's view, science exemplified the core ideal of a broader process of professionalization: the adoption of technical standards and methodological procedures that, he asserted, naturally directed effort in every area of specialized labor toward the good of all. Science embodied perfectly the modern ideal of "genuine individuality," in which "the complete emancipation of the individual" issued in "complete disinterestedness" – the single-minded focus of each citizen on "the excellence of his work," rather than pecuniary gain. Once all individuals, in all stations of life, sought prestige solely by fulfilling the technical standards of their professions, a thoroughly functional society in which "men were divided from one another by special purposes, and reunited in so far as these individual purposes were excellently and successfully achieved," would emerge. Croly insisted that the "authoritative technical tradition associated with any one of the arts of civilization" neatly connected individual striving to the collective good. Each such standard codified "the accumulated experience of mankind in a given region," indicating how to produce the outcome most favorable to all. In fact, insofar as a genuine standard of this variety existed in a field, Croly declared that society was "justified in imposing it." He

---

[47] Herbert Croly, *The Promise of American Life* (New York: Macmillan, 1909), 405–406, 418. Helpful treatments include David W. Levy, *Herbert Croly of the New Republic: The Life and Thought of an American Progressive* (Princeton: Princeton University Press, 1985); Kevin C. O'Leary, "Herbert Croly & Progressive Democracy," *Polity* 26, no. 4 (Summer 1994): 533–552; Gillis J. Harp, *Positivist Republic: Auguste Comte and the Reconstruction of American Liberalism, 1865–1920* (University Park: Pennsylvania State University Press, 1995), 183–209; Claudio J. Katz, "Syndicalist Liberalism: The Normative Economics of Herbert Croly," *History of Political Thought* 22, no. 4 (2001): 669–702; and John Allphin Moore, *Herbert Croly's The Promise of American Life At Its Centenary* (Newcastle upon Tyne: Cambridge Scholars Press, 2009). Dewey likewise found guild socialism attractive during the 1910s: Westbrook, *John Dewey and American Democracy*, 225–226, 244–252.

believed that the economic, social, and political apparatus would promote its proper end of "human amelioration" when Americans aligned their individual motives with communally determined professional standards, and thus competed solely to excel one another in social serviceability.[48]

The political vision set forth in Walter Lippmann's *A Preface to Politics* (1913) and *Drift and Mastery* (1914) demanded much more of the professional middle class than did those of Croly and Weyl. Those two urged middle-class Progressives to openly embrace their own, putatively non-capitalist values and work to spread these throughout society. The "better American individual," wrote Croly, "needs to do what he has been doing, only more so, and with the conviction that thereby he is becoming not less but more of an American." In his vision, the new middle class would drive social reconstruction by promoting the cause of "personal efficiency," making every occupation hospitable to the "comparatively zealous and competent individual performer."[49] Weyl likewise called on middle-class reformers to remake the public in their own image. Lippmann, by contrast, applied the pragmatists' deflation of absolute truth to the political sphere, urging Americans to submit their beliefs and values to the stern test of human needs. Although his conception of politics produced many of the same policy orientations as did that of Croly and Weyl, he diverged from both of them on the public role of science. Lippmann's writings foreshadowed the profoundly anti-utopian liberalism that would dominate the American academy in the 1950s, as well as the arguments about science and epistemology that anchored it.

Adding Freud's insights to those of Dewey, Lippmann also urged a franker acceptance of human desires and hewed to a darker view of humanity than did the mystical Croly and the sunny Weyl, both almost a generation his senior and steeped in Victorian ideals. He came of age on the margins of Greenwich Village cultural modernism and echoed its critique of Victorian values. Lippmann articulated a relativistic definition of culture paralleling that of Sumner and Boas. Yet, despite a romantic streak that would linger for a few more years, Lippmann stood closer, temperamentally and intellectually, to the realist Ross than to any of the other figures described in this chapter. Like Ross, Lippmann viewed the cultural process not as a matter of individuals gaining reliably social natures but rather as an ongoing struggle to keep a lid on a roiling sea of emotions through various modes of cultural discipline. He argued that modern psychology's recognition of the power of irrational motives made possible – and required – a political process that redirected individual needs and desires into socially constructive channels, through a process akin to Freud's "sublimation."[50]

Lippmann sought to move conservatives and their radical critics onto the middle ground of Progressivism, which he identified as the needed "human

[48] Croly, *The Promise of American Life*, 411, 417, 433, 210, 434.
[49] *Ibid.*, 430–431.
[50] Walter Lippmann, *A Preface to Politics* (New York: M. Kennerley, 1913), 49, 306–307.

politics." Having found American workers distinctly non-revolutionary during his brief flirtation with socialism in 1912, Lippmann now reassured a fearful middle class that socialists, feminists, and the like wanted nothing more than the goal they themselves pursued: namely, the "civilizing opportunities" and "civilized environments" necessary for true self-realization. Explaining that a hungry man "isn't the less hungry because he asks for the wrong food," he identified these social movements as, at bottom, merely demands for a human politics that would align industrial production with the material and psychological needs of the entire population. Like all individuals, Lippmann argued, the radical critics of the day ultimately aimed at "a frank recognition of desire, disciplined by a knowledge of what is possible, and ordered by the conscious purpose of their lives."[51]

Lippmann's vision required substantial changes in American political culture. On the side of government, he stressed the need for a true statesman in the vein of Theodore Roosevelt, a "political creator" or "political inventor" who could draw out the deep political meaning in each emerging social movement. A skilled "translator of agitations," Lippmann's political inventor would redirect and harmonize the ever-changing "compound of forces" that constituted society, after discerning the human core behind the overtly stated goals of such movements. Meanwhile, the public, imbued with a "wilful humanistic culture" by critics acting in accordance with modern psychology, would firmly tether the political creator to the public will by putting into place political mechanisms ensuring "that no leader's wisdom can be applied unless the democracy comes to approve of it." Like Weyl and Croly, Lippmann insisted that democratic institutions would "not work without a people to work them."[52]

But Lippmann leaned more heavily on science, and in a different way, than did Weyl and Croly. At the end of *A Preface to Politics*, and throughout *Drift and Mastery* – perhaps the most widely read formulation of scientific democracy ever published – Lippmann drew a powerful analogy between science and politics. This analogy did not paint political action as the equivalent of solving an engineering problem. Although Lippmann may seem to have authorized experts to take the reins of society, he actually had something quite different in mind. Far from symbolizing a body of technical standards, as it did for Croly, science represented for Lippmann an embodiment of the fundamental insight of human fallibility and an illustration of the limited but real accomplishments that followed from grasping this insight. Identifying policy proposals as hypotheses designed to accommodate the "facts" of the political process – namely, the deepest urges of the people – Lippmann insisted that democratic governance served to keep the self-aggrandizing tendencies of

---

[51] Lippmann, *A Preface to Politics*, 269–270, 89, 88, 96–97; Lippmann, *Drift and Mastery: An Attempt to Diagnose the Current Unrest* (New York: M. Kennerley, 1914), 148–149. On Lippmann and socialism, see Mark Pittenger, *American Socialists and Evolutionary Thought, 1870–1920* (Madison: University of Wisconsin Press, 1993), 228–232.

[52] Lippmann, *A Preface to Politics*, 98–99, 116, 306–307.

leaders in check by forcing them to test their proposals in the arena of popular opinion, in the political equivalent of a scientific experiment. Only those
proposals that passed democratic muster were considered "true," Lippmann
explained, just as scientific hypotheses retained adherents only so long as they
were not disconfirmed by evidence. He identified science's primary contribution to democracy as its philosophical fallibilism, not its technical findings.
To Lippmann, the cultural and political meaning of modern science stemmed
from the reflexive self-consciousness of its method, its development of a technique for moderating the natural human tendencies toward fatuous credulity
and self-aggrandizement.[53]

Yet, in Lippmann's view, science also revealed the areas of practical success
that opened up when one adopted a fallibilistic, realistic orientation, abandoning pretensions to transcendent truth or absolute perfection. Just as scientists
achieved considerable predictive accuracy, so, too, would political inventors
who heeded science's central lesson: that they could come closest to objectivity by acknowledging their own subjectivity. A statesman who understood the
deep, human needs underlying prevailing social impulses could attain considerable success in accommodating those needs, just as scientists gained the ability
to guide and channel natural forces by carefully tracing their contours rather
than wishing them to be other than what they were. A policy built on the raw
data of human desires could attain something like the solidity of a scientific
theory, at least until the balance of social forces shifted. Meanwhile, a scientific
public, capable of bracketing its utopian fantasies and keeping its sights trained
on legitimate human needs, would complete the democratic picture by consistently returning political inventors rather than charlatans to office.

In the end, Lippmann's term "mastery" did not mean the capacity to engineer the social environment or the ability to ruthlessly repress one's emotions, but rather a different kind of self-control: the habit of recognizing and
checking in oneself the universal human pretension to infallibility. Science, to
Lippmann, embodied a deep awareness of the irrationality and partiality of all
human beings, which led its possessors to oppose the will to dominate wherever it appeared. Nothing was more dangerous to society than actors who
were unaware of their own subjectivity, Lippmann insisted.[54] He accused
both *laissez-faire* thinkers and socialists of fostering this error, violating the
terms of scientific politics by treating the will of a single actor – the private

[53] Lippmann, *Drift and Mastery*, 150, 15–16. The standard, managerial reading appears in Ronald
Steel, *Walter Lippmann and the American Century* (Boston: Little, Brown and Company, 1980),
77; and Jordan, *Machine-Age Ideology*, 72–75. Alternative accounts include David A. Hollinger,
"Science and Anarchy: Walter Lippmann's *Drift and Mastery*," in *In the American Province:
Studies in the History and Historiography of Ideas* (Baltimore: Johns Hopkins University Press,
1989); D. Steven Blum, *Walter Lippmann: Cosmopolitanism in the Century of Total War* (Ithaca:
Cornell University Press, 1984), 31–39; and James T. Kloppenberg, *Uncertain Victory: Social
Democracy and Progressivism in European and American Thought, 1870–1920* (New York:
Oxford University Press, 1986), 318–320.
[54] Lippmann, *A Preface to Politics*, 205–206. Cf. Ross, *Social Control*, 51–52.

individual or the strong leader – as an absolute truth for the purposes of policy formation. By contrast, Lippmann argued that true democrats, like scientific thinkers, protected themselves from authoritarian dreams of omnipotence by habitually hurling their cherished ideals against the hard rock of reality. When constantly tested against political facts in this manner, he argued, the very force that most threatened social order – the "creative imagination," with its relentless "attempts to bridge the gap between what we wish and what we have" – became its greatest protector. Lippmann held that the emergence of "a disciplined love of the real world," given institutional form in the ongoing replacement of putatively objective political authorities by a humbler, democratic utility akin to that of science, represented "the profoundest change that has ever taken place in human history."[55]

In the final instance, Lippmann placed historical agency squarely in the hands of a democratic public. "[W]hat thwarts the growth of our civilization," he wrote, "is not the uncanny, malicious contrivance of the plutocracy, but the faltering method, the distracted soul, and the murky vision of what we call grandiloquently the will of the people. If we flounder, it is not because the old order is strong, but because the new one is weak."[56] Most Progressive theorists saw the old order as far more powerful than Lippmann suggested, stubbornly persisting in the habitual beliefs and practices of ordinary Americans and powerfully reinforced by a barrage of media messages. None of these figures, however, adopted the mindset of a would-be social engineer, viewing social problems as technical questions to be solved through top-down political manipulation guided by empirical inquiry. In fact, rather than seeking to redesign the state, they sought to alter prevailing understandings of democracy and the cultural basis on which it rested.

This is true, in large part, because most members of the older generation of Progressive theorists – those who shaped the conversation before the likes of Bernard and Lippmann came onto the scene – viewed centralized state administration as a fading vestige of an authoritarian regime in which social order had been maintained by physical force. But the conceptions of science these figures employed reinforced their suspicion of coercive state power. Of the academic figures treated in this chapter, only Bernard and Sumner viewed scientific knowledge in the now-traditional manner as the product of individual investigators who willfully suppressed their emotional commitments and thereby gained an unclouded view of external reality. The other scholars, like Lippmann, identified science as the result of a dialogic process of mutual criticism. Science, to them, entailed individuals submitting their imaginative, emotionally inflected interpretations of human experience to the twin tests of empirical investigation and mutual criticism. Here, scientific rationality remained a matter of emotions

---

[55] Lippmann, *Drift and Mastery*, 171, 139. Cf. Westbrook, *John Dewey and American Democracy*, 193–194.
[56] Lippmann, *Drift and Mastery*, 15–16.

and values, albeit discursively and empirically disciplined ones, rather than stemming from their absolute elimination.

The resulting writings represented something even more important than attempts to grapple theoretically with the nature of social relations and organization, for these texts – here joined by those of Bernard and Sumner – also powerfully intervened in a national discourse about the nature of the American polity. The concepts of social control, the social self, folkways and mores, and culture implicitly ruled out alternative understandings of political order as the product of economic relations, legal strictures, or other non-psychological factors. They traced the success of democracy to its reliance on a dense network of horizontal, person-to-person communication, which provided a non-coercive, non-hierarchical mode of social coordination. These Progressive theorists described all social institutions, including the state, as outgrowths of public opinion, manifestations of the deeper cultural substrate. Finally, they assumed that the boundaries of American culture coincided with those of the political nation. The net effect was to portray the United States as a pure deliberative polity, its institutions resting on a foundation of interpersonal discourse.

This image of a society rooted in deliberation defined a professionally comfortable and apparently consequential role for politically committed scholars in the human sciences: they could change the polity indirectly by influencing the cultural discourse out of which American institutions grew. By augmenting the technological innovation attendant on progress in the natural sciences with a steady flow of reliable social knowledge, human scientists could resolve the political crisis of industrialization and forestall any further crises by ensuring cultural flexibility. In addition to revealing the outlines of the social organism, the human sciences would also enable that organism to flourish as never before, despite the profoundly dangerous environment into which natural scientists had guilelessly led it.

The writings discussed here also produced another, more ephemeral political effect: they enabled scholars to believe, at least for a time, that their own favored principles underlay the political initiatives of the Roosevelt, Taft, and Wilson administrations, as well as the reform movements of workers, women, and perhaps even progressive business leaders. All of these groups, it appeared, merely expressed in distorted forms a swelling popular demand for Lippmann's "human politics" and the scientific knowledge on which it would be based.

Scientific democrats holding this hopeful view of American opinion would get a rude shock in the era of World War I. But while the war caused Progressive scholars to drastically revise their understanding of the public's commitments, it left their ideal vision essentially untouched. The core tenets of Progressive social theory underwrote a wide range of scholarly endeavors and political projects between the wars. Still, the Progressive theorists' analysis of social order and social change left important questions about social inquiry unanswered. What was the exact status, in terms of both scope and authority, of the factual and moral judgments offered by human scientists? Whose experience, or rather reports of experience, needed to be included when developing generalized guidelines for behavior – those of trained experts only, or those of all citizens?[57] Likewise, were experts or citizens the proper audience in justifying these guidelines? In short, should citizens somehow be included in the process of social inquiry, or were the judgments of professional scholars a sufficient guide to social action? These questions would come out into the open in the 1920s, as post-Progressive scientific democrats worked to build up the cultural sciences and connect their scholarship more firmly to public discourse.

---

[57] On the deep history of this question, see Steven Shapin, *A Social History of Truth: Civility and Science in Seventeenth-Century England* (Chicago: University of Chicago Press, 1994).

5

The Biology of Culture

Before World War I, as we have seen, leading scientific democrats traced both social order and social change to a psychological realm of "society" or "culture": a vast network of interpersonal communications that provided the foundation for all social institutions, including the state. Although the question of institutional mechanisms for social discipline preoccupied these theorists, others began to focus on the conceptions of the human mind that such an understanding of society might entail. Why were people so easily influenced by the arguments and judgments of others? Why did public opinion represent such a powerful force in human behavior? These were hardly idle questions to the scientific democrats. Committed to the project of changing their fellow citizens' beliefs and values, they needed to understand better the processes of believing and valuation.

By the early 1920s, a new set of debates about the nature of the mind roiled the disciplines of psychology, biology, and philosophy, cutting freely across the barriers between these fields and revealing the shared cultural goals that united many of their practitioners. These debates have largely been obscured in standard histories of these fields, which suggest that psychology was dominated in the era by John B. Watson's behaviorism, biology by T. H. Morgan's program in genetics, and philosophy by John Dewey's remarkable works of the 1920s. From the standpoint of scientific democracy, however, Morgan's research was largely irrelevant, while Watson's struck many as a powerful threat to the project of improving democracy by rendering American culture more scientific. In philosophy, meanwhile, many other schools of thought proved just as compatible with scientific democracy as Dewey's instrumentalism. A missing element in the history of all these disciplines is the concept of emergent evolution, which seemed to resolve key theoretical issues facing scientific democrats of many different stripes – most notably, the apparent contradiction between scientific naturalism and free will. This chapter explores the development of emergent evolutionism and several related schools of thought, showing that these were closely tied to scientific democracy in the 1920s and early 1930s.

DETERMINISM AND EMERGENCE

It is often said that the American social sciences liberated themselves from biology in the 1920s, a decade that saw biological determinism decisively routed as an explanation of human behavior in those fields of study. In fact, this liberation often amounted to a shift of allegiance from a deterministic theory of biology's influence over behavior to a non-deterministic, yet still biologically grounded, conception of human behavior. Biologists, philosophers, and psychologists explored several ways of elevating subjectivity to the status of a biological fact, a move that embedded the mind in nature and created a firm foundation in the natural sciences for a set of autonomous and non-deterministic social sciences. No longer would scientific democrats need to square their social theories with the data of evolutionary biology – or, worse yet, physics – in order to demonstrate their scientific credentials or to describe science as the basis for a modern culture.[1]

The specter of determinism haunted the leading American biologists of the 1920s, as it had William James before them. Although Morgan, at Columbia, emerges as the key figure when the period is viewed in retrospect, it was actually men such as C. Judson Herrick, Edwin Grant Conklin, and Herbert Spencer Jennings who constituted the mainstream of the discipline at the time. They pursued a very different quarry than a full understanding of genetic inheritance: namely, a scientific account of purposive behavior. These biologists firmly embraced empiricism. Yet they held that science's ultimate social value lay in its contributions to the wider culture, which shaped social institutions. They set out to combat the apparently pessimistic implications of scientific determinism by establishing an empirical basis for free will and moral choice.[2]

This was not an abstract problem in the 1920s. To the many scientific democrats who believed that the creation of a scientific culture would instantiate the core ethical values promoted by liberal Protestants, the developments of the 1920s presented a rude shock. The nationwide push for Progressive political reforms ended with a whimper in 1920, with the election of the genial nondescript Warren G. Harding to the presidency. Shortly thereafter, a significant fraction of the American people – rural residents, especially – rose up in anger

---

[1] Classic accounts of the decline of biological determinism include Hamilton Cravens, *The Triumph of Evolution: American Scientists and the Heredity-Environment Controversy, 1900–1941* (Philadelphia: University of Pennsylvania Press, 1978); Rosalind Rosenberg, *Beyond Separate Spheres: The Intellectual Roots of Modern Feminism* (New Haven: Yale University Press, 1982); and Carl N. Degler, *In Search of Human Nature: The Decline and Revival of Darwinism in American Social Thought* (New York: Oxford University Press, 1991).

[2] Philip J. Pauly, *Biologists and the Promise of American Life, from Meriwether Lewis to Alfred Kinsey* (Princeton: Princeton University Press, 2000), 198–201; Herbert Spencer Jennings, *Some Implications of Emergent Evolution: Diverse Doctrines of Evolution – Their Relation to the Practice of Science and of Life* (Hanover: Sociological Press, 1927), 1–11; Edwin Grant Conklin, *The Direction of Human Evolution* (New York: Charles Scribner's Sons, 1922), 122–123.

against the teaching of evolutionary theory in the schools, insisting that parents rather than scientists should set the content of curricula. In the big cities, meanwhile, a noticeable slackening of morals went hand in hand with the rise of a culture of consumption and an intense popular interest in psychological theories that identified human behavior as the result of subrational motives. The nightmare scenario was spectacularly on display in the celebrity attorney Clarence Darrow's 1924 claim that the teenagers Nathan Leopold and Richard Loeb had murdered their neighbor due solely to biological and social forces. This was hardly what most scientific democrats had in mind when they thought of cultural change.[3] Herrick and other leading biologists explicitly addressed the citizen who would conclude from modern scholarship "that I have no control of myself, my environment or my fellow-men," and on that basis disclaim all moral responsibility. The individual, they insisted, was much more than a "bit of froth floating passively on the stream of circumstance."[4]

Although Freud's conception of human behavior in terms of unconscious (and largely sexual) forces played a key role in the popular psychology boom of the 1920s, the scientific democrats in the American discipline of biology had a target closer at hand, namely Watson's profoundly anti-rationalist system. Behaviorism, which Watson formally launched in 1913, excised all references to a "mind" or "consciousness" intervening between environmental stimuli and behavioral responses. The responses, he suggested, could be explained entirely by reference to the stimuli. Although Watson allowed that certain positive and negative emotions inhered in the individual, he insisted that these vague impulses found expression only in forms dictated by the social environment. Different sets of stimuli produced different responses, and thus different kinds of people.[5]

Politically, Watson's writings pointed toward a model similar to that of Luther Lee Bernard. Like Bernard, Watson proposed to mold individual desires in accordance with the functional needs of society. Bernard suggested that social engineers could do this by associating socially valuable actions with positive feelings. Watson's description of stimulus-response mechanisms explained why and how they could create such associations. His emphasis on the shaping power of external stimuli allowed him to suggest that a modern society could produce any type of individual it liked, simply by modifying social structures and environments. Watson famously boasted, "Give me a dozen healthy infants, well-formed, and my own specified world to bring them up in and I'll guarantee to take any one at random and train him to become any type of specialist I might select."[6]

[3] Jennings, for example, described himself as a "hopeful uplifter" rather than an enemy of traditional morality: *Some Implications of Emergent Evolution*, 1.
[4] C. Judson Herrick, *Fatalism or Freedom: A Biologist's Answer* (New York: Norton, 1926), 95, 17.
[5] Kerry W. Buckley, *Mechanical Man: John Broadus Watson and the Beginnings of Behaviorism* (New York: Guilford, 1989).
[6] Benjamin Harris, "'Give Me a Dozen Healthy Infants': John B. Watson's Popular Advice on Childrearing, Women, and the Family," in *In the Shadow of the Past: Psychology Portrays the Sexes*, ed. Miriam Lewin (New York: Columbia University Press, 1984): 126–154.

To those favoring a more robust conception of human freedom, Watson's work represented a major theoretical challenge. On the one hand, his behaviorism potentially reinforced their identification of the intersubjective realm of culture as the basis for social order. But Watson pushed toward a stricter form of determinism. This raised the alarming prospect that scientists might not be able to infuse public culture with new ideas. If behavior was entirely conditioned, then scientists could change the actions of their fellow citizens only by reconditioning them, not by persuading them. Watson's theory ran directly counter to the desire of most scientific democrats to make room in nature for the human mind, and thereby to create a naturalistic system of morality. No such thing as the mind existed, Watson declared forthrightly. An environmental stimulus produced a direct behavioral response, with no intervening choice or even reflection on the part of the individual involved. To critics, this view rendered human beings utterly impotent to effect change in the world – or even to influence individual behavior, as Watson proposed – by adopting new ways of thinking and new institutions.[7]

Watson's work presented other problems as well. For example, he sought to establish laboratory studies as the gold standard for assessing theoretical claims in psychology. He viewed psychology as a natural, and therefore experimental, science. This methodological imperative was tied up with Watson's insistence on banishing from psychological discourse all entities that could not be directly observed via the human senses. Additionally, Watson gained a high public profile in the psychology-obsessed popular culture of the 1920s. Fired from Johns Hopkins in 1920 for having an affair with his graduate assistant, Watson spent the rest of the decade guiding advertising campaigns and writing for the public, urging Americans to reengineer themselves. In *Behaviorism* (1924) – published by the People's Institute of New York, a major center for popular lectures and discussion groups – he used straightforward language and employed a personal idiom of "you" and "we," even in the most technical passages.[8] Watson wrote frequently for the numerous parenting magazines that popped up in the 1920s and jousted with psychological defenders of idealism and introspection in high-profile debates.[9] These activities served to strengthen the public's association of science with materialism and immorality. Behaviorism threatened to take human purposes off the scientific map entirely.

Watson's attempt to eliminate the mind sent many scientific democrats scrambling to uphold an image of citizens as, in the final instance, guided by

[7] Sharon E. Kingsland, "A Humanistic Science: Charles Judson Herrick and the Struggle for Psychobiology at the University of Chicago," *Perspectives on Science* 1 (1993), 463; Jennings, *Some Implications of Emergent Evolution*, 7–10; Roy Wood Sellars, "Why Naturalism and Not Materialism?" *Philosophical Review* 36 (1927), 218–224; Herrick, *Fatalism or Freedom*, 91.

[8] E.g., John B. Watson, *Psychology from the Standpoint of a Behaviorist* (Philadelphia: Lippincott, 1919), 9–10.

[9] Franz Samelson, "Struggle for Scientific Authority: The Reception of Watson's Behaviorism, 1913–1920," *Journal of the History of the Behavioral Sciences* 17 (1981): 399–425; John B. Watson and William McDougall, *The Battle of Behaviorism: An Exposition and an Exposure* (London: K. Paul, Trench, Trubner & Co., 1928).

conscious decisions and freely chosen moral commitments. His critics sought a conception of human behavior that neutralized the challenge of determinism by allowing for the possibility of mental, and thus moral, freedom. They hoped to forestall a growing public fear – lingering in many of their own minds as well – that the rise of science meant the decline of any conception of moral responsibility. Their efforts lent biological authority to the postulation of an autonomous realm of culture and thus helped pave the way for the full flowering of the human sciences between the wars.

The theory of evolutionary emergence became a favored means of undermining determinism in the 1920s. Lester Frank Ward had offered the basic insight of emergentism four decades earlier, arguing that at key points in the course of evolutionary history, qualitatively new phenomena had appeared: aggregations exhibiting laws that were distinct from, and not predictable by or reducible to, the laws governing the behavior of their constituent parts. In particular, Ward stressed that the human mind was such an emergent – literally, a *"new power"* in the universe, possessing qualities not found on the molecular level. He used his rudimentary theory of emergence to argue that the advent of intelligence had switched human progress onto a new track, away from the merely biological evolution of earlier epochs and toward the vastly more effective processes of cultural evolution. As the theorists of the 1920s recognized, emergentism provided a solid foundation in evolutionary biology for the human sciences, which concerned phenomena that most scientific democrats viewed as the products of mental freedom. Many interwar thinkers relied on emergent evolution to demonstrate the compatibility of their visions of democracy with Darwinian biology. Emergentism authorized the human sciences by describing reality as both fully naturalistic and organized into a series of levels of ascending complexity, each featuring its own distinct laws and thus requiring a discrete – though fully scientific – form of inquiry. Meanwhile, its adherents could also wield it against determinism. They argued, after the fashion of Ward, that the deterministic laws of the physical world actually represented the precondition for mental freedom, rather than ruling this phenomenon out of existence.[10]

The British psychologist C. Lloyd Morgan developed the first scientifically persuasive theory of emergentism in the years before World War I. Lloyd Morgan was one of many Western thinkers attempting to square Darwin's seemingly amoral account of the struggle for survival with traditional conceptions of moral purpose and, ideally, moral progress. Previously, the leading candidate for this harmonizing task had been vitalism, championed among natural scientists by the German embryologist Hans Driesch and in philosophical

[10] Lester Frank Ward, "Mind as a Social Factor," *Mind* 9, no. 36 (October 1884), 565; David Blitz, *Emergent Evolution: Qualitative Novelty and the Levels of Reality* (Boston: Kluwer, 1992). The French sociologist Émile Durkheim also espoused a version of emergentism in *The Rules of Sociological Method* (New York: Free Press, 1964 [1895]).

circles by the French thinker Henri Bergson. Vitalism, which was well known to educated Americans – Bergson's writings engendered a veritable craze in the United States and led an enthusiastic William James to declare him a "genius" – postulated an immaterial, undetectable force lying behind all chemical and biological processes. This *élan vital*, according to Bergson, drove evolutionary change and thus accounted for the appearance of new phenomena such as the mind.[11]

Although Bergson's vitalism captured the public imagination, many working biologists felt that it verged on supernaturalism. Lloyd Morgan's work did not generate such opposition, despite his insistence that a supreme intelligence was at work in the evolutionary process. This postulate could be excised relatively easily from Lloyd Morgan's main conceptual innovation: the claim that evolution proceeded in successive stages, each inaugurated by the appearance of a new mode of interaction between existing entities. He offered multiple versions of his theory over the years, dividing the evolutionary process into stages in various ways. One version proposed that physico-chemical interactions had been joined by organic interactions and later by cognitive interactions. But Lloyd Morgan's most important point was that each such mode of interaction – each level of reality – was perfectly natural yet featured its own laws, which could not be formulated in terms of the laws governing earlier or lower levels of reality. Most importantly, the mind did not follow the physico-chemical laws of the molecular world. Its constitution by cognitive relations gave the mind additional properties and laws that formed the stuff of psychology, Lloyd Morgan argued.[12]

For a number of years, American philosophers had been working toward a similar theory from less theistic angles. In fact, these philosophers had provided some of the key materials from which Lloyd Morgan built his theory. Particularly important were the New Realists E. G. Spaulding and Walter T. Marvin. In a 1912 essay, Spaulding articulated certain aspects of emergentism in the course of defending science as a legitimate mode of inquiry. Many idealists had long questioned science's validity by challenging the analytic method: the process of breaking down wholes into smaller parts for ease of investigation. This would not do, the critics argued, because parts could not be understood in isolation from wholes. Idealists found evidence for this claim in the fact that the isolated parts typically featured different characteristics than the encompassing wholes. Spaulding defended scientific analysis by responding that the properties of a whole not found in its constituent parts could be traced to the *relations* between these parts. In other words, the distinctive properties

[11] Horst H. Freyhofer, *The Vitalism of Hans Driesch: The Success and Decline of a Scientific Theory* (Frankfurt am Main: P. Lang, 1982); Peter J. Bowler, *The Eclipse of Darwinism: Anti-Darwinian Evolutionary Theories in the Decades Around 1900* (Baltimore: Johns Hopkins University Press, 1983); Thomas Quirk, "Bergson in America," *Prospects* 11 (October 1986): 453–490.
[12] Blitz, *Emergent Evolution*, 59, 78.

of wholes were structural or organizational in origin, stemming not from the fact that they were wholes *per se*, but rather from certain of their structural properties.[13]

Marvin, too, divided nature into separate levels that had emerged sequentially in time. Replacing the *élan vital* with pure chance, Marvin described all of the basic components of the universe as generators of novelty, their behavior stemming from both lawful causes and random, spontaneous events. He added that reality could be viewed (at least for heuristic purposes) as a series of layers or "logical strata," each higher level featuring a smaller number of entities with more complex characteristics. Marvin's theory held that science provided reliable knowledge on which to base most forms of action, even though the future remained partly undetermined because individual behavior had a random component and because a new stratum of reality could appear at any moment.[14]

Although the New Realists participated most actively in its development, emergentism also fit well with the pragmatists' emphasis on moral freedom. Interpreters have discerned emergentist themes in the writings of James, Dewey, and Mead, who took the universe to be dynamic, incomplete, and subject to human intervention, rather than tightly bounded by deterministic laws.[15] In fact, Lloyd Morgan's theory, stripped of its theistic framework, solved a whole series of problems associated with the broad political program of the post-Progressive scientific democrats. For one thing, it seemed to validate the possibility of a middle ground between pure statism and pure individualism. Assuming that a single model characterized all part-whole relations, early-twentieth-century philosophers struggled to avoid treating either society or the individual as a mere reflex of the other. With emergentism in hand, they could distinguish an organically integrated society that nevertheless preserved a broad sphere of freedom from both a purely "mechanical" collection of individuals and a monolithic social whole that utterly negated the individual. In addition, emergentism seemed to reinforce the ethical economists' assumption that the economic laws of an industrial society differed radically from those characterizing the merely "aggregative," preindustrial phase. Finally, emergentism gave

[13] Edward Gleason Spaulding, "A Defense of Analysis," in Edwin B. Holt et al., *The New Realism: Coöperative Studies in Philosophy* (New York: Macmillan, 1912), 158, 160, 238. Spaulding elaborated on this view in *The New Rationalism* (New York: H. Holt: 1918), identifying physico-chemical, biological, and ethical levels of natural phenomena.

[14] Blitz, *Emergent Evolution*, 89. Blitz notes that Marvin's levels included the logical, the physical, the biological, and the mental or social (90). His theory can be found in *A First Book in Metaphysics* (New York: Macmillan, 1912).

[15] Daniel S. Malachuk, "'Loyal to a Dream Country': Republicanism and the Pragmatism of William James and Richard Rorty," *Journal of American Studies* 34 (2000), 95–96; John Herman Randall Jr., "The Changing Impact of Darwinism on Philosophy," *Journal of the History of Ideas* 22 (1961), 437; Robert B. Westbrook, *John Dewey and American Democracy* (Ithaca: Cornell University Press, 1991), 334–335; Murray J. Leaf, "Mead, George Herbert," in *Thinkers of the Twentieth Century, Second Edition*, ed. Roland Turner (Chicago: St. James Press, 1987), 523. A good illustration of the tendency appears in George Herbert Mead, *The Philosophy of the Present*, ed. Arthur E. Murphy (Chicago: Open Court, 1932), 32–47.

a biological status to the supra-individual phenomena of society and culture described by figures such as Ross, Cooley, Sumner, and Boas. Emergentism thus validated the extension of scientific methods into the realm of the social. It suggested that social or cultural phenomena were fully natural in a metaphysical sense, yet irreducible to the simpler terms of physics and chemistry.[16]

All these aspects of emergentism appealed to scientific democrats. So, too, did its ability to combat the widespread public perception that Darwin's theory amounted to materialism, undermining morality and authorizing a ruthless competition for resources. World War I presented both a threat and an opportunity in this regard. On the one hand, German officials often equated Darwinism with a vicious struggle for survival, presenting the international sphere as a proving ground of national strength. On the other hand, this phenomenon opened room for American thinkers to read Darwin's theory in more cooperative terms and to identify it with democracy. More generally, German arguments reinforced American scholars' already strong tendency to interpret the war as a battle between competing philosophies. Biologists and philosophers disagreed on the ultimate source of Germany's theoretical error, variously identifying Nietzsche, Haeckel, Hegel, Kant, and Fichte as the root of its philosophical ills. But they agreed that the conflict centered on fundamental worldviews, and that democracy went hand in hand with a correct reading of biology's social meaning on the Allied side.[17]

Biologists outdid even philosophers in framing the war as a conflict between philosophies of nature. They stood at a distinct disadvantage in the war effort relative to their colleagues in physics and chemistry, who could offer specific points of technological assistance. Biologists did occasionally propose material contributions to the war effort, but improved control of crop-destroying insects paled in comparison to the tanks, submarines, and poisonous gases created by physical scientists. Thus, biologists typically joined philosophers and social scientists in waging the ideological phase of the war, primarily by reading evolutionary theory through the lens of democratic theory. Figures such as Conklin and Raymond Pearl attacked German interpretations of the "survival of the fittest" and lauded the superiority of self-government. They insisted that the most cooperative, integrated societies, not those paralyzed by internal competition, would win out in the international sphere, into which they would then extend the principle of cooperation.[18]

---

[16] The cultural critic Joseph Wood Krutch discussed "the very modish psychological theory of emergence" in his bestseller *The Modern Temper: A Study and a Confession* (New York: Harcourt Brace, 1929), 141–146.

[17] Bruce Kuklick, *The Rise of American Philosophy, Cambridge, Massachusetts, 1860–1930* (New Haven: Yale University Press, 1977), 435–447; James Campbell, "Dewey and German Philosophy in Wartime," *Transactions of the Charles S. Peirce Society* 40, no. 1 (Winter 2004): 1–20; Campbell, *A Thoughtful Profession: The Early Years of the American Philosophical Assoociation* (Chicago: Open Court, 2006), 199–243.

[18] Gregg Mitman, "Evolution as Gospel: William Patten, the Language of Democracy, and the Great War," *Isis* 81 (1990), 452–453, 446; Mitman, *The State of Nature: Ecology, Community, and American Social Thought, 1900–1950* (Chicago: University of Chicago Press, 1992), 58–62.

Headquarters Nights (1917), by the Stanford entomologist Vernon L. Kellogg, was the best-known contribution to the biologists' war effort. Originally serialized in the *Atlantic Monthly*, the book offered an exposé of the cultural sway over Germans of a belligerent reading of Darwinism holding that "war is natural, it is inevitable, and is, indeed, to be welcomed as the necessary final test of the value of the different lines of development and organization of human life and society represented by various existing human groups." But Kellogg's effort to counter competitive readings of evolutionary theory backfired. In the early 1920s, the surging popular movement against teaching evolution in the public schools took *Headquarters Nights* as one of its main texts, finding in it evidence that Darwinism led directly to cutthroat struggle.[19]

In response, biologists and philosophers stepped up their efforts to justify and disseminate a more cooperative understanding of evolution that left room for moral freedom as well as social harmony. The search took some evolutionary theorists back to Jean-Baptiste Lamarck's pre-Darwinian theory that offspring could inherit characteristics their parents had acquired through repeated use. In the classic example, Lamarck postulated that giraffes had slowly lengthened their necks over time by stretching for leaves and passing the incremental gains on to their children. Many American biologists in the 1920s were drawn to the work of modern-day Lamarckians such as the Italian zoologist Eugenio Rignano.[20]

But emergent evolution, unlike Lamarckism, intersected harmoniously with Darwin's theory of natural selection while still undermining a materialistic reading of human action. Aiming at a theoretical gap in Darwin's framework that Mendelian genetics had not yet filled, emergentism accounted for the development of important new traits – most importantly, the unprecedented capacity for conscious reason that virtually all commentators took to differentiate human beings from other animals.[21] Natural selection, emphasized Dartmouth's William Patten, "sifts life ... *but it never creates the thing it sifts.*" With Patten, many American biologists folded emergentism into theories that identified constructive principles of growth in the evolutionary process. They typically discerned an upward, progressive movement in the course of biological history, leading away from individual competition toward harmonious cooperation.[22]

[19] Mitman, "Evolution as Gospel," 446, 454–456; Vernon L. Kellogg, "War and Evolution: Germanized," *North American Review* 207, no. 748 (March 1918), 365; Mark Aaron Largent, "'These are Times of Scientific Ideals': Vernon Lyman Kellogg and Scientific Activism, 1890–1930" (PhD dissertation, University of Minnesota, 2000).

[20] Bowler, *The Eclipse of Darwinism*; William Morton Wheeler and Thomas Barbour, *The Lamarck Manuscripts at Harvard* (Cambridge, MA: Harvard University Press, 1933). Charles S. Peirce had favored Lamarckism, which he found compatible with the ethical teachings of Christ, over Darwinism, which he viewed as a projection onto nature of "politico-economic views of progress": *Chance, Love, and Logic* (New York: Harcourt, Brace, 1923), 275.

[21] Bowler, *The Eclipse of Darwinism*, 275.

[22] Quoted in Mitman, "Evolution as Gospel," 450. Cf. Patten, *The Grand Strategy of Evolution: The Social Philosophy of a Biologist* (Boston: Richard G. Badger, 1920). These visions of evolutionary

Not all of these biologists assigned the human mind a leading role in the creation of the cooperative commonwealth. A strikingly anti-rationalist account came from the pen of Harvard's William Morton Wheeler, a leading proponent of emergent evolution in the 1920s. Lloyd Morgan's description of the levels of reality had jumped directly from the individual mind to God. Wheeler and the philosopher Roy Wood Sellars both contributed powerfully to the discourse of emergentism by adding society as a separate level of existence. Unusually, however, Wheeler's system omitted the individual mind as a distinct phase of reality. An entomologist by specialty, he argued that human societies took their shape from exactly the same instinctual forces driving social insects such as ants and termites to form societies. Wheeler drew distinctly unappetizing lessons from his social insects, mordantly raising the prospect of "a society of very low intelligence combined with an intense and pugnacious solidarity of the whole."[23]

Other biologists of the era hewed to Wheeler's emergentism without his alarming suggestion that social cohesion stemmed solely from innate instincts. They stressed the free operation of human reason instead. Princeton's Edwin Grant Conklin, one of the highest-profile American biologists of the interwar years and an inveterate defender of harmony between science and religion, used the concept of emergent levels to argue that, although nature always selected the fittest specimens, fitness itself meant different things at different evolutionary stages. Physically, he explained, "the fittest is the most viable," but "intellectually, the fittest is the most rational," and "socially, the fittest is the most ethical." Conklin coupled this claim with the anti-Spencerian principle that in "every line of progressive evolution there comes a time when specialization can go no farther without interfering with the harmonious interrelation of parts and thus breaking down co-operation." On this basis, he predicted that mankind's future evolution would proceed on the social rather than the intellectual or physical plane. He looked forward to the gradual "elimination of the socially unfit," those who preferred their own selfish interests to "the freedom of fellowship, common service, and mutual esteem." In fact, Conklin saw a larger purpose at work in the course of evolution, a teleological pull toward "a Society of Nations, a 'Federation of the World.'"[24]

progress found their highest interwar expression in a popular Anglo-American collaboration: Frances Mason, ed., *Creation By Evolution: A Consensus of Present-Day Knowledge as Set Forth by Leading Authorities in Non-Technical Language that All May Understand* (New York: Macmillan, 1928).

[23] Blitz, *Emergent Evolution*, 122, 126; William Morton Wheeler, *Emergent Evolution and the Social* (London: Kegan Paul, Trench, Trubner & Co., 1927), 36; Charlotte Sleigh, "Brave New Worlds: Trophallaxis and the Origin of Society in the Early Twentieth Century," *Journal of the History of the Behavioral Sciences* 38, no. 2 (Spring 2002), 146.

[24] Conklin, *The Direction of Human Evolution*, 85, 61, 76, 124, 75. On Conklin's public engagements, see J. W. Atkinson, "E. G. Conklin on Evolution: The Popular Writings of an Embryologist," *Journal of the History of Biology* 18 (1985): 31–50; and Kathy J. Cooke, "A Gospel of Social Evolution: Religion, Biology, and Education in the Thought of Edwin Grant Conklin" (PhD dissertation, University of Chicago, 1994).

While Wheeler and Conklin stressed the holistic side of emergentism, some younger figures saw it as validating individual freedom. Herbert Spencer Jennings picked up from William E. Ritter of the Scripps Institution the idea that each individual person might be an emergent, "a thing set off from all others, in some respects unique." Such a being, Jennings wrote, was "free from the tyranny of general law," free "to act in accordance with its own nature alone; and yet in its acts there is no breach of experimental determinism." Jennings, a former undergraduate student of Dewey's at the University of Michigan, argued that emergent evolution called for an experimental approach to policy formation in which reason informed practice but remained a thoroughly fallible guide, guaranteeing no certainty of results.[25]

Philosophers, too, looked to biology for lessons concerning the nature and future of human behavior. Well before some of the New Realists began to work toward emergentism in the 1910s, leading American thinkers had sought a way to firmly embed the mind in nature. Four figures, especially – Harvard's William James and George Santayana and the Columbia duo of John Dewey and Frederick J. E. Woodbridge – prepared the ground for emergentism's rapid spread in the 1920s. Although important disagreements divided these men, their work carved out a broad space of philosophical naturalism that defined the mind as a natural fact. Eschewing the traditional dualism that deemed minds and bodies entirely distinct entities, James, Santayana, Dewey, and Woodbridge each insisted that the human mind, though free to choose and capable of generating its own purposes and ideals, was essentially continuous with the rest of nature. In the early 1920s, this position appeared most influentially in Dewey's *Human Nature and Conduct* (1922) and *Experience and Nature* (1925). Just as many biologists lent the authority of empirical research to emergentism, so too did leading philosophers give their imprimatur to the view that, as Conklin put it, "structure and function, body and mind, brain and consciousness appear to be two aspects of one thing – namely, organization or life – and neither can be fully explained in terms of the other."[26]

[25] Jennings, *Some Implications of Emergent Evolution*, 3–6, 11. Ritter equated emergentism with what he had previously labeled the "organismal conception of life": William E. Ritter and Edna W. Bailey, "The Organismal Conception," *University of California Publications in Zoology* 31, no. 14 (1928), 334.

[26] Bruce Kuklick, *A History of Philosophy in America, 1720–2000* (New York: Oxford University Press, 2001), 190–191; Westbrook, *John Dewey and American Democracy*, 287–293, 321–346; Conklin, *The Direction of Human Evolution*, 189. Cf. C. F. Delaney, *Mind and Nature: A Study of the Naturalistic Philosophies of Cohen, Woodbridge and Sellars* (Notre Dame: University of Notre Dame Press, 1969), 226. Helpful introductions to Santayana include Victorino Tejera, *American Modern, The Path Not Taken: Aesthetics, Metaphysics, and Intellectual History in Classic American Philosophy* (Lanham: Johns Hopkins University Press, 1996), 79–116; Kuklick, *A History of Philosophy in America*, 207–210; and John Lachs, *On Santayana* (Belmont: Thomson Wadsworth, 2006). On the metaphysical (though not political) harmony between Santayana and Dewey's circle, see also Glenn Tiller, "The Unknowable: The Pragmatist Critique of Matter," *Transactions of the Charles S. Peirce Society* 42, no. 2 (Spring 2006): 206–228. Woodbridge's thought is covered in William Frank Jones, *Nature and Natural Science: The*

The University of Michigan's Roy Wood Sellars played a central role in integrating emergentism into the broad current of philosophical naturalism established by James, Santayana, Dewey, and Woodbridge. Sellars, who joined Wheeler in adding the social level to emergentism, used the theory to differentiate modern naturalism from nineteenth-century materialism. He explained that naturalists shared with materialists a belief that all experienced phenomena occupied "one great physical theatre." Naturalists, however, denied that "the movement of atoms in a mechanical way" could explain human affairs, he continued. Social institutions, for example, emerged through the operations of mind – a natural phenomenon, to be sure, but not a purely physical (or even a spatiotemporal) one. "With the admission of levels in nature and the efficacy of mind as a living kind of organization," Sellars declared, modern naturalists could confidently affirm "the significance of values and ideals as effective elements in the functioning of a human organism." On the emergentist theory, he summarized, "[e]thics and sociology become natural sciences resting on psychology, just as psychology becomes continuous with biology."[27] Here, as in much scholarly thought of the 1920s, the move was not to leave biology behind, as our histories would have it, but rather to reconcile biology with the human sciences by changing the former rather than the latter – by developing a non-reductive biological understanding that accounted for phenomena more complex and less deterministic than physico-chemical relations.

PSYCHOBIOLOGY

The theoretical clash over behaviorism and determinism was not a merely academic affair. In 1924, Watson squared off against the British-born Harvard psychologist William McDougall, a doughty champion of introspective methods and inherited traits, in the "Battle of Behaviorism," a public debate in Washington that drew more than a thousand spectators and which many interpreters viewed as a clash between religion and science – in other words, morality and materialism – after the fashion of the following year's Scopes Trial. Out of the public eye, meanwhile, the debate came to a head at the University of Chicago. There, Herrick and a group of fellow "psychobiologists" squared off against Karl Lashley, a student of Watson. Working closely together, despite

Philosophy of Frederick J. E. Woodbridge (Buffalo: Prometheus Books, 1983) and William M. Shea, *The Naturalists and the Supernatural: Studies in Horizon and an American Philosophy of Religion* (Macon, GA: Mercer University Press, 1984), 143–170.

[27] Sellars, "Why Naturalism and Not Materialism?", 223, 222. Also see Sellars' *Evolutionary Naturalism* (Chicago: Open Court Publishing, 1922). Treatments of Sellars' thought include Delaney, *Mind and Nature*, 145–208; William Preston Warren, *Roy Wood Sellars* (Boston: Twayne, 1975); Edmond Wright, "A New Critical Realism: An Examination of Roy Wood Sellars' Epistemology," *Transactions of the Charles S. Peirce Society* 30, no. 3 (1994): 477–514; Pouwel Slurink, "Back to Roy Wood Sellars: Why His Evolutionary Naturalism is Still Worthwhile," *Journal of the History of Philosophy* 34, no. 3 (July 1996): 425–449; and Kuklick, *A History of Philosophy in America*, 211–214.

their philosophical and political differences, these men undertook sophisticated empirical investigations to determine whether a naturalistic outlook should make room for freedom and purpose or toss these terms onto the trash heap of science with phlogiston and other discredited concepts.[28]

Like Watson, Lashley was a transplanted Southerner. A firm "believer in *things as they are*," he harbored a virulent racism, appending "Heil Hitler and Apartheit!" to private letters as late as 1956. He scorned all talk of consciousness, insisting that human beings possessed no qualities or capacities that set them apart from lower animals and declaring that "the inclusion of 'mind' will add nothing to scientific psychology." Lashley also joined Watson in making the behaviorist's key methodological move: studying white rats in order to draw conclusions about human behavior.[29]

Nonsense, Herrick replied; "men are bigger and better than rats." Herrick's 1926 book *Brains of Rats and Men* directly challenged the behaviorists' denial of qualitative differences between the two organisms. Downgrading physiology in favor of neurology, comparative anatomy, and embryology, Herrick and his fellow psychobiologists sought a way of grounding psychology in biology that left room for the distinctive, purposive character of human action. They wanted to buttress the common-sense observation that individuals could reshape their thoughts and values, "socialize" their ideals, "learn to lay out a deliberate program of self-culture and to subordinate other motives to this purpose," and, not least, "evaluate experience in terms of both personal enjoyment and the happiness of others" in order to "strike a balance between these which is good for all of us."[30]

Herrick's University of Chicago colleague Charles Manning Child developed one of the key concepts of psychobiology, though its social implications remained somewhat unclear. Child took his research on starfish and sea urchins to indicate that changes in the physical and social environment caused differentiation and structuration within an organism's protoplasm – the material filling its cells – by forming "gradients" between areas of higher and lower metabolic activity. He believed that he had discovered a means by which external stimuli permanently altered an organism's physiological structure. Both Child and Herrick used the gradient theory to portray the organism as self-consciously

[28] Kingsland, "A Humanistic Science," esp. 464–470; Franz Samelson, "Organizing for the Kingdom of Behavior: Academic Battles and Organizational Policies in the Twenties," *Journal of the History of the Behavioral Sciences* 21, no. 1 (1985): 33–47; Nadine C. Weidman, "Psychobiology, Progressivism, and the Anti-Progressive Tradition," *Journal of the History of Biology* 29 (1996): 267–308; Weidman, *Constructing Scientific Psychology: Karl Lashley's Mind-Brain Debates* (New York: Cambridge University Press, 1999).

[29] Quoted in Weidman, "Psychobiology, Progressivism, and the Anti-Progressive Tradition," 302–304; Karl S. Lashley, "The Behavioristic Interpretation of Consciousness. II," *Psychological Review* 30, no. 5 (1923), 352.

[30] Quoted in Weidman, "Psychobiology, Progressivism, and the Anti-Progressive Tradition," 298; Kingsland, "A Humanistic Science," 449, 460; C. Judson Herrick, *The Thinking Machine* (Chicago: University of Chicago Press, 1929), 364.

altering its own structure, although it could certainly comport with the behaviorists' stimulus-response dynamic as well. Child also applied organic analogies to democratic societies, suggesting that at all levels and in all environments, democratically structured organizations proved the most effective. Herrick, meanwhile, insisted that physical evolution had largely ended, leaving cultural evolution as the leading influence on human behavior. He declared that the latter process moved constantly in the direction of greater cooperation.[31]

Herrick also built on the ideas of his friend George E. Coghill, of Philadelphia's Wistar Institute of Anatomy and Biology. Coghill agreed with Child that an organism's responses to its environment could profoundly alter its physiology. In fact, he traced the success of education to a deep-seated process of "protoplasmic learning" within each cell. Coghill's most distinctive contribution to psychobiology, however, was an account of evolution holding that the nervous system evolved as an integrated whole, rather than originating as a collection of discrete, isolated reflexes. Coghill insisted that behavioral responses took a holistic form; these enlisted the entire organism and were profoundly inflected by its chosen goals. He described the nervous system as an emergent, exhibiting a property – purposiveness – that did not inhere in its component parts. Every organism "does creative work," Herrick summarized. "Some of them even create themselves, as in species formation in the evolutionary series and character building by human beings."[32]

Herrick drew on both of these theorists' ideas, as well as the theory of emergent evolution, in his attempt to reconcile scientific naturalism with the everyday experience of moral choice and purpose-driven behavior. All of nature could be understood in terms of scientific laws, Herrick allowed. But human beings could never fully understand those laws, once and for all, because qualitatively new phenomena following entirely new kinds of laws could arise at any time. According to Herrick, this meant both that scientists could never fully predict the future and that any given system of scientific laws held true only for a limited group of phenomena. His view gave science an irrevocable historical dimension: "When an entirely new configuration of things arises naturally, or first comes to our knowledge by observation," as with the discovery of radium, "we cannot predict what it will do next until we learn the laws of its behavior by further observation and experiment." Existing scientific laws would continue to hold true within their sphere of relevance, but they would not necessarily apply in the new realm of observed behavior. Herrick believed that this inability to fully predict the future left considerable room for moral freedom, as individuals would always face numerous unforced choices. He thus declared strict natural causation fully compatible with "feeling, thinking, willing, purpose, self-culture, moral sanctions, choice, freedom, faith, hope and charity."[33]

---

[31] Kingsland, "A Humanistic Science," 455–456.
[32] Kingsland, "A Humanistic Science," 454–456, 465–466; Weidman, "Psychobiology, Progressivism, and the Anti-Progressive Tradition," 284, 278; Herrick, *The Thinking Machine*, 302.
[33] Herrick, *The Thinking Machine*, 304; Herrick, *Fatalism or Freedom*, 92.

The emergentist assumption that the universe produced real novelty enabled Herrick to reconcile the apparent discord – a source of frustration for scientific thinkers since Kant's days – between the pictures of the individual mind as a natural phenomenon, on the one hand, and a locus of creative thought and action, on the other.

SCIENCES OF SUBJECTIVITY

Emergentism's validation of human purposes and a naturalistic approach to moral and social phenomena rendered it highly attractive to interwar scientific democrats. From the side of biology, Jennings argued that the theory allowed the scientist to finally make peace with his humanistic colleagues who studied the moral life of mankind. Many philosophers, too, welcomed the prospect of a scientific account of freedom. Arthur O. Lovejoy defended emergentism alongside Wheeler at a special panel of the 1926 World Congress of Philosophy. In Vienna, a youthful Karl Popper also drew on Lloyd Morgan's theory of emergent evolution in the 1920s. By 1945, when John Dewey, Sidney Hook, and Ernest Nagel issued a joint rebuttal to a growing chorus of claims that modern naturalism amounted to nothing more than nineteenth-century materialism, they took for granted a scholarly consensus on emergentism.[34]

The new conception of evolution also found a warm reception elsewhere in the cultural sciences. In a nice irony, this development among biological thinkers gave cultural scientists license to assert the reality and the irreducibility of non-material entities, and thus to declare their own independence, on a day-to-day basis, from biology. Emergentism's postulation of multiple layers of reality, each possessing its own laws, authorized the development of new disciplines that were fully scientific, yet methodologically distinct from physics and even biology.[35] In fact, the theory seemed to positively require a series of separate sciences dealing with the various extra-personal concepts that had long been the stuff of social theory – the state, society, and what came to be widely called culture in the 1920s. If each of these emergent entities had its own distinct laws, then each required a separate form of analysis.[36]

[34] Jennings, *Some Implications of Emergent Evolution,* 11; Arthur O. Lovejoy, "The Meanings of 'Emergence' and Its Modes," *Journal of Philosophical Studies* 2, no. 6 (April 1927): 167–181; Blitz, *Emergent Evolution,* 122, 126; Michel Ter Hark, "The Psychology of Thinking, Animal Psychology, and the Young Karl Popper," *Journal of the History of the Behavioral Sciences* 40, no. 4 (Fall 2004): 375–392; John Dewey, Sidney Hook, and Ernest Nagel, "Are Naturalists Materialists?" *Journal of Philosophy* 42 (September 13, 1945), 519. See also Lovejoy, "The Discontinuities of Evolution," *University of California Publications in Philosophy* 5 (1924): 173–220; and Stephen C. Pepper, "Emergence," *Journal of Philosophy* 23, no. 9 (April 29, 1926): 241–245.
[35] Arthur O. Lovejoy called this component of the theory "functional emergence," as distinct from "existential emergence," which is an account of the historical evolution of reality: "The Meanings of 'Emergence' and Its Modes," 174.
[36] The traffic between biology and the human sciences also ran the other way, at a time when disciplinary borders remained fairly permeable. Herrick, for example, drew on the sociologist

Moreover, as figures such as Ralph Barton Perry and the political scientist Harold D. Lasswell recognized, emergentism also rendered plausible the claim of progressives that individuals would actually be freer in a highly coordinated society.[37] Theorists such as Richard T. Ely and John Dewey had developed a concept of "positive freedom" somewhat distinct from that later defined by Isaiah Berlin. Here, freedom referred to the ability to realize one's impulses, not simply the absence of legal or physical obstacles. Thus, an individual stranded alone in a desert was not free to eat a solid meal, whereas the inhabitant of a modern city was free to do so – if given the proper income. On some readings, emergentism suggested that the individual components of a system retained all of their prior freedom and gained further advantages from the advent of a new form of organization.[38]

Mainstream biologists also explored other conceptions of freedom in the 1920s. Conklin employed a full-blown organic analogy, defining society as an organism and individual citizens as its component cells. This image worked well for combating self-seeking, which Conklin compared to the destructive action of cancer cells. People did not want to be malignant, he averred. Instead, they naturally sought to function like "the normal cells of the body, each of which is a unit, preserving its own individuality, and to a certain extent its own independence, and free to do the work for which it is fitted under the control of the body as a whole."[39]

But body cells were not free to move around or to change functions, a fact that cut against the use of the organic analogy by many scientific democrats. Herrick offered a different strategy for grounding freedom in biology, one that minimized references to the social whole. All around him in the 1920s, Herrick saw a long-overdue embrace of "natural freedom," which he described as "freedom to live and grow, to know that we do it and to enjoy this knowledge, to forecast the probable or possible future course of life and of personal development, and consciously to participate in the control of the future development." Herrick identified this conception of natural freedom as a more powerful reply to the moral skepticism of Clarence Darrow than any religious lessons or preachy moralizing. Herrick's biological rendering of self-realization, however, confined natural freedom within the boundaries set by scientific laws. The

Frank H. Hankins' "Individual Freedom with Some Sociological Implications of Determinism," *Journal of Philosophy* 22 (1925): 617–634. He also published alongside social scientists in interdisciplinary texts; see "Self-Control and Social Control," in Jane Addams *et al.*, *The Child, The Clinic and the Court* (New York: New Republic, 1925): 156–177 and "The Scientific Study of Man and the Humanities," in *The New Social Science*, ed. Leonard D. White (Chicago: University of Chicago Press, 1930): 112–122.

[37] Heinz Eulau, "The Maddening Methods of Harold D. Lasswell: Some Philosophical Underpinnings," *Journal of Politics* 30 (1968), 9. For Perry's use of the concept, see the discussion that follows.

[38] Dewey spelled out this concept of positive freedom in *Outlines of a Critical Theory of Ethics* (1891), in Jo Ann Boydston, ed., *The Early Works of John Dewey, 1882–1898, Volume 3* (Carbondale: Southern Illinois University Press, 1967), 343–344. For Ely's analysis of self-realization, see Chapter 2.

[39] Conklin, *The Direction of Human Evolution*, 125.

"normal, healthy man," he summarized, "guides his life as the pilot guides his plane," either following "the rules of the game" or going down in flames. Like Ward some forty years earlier, Herrick argued that individuals could "act in view of possible future contingencies" only if science revealed the possibility of doing so. Freedom required the existence of deterministic laws in the physical world.[40]

Meanwhile, Perry drew out the consequences of emergentism for the question of self and society. Building on his teacher James' emphasis on the generative possibilities of human "federation," Perry argued in 1926 that social coordination gave each human desire the fullest scope for realization. He displayed a somewhat wobbly grasp of emergentism, however, for he employed not only the common illustration of chemical synthesis, which produced an entity exhibiting qualities not found in the original parts, but also the model of a resultant of two forces, which simply pointed in a different direction than the original forces while incorporating both of them fully. The resultant-of-forces image figured heavily in Perry's analysis of collective decision-making. He proposed that for any social conflict, there existed a truly integrative solution in which "no factor operates save the two interests themselves, each in turn taking account of the other, and both contributing dynamically to the outcome." In the resulting plan of action, Perry emphasized, "both interests may be said to be wholly satisfied." As a matter of pure logic, an integrative solution could not abridge either interest, because if one or the other were not "fully operative," then the outcome would change. Perry argued that a highly organized society could foster the full and harmonious realization of every existing interest, without any serious loss, by adopting "inclusiveness" as its supreme ethical standard and then crafting integrative solutions to all conflicts.[41]

The widespread interest in emergent wholes among cultural scientists in the 1920s paved the way for a ready acceptance of *Gestalt* psychology in the American disciplines. This school of thought was developed during the 1910s and 1920s by a group of German figures, led by Kurt Koffka, Wolfgang Köhler, and Max Wertheimer, who subsequently relocated to United States. *Gestalt* theorists defined the mind as natural and entirely accessible to science, but they firmly rejected a mechanistic understanding of nature such as that found in Watson's behaviorism. Instead, they stressed the patterned, holistic nature of human responses to environmental stimuli. In a series of classic tests, for example, they demonstrated that human beings perceived visual patterns directly as wholes rather than building them up inductively from their constituent dots, which of course lacked all of the characteristics of the larger patterns. Proponents of emergentism such as Sellars and Wheeler recognized its essential affinity with the *Gestalt* approach.[42]

[40] Herrick, *Fatalism or Freedom*, 88, 95, 87, 84–85.
[41] Ralph Barton Perry, *General Theory of Value: Its Meaning and Basic Principles Construed in Terms of Interest* (New York: Longmans, Green, 1926), 664, 660–661, 655.
[42] Ernest Keen, *A History of Ideas in American Psychology* (Westport: Praeger, 2001), 89–98; Michael Sokal, "The Gestalt Psychologists in Behaviorist America," *American Historical Review*

The *Gestalt* theorists became a major presence in the United States during the 1930s. Koffka came to Smith in 1927, and most of the other major figures visited the U.S. frequently in the late 1920s. During that period, their writings served as one of the major vectors through which the holistic, non-reductive outlook also associated with emergentism made its way into the other cultural sciences. Boasian anthropologists, for example, picked up on *Gestalt* ideas. Margaret Mead read Koffka's *The Growth of the Mind* in 1925, gave it to Edward Sapir, and later discussed it with Ruth Benedict. Perry also welcomed the movement in that same year. "The *gestaltists* in psychology have taught us to regard personalities as functioning wholes," explained the sociologist Charles A. Ellwood in 1931. In the 1930s, Wertheimer taught an influential seminar on social science methodology at the New School for Social Research's University in Exile, which served as the home for many *Gestalt* theorists and other Central European émigrés.[43]

Mainstream American psychologists, including even self-proclaimed behaviorists, were also receptive to *Gestalt* concepts in the movement's early days. For one thing, many of the thinkers who in the 1920s labeled themselves "behaviorists" went on invoking the very terms, such as "consciousness," "purpose," and "ideals," that Watson sought to ban. Some sought to give the label "behaviorism" a wider meaning, while others simply missed the radically reductive nature of Watson's project.[44] Watson's efforts notwithstanding, the philosophical framework of behaviorism – a term also employed by the Harvard philosopher Edwin B. Holt in the 1910s – remained fairly capacious. Additionally, the rigorous, experimental approach of the *Gestalt* theorists allowed American psychologists to recognize the movement as a contribution to science even on Watson's strict methodological terms. Thus, even such a dedicated Watsonian as Lashley embraced key *Gestalt* concepts, such as configurations and fields. Meanwhile, Harvard's Edwin G. Boring, a vigorous proponent of the experimental method, told Köhler in 1925, "I am not a *Gestalt* psychologist but merely a scientist. *Gestalt* psychology seems to me to be nothing more than the introduction of science into psychology."[45]

89 (December 1984): 1240–1263; Sellars, "Why Naturalism and Not Materialism?" 223; Wheeler, *Emergent Evolution and the Social*, 14.

[43] Sokal, "The Gestalt Psychologists in Behaviorist America"; Margaret Mead, "Patterns of Culture: 1922–1934," in Mead, ed., *An Anthropologist at Work: Writings of Ruth Benedict* (Boston: Houghton Mifflin, 1959), 207; James T. Kloppenberg, "The Place of Value in a Culture of Facts: Truth and Historicism," in *The Humanities and the Dynamics of Inclusion Since World War II*, ed. David A. Hollinger (Baltimore: Johns Hopkins University Press, 2006), 129; Ellwood, "Scientific Method in Sociology," *Social Forces* 10 (October 1931), 19; Peter M. Rutkoff and William B. Scott, *New School: A History of the New School for Social Research* (New York: Free Press, 1986), 123–127.

[44] In a good example of the former response, Ralph Barton Perry lambasted critics who would "make a present of the label 'behaviorism' to Professor Watson": "Reply to Professor Calkins," *Journal of Philosophy* 24, no. 25 (December 8, 1927), 685.

[45] Quoted in Sokal, "The Gestalt Psychologists in Behaviorist America," 1246.

However, by the time several more of the *Gestalt* theorists immigrated to the United States in the 1930s, tensions with American behaviorists had increased. Despite their experimental bent, these theorists employed a sharp, openly philosophical critique of mechanistic naturalism that irritated many Americans – most prominently Boring, who flatly disavowed philosophy. When Köhler wrote *Gestalt Psychology* in 1929 to introduce English speakers to the school, he opened his book by contrasting *Gestalt* concepts to a narrow, caricatured behaviorism. The book led several American psychologists to turn against *Gestalt* concepts, declaring them overly mystical. Boring, for example, underplayed his early support for *Gestalt* principles in his influential 1929 *A History of Experimental Psychology*. Personal friction between Köhler and Boring deepened the split, as did the Gestaltists' outspoken attacks on positivism ("an evil that destroys young energies as surely as does tuberculosis," according to Köhler) in the ideologically contentious 1930s. In the mid-1930s, Boring and his psychological allies managed to prevent Perry and other philosophers from appointing Köhler to the department they shared at Harvard. The *Gestalt* theorists held that a true understanding of democracy itself required the articulation of non-reductive science and the rejection of behaviorism, which embodied "the wrong scientific attitude."[46] For this reason, of course, *Gestalt* theories appealed to Ellwood, a dogged critic of value-neutral approaches, and the many other scientific democrats of the 1930s who likewise interpreted behaviorism narrowly and held up an ethically engaged, philosophically informed, and frankly normative science as the foundation for democracy. The tussle over *Gestalt* psychology represented an early episode in the grand epistemological battle around "science and values" that would consume many scientific democrats in the 1930s.

But the basic insights of emergentism and *Gestalt* theory also made their way into the cultural sciences independently of Köhler's strident philosophical critique. The Berkeley psychologist Edward C. Tolman, who was more ecumenical than his Harvard counterpart Boring, explicitly incorporated *Gestalt* concepts into his influential formulation of "neobehaviorism" (he dubbed it "purposive behaviorism") in the early 1930s. Tolman endorsed two major components of Watson's program for psychology. First, he assumed that all behavior stemmed from a basic physiological drive to replace a state of "disturbance" with one of "quiescence." Second, he declared that scientific proof required observable evidence of behavior in a laboratory setting. But he insisted that something like "consciousness" stood between stimuli and responses, and he argued that experimentation clearly demonstrated the existence of such an entity, even if

---

[46] *Ibid.*, 1246; Wolfgang Köhler, "The Naturalistic Interpretation of Man (The Trojan Horse)," in *The Selected Papers of Wolfgang Köhler*, ed. Mary Henle (New York: Liveright, 1971), 344; Kloppenberg, "The Place of Value in a Culture of Facts," 131; Kurt Koffka, "Why Psychology?" in *Principles of Gestalt Psychology* (New York: Harcourt Brace, 1935), 9. Cf. Wertheimer's "On the Concept of Democracy" and "A Story of Three Days," in *Documents of Gestalt Psychology*, ed. Mary Henle (Berkeley: University of California Press, 1961): 42–51, 52–64.

it could not itself be perceived directly from the outside. Consciousness was a hypothesis that fit the observed facts. For Tolman, as for many other self-proclaimed behaviorists, that label indicated allegiance to an experimental methodology, not a specific theory of the mind.[47]

In fact, Tolman explicitly preferred the approach of the *Gestalt* theorists, especially Kurt Lewin, to that of Watson. He had studied not only with the real-ist philosophers Perry and Edwin B. Holt at Harvard, but also with Koffka at Giessen. Tolman agreed with Watson that an organism's behavior amounted to a series of responses to the "internal physiological disequilibriums" caused by stimuli, and that these responses were conditioned by past experience. Tolman insisted, however, that learning involved more than trial and error. Experiential situations never recurred in exactly the same form, he reasoned, and thus trial and error was largely useless as a guide to behavior. But human beings also learned through what Tolman called "inventive ideation," a process of using mental symbols – shorthand for experience – to predict the effects of broad categories of stimuli and thus to address situations that had never before been experienced in their particulars. Inventive ideation was possible, according to Tolman, because mental processes were symbolic or conceptual in nature, involving "sign-gestalt formation, refinement, selection or invention."[48]

Epistemologically, Tolman followed William James – by way of Tolman's teachers Perry and Holt, both students of James – in arguing that immediately given experience was both objective and subjective, "an initial, common matrix out of which both physics and psychology are evolved." While this immedi-ate experience was "rich" and "qualitied," it was largely "ineffable, that is, logically incommunicable from one sentient being to another." Nor could it serve as material for inventive ideation, which entailed the manipulation of abstracted concepts – literally, the communication of concepts from a pre-sent being to its future self. Tolman wrote that external objects made them-selves known "only in their guise as possible *behavior-supports* for the given organism," because "men know the world only for the purposes of behaving to it." Knowledge was inevitably "strained through the behavior-needs and the behavior-possibilities of the particular organisms who are gathering that knowledge." In short, Tolman adopted a pragmatic conception of the know-ing process. That "knowledge is 'true,'" he wrote, "which 'works,' given the

[47] Sokal, "The Gestalt Psychologists in Behaviorist America"; Edward C. Tolman, *Purposive Behavior in Men and Animals* (New York: Century, 1932), 415, 418; Keen, *A History of Ideas in American Psychology*, 125. Tolman's reformulation of the behaviorist position is outlined in Laurence D. Smith, *Behaviorism and Logical Positivism: A Reassessment of the Alliance* (Stanford: Stanford University Press, 1986), 67–146 and Keen, *A History of Ideas in American Psychology*, 123–134.

[48] Smith, *Behaviorism and Logical Positivism*, 69–81; Edward C. Tolman, "Psychology versus Immediate Experience," *Philosophy of Science* 2 (July 1935), 365; Tolman, *Purposive Behavior in Men and Animals*, 219, 372. On the thought of Perry and Holt, see Kuklick, *The Rise of American Philosophy*, 338–350, 417–434; and Kuklick, *A History of Philosophy in America*, 202–207. In the former book, Kuklick notes the congruity with Tolman (431).

particular behavior-needs and the particular behavior-capacities of the type of organism gathering such knowledge."[49]

Tolman defined science as nothing more than a system of concept formation that had proven highly productive for the purposes of communication and inventive ideation. Its concepts were not "re-livings of immediate experience," but rather "logical constructs," a set of "rules and equations whereby we are aided in finding our way about from one moment of immediate experience to another." Rather than creating "a complete replica of reality," which was impossibly complex and in any case incapable of communication, science facilitated future behavior by offering a simplified model, "a short-hand for finding one's way about from one moment of reality to the next."[50]

Tolman's epistemological analysis opened behaviorism to any entity that was not immediately observable but could be employed as a hypothetical guide to behavior. He proposed to represent the mental processes that could be "inferred 'back' from behavior," though not observed directly, by a series of "intervening variables" that represented an organism's range of responses to similar stimuli. Tolman presented his theory as an elaboration of Watson's behaviorism, not a substitute for it. He explained that the "arch-behaviorist" Watson and many others missed these intervening variables because they were concerned with "molecular or microscopic concepts," seeking to reduce psychology to physiology by creating "a complete neural, glandular and visceral picture." As scholars such as Holt, Perry, Lashley, Theodore de Laguna, and Albert Paul Weiss had shown, the behaviorist could equally well seek "molar or macroscopic concepts" that described "behavior-readiness" in terms of the holistic, goal-oriented responses of the entire organism that everyday speakers called "'demands,' 'intentions,' 'expectations' and 'attainments.'"[51]

At the molar level, Tolman explained, behavior was "gestalted"; each type of response represented "an 'emergent' phenomenon that has descriptive and defining properties of its own," in addition to its lower-level, physiological features. Still, these types were "objectively definable," and thus proper subjects for scientific analysis. Tolman expected to find, through experimentation, "laws of sensory capacity, memory, learning and reasoning." These, he believed, would show that the human animal, unlike the rat, exhibited "an ultimate demand for harmonious sense-impressions, an ultimate demand for the company of one's fellows and perhaps an ultimate demand for the social approval of one's fellows." In turn, such ultimate demands produced subordinate or conditional demands, including "(a) demands for specific types and instance of goal-objects and (b) demands for specific means-objects." Such subordinate demands, which came

[49] Tolman, "Psychology versus Immediate Experience," 359–360, 363; Tolman, *Purposive Behavior in Men and Animals*, 429–430.

[50] Tolman, "Psychology versus Immediate Experience," 380, 359–360; Tolman, *Purposive Behavior in Men and Animals*, 424–425.

[51] Tolman, *Purposive Behavior in Men and Animals*, 414; Tolman, "Psychology versus Immediate Experience," 379, 365.

to function as ends in themselves rather than means to ends, made up "the very breath and substance of life," Tolman explained. They comprised most of "what the anthropologists find in the way of specific culture-patterns and what psychologists find in the way of individual idiosyncrasies."[52]

Tolman's neobehaviorism offered a way for psychologists and other cultural scientists to embrace human purposes and ideals while still thinking of themselves as rigorously scientific, in the positivist sense of eschewing philosophical speculation. Tolman differentiated science from philosophy along three axes. First, science sought "objectively stateable laws and processes governing behavior," whereas metaphysicians attempted to communicate directly the "immediately given pre-analytical complex" as it "appears to the naïve man." Tolman also identified science with the reductive method of analysis. "However falsifying analysis may be," he argued, both communication and prediction required the reduction of experiential phenomena to a "barren and 'unfelt' array of functionally defined immanent determinants and behavior-adjustments." The emotional unreality of concepts was simply the price that human beings had to pay for the capacity to think and speak intersubjectively. Finally, Tolman associated philosophy with an ontological dualism that posited a realm of pure mentality standing entirely apart from that of material reality. This he flatly rejected. If there was any divide in experience, Tolman insisted, it ran "between reality and its maps and not between two types of reality." On all of these grounds, Tolman insisted, the scientific psychologist gave speculative philosophy a wide berth – yet still avoided Watson's reductive view of the mind.[53]

What, then, of eugenics, by far the best-known effort of the 1920s to draw lessons regarding human affairs from biology? Few of the figures discussed in this chapter denied, in principle, that parents could influence the specific physical and mental characteristics of their offspring by choosing their mates with an eye to desirable, heritable traits. But their assumption that culture amounted to a distinct and uniquely powerful track of evolutionary change, rather than a reflex of biological traits, rendered eugenics largely extraneous. Patten was unusual in openly questioning the very possibility of knowing what to breed for: "Who shall decide what is 'good human stock' and what are 'good environments' for it? No one surely knows what those 'good things' are." But many of the other mainstream biologists, including Conklin and the psychobiologists, argued that education represented a far more effective means to the desired end of harmonious cooperation than selective breeding.[54]

[52] Tolman, *Purposive Behavior in Men and Animals*, 418, 4; Tolman, "Psychology versus Immediate Experience," 368–370, 379.

[53] Tolman, "Psychology versus Immediate Experience," 356, 359–360; Tolman, *Purposive Behavior in Men and Animals*, 420, 424.

[54] William Patten, *Growth: An Introduction to the Study of Evolution* (Hanover: Dartmouth Press, 1929 [1923]), 36; Kathy J. Cooke, "Duty or Dream? Edwin G. Conklin's Critique of Eugenics and Support for American Individualism," *Journal of the History of Biology* 35, no. 2 (2002): 365–384; Cooke, "A Gospel of Social Evolution," 198; Miriam Reumann and Anne Fausto-Sterling,

Meanwhile, human scientists outside the discipline of biology eagerly embraced emergentism, *Gestalt* theory, and other systems promising the best of both holistic and reductive approaches. Scientific democrats in the 1920s identified all manner of immaterial entities – especially the personality, society, and culture – as emergent phenomena, rendering these completely natural, yet not reducible to physics or even biology. Ironically, then, the cultural sciences proclaimed their independence from biology by embracing a biological theory, albeit one that lasted far longer outside the circle of professional biologists than within it. Moreover, as Chapter 8 reveals, many cultural scientists followed Tolman in drawing on the new view of nature as a complex, multilayered affair to define scientific knowledge as a human construct, not a mere reflection of an external world.

"Notions of Heredity in the Correspondence of Edwin Grant Conklin," *Perspectives in Biology & Medicine* 44, no. 3 (Summer 2001): 414–425; Weidman, "Psychobiology, Progressivism, and the Anti-Progressive Tradition," 273. Jennings and Pearl are usually classed with the left-leaning, largely British "reform eugenicists" of the 1930s: Diane Paul, "Eugenics and the Left," *Journal of the History of Ideas* 45 (1984): 567–590; Daniel J. Kevles, *In the Name of Eugenics: Genetics and the Uses of Human Heredity* (Berkeley: University of California Press, 1985), 164–175; Michael Mezzano, "The Progressive Origins of Eugenics Critics: Raymond Pearl, Herbert S. Jennings, and the Defense of Scientific Inquiry," *Journal of the Gilded Age & Progressive Era* 4, no. 1 (January 2005): 83–97. See also Elazar Barkan, *The Retreat of Scientific Racism: Changing Concepts of Race in Britain and the United States Between the World Wars* (New York: Cambridge University Press, 1992), 189–220.

6

The Problem of Cultural Change

If the events of World War I and its immediate aftermath fueled a surge of philosophical reflection among American biologists and psychologists, their impact on other areas of the cultural sciences proved even more transformative. The monumental irrationality of the war years hovered over the theoretical work of the 1920s. Innumerable developments of that era struck scientific democrats as disastrous: the intense outbursts of anti-immigrant and antiradical sentiment on the home front; the public's rejection of the League of Nations and then Progressive reform itself; and, of course, the very fact of armed conflict – the inability of civilized Western nations to prevent minor disagreements from escalating into years of mass slaughter. No doubt a factor as well, if perhaps only subconsciously, was the fact that leading American scholars had abetted the cataclysm by urging their fellow citizens to enter the war, in the more idealistic days of 1917. It seemed to the scientific democrats that all of the constructive cultural work they had done since the 1890s – and much more – had been torn down in a cataclysm of sheer savagery.

In the face of this massively widened gap between ideal and reality, many scientific democrats adopted new strategies for producing cultural change. A few gave up on the public altogether and thus moved beyond scientific democracy, calling for government by an expert-guided elite. Most followed their Progressive forebears in seeking to establish social control, in Edward A. Ross' sense of the phrase: a system of coordination designed to promote the public good and rooted primarily in horizontally, informally maintained norms and beliefs. These post-Progressive scientific democrats preserved the hopeful assumption that social and political institutions took their outlines from the underlying cultural substrate and would shift in tandem with the latter. But they discovered that decades of research and theorizing had failed to appreciably dent the public mind, especially in the crucial area of social ethics. The scientific democrats thus ratcheted up their efforts to shape culture directly.

Whereas Chapter 7 describes a series of practical initiatives undertaken by scientific democrats seeking to reach the American people more effectively with their ethical and political message, this chapter demonstrates that scientific democrats proceeded not only by undertaking practical changes, but also by looking more closely at the object they hoped to change: the public mind itself. Going far beyond the problem of free will and determinism – and often taking the emergentist resolution of that problem for granted – these cultural scientists devoted considerable attention to the social role of academic ideas and the mechanisms of their diffusion. The spread of psychological and anthropological concepts through the human sciences, a familiar theme to historians of the 1920s, took much of its impetus from scientific democrats' desire to understand better the nature of public opinion. The political function of expertise served as a focal point for much of this work, as questions about the proper relationship between experts and citizens echoed through the disciplines. One conclusion, at least, produced virtual unanimity: the cultivation of the public mind could not be left to chance, or to the trickling down of academic ideas and values to the public through the personalities of college graduates. The scientific attitude would not spread through the population like a contagion; it would need to be carefully cultivated through hard, focused work.[1]

PARTICIPATION AND EXPERTISE

Many cultural scientists set out to bridge the gap between academic knowledge and the public by urging experts to engage closely and respectfully with their fellow citizens. No one advocated this approach as forcefully and as influentially as John Dewey. Dewey, more than any other single individual, set the tone for the cultural sciences in the 1920s. In a remarkable series of books published during that decade, Dewey fleshed out his conceptions of philosophy's public role (*Reconstruction in Philosophy*), social psychology (*Human Nature and Conduct*), metaphysics (*Experience and Nature*), and epistemology (*The Quest for Certainty*).[2] But *The Public and Its Problems* (1927), Dewey's most sustained endeavor in political theory, revealed most clearly how he expected scientific scholars to operate in the democracy of the future. Dewey took issue with a disillusioned Walter Lippmann's claim in the early 1920s that citizens wanted little more from their government than material prosperity, and that its provision could be left to expert-guided bureaucracies. Dewey argued that all human beings naturally sought to participate in collectively controlling their environment, and that no political system that failed to provide an outlet for that need could long stand. He also worried that attaching social scientists to government, even in an advisory role, would distance them from citizens and corrupt their findings.

[1] In the wake of World War I, Stuart Chase later claimed, reformers threw off the "ancient theory that the seed of truth, once planted, would surely sprout": *The Tyranny of Words* (New York: Harcourt, 1938), 6.
[2] Robert B. Westbrook, *John Dewey and American Democracy* (Ithaca: Cornell University Press, 1991), 275–373; Bruce Kuklick, *A History of Philosophy in America, 1720–2000* (New York: Oxford University Press, 2001), 185–197.

In his inimitable, meandering way, Dewey proposed that social scientists should avoid these pitfalls by working with artists and journalists to help citizens constitute themselves as what he idiosyncratically called "publics." Dewey defined a public as a group of people who found themselves affected, for good or for ill, by a certain form of social action, whether that action was organized as an institution or a mere matter of unorganized, private behavior. He then invoked the Progressive conception of organized politics as a mere reflex of the underlying culture, identifying the ideal state as simply the sum of all proposed restraints on social action put in place by all extant publics. Thus, citizens themselves would play the leading role in shaping the state, as Dewey imagined it. They would not exercise control directly, but they would carefully rein in their representatives, confining the latter strictly to agendas set by the prevailing publics. Yet social scientists would in turn guide the citizens, by disseminating knowledge of the causal relationships between actions and their consequences under current social conditions. These experts would help reveal the contours of latent publics and empower the members of each such public to recognize themselves as such, so that they could make the proper governmental changes. Armed with reliable social knowledge, the public as a whole would bring the existing state into line with its manifest needs and then watch for changes in the conditions of social action, modifying the state accordingly.[3]

The organization theorist Mary Parker Follett, who worked with Dewey and Herbert Croly in the postwar National Conference on the Christian Way of Life, called for an even closer relationship between experts and citizens in *The New State* (1918) and *Creative Experience* (1924). Dewey outlined a popularization model, asserting that "methods which make presentation of the results of inquiry arresting and weighty" would lead the public to "[buy] the facts" produced by scholars, and that keeping the latter out of positions of authority would preserve their ability to effectively address public questions. Follett, by contrast, presumed that ideas and practices spread only through personal contact. Somehow the intimacy of face-to-face interchange would need to be re-created in complex modern societies. Building on philosophical idealism and, later, the New Realist Edwin B. Holt's version of behaviorism, Follett combined a deliberative theory of politics with a decentralized conception of expertise and arrived at "a scientific technique of evolving the will of the people." She would organize society into small local groups, thereby embedding the "relation of neighbors one to another" in the very "substance of the state." In a democracy, she insisted, "the good and the wise" could make their mark only through "a subtle permeation," not direct coercion.[4]

[3] John Dewey, *The Public and Its Problems*, in *John Dewey: The Later Works, 1925–1953, vol. 2*, ed. Jo Ann Boydston (Carbondale: Southern Illinois University Press, 1988); Westbrook, *John Dewey and American Democracy*, esp. 300–318; Laura M. Westhoff, "The Popularization of Knowledge: John Dewey on Experts and American Democracy," *History of Education Quarterly* 35 (1995), 44.
[4] William Graebner, "The Unstable World of Benjamin Spock: Social Engineering in a Democratic Culture, 1917–1950," *Journal of American History* 67 (1980), 626; John Dewey, "Practical

Make their mark they would, however. Follett waxed mystical about the creative power of intersubjective relations, systematically drawing out the assumptions that George Herbert Mead, Ralph Barton Perry, and many others had used to dissolve apparent conflicts of interest in society. Thanks to the "fundamentally blessed relation between self and circumstance," Follett wrote, individuals could resolve their disagreements in an additive way by entering into truly integrative relations in which nothing was lost, and much gained, on both sides. Through the "revaluation" of combatants' interests, she explained, all could get what they sought without sacrifice. Follett proposed a similarly integrative relationship between experts and citizens. She called for "experience meetings" in which citizens would "unite our various experiences, one with the other and with the material provided by the expert." Meanwhile, she continued, experts participating in these groups would assimilate into their knowledge the less technical, more personal experiences of their fellow citizens. Follett's neighborhood groups would function officially as a branch of government, sending policy recommendations upward to the central coordinating agency of the state and getting back policy decisions based on the integration of all such recommendations. News and other information would also flow up and down the chain. Once expert judgments had been leavened by popular experience, and each citizen shown the "relation of his own activity to the satisfaction of his desires," Follett wrote, corporate power would dissolve and the true "World State" would arise.[5]

Many cultural scientists in the 1920s shared Follett's belief that integrative solutions gave much more to each party than did negotiated compromises. Some also joined her in proposing systems of face-to-face discussion groups to anchor what we would now call "deliberative democracy."[6] This emphasis on direct public participation, however, did not necessarily entail a rosy view of the average citizen's political capacity. The University of Kansas sociologist Seba Eldridge outlined a remarkably elitist version of deliberative politics,

Democracy," in *The Later Works, Volume 2*, 220; Mary Parker Follett, *The New State: Group Organization the Solution of Popular Government* (New York: Longmans, Green, 1918), 160, 157. On Follett's conceptions of expertise and democratic deliberation, see Kevin Mattson, *Creating a Democratic Public: The Struggle for Urban Participatory Democracy During the Progressive Era* (University Park: Pennsylvania State University Press, 1998), 87–104. Follett is best known today for her contributions to management theory and practice: e.g., James Hoopes, *Community Denied: The Wrong Turn of Pragmatic Liberalism* (Ithaca: Cornell University Press, 1998), 145–163.

[5] Follett, *The New State*, 7, 11; Follett, *Creative Experience* (New York: Longmans, Green, 1924), 131–132, 212–213.

[6] The sociologist Hornell Hart dubbed this the "technique of accommodation," arguing that it led to solutions based on the "highest common denominator": "The History of Social Thought: A Consensus of American Opinion," *Social Forces* 6, no. 2 (December 1927), 630, 632. For a sharp critique of the small-group model, see William Graebner, *The Engineering of Consent: Democracy and Authority in Twentieth-Century America* (Madison: University of Wisconsin Press, 1987); cf. Nikolas Rose, *Inventing Ourselves: Psychology, Power, and Personhood* (New York: Cambridge University Press, 1996), 136–148.

more in line with ancient republican models than with the era of mass suffrage. Like Follett, Eldridge called for the creation of a system of neighborhood discussion groups that would include experts and insisted that these experts could not determine the interests of the citizens. He, too, sought to solve the problem of expertise by instituting "continuous, intensive, systematic study of the citizen's problems, including relevant 'scientific' principles, by the citizen himself." Eldridge suggested expansively that such a pyramidal structure of discussion groups might even win the religious loyalties formerly directed toward the churches, but he was hardly idealistic about the capacities of the public or its current leaders. Even if existing religious and political divisions could be bridged, Eldridge suggested, a "very large percentage" of citizens might turn out to lack the intelligence or interest needed for constructive civic participation.[7]

Eldridge broached the possibility of a "functionally selected electorate," a small group whose distinctive characteristics qualified them for the tasks of the citizen. He further proposed to subject these individuals to a rigorous course of education. After completing a two-year, part-time "orientation course in the social sciences," followed by a similarly structured "general survey course in the major problems of the citizen – local, national and international," Eldridge's citizen-specialists would continue to spend at least two hours a day studying and solving social problems. Only under these conditions, he believed, would true self-government arise.[8]

The political proposals put forth by the University of Chicago political scientist Harold D. Lasswell likewise reveal that Follett's emphasis on the potency of integrative solutions to human problems hardly entailed a commitment to widespread popular participation. Lasswell stood about as far as possible from Follett on the structure of the expert-citizen relationship. Arguing that experts could determine the public's needs better than it could, Lasswell flatly denied that "social harmony depends upon discussion" or that "discussion depends upon the formal consultation of all those affected by social policies." But in *Psychopathology and Politics* (1930), his pioneering application of Freudianism to political science, Lasswell cited Follett approvingly on the need to resolve each social conflict through a "synthesis" rather than a "trade." Echoing Lippmann's early work, Lasswell defined the true political leader as "a discoverer of inclusive advantages" who, when faced with a social conflict, offered a "reinterpretation of the situation" that "renders the old line of battle, the older definition of interest, irrelevant."[9]

---

[7] Seba Eldridge, *The New Citizenship: A Study of American Politics* (New York: Thomas Y. Crowell, 1929), 258, 295, 332.

[8] *Ibid.*, 300, 262.

[9] Harold D. Lasswell, *Psychopathology and Politics* (Chicago: University of Chicago Press, 1930), 196, 48. On Lasswell's work, see Mark C. Smith, *Social Science in the Crucible: The American Debate Over Objectivity and Purpose, 1918–1941* (Durham: Duke University Press, 1994), 212–252; and William Ascher and Barbara Hirschfelder-Ascher, *Revitalizing Political Psychology: The Legacy of Harold D. Lasswell* (Mahwah, NJ: Lawrence Erlbaum, 2005).

Lasswell combined Follett's faith in the possibility of win-win solutions with a remarkably manipulative view of the political future, wherein cultural leaders would program citizens not to ask anything of the government so long as its experts satisfied their basic needs. Lasswell espoused a "preventive politics" that would keep the social "tension level" low through a "rigorous audit of the human consequences of prevailing political practices." He called on social scientists to use any means necessary to protect the public from itself. "Inform, cajole, bamboozle and seduce in the name of the public good," he urged them: first "find the good," then discover "how to make up the public mind to accept it." According to Lasswell, the future would go to those best able to employ the tools of modern propaganda. He fervently hoped that it would be those public-minded experts who recognized citizens' innate need "to be ruled by the truth about the conditions of harmonious social relations," yet hewed to the democratic ideal of a "public order in which coercion is at a minimum and reliance is upon persuasion." Like Luther Lee Bernard, he combined extravagant assertions of science's cultural authority with a steady confidence in the possibility of harmonious solutions to social conflicts and a deep aversion to formal, political means of social coordination.[10]

FOUNDATIONS AND VALUE-NEUTRALITY

Few cultural scientists in the 1920s could match the intensity of Lasswell's desire to expunge all traces of emotion from the political sphere. Many, however, pushed for a form of democratic politics grounded more firmly in academic expertise. "It is not important or desirable that the political scientists should govern the world," wrote Lasswell's University of Chicago mentor Charles E. Merriam, "but it is fundamental that they be heard before decisions are made on broad issues, and that the scientific spirit be found in the governors and the governed as well." Under the guidance of Merriam and Rockefeller executive Beardsley Ruml, the Social Science Research Council (SSRC) sought to advance this vision of scientific democracy by channeling tens of millions of Rockefeller dollars into social-scientific research in the 1920s.[11]

[10] Harold D. Lasswell, *Propaganda Technique in the World War* (London: K. Paul, Trench, Trubner and Co., 1927), 5; Lasswell, *Psychopathology and Politics*, 197, 173, 245, 310, 200, 198.

[11] Charles E. Merriam, *New Aspects of Politics* (Chicago: University of Chicago Press, 1925), 232; Dorothy Ross, *The Origins of American Social Science* (New York: Cambridge University Press, 1991), 400–402. Recent treatments of the Rockefeller programs typically draw on Barry D. Karl and Stanley N. Katz, "The American Private Philanthropic Foundation and the Public Sphere, 1890–1930," *Minerva* 19 (1981): 236–270 and several studies by Martin Bulmer, especially "The Early Institutional Establishment of Social Science Research: The Local Community Research Committee at the University of Chicago, 1923–30," *Minerva* 18 (1980): 51–110; "Support for Sociology in the 1920s: The Laura Spelman Rockefeller Memorial and the Beginnings of Modern, Large-Scale, Sociological Research in the University," *American Sociologist* 17, no. 4 (November 1982): 185–192; *The Chicago School of Sociology: Institutionalization, Diversity, and the Rise of Sociological Research* (Chicago: University of Chicago Press, 1984); and, with Joan Bulmer, "Philanthropy and Social Science in the 1920s: Beardsley Ruml and the Laura

As numerous historians have noted, this flood of SSRC money placed a premium on formally nonpolitical language in grant proposals and the ensuing publications. Rockefeller and other foundations faced intense political criticism for their apparent ability to subvert the democratic process by using huge amounts of cash to sway public opinion. They responded by claiming to be nonpartisan and favoring studies framed in a neutral language and unaccompanied by overtly ideological recommendations for action. Grant applicants, many already inclined toward a politically unaffiliated stance, quickly learned to adapt their language to this image of nonpolitical expertise.[12]

Detachment from party politics hardly meant disengagement from cultural questions, however. Rather than celebrating the status quo, SSRC-sponsored scholars simply tended to frame the cultural and political changes they sought – including, centrally, the move beyond party politics itself to a polity that employed scientific knowledge to guide social change – as imminent and inevitable, rather than merely valuable and desirable. Merriam exemplified the combination of empiricism and commitment to cultural change characteristic of SSRC-sponsored research.[13] On the one hand, his *New Aspects of Politics*, published in 1925, codified the call for a more empirical approach in the field of political science. Merriam also worked diligently to find sources of funding for empirical work, becoming a major institutional player. And he hewed to one of the key props of the value-neutral approach: a distinction, systematized by his Columbia teacher Frank Goodnow, between "politics" and "administration," the former involving the expression of the popular will and the latter – the means to political ends determined by the public – representing the province of experts. But Merriam did not seek to eliminate the realm of politics proper, after the fashion of Lasswell. A longtime Chicago alderman and a serious contender for mayor in 1911, Merriam combined the typical Progressive's distaste for backroom deals and blind partisanship ("jungle politics") with a profound commitment to political means of change. Like Dewey, he favored democratic participation for its ability to give expression to the "political possibilities in human nature." Merriam's call for neutrality stemmed from his belief that social scientists should stand ready to carry out any tasks assigned to them by a democratic public.[14]

Spelman Rockefeller Memorial, 1922–29," *Minerva* 19, no. 3 (1981): 347–407. See also Donald Fisher, "Philanthropic Foundations and the Social Sciences: A Response to Martin Bulmer," *Sociology* 18, no. 4 (1984): 580–587.

[12] Smith, *Social Science in the Crucible*, 27.

[13] Robert Adcock finds a pairing of "low-key empiricism with reformist sensibilities" between the wars: "Interpreting Behavioralism," in Adcock, Mark Bevir, and Shannon C. Stimson, eds., *Modern Political Science: Anglo-American Exchanges Since 1880* (Princeton: Princeton University Press, 2007), 187.

[14] Merriam, *New Aspects of Politics*, viii; Smith, *Social Science in the Crucible*, 101–105, 84. Smith notes that Merriam privately upheld the right of Tennesseans to ban evolution from their classrooms in 1925 (200). Recent treatments include Dorothy Ross, "Anglo-American Political Science, 1880–1920," in Adcock, Bevir, and Stimson, *Modern Political Science*, 29–30; Mark C. Smith, "A Tale of Two Charlies: Political Science, History, and Civic Reform, 1890–1940," in

Ideally, of course, the public would define these tasks only after it had digested the findings of experts. To ensure that outcome, Merriam hoped to form in each American child, through the schools, "the constructive type of political mind." All citizens, in his view, should learn to act as "intelligent critics, analysts, adaptors, and reconstructors of government," constantly assimilating and applying the findings of the social sciences in keeping with changing conditions. Like so many other scientific democrats of his day, Merriam expected the findings of social scientists to operate primarily in the public sphere. Their theories would reshape the beliefs and values of citizens and their representatives, who would then reshape governance directly.[15]

While Merriam was working to direct the attention of political scientists toward problems of administration and the attention of the public toward the value of expertise, sociologists were also taking advantage of the influx of Rockefeller money. Here, the new funding source gave a major boost to a group of scholars who argued that quantitative, statistical methods offered a uniquely powerful means of correcting for human biases. At Columbia and later Chicago, William F. Ogburn, a fervent advocate of both value-neutrality and quantification as a means of achieving it, helped Rockefeller dollars find their way into the pockets of like-minded sociologists, men such as Luther Lee Bernard and F. Stuart Chapin.[16]

Ogburn by no means rejected the cultural project of scientific democracy, however. He differed from Merriam, first, in downplaying the importance of direct participation by citizens in an ideal system of governance, and second, in presenting the emergence of such a system as a foregone conclusion. Ogburn's formulation of technological determinism and his concept of "cultural lag" became central to the mid-1920s discourse on the wider meaning of science. Much more than Merriam, Ogburn stressed the inevitability of political change – the impending arrival of a public culture tuned to what he saw as the innate logic of industrial technology. His claim that technological evolution proceeded independently of human choice and guidance represented a particularly sharp formulation of the Progressives' sense that the social organism

---

ibid., 118–136; and Michael T. Heaney, "The Chicago School That Never Was," *PS: Political Science & Politics* 40, no. 4 (October 2007): 753–758.

[15] Merriam, *New Aspects of Politics*, 208, 206. Merriam's book prefigured Lasswell's call for the creation of properly democratic personalities. "We can measure the patterns of conservative and radical," he wrote, "and, I venture to say, manufacture as many of either as we like by proper treatment, physical, psychical, or economic" (228).

[16] For the impact of SSRC money on sociology, see Stephen P. Turner and Jonathan H. Turner, *The Impossible Science: An Institutional Analysis of American Sociology* (Newbury Park: Sage Publications, 1990), 41–57. On Ogburn: Robert C. Bannister, *Sociology and Scientism: The American Quest for Objectivity, 1880–1940* (Chapel Hill: University of North Carolina Press, 1987), 161–187; Dennis Smith, *The Chicago School: A Liberal Critique of Capitalism* (Houndmills, Basingstoke: Macmillan Education, 1988), 167–183; and Rudi Volti, "Classics Revisited: William F. Ogburn, Social Change with Respect to Culture and Original Nature," *Technology and Culture* 45, no. 2 (April 2004): 396–405.

had entered a new industrial "environment" that worked, as it were, from the outside to reshape fundamental institutions.[17]

Ogburn's influential 1922 book *Social Change, With Respect to Culture and Original Nature* conceptually restricted the scope of political action by declaring it simply one among many means of adjusting to an overarching, self-moving technological process. He divided the cultural realm into a sphere of "material culture," comprising techniques, inventions, and the built environment, and a much slower-changing "adaptive culture," encompassing the whole array of social institutions – including the family, government, and much else – whose contours, he believed, depended in part or in whole on material conditions. Ogburn downplayed the needed adjustments in the adaptive culture, calling them "relatively minor changes." But they added up to a fundamental overhaul of virtually all aspects of public and private behavior. Ogburn called, not just for shorter working hours and new forms of education, but also for altered family structures, "modification of social codes," and "recognition of boundaries to selfishness" across the full range of institutions. Ogburn hoped his statistical analyses would guide a massive mobilization of reform energies by identifying the many points of maladjustment between industrial production and other culture elements.[18]

Later in the decade, this reformist brand of quantitative analysis fueled the emergence of the "attitude" concept in the new field of social psychology. Like Edward Tolman, attitude theorists postulated contingent but relatively stable internal structures – attitudes – that served as the link between stimuli and responses in human behavior. An attitude represented an outcropping of cultural experience within the mind of the individual, directing him or her to respond in a certain manner when faced with a given stimulus. As such, it bridged evaluation and cognition, reinforcing the view of Dewey and many other scientific democrats that knowledge and values were inextricably intertwined. Yet many social psychologists believed that they could, via survey instruments and statistical analysis, study such attitudes in a relatively neutral fashion.[19]

Not surprisingly, early attitude theorists such as L. L. Thurstone, Emory S. Bogardus, Read Bain, and Kimball Young concerned themselves above all with social and political attitudes. Few of the pioneers in the field cared what Americans thought about, say, a particular brand of soap. Instead, they sought to assess (and influence) the public mind on matters of particular concern to science-minded reformers: tolerance of ethnic and religious differences, perceptions of economic causation, openness to change in general, and above all, attitudes toward social institutions. Revealingly, one of the most important

[17] William F. Ogburn, *Social Change, With Respect to Culture and Original Nature* (New York: B. W. Huebsch, 1922), 364–365.

[18] Ogburn, *Social Change*, 364–365, 202–203, 212–213.

[19] Donald Fleming, "Attitude: The History of a Concept," *Perspectives in American History* 1 (1967): 287–365, esp. 358.

early studies in the field, by the radical psychologist Theodore M. Newcomb, explored the effects of collegiate education on students' political attitudes. Prominent targets of measurement by attitude theorists in the late 1920s and early 1930s included traits such as "fair-mindedness" and, inevitably, "the scientific attitude."[20]

Not all researchers claiming the ability to distance themselves from prevailing political ideologies employed statistical methods. Plenty of Rockefeller funds also flowed to figures who adopted a reportorial mode of analysis, seeking to document both the qualitative and quantitative characteristics of life in the modern world. For example, Rockefeller's Institute for Social and Religious Research originally underwrote Robert S. Lynd and Helen M. Lynd's massive, mixed-methods study of Muncie, Indiana, published in 1929 as *Middletown.* A striking feature of the book is the Lynds' concern with what Eldridge had called "the failure of the citizen." To be sure, they were hardly enthusiasts of what they described (in quotes) as "democratic government," which they equated with the absurd theory that the voters would automatically choose the best person for each public post. In fact, the Lynds found Middletown a political disaster area. Comparing the town with its 1890s predecessor, as was their usual method of indirect critique, they noted precipitous drops in voter turnout over the years. Moreover, even the most politically active citizens of Middletown fell far short of what their institutions, "framed to function in a simpler culture," required them to know about local affairs and candidates for office. As a result, Middletown's residents, while habitually complaining about matters of policy, no longer elected the "best citizens" as a matter of course. In politics, as in religion and education, the Lynds noted ruefully, "Middletown people are tending increasingly to delegate their interests" while concerning themselves primarily with economic activities.[21]

A similar mixed-methods approach characterized the numerous urban studies produced by the "Chicago School" of sociologists, led in the 1920s by Robert E. Park. A former journalist who had worked with John Dewey on the abortive "Thought News" project, Park imparted to his sociological students the reporter's combination of balanced coverage and an eye for the juicy or telling detail. His journalistic background may also explain the fact that he

---

[20] Theodore M. Newcomb, *Personality and Social Change: Attitude Formation in a Student Community* (New York: Holt, Rinehart and Winston, 1943); Goodwin B. Watson, *The Measurement of Fair-Mindedness* (New York: Teachers College, 1925); Victor H. Noll, "Measuring the Scientific Attitude," *Journal of Abnormal and Social Psychology* 30 (1935): 145–154.

[21] Eldridge, *The New Citizenship*, 5; Robert S. Lynd and Helen Merrell Lynd, *Middletown: A Study in American Culture* (New York: Harcourt, Brace and Co., 1929), 413–414, 417, 425, 421, 434. Recent treatments include Rita Caccamo, *Back to Middletown: Three Generations of Sociological Reflections* (Stanford: Stanford University Press, 2000), 1–54; and Sarah E. Igo, *The Averaged American: Surveys, Citizens, and the Making of a Mass Public* (Cambridge, MA: Harvard University Press, 2007), 23–102. On Lynd's career, see Smith, *Social Science in the Crucible*, 120–158.

took the city, rather than the nation, as his unit of analysis. Combining up-to-date concepts from population ecology and psychobiology with core tenets of Progressive sociology, Park identified the city as a whole and each of its constituent neighborhoods as a kind of emergent, the spontaneous product of statistical patterns in individual behavior.[22]

Challenging managerial conceptions of politics, Park identified cultural change rather than improved administration as the key to preserving democracy – indeed, social order itself – under the set of urban conditions that today's theorists mark with the term "modernity." In his seminal 1915 essay "The City," Park insisted that Sumner's mores could no longer perform their integrative task in a polyglot industrial city, wherein individuals and groups depended on one another for their existence but were "widely removed in sympathy and understanding." Absent the community of sentiment on which so many Progressives had pinned their hopes, Park concluded, the invisible tendrils of custom gave way among city dwellers to "fashion," "public opinion," and "positive law" as the dominant means of social control. To his mind, this was no loss. A more dedicated cosmopolitan than most of his contemporaries – not to mention his major donors – Park portrayed the city as a wondrous entity whose patterns of "vast unconscious co-operation" produced "the diversity of interests and tasks" needed to give every individual "the opportunity to choose his own vocation and develop his peculiar individual talents." Every element of city life, in fact, served "to select and emphasize individual differences," even as the overall process created a "community of interests" uniting the whole. The city dweller needed only to possess the openness to embrace this initially disconcerting mode of life.[23]

Park sought to refute the assumption of both moralists and planning enthusiasts that urban life could be controlled from above through the application of centralized force. He called the city "a product of the artless processes of nature and growth," produced by the constant interaction of individuals with the structures created by such interactions in the past. Park and other Chicago School thinkers sought, like all scientific investigators, to discover regularities in these social processes. Yet the most one could find, they argued, was statistical patterns in individual behavior. The only way to change these patterns was to

---

[22] Robert E. Park, "The City: Suggestions for the Investigation of Human Behavior in the City Environment," *American Journal of Sociology* 20 (1915), 585, 587. Cf. Roderick McKenzie, "The Ecological Approach to the Study of the Human Community," *American Journal of Sociology* 30 (1924), 289. Discussions of the school's conceptions of knowledge include Emanuel Gaziano, "Ecological Metaphors as Scientific Boundary Work: Innovation and Authority in Interwar Sociology and Biology," *American Journal of Sociology* 101 (1996): 874–907; and Tony Burns, "The Theoretical Underpinnings of Chicago Sociology in the 1920s and 30s," *Sociological Review* 44, no. 3 (1996): 474–494.

[23] Park, "The City," 597, 605, 610, 608, 585, 587. On Park's politics, see Ross, *The Origins of American Social Science*, 304–305; for his sleight of hand with donors, William H. McNeill, *Hutchins' University: A Memoir of the University of Chicago, 1929–1950* (Chicago: University of Chicago Press, 1991), 10.

change the behavior of individuals. In the view of Chicago School sociologists such as William I. Thomas and Florian Znaniecki, reformers should use the schools to foster a type of personality – "Creative Man" – that would embody the meta-level attitude of openness to new experience, and thus would be capable of reorganizing social institutions from within.[24]

Over in economics, however, interwar "institutionalists" embraced new modes of economic planning. Although they were centrally concerned with the social dimension of economic behavior, many of these theorists ignored the question of human scientists' relation to individual citizens – likely because they felt sure that planners could grasp and implement the common good. Building on the writings of the ethical economists and Thorstein Veblen, a biting critic of the cultural dominance of pecuniary values, institutionalists such as Columbia's Wesley C. Mitchell and Wisconsin's John R. Commons (a student of Richard T. Ely) firmly embedded economic behavior within a broader social and cultural matrix. Resolutely statistical, though not dogmatically so, Mitchell and Commons hoped to create the knowledge base for a state that managed economic affairs in order to smooth out business cycles. Other institutional economists, including Texas' Clarence E. Ayres and Yale's Walton H. Hamilton, took additional inspiration from Cooley, Dewey, and the like and attended directly to the task of establishing a foundation of public legitimacy for such a state. Still, the writings of the institutional economists, far more than most contributions to the human sciences in the 1920s, pointed toward the policy-oriented social science undertaken by New Deal agencies in the following decade. Of all the theoretical innovations of the 1920s, it was institutional economics that would be taken up most systematically by the Roosevelt administration in its search to right the nation's economic ship.[25]

[24] Park, "The City," 578; McKenzie, "The Ecological Approach to the Study of the Human Community," 289, 292; Fleming, "Attitude: The History of a Concept," 329–330.

[25] Helpful introductions to Veblen's work include Ross, *The Origins of American Social Science*, 204–216; John P. Diggins, *Thorstein Veblen: Theorist of the Leisure Class* (Princeton: Princeton University Press, 1999); Guglielmo Forges Davanzati, *Ethical Codes and Income Distribution: A Study of John Bates Clark and Thorstein Veblen* (New York: Routledge, 2006); Howard Brick, *Transcending Capitalism: Visions of a New Society in Modern American Thought* (Ithaca: Cornell University Press, 2006), 47–50, 63–64; and Rick Tilman, *Thorstein Veblen and the Enrichment of Evolutionary Naturalism* (Columbia: University of Missouri Press, 2007). On institutionalism's younger proponents, see Ross, *The Origins of American Social Science*, 371–386, 410–420; Smith, *Social Science in the Crucible*, 49–83; Ross, "The Many Lives of Institutionalism in American Social Science," *Polity* 28, no. 1 (Autumn 1995): 117–123; David A. Moss, *Socializing Security: Progressive-Era Economists and the Origins of American Social Policy* (Cambridge, MA: Harvard University Press, 1996); Malcolm Rutherford, "Understanding Institutional Economics: 1918–1929," *Journal of the History of Economic Thought* 22, no. 3 (2000): 277–308; Rutherford, "On the Economic Frontier: Walton Hamilton, Institutional Economics, and Education," *History of Political Economy* 35, no. 4 (Winter 2003): 611–653; Brick, *Transcending Capitalism*, 65–82; Mary O. Furner, "Structure and Virtue in United States Political Economy," *Journal of the History of Economic Thought* 27, no. 1 (March 2005): 13–39; Clive Lawson, "Ayres, Technology and Technical Objects," *Journal of Economic Issues* 43, no.

CULTURE AND GOVERNANCE

It would be easy to assume that, while scholars in the more established fields of political science, sociology, and economics debated the relationship between scientific experts and the public, those in the less overtly scientific fields of anthropology, history, and philosophy dealt with an entirely separate range of phenomena. In fact, however, questions, concepts, and theories circulated freely between these two wings of the interwar cultural sciences, as well as the third, biological and psychological wing discussed in Chapter 5. Although scientific democrats in anthropology history, and philosophy spoke in terms of the power of ideas and values in cultural change, rather than the social role of experts, and were less concerned to frame their studies in politically nonpartisan terms than were scientific democrats in the other human sciences, they shared a broad political vision with their counterparts and faced the same strategic dilemmas: How could the ideas developed by cultural scientists – especially ideas about the proper cultural and political response to industrialization – make their way into the public mind? What were the sources of resistance to new ideas? What resources could scholars employ to overcome such resistance? Across these fields, the Boasian concept of culture and various equivalents drew substantial scholarly attention in the 1920s, because these referred to what many scientific democrats identified as the locus of resistance to their program.

The writings of Boas and his students neatly explained why social beliefs persisted so strongly after the economic conditions supporting them had vanished – why, for example, Americans still inclined toward rugged individualism and *laissez-faire* decades after the closing of the frontier and the emergence of massive industrial corporations seemed to have rendered those principles destructive of freedom. "Culture" provided a name and an explanation for the drag on social progress, the missing factor that revealed why public opinion lagged so far behind the era's reform proposals: the prevailing culture naturalized or reified socially contingent preferences, projecting the dominant values and beliefs of a bygone era onto the very structure of reality itself. A fundamental tenet of Boasian anthropology – one that resembled William James' view of learning – held that people were fundamentally conservative in their interpretive habits, viewing new phenomena through deeply ingrained perceptual molds.[26]

---

3 (2009): 641–659; and Rutherford, "Towards a History of American Institutional Economics," *Journal of Economic Issues* 43, no. 2 (June 2009). Some interpreters consider Cooley part of the founding generation of institutional economists: e.g., Yuval P. Yonay, *The Struggle Over the Soul of Economics: Institutionalist and Neoclassical Economists in America Between the Wars* (Princeton: Princeton University Press, 1998), 52. Cf. Rick Tilman and Terry Knapp, "John Dewey's Unknown Critique of Marginal Utility Doctrine: Instrumentalism, Motivation, and Values," *Journal of the History of the Behavioral Sciences* 35, no. 4 (1999): 391–408.
[26] George W. Stocking Jr., "Introduction: The Basic Assumptions of Boasian Anthropology," in *The Shaping of American Anthropology, 1883–1911: A Franz Boas Reader*, ed. Stocking (New York: Basic Books, 1974), 6. On the Boasians in the 1920s, see also Sydel Silverman, "The Boasians and the Invention of Cultural Anthropology," in Fredrik Barth et al., *One Discipline, Four Ways:*

However, the proliferating versions of cultural anthropology that appeared in the 1920s also highlighted the ability of science to break through the cake of custom. Within the broad frame of Boasian cultural relativism, theorists employed many different assumptions about what human beings needed and sought to achieve, how cultures operated to frustrate or facilitate those needs and impulses, and how scientific scholars might conceivably change the existing American culture. Whereas Boas stressed the ability of a culture to facilitate or block the pursuit of material needs by individuals engaged in a struggle with nature, Edward Sapir focused instead on the interaction between a culture and the expressive needs of individuals. Meanwhile, another Boasian, Berkeley's Robert H. Lowie, presented society as the locus of pitched material conflicts between individuals, in which the winners used culture elements to rationalize their dominance and the rest sought merely to rationalize their disadvantaged position. Yet Lowie, like so many other scholars of the era, found in science a ray of hope for the modern world.

Boas' popular 1928 book *Anthropology and Modern Life* identified the present as a moment of social transition – and thus cultural lag – in which scientific intellectuals needed to actively lead the public. Normally, he explained, intellectuals were the most conservative of social groups, by a wide margin. Segregated from the wider culture by their privileged status, steeped in the ideas of past generations through their long training, and no more intelligent or free from convention than the lowliest worker, intellectuals generally took their cues from their "class interest," whereas the masses instinctively preserved the "moral obligation of altruistic behavior" that Boas found underlying all cultures. That obligation, he argued, took on an increasingly universalistic form in the contemporary world, because of the contact of cultures and advances in knowledge. In fact, Boas contended that all of history represented a movement toward "a *human* ideal as opposed to a *national* ideal." At every point in that process, he emphasized, "the desires of the masses are in a wider sense human than those of the classes," who, under typical circumstances, merely enforced the existing forms of society.[27]

Yet, Boas continued, action was a matter of knowledge and its associated habits, not just ethical ideals. And there were moments, such as the present, when intellectuals temporarily attained greater freedom of thought than the masses, as part of the intertwined processes of scientific advancement, social progress, and cultural upheaval. At these times, intellectuals needed to transcend their class interest and show ordinary citizens how to implement their altruistic ideal, even if this meant directly challenging the prevailing beliefs of the masses. It was the task of such transitional intellectuals, he wrote, "not only to free ourselves of traditional prejudice, but also to search in the heritage of the past for what is useful and right, and to endeavor to free the mind of

British, German, French, and American Anthropology (Chicago: University of Chicago Press, 2005); and Brick, *Transcending Capitalism*, 88–98.
[27] Franz Boas, *Anthropology and Modern Life* (New York: Norton, 1928), 191, 195.

future generations so that they may not cling to our mistakes, but may be ready to correct them." In Boas' view, scientific thinkers needed to call attention to the contingency of the prevailing culture and also indicate the path beyond it, by tracing the outline of a new culture better adapted to the needs of the present.[28]

What were those needs? As in his earlier writings, Boas identified a successful culture as one that enabled its members to ensure their material welfare under the present environmental conditions, while retaining sufficient flexibility to adapt to future changes in the environment or to adopt from other cultures techniques better suited to extant conditions. In his 1928 book, he described the Eskimo as utterly unable to take advantage of cultural innovations, due solely to the "automatic reactions" embedded in their highly rigid culture rather than direct institutional constraints. Lacking any center of formal authority, Boas explained, the Eskimo were nevertheless shackled to prevailing practices and thus were profoundly unfree: "New inventions are rare and the whole industrial life of the people runs in traditional channels." Throughout the book, Boas stressed the need for scientists to help increase the flexibility of existing cultures, both to enhance material well-being and to give the altruistic sentiment a universal object, in keeping with the growing interdependence of the peoples of the world. With the help of a cosmopolitan embrace of other cultures and an ever-expanding body of scientific knowledge, humanity could steadily widen the circle of ethical obligation to include the entire globe, while making available to all peoples the innovative material techniques developed by the modern, scientific societies.[29]

Whereas Boas saw the promise of worldwide prosperity in modern modes of production, his student Edward Sapir worried about industrial society's deadening effects on individual creativity. A specialist in American Indian languages who taught at Chicago in the late 1920s and Yale in the 1930s, and a widely published poet during the early part of that period as well, Sapir distinguished between "genuine" and "spurious" cultures. Genuine cultures, he explained, drew out the creative impulses of aesthetically and ethically sensitive individuals, whereas spurious cultures discouraged their expression. Sapir held that all people possessed an innate urge to sustain formal standards of cultural authority and live in accordance with them. Freedom could not mean the absence of cultural standards, an impossible condition for any human group. Instead, a genuine culture provided inspiring, inspiriting standards that individuals experienced as a "gentle embrace" rather than "a burden and a tyranny." On this score, Sapir favorably contrasted the situation of "[t]he American Indian who solves the economic problem with salmon-spear and rabbit-snare" to that of the telephone operators and other alienated laborers of his own time. He sought a culture in which each individual's mode of employment would "directly satisfy his own creative and emotional impulses." This would allow modern societies

[28] *Ibid.*, 195–197.
[29] *Ibid.*, 181–182, 220.

to produce cultural leadership of the caliber found in the great pre-industrial civilizations.[30]

For his part, Lowie painted a portrait of contemporary industrial life even darker than Sapir's image of widespread creative frustration. In fact, he moved beyond the boundaries of scientific democracy altogether, declaring that science could offer no moral lessons whatsoever. In the widely read *Primitive Society* (1920) and especially *Are We Civilized?* (1929), Lowie voiced a skeptical, cosmopolitan realism that described society as the scene of sharp clashes between individuals and groups who mobilized intellectual beliefs as part of the struggle. "Life is grim," he stated bluntly. "The savage who believes in sinister forces lowering on every side expresses everyday experience more accurately than the philosophers of optimism."[31]

A maverick socialist raised in New York by German-speaking émigrés from the Enlightened Jewish community of Vienna, Lowie stood on the margins of American academic culture. Like Émile Durkheim, he identified all moral claims as nothing more than reflections of the irreducible fact of social integration itself. Thus, whereas Sapir contrasted genuine to spurious cultures, Lowie saw in the course of history only "a transfer of power from one mystically sanctified source of authority to another," as "from a church to a book, from a book to a state, or to an intangible public opinion." He identified the individuals who undertook such struggles to win and then sustain a position of cultural dominance as utterly ruthless in pursuing their "supposed interests" and "imposing their own ideals on others." At each stage, meanwhile, the oppressed masses, desiring only "peace, security, and relaxation," found solace in a system of theodicy – a "complete world-view" providing an absolute, incontrovertible explanation of why bad things happened to good people. On all sides, in Lowie's view, reason served primarily to rationalize the hard fact of human domination.[32]

Lowie's skepticism about the potency of human reason extended both to politics, where he feared large-scale programs of social engineering in the name of the common good, and to science. "Science is a part of culture," he summarized. "It is not something apart floating in an ether of pure reason," but rather

---

[30] Edward Sapir, "Culture, Genuine and Spurious," *American Journal of Sociology* 29 (January 1924): 401–429. On Sapir, see Regna Darnell, *Edward Sapir: Linguist, Anthropologist, Humanist* (Berkeley: University of California Press, 1990); Darnell, *Invisible Genealogies: A History of Americanist Anthropology* (Lincoln: University of Nebraska Press, 2001), 105–135; and Richard Handler, *Critics Against Culture: Anthropological Observers of Mass Society* (Madison: University of Wisconsin Press, 2005), 49–122.

[31] Robert H. Lowie, *Are We Civilized? Human Culture in Perspective* (New York: Harcourt, Brace, 1929), 291. On Lowie's life and work, see Robert F. Murphy, *Robert H. Lowie* (New York: Columbia University Press, 1972); Mark Pittenger, *American Socialists and Evolutionary Thought, 1870–1920* (Madison: University of Wisconsin Press, 1993), 233–238; and Regna Darnell, "Robert H. Lowie," in *Celebrating a Century of the American Anthropological Association: Presidential Portraits* (Arlington: American Anthropological Association, 2002), 69–72.

[32] Lowie, *Are We Civilized?* 284; Lowie, *Primitive Society* (New York: Boni and Liveright, 1920), 440.

a product of the irrational drives that fueled all other social pursuits. Still, Lowie believed that scientists did make genuine discoveries, if only through "the quite unreasonable clash of conflicting temperaments," and typically in unintended directions. Although he built primarily on the positivism of Ernst Mach, Lowie's understanding of science sounded much like that of Charles S. Peirce. He argued that competition among irrational, self-seeking individuals produced spontaneous order in the sciences. A similar process operated in the technological realm, generating material advances out of the clash of interests. In the end, Lowie wrote, while men were still savages, "the word loses its sting when we recall what savages have achieved." Although civilization was everywhere a "planless hodgepodge," a "thing of shreds and patches" with no coherent design, Lowie's modern savages still managed to eke out advances in material technique and abstract knowledge. For him, as for Boas – though less so for Sapir – scientists and technologists represented the leading force for cultural change, even if Lowie thought they produced such change through self-interested behavior.[33]

In the discipline of history, the "New Historians" James Harvey Robinson and Charles A. Beard combined their own skepticism about human beings in the aggregate with the classic nineteenth-century image of a small, selfless group of scientific thinkers dragging the unthinking mass of humanity down the path of progress. Both men left Columbia during World War I and wrote as independent scholars in the 1920s, having long worked to shift the field's focus from the national politics and military campaigns that dominated traditional histories to deeper structural forces: economic shifts, broad cultural changes, and the inexorable progress of science and technology. As Robinson noted in 1921, the New Historians sought to replace the "conventional chronicle of remote and irrelevant events" with "a study of how man has come to be as he is and to believe as he does." Such histories, they believed, would give the general public causal knowledge it could use to transform the political situation.[34]

According to Robinson, the most important fact was that the public's conception of human relations lagged a good 2,000 years behind its grasp of

[33] Lowie, *Are We Civilized?*, 270, 274, 280, 294–295; Lowie, *Primitive Society*, 441.
[34] James Harvey Robinson, *The Mind in the Making: The Relation of Intelligence to Social Reform* (New York: Harper, 1921), 5–6. On the New Historians, see Ellen Nore, *Charles A. Beard: An Intellectual Biography* (Carbondale: Southern Illinois University Press, 1983); Peter Novick, *That Noble Dream: The "Objectivity Question" and the American Historical Profession* (New York: Cambridge University Press, 1988), esp. 89–105; Clyde W. Barrow, *More Than a Historian: The Political and Economic Thought of Charles A. Beard* (New Brunswick, NJ: Transaction, 2000); Ellen F. Fitzpatrick, *History's Memory: Writing America's Past, 1880–1980* (Cambridge: Harvard University Press, 2002), 52–53, 69–75 and Smith, "A Tale of Two Charlies." Ernst A. Breisach views the pair in an international context: *American Progressive History: An Experiment in Modernization* (Chicago: University of Chicago Press, 1993). On Beard's frequently overlooked writings on technology, see John M. Jordan, *Machine-Age Ideology: Social Engineering and American Liberalism, 1911–1939* (Chapel Hill: University of North Carolina Press, 1994), 214–221; and Eric Schatzberg, "'Technik' Comes to America: Changing Meanings of 'Technology' Before 1930," *Technology & Culture* 47, no. 3 (July 2006), 508–511.

natural science. Drawing heavily on Dewey and Veblen, he argued that "almost all that passed for social science, political economy, politics, and ethics in the past may be brushed aside by future generations as mainly rationalizing." Yet the knowledge base for vast institutional improvements now stood at the ready in the social sciences, Robinson insisted. "If the majority of influential persons held the opinions and occupied the point of view that a few rather uninfluential people now do," he contended, there would be "no likelihood of another great war; the whole problem of 'labor and capital' would be transformed and attenuated; [and] national arrogance, race animosity, political corruption, and inefficiency would all be reduced below the danger point."[35]

Like Dewey in philosophy and Boas in anthropology, Robinson sought to employ his particular disciplinary expertise to indirectly but powerfully advance the cause of scientific democracy. These Columbia-based icons of post-Progressive thought hoped to break Americans out of a cultural impasse that, in their view, scotched all efforts by political scientists, sociologists, and economists to create more cooperative, humanly satisfying institutions – namely, the prevalence of a belief that social truths and values were fixed for all time, rather than pegged to an ever-changing environment and subject to empirically grounded criticism on the basis of their fit with that environment. Robinson thus argued that Americans would need to recognize the fungibility and the empirical character of social ethics before they would embrace the new political forms latent in industrial enterprise. He added a strategic consideration as well, suggesting that this indirect form of criticism would forestall the repressive reaction by conservatives that would inevitably greet a direct assault by human scientists on prevailing institutions. "It is premature to advocate any wide sweeping reconstruction of the social order, although experiments and suggestions should not be discouraged," Robinson wrote. "What we need first is a change of heart and a chastened mood which will permit an ever increasing number of people to see things as they are, in the light of what they have been and what they might be." Historians could nurture the needed outlook by joining anthropologists in stressing the contingency of thought and demonstrating that widely held social beliefs had concrete, material origins rather than reflecting a timeless realm of Platonic essences. When ordinary citizens saw social institutions and principles as "half-solved problems" rather than sacred cows, Robinson averred, the needed "moral and economic regeneration" would begin. Like Dewey and Boas, Robinson hoped to clear the way for thinkers in the more established human sciences to create, through fully democratic processes of persuasion, the base of cultural legitimacy for a new social and political order.[36]

Beard, who straddled the disciplines of history and political science, likewise directly attacked the prevailing views of modern society in the 1920s. He first gained fame for his controversial 1913 book *An Economic Interpretation of*

[35] Robinson, *The Mind in the Making*, 47, 157, 3.
[36] *Ibid.*, 217, 220, 4.

the Constitution of the United States, which advanced the cause of Progressive politics – and helped codify the Progressive school of historiography – by reading the contemporary struggle between "the people" and "the interests" back into the founding period. In his writings of the late 1920s, however, he turned his attention to the present day and, like Ogburn, described historical change as a combination of inexorable technological progress and halting advances by other cultural elements, which lagged behind technology to a greater or lesser degree. Technology, he wrote, driven by unquenchable curiosity and an "acquisitive passion" that would long outlive capitalism, "marches in seven-league boots from one ruthless, revolutionary conquest to another." Yet Beard portrayed academic scholarship as a crucial instrument for minimizing cultural lag, if the formidable obstacles to creative thought could be removed.[37]

As interpreters have noted, Beard changed his rhetorical strategy as well as his thematic focus and his account of the dynamics of historical change in the 1920s. He had adopted a detached, scholarly tone in his 1913 book and other writings from the era, claiming only to give his readers the facts of the matter. After World War I, however, Beard concluded that Progressive initiatives had failed to dent the shell of popular conservatism and that intellectuals needed to take a more interventionist approach to public opinion. He now sought to change American public culture by presenting facts about the present rather than the past and by accompanying these facts with explicitly normative conclusions about their policy implications and wider cultural meaning.[38]

Philosophers as well as anthropologists and historians could advance the cause of building a more interdependent, cooperative society. But the philosophers of the 1920s did not see eye to eye on the proper means of doing so, even though shared ethical values tended to knit them together more closely than might otherwise have been the case. One group of interwar philosophers, led by Dewey and heavily populated by his students, sought to bolster the modern social sciences both indirectly, by targeting belief in the absolute character of social ethics, and directly, by articulating what they took to be the system of social ethics appropriate to the industrial era. A second group of naturalists, pragmatists, and realists – most trained or influenced by William James – aligned themselves instead with the biological-psychological wing of scientific democracy, addressing the questions about natural science and individual freedom discussed in Chapter 5. And a third group of idealists took up the parallel campaign, led by Harvard's Josiah Royce during the Progressive Era, to develop idealist accounts of metaphysics and epistemology consistent with both the natural sciences and human purposes.

[37] Charles A. Beard, "Time, Technology, and the Creative Spirit in Political Science," *American Political Science Review* 21, no. 1 (February 1927), 5.
[38] Raymond Seidelman, with the assistance of Edward J. Harpham, *Disenchanted Realists: Political Science and the American Crisis, 1884–1984* (Albany: State University of New York Press, 1985), 95–96.

Thus, interwar philosophers divided, with some seeking to aid social scientists in their cultural initiatives and others attempting to rescue the natural sciences from overly zealous defenders who seemed to deny the reality of purposive, ideal-driven behavior. Whereas those in Dewey's orbit sought to link philosophy to the social sciences, philosophers following the leads of James or Royce largely ignored those disciplines, seeking to draw social meanings out of modern science without attending to empirical analyses of recent economic and social changes. Although the problems associated with industrialization and consumerism clearly formed the backdrop to interwar philosophical discussions, few outside of Dewey's circle identified the accounts of such phenomena developed by their social-scientific contemporaries as important contributions to the solution of social ills. Instead, they followed the philosophically inclined natural scientists of the time in seeking to foster cultural change by proving that free will, purposive behavior, and altruism had firm foundations in nature.

On one side of this implicit divide, the group of "Columbia naturalists" led by Dewey saw the most consequential element of cultural lag in modern America as the stubborn allegiance of the commercial middle class to a faulty social-scientific theory, namely orthodox political economy. They believed that the nation would realize its core values – by allowing workers and other downtrodden groups to share in the benefits of industrial production – only when the culturally dominant portions of the middle class embraced the basic outlines of ethical economics, with its understanding of economic principles as contingent and changing. In particular, Dewey and younger naturalists such as John Herman Randall Jr. and Herbert W. Schneider targeted what they viewed as the two key props of *laissez-faire* theory: a Lockean, empiricist conception of knowing as a matter of generalizing from discrete sensory experiences and a transcendental, Kantian conception of ethics as an entirely non-empirical affair. They held that each of these positions rendered prevailing principles of social ethics immune from empirical criticism and revision – empiricism by isolating the mind from the world and ethical transcendentalism by isolating evaluative judgments from factual ones. The Columbia naturalists identified these two tenets as the leading ideological weapons of the commercial middle class. They also saw a lingering ethical transcendentalism as the theoretical weak spot that prevented liberal Protestant critics from striking at the ideological core of bourgeois rule.[39]

---

[39] On Dewey's critique of empiricism, see Alan Ryan, *John Dewey and the High Tide of American Liberalism* (New York: Norton, 1995), 126. "New York naturalism" is summarized in Kuklick, *A History of Philosophy in America*, 190–196; see also William M. Shea, *The Naturalists and the Supernatural* (Macon, GA: Mercer University Press, 1984); George Cotkin, "Middle-Ground Pragmatists: The Popularization of Philosophy in American Culture," *Journal of the History of Ideas* 55 (1994): 283–302; John Ryder, ed., *American Philosophic Naturalism in the Twentieth Century* (Amherst, NY: Prometheus, 1994); Victorino Tejera, *American Modern, The Path Not Taken: Aesthetics, Metaphysics, and Intellectual History in Classic American Philosophy* (Lanham, MD: Rowman & Littlefield, 1996); and John P. Anton, *American Naturalism and Greek Philosophy* (Amherst, NY: Prometheus, 2005).

Although it took the form of an empirical intellectual history, Randall's *The Making of the Modern Mind* (1926) neatly captured additional elements of the political critique that underlay Deweyan naturalism. Widely used as a textbook for decades, Randall's narrative described mechanistic versions of social science as outmoded vestiges of the commercial middle class's struggle for political dominance. The ideologists of the Enlightenment era had not been entirely wrong, he allowed, particularly insofar as they turned toward a pragmatic understanding of social ethics. Most importantly, the utilitarian Jeremy Bentham had correctly recognized that "what is socially useful can stand on its own feet, without the additional support of roots in the natural and divine order." But Randall insisted that a naturalistic politics could work only if it recognized the enormous variability of persons and the influence of social groups on individual behavior – only if it replaced Bentham's calculating machine with modern psychology's "infinitely more complex creature of impulse and passion and emotional preference."[40]

Randall and the other Deweyan naturalists identified the rational, self-maximizing *homo economicus* of orthodox economics as the long cultural arm of the once-revolutionary bourgeoisie, keeping workers down materially while preventing them from staking a claim to the non-material goods monopolized by the wealthier members of society. By contrast, they held that the more nuanced, less rationalistic conception of human motivation characteristic of contemporary social psychology – along with the contextual, value-laden understanding of science that the Deweyans saw implicit in that framework – represented the political basis for a fully inclusive, egalitarian polity. This polity would fulfill the whole range of needs for all citizens, in part by allowing them to exercise their innate impulse to participate in controlling their own affairs. Human nature, said Randall, inevitably rejected the "benevolent despotism" of "'liberty' handed down from above." He mobilized the authority of modern psychology against what the Deweyans took to be the main tenets of the nineteenth-century ideology of science: the image of scientific knowledge as the product of morally and socially detached individuals and the accompanying technocratic dream of a "rational and scientific politics administered by an expert."[41]

Other philosophical naturalists instead viewed the contest over the status of human thought and values in the natural sciences as the theoretical bottleneck preventing the emergence of a more cooperative social order. Thus, they focused their attention on the metaphysical and epistemological implications of evolutionary biology and modern physics. As we saw in Chapter 5, Roy

[40] John Herman Randall Jr., *The Making of the Modern Mind: A Survey of the Intellectual Background of the Present Age* (Boston: Houghton Mifflin, 1926), 341, 504. One commentator has credited Randall with introducing American historians to the concept of "the Enlightenment" as a distinct period: R. R. Palmer, "A Century of French History in America," *French Historical Studies* 14, no. 2 (1985), 169–170.
[41] Randall, *The Making of the Modern Mind*, 339.

Wood Sellars developed a theory of emergence that seemed to reconcile the existence of mental and cultural phenomena with the materialistic bent of evolutionary theory. Meanwhile, Wisconsin's Max C. Otto concerned himself with the place of human purposes in a scientific outlook. In a pair of popular 1920s books, Otto reassured anxious moderns that their longstanding ideals held up perfectly well in the new world of the biologists. Like Sellars, Otto sought to authorize the application of scientific reasoning to human affairs by combating the popular perception that science entailed determinism, not by taking on Dewey's target, the belief that social structures and ideals could not change.[42]

Several other naturalists, pragmatists, and realists likewise carried forward the view of the mind as a natural fact that underpinned Dewey's work and had also been developed by James, George Santayana, and Frederick J. E. Woodbridge. At Harvard, a number of younger figures carried the torch for a broad, ethically engaged naturalism after James died in 1910 and Santayana returned to Europe in 1912. In *Mind and the World-Order* (1929), Josiah Royce's student C. I. Lewis brought a Deweyan conception of science into dialogue with the studies of symbolism described in Chapter 8 by highlighting the centrality of conceptual abstractions in empirical inquiry. Coming to epistemology from the side of symbolic logic, then just beginning to make its presence felt in American philosophical life, Lewis tried to refine and strengthen Dewey's claim that, although knowledge emerged out of practical activities, it issued in abstractions rather than particular or concrete statements. Lewis stressed the ability of science's "objective and impersonal" concepts to foster coordinated action, if not outright agreement, among subjective individuals. He drew out pragmatism's implication that acting together – after all, concepts were merely programs for future action – did not require uniformity of motives, emotions, or underlying interpretations. By describing scientific abstractions as sites and expressions of what we might call today "overlapping consensus," Lewis suggested that diverse individuals could cooperate socially without agreeing on why it was a good idea to do so. Like Dewey's formulation of pragmatism, Lewis' version reconciled natural science with human purposes by describing scientific knowledge itself as a set of tools that individuals used to pursue their ends.[43]

[42] Kuklick, *A History of Philosophy in America*, 170–171, 211–214; Roy Wood Sellars, *Evolutionary Naturalism* (Chicago: Open Court Publishing, 1922); Max C. Otto, *Natural Laws and Human Hopes* (New York: H. Holt, 1926), esp. 48–49; Otto, *Things and Ideals: Essays in Functional Philosophy* (New York: H. Holt, 1924). See also Edwin A. Burtt, *The Metaphysical Foundations of Modern Physical Science: A Historical and Critical Essay* (New York: Harcourt, Brace, 1925) and Durant Drake, *Mind and Its Place in Nature* (New York: Macmillan, 1925).

[43] C. I. Lewis, *Mind and the World-Order: Outline of a Theory of Knowledge* (New York: C. Scribner's Sons, 1929), 70; Bruce Kuklick, *The Rise of American Philosophy, Cambridge, Massachusetts, 1860–1930* (New Haven: Yale University Press, 1977), 533–562; Kuklick, *A History of Philosophy in America*, 214–220. See also Murray G. Murphey, *C. I. Lewis: The Last Great Pragmatist* (Albany: State University of New York Press, 2005); and Sandra B. Rosenthal, *C. I. Lewis in Focus: The Pulse of Pragmatism* (Bloomington: Indiana University Press, 2007).

Meanwhile, Lewis' Harvard colleague Ralph Barton Perry crafted his massive, influential *General Theory of Value* (1926). There, Perry defined a value as "any object of any interest," rendering the concept of value a universal currency with which to relate and compare the whole range of ethical, aesthetic, economic, and religious impulses. Perry believed that correctly grasping and applying the concept of value would facilitate the resolution of the three great social-philosophical disputes racking modern society in the wake of natural science's rise: the conflict between capital and labor; the struggle between those who saw the key to social progress in technological innovation and those who looked to the arts instead; and the debate over religion's role in the modern world. Combining his definition of value with the standard of "inclusiveness" described in Chapter 5, Perry re-grounded a set of traditionally Christian ethical principles and suggested that social integration was inherently good for – and naturally desired by – self-interested individuals. A generation older than Sellars, Otto, and Lewis, Perry had already grappled with the natural status of the human mind in his earlier, epistemological contributions to the controversy over the New Realism. Now, he worked his way back toward the concern with social ethics that drove Dewey and his students.[44]

Most philosophical idealists, by contrast, sought to validate nonempirical, religious ways of knowing as the needed ethical complement to scientific facts. At Harvard, William Ernest Hocking carried on in the vein of his mentor Royce, referencing individual purposes to an overarching "invisible unity" that infused all of reality with a meaning accessible through "the mystic's worship." Meanwhile, Princeton's Warner Fite and a number of other philosophers and theologians developed a more individualistic, pluralistic formulation of idealism, rooted in the work of the nineteenth-century Boston University philosopher Borden Parker Bowne, that went by the name of "personalism." These figures stood beyond the boundaries of scientific democracy, sympathetic though they were to the cognitive claims of the modern sciences in their proper spheres.[45]

But the Cambridge-trained Alfred North Whitehead, another key figure at Harvard between the wars, articulated a third, highly idiosyncratic, system of philosophical idealism that interfaced more easily with versions of scientific democracy. A pioneer in the development of analytic philosophy during his days in Britain, Whitehead had shifted his focus to metaphysics by the time he arrived in the United States in 1924. *Science and the Modern World* (1925) announced his intention to articulate a worldview compatible with modern

[44] Ralph Barton Perry, *General Theory of Value: Its Meaning and Basic Principles Construed in Terms of Interest* (New York: Longmans, Green, 1926), 115, 655, 7. Kuklick discusses this book in *The Rise of American Philosophy*, 508–515.

[45] William Ernest Hocking, *The Self: Its Body and Freedom* (New Haven: Yale University Press, 1928), 171, 173; Kuklick, *The Rise of American Philosophy*, 481–495; Warner Fite, *Moral Philosophy: The Critical View of Life* (New York: L. MacVeagh, 1925); Gary J. Dorrien, "Making Liberal Theology Metaphysical: Personalist Idealism as a Theological School," *American Journal of Theology & Philosophy* 24, no. 3 (September 2003): 214–244.

biology and physics, which now spoke of interacting organisms and fields rather than discrete particles in motion. He insisted that the world was constituted by a vast range of natural existents standing in multiple, overlapping relations of mutual influence, rather than a series of impermeable billiard-ball atoms caroming through empty space.[46]

Whitehead shared Randall's goal of using the history of ideas to advance a particular reading of the beliefs and values appropriate to the contemporary age. He proceeded by way of a historical account of changing views of nature, describing a kind of cultural lag between the metaphysical conceptions that, he averred, had implicitly guided research scientists since the late nineteenth century and those currently bandied about by philosophers. But Whitehead centered his account on biological and physical theories rather than views of man and society as such. Nor did he, after Randall's fashion, address the specific political implications of his narrative. Instead, Whitehead hoped to influence the culture of his day by establishing the more general claim that sociality, not individuality, was the primary characteristic of reality.

In the end, Whitehead and other idealists stood at arm's length from the main current of scientific democracy by the late 1920s. They were too committed to philosophical and religious modes of knowledge to condone the increasingly naturalistic tenor of the other human sciences. On this score, it is revealing that the nascent humanist movement of the late 1920s and 1930s enlisted pragmatists, naturalists, and realists, but no idealists. Far more self-consciously than most of their peers in the human sciences, philosophical advocates of humanism sought to establish a quasi-religious framework of meaning and practices that could perform the cultural functions hitherto carried out by traditional forms of theism.[47] This, however, was precisely the role that philosophical idealists had long hoped to fill themselves. The emergence of the humanist movement reflected the end of idealism's dominance over American philosophy. By the 1920s, most idealists identified themselves closely with the Protestant churches. To take just one example, as Dewey, Randall, and Sellars were helping to craft the *Humanist Manifesto* of 1933, which touted a purely naturalistic ethics, Hocking was putting the finishing touches on his two-year

---

[46] Alfred North Whitehead, *Science and the Modern World* (New York: Macmillan, 1925). Cf. Roy Wood Sellars, "Why Naturalism and Not Materialism?" *Philosophical Review* 36 (1927), 218–219, 221–223. Westbrook finds similar metaphysical themes in Dewey's writings of the 1920s: *John Dewey and American Democracy*, 321–361. A later installment of Whitehead's project, *Process and Reality* (1929), launched the important traditions of "process philosophy" and "process theology": Lewis S. Ford, *The Emergence of Whitehead's Metaphysics, 1925–1929* (Albany: State University of New York Press, 1984).

[47] Donald H. Meyer, "Secular Transcendence: The American Religious Humanists," *American Quarterly* 34 (1982): 524–542; Stephen P. Weldon, "The Humanist Enterprise from John Dewey to Carl Sagan: A Study of Science and Religion in American Culture" (PhD dissertation, University of Wisconsin-Madison, 1997), 59–98; John R. Shook, "John Dewey and Edward Scribner Ames: Partners in Religious Naturalism," *American Journal of Theology & Philosophy* 28, no. 2 (May 2007): 178–207.

effort to modernize the theological basis for the Protestant missionary enterprise.[48] The dominant scientific outlooks of the interwar years struck philosophical idealists as inadequate bases for a modern culture.

Across the human sciences in the 1920s, in a range of value-neutral and normatively engaged idioms, scholars worked to lay the theoretical groundwork for a scientific transformation of American culture. Despite the very real differences between them, these figures shared a common goal: to articulate and disseminate what they took to be the social meaning of modern science, above all its revelation of the need for citizens to adopt a greater sense of social obligation and mutuality. But when the rampant consumerism of the 1920s gave way to the economic collapse of the 1930s, the intellectual divisions within the human sciences would harden and generate sustained conflict. Facing both a social crisis and the prospect – suddenly very real – of extensive state intervention in the economy, Depression-era scholars would bitterly contest the proper means of pursuing their shared cultural goals.

Before those divisions opened up, however, scientific democrats undertook by various practical means to spread their ethical and political outlooks throughout the American population. Within the universities, they developed a series of curricular innovations aimed at spelling out science's cultural meaning. Beyond those institutions, they sought to teach the same lessons to a far wider audience. Throughout the 1920s and into the 1930s, scientific democrats worked to turn their popular writing and undergraduate classes into powerful, focused tools of cultural change.

---

[48] Edwin H. Wilson, *The Genesis of A Humanist Manifesto* (Amherst, NY: Prometheus, 1995); William Ernest Hocking et al., *Re-thinking Missions: A Laymen's Inquiry after One Hundred Years* (New York: Harper & Brothers, 1932).

7

Making Scientific Citizens

In the 1920s and early 1930s, scientific democrats engaged in a range of efforts to codify and distribute the results of new research on the human person and the sphere of interpersonal discourse that constituted society. The image of a democratic society as an organized network of communicating and resonating nodes, its structures both revealed and given flexibility by the cultural sciences, found new curricular and institutional expressions in the universities. Many of its advocates reached out directly to the public, seeking in a variety of ways to engage American citizens *en masse* and extend their influence beyond the still small circle of college graduates. The period since 1900 had brought an explosion of specialized knowledge production, often accompanied by a belief that merely augmenting the stock of scientific knowledge would automatically drive moral progress. But with the apparent setbacks of the 1920s, growing discontent with a *laissez-faire* approach to the distribution of the scientific attitude blossomed into a series of sustained initiatives to actively cultivate that attitude among citizens. Curricular reformers and popularizers tried to tame a sprawling body of scientific work and put its core findings and values into the hands of the masses.

On the curricular side, two interconnected phenomena – the system of free election in undergraduate education and the increasingly narrow, specialized character of advanced knowledge – bore the brunt of criticism from scientific democrats. Not far from the surface, however, were concerns about students' values and personal traits. The student body was increasingly vocational in outlook and Jewish in composition in the 1920s, as second-generation immigrants who had gained a precarious economic foothold took advantage of burgeoning opportunities for free or low-cost collegiate education. But the widespread concern with students' character in American higher education in the 1920s can hardly be reduced to anti-Semitism alone, prominent though that impulse was.[1] Many scientific thinkers, moreover, proved receptive to the educational attainments of Jews in a way that their counterparts in the humanities often

[1] On this point, see Julie A. Reuben, *The Making of the Modern University: Intellectual Transformation and the Marginalization of Morality* (Chicago: University of Chicago Press, 1996), 262–264.

were not. But most scientific democrats in the cultural sciences feared that the disconnected nature of modern knowledge prevented students from grasping and internalizing the true cultural meaning of science. They believed that undergraduates were learning technical, vocational skills without the ethical virtues appropriate to an industrial age. Coupled with other streams of curricular criticism – from humanists who found science itself irretrievably materialistic and from scientists who charged that the elective system failed to prepare incoming graduate students for advanced research – scientific democrats' frustration with the prevailing means of distributing knowledge helped to fuel a range of educational reforms, most prominently the establishment of survey and orientation courses associated with programs of general education.

Many scientific democrats also reached out beyond the university's borders after World War I, attempting to introduce the general population to the scientific attitude. For many of its proponents, the interwar campaign to take science to the people rested on a belief that the scholar's professional tasks included both producing knowledge and making its wider spirit and meaning available to the masses. A smaller but highly visible group also presumed that the cultural sciences both delineated the institutional contours of a genuine industrial democracy and carried its ethical foundations in their very marrow.

But multiple visions of science and of scientific democracy jockeyed for position in the emerging adult education movement and the bustling marketplace of popular publishing during the 1920s. As scientific democrats pushed beyond theorizing culture in general toward seeking to change American culture in particular, they ran up against much more powerful competitors – and implicitly cut against one another's programs. For the first time since they had attempted to remake the American universities in accordance with their ideals, the various groups of scientific democrats came into direct, institutional conflict with those harboring different visions for the nation's future.

CURRICULAR REFORM

This opposition emerged in a vigorous interwar debate over the cultural role of the universities, a debate that centered on competing proposals to replace or limit the elective system. That curricular approach, which allowed students to choose freely from among a vast array of courses, had always been associated with science in the minds of both its critics and its now dwindling body of supporters. Harvard's widely emulated transition to free election began well before Charles W. Eliot arrived and remained incomplete until 1894, when the system was finally extended to freshmen. But Eliot's impassioned defense of free election strongly identified that system with the cause of scientific reform. Eliot and many other early scientific democrats hoped that the new curricular approach would allow the modern subjects to establish themselves from the inside out by demonstrating student demand in a competitive academic marketplace and obviating the need to impose changes in a top-down manner. Yet the elective model remained subsidiary to the larger end of adding modern courses and fostering open communication between science-minded faculty and students. In fact, by the early twentieth century even Eliot came to doubt the wisdom of his signature reform. Students shied away from courses in the sciences and other modern subjects, while the wider culture seemed untouched by their ethical lessons.[2]

In 1909, A. Lawrence Lowell, a political scientist committed to rendering his discipline more objective, succeeded Eliot as Harvard's president largely on the strength of his critique of free election. But humanists, then in the process of putting their own disciplines on firm professional footing, raised the anti-elective cry most loudly before World War I. Drawing on the work of the British cultural critic Matthew Arnold, figures such as Harvard's Charles Eliot Norton, a cousin of Eliot but a sharp critic of the elective system, argued that the "Western tradition" of art and literature served as the proper basis for a common undergraduate education, even – indeed, especially – in a bustling industrial democracy.[3]

A fourth Harvard figure, Irving Babbitt, took the helm of the humanistic campaign against free election in the years before Lowell's inauguration. His opposition to that system reflected a deeper critique of the science-centered modern university. In *Literature and the American College* (1908), Babbitt deplored the political effects of a decentralized, user-driven model of higher education, complaining of the "comparative indifference to clearness and consistency of thought" exhibited by modern Americans. The problem, in his view, was not political democracy, but rather cultural democracy – the misguided belief that all ideas and all subjects were equally valuable, perhaps even equally true. A genuine college, Babbitt argued, would "check the drift toward a pure democracy" and help create "an aristocratic and selective democracy." He wanted to confine the humanitarian ideal of "uplift for the many" to the pre-collegiate institutions and assign the creation of "encyclopedic knowledge" to the graduate schools, reserving undergraduate training for "the creation of a social élite" imbued with "high and objective standards of human excellence."

[2] Richard Hofstadter and Walter P. Metzger, *The Development of Academic Freedom in the United States* (New York: Columbia University Press, 1955), 360; W. B. Carnochan, *The Battleground of the Curriculum: Liberal Education and the American Experience* (Stanford: Stanford University Press, 1993), 53; Reuben, *The Making of the Modern University*, 63, 201–202, 230–243.

[3] James Turner, *The Liberal Education of Charles Eliot Norton* (Baltimore: Johns Hopkins University Press, 1999); Caroline Winterer, *The Culture of Classicism: Ancient Greece and Rome in American Intellectual Life, 1780–1910* (Baltimore: Johns Hopkins University Press, 2002), 99–183; Linda Dowling, *Charles Eliot Norton: The Art of Reform in Nineteenth-Century America* (Durham: University of New Hampshire Press, 2007). On the rise of the humanities, see also Laurence R. Veysey, "The Plural Organized Worlds of the Humanities," in *The Organization of Knowledge in Modern America, 1860–1920*, ed. Alexandra Oleson and John Voss (Baltimore: Johns Hopkins University Press, 1979): 51–106; Bruce Kuklick, "The Emergence of the Humanities," *South Atlantic Quarterly* 89, no. 1 (1990): 195–206; Reuben, *The Making of the Modern University*, 211–229; and Jon H. Roberts and James Turner, *The Sacred and the Secular University* (Princeton: Princeton University Press, 2000).

Building on "the sifted experience of generations," rather than discarding it, Babbitt's college would create "men of quality" for "a quantitative age." He equated true democracy with the "disciplined and selective sympathy" of the humanist, not the scientist's shallow materialism and watery relativism.[4]

Babbitt had no problem with the natural sciences, but he deplored the "invasion" of science's "hard literalness" into the study of man. "We have invented laboratory sociology, and live in a nightmare of statistics," he exclaimed. Equating positivism with pragmatism, Babbitt argued that both "forego the discipline of a central standard, and make of the individual man and his thoughts and feelings the measure of all things." Worse still, he continued, a scientific approach flatly denied the purposive character of human behavior. Such an understanding of man did not provide an adequate basis for political organization. Babbitt specifically rejected many human scientists' identification of the emotions as the glue holding society together, seeing in this another version of Jean-Jacques Rousseau's baseless claim that "unbounded brotherhood" could offset "unbounded self-assertion." Nor could the enlightened self-interest of the classical political economists do the trick, said Babbitt. Only a "religious restraint" could prevent modern cultures from swinging back and forth between "an anarchical individualism" and "a utopian collectivism," producing an "inevitable drift toward imperialistic centralization." Babbitt and fellow "New Humanists" such as Paul Elmer More sought to fill the cultural void in modern society with a secularized version of traditional Christian dualism: a view of the human person as possessing a higher self, animated by reason and altruistic ideals, that properly restrained the lower self of passion and self-seeking interests.[5]

But the New Humanists hardly stood alone in believing that the nation needed new cultural standards to offset the ethical erosion fueled by new scientific techniques. As we saw in Chapter 3, philosophers drawn to science proved particularly likely to worry about the moral impact of the modern curriculum before World War I. Thus, Morris R. Cohen welcomed "the recent reassertion of the old ideal of culture as the aim of college training" and nominated a truly integrative philosophy as the proper complement to specialized knowledge. The following year, Frederick J. E. Woodbridge declared the oft-invoked ideals of service, character, and efficiency inadequate for a modern education. According to Woodbridge, the colleges should instead concentrate on "educating the emotions to act rationally." Cohen, Woodbridge, and other science-minded philosophers articulated the longstanding assumption that scientific knowledge represented the leading edge of a comprehensive scientific culture.

---

[4] Irving Babbitt, *Literature and the American College: Essays in Defense of the Humanities* (Boston: Houghton Mifflin, 1908), 1–2, 78, 80–81, 74–75, 85, 87, 8. Recent studies include Milton Hindus, *Irving Babbitt, Literature, and the Democratic Culture* (New Brunswick: Transaction, 1994); and George A. Panichas, *The Critical Legacy of Irving Babbitt: An Appreciation* (Wilmington: ISI Books, 1999).

[5] Babbitt, *Literature and the American College*, 89, 26–27, 61–62; J. David Hoeveler Jr., *The New Humanism: A Critique of Modern America, 1900–1940* (Charlottesville: University Press of Virginia, 1977), 125.

Even their contributions to the campaign for academic freedom, embodied in the 1915 creation of the American Association of University Professors, sought to open more cultural space for the findings and outlook of the sciences.[6]

Frustration with the elective system spread rapidly through the ranks of scientific democrats in the wake of World War I. These critics shared with the New Humanists a desire to embed the industrial economy and the bureaucratic state – still growing rapidly, despite the free-market rhetoric of a series of Republican presidents – in a cushion of ethical commitments shared by the entire population, and they too sought a partially or wholly prescribed curriculum that could spread the needed outlook. But the two groups clashed directly over the content and justification of such a curriculum. The New Humanists declared that the cultural sciences embodied modern materialism and elevated the animal side of human nature over its higher, spiritual element. Scientific thinkers, they charged, refused to acknowledge the need for inner discipline in the name of common interests and a higher, objective law. To the New Humanists, the human sciences validated outright hedonism. By contrast, many scientific democrats argued that those disciplines embodied the core ethical lesson of the modern era – the need for a more collective, altruistic orientation – without trafficking in the outworn, prescientific language of dualism and higher selves. They also rejected the New Humanists' assumption that social truths held true for all ages and emerged through "the selection of time."[7]

But another group, wielding considerably more power than the scientific democrats, also spoke with the voice of science in curricular debates. Abraham Flexner emerged as the leader of a group of research scientists who viewed the undergraduate college as primarily an adjunct to the university's research arm – a feeder for the graduate programs that they saw as the institution's lifeblood. Flexner, the author of an influential 1910 study that had reshaped American medical education, targeted the undergraduate curriculum in *A Modern College and a Modern School* (1923) and its 1930 sequel *Universities: American, English, German*. In a phrase picked up by many subsequent reformers, Flexner dismissed the existing universities as mere "service stations" that failed miserably at their appointed task: creating a true "cultural tradition" in the United States by intervening to alter the nation's "pioneer quality."[8]

[6] Morris R. Cohen, "The Conception of Philosophy in Recent Discussion," *Journal of Philosophy, Psychology and Scientific Methods* 7 (July 21, 1910), 409; quoted in Thomas Le Duc, *Piety and Intellect at Amherst College, 1865–1912* (New York: Arno Press, 1969 [1946]), 150; George M. Marsden, *The Soul of the American University: From Protestant Establishment to Established Nonbelief* (New York: Oxford University Press, 1994), 296–312; Reuben, *The Making of the Modern University*, 193–201.
[7] Babbitt, *Literature and the American College*, 82.
[8] Quoted in David O. Levine, *The American College and the Culture of Aspiration, 1915–1940* (Ithaca: Cornell University Press, 1986), 104; Abraham Flexner, *A Modern College and a Modern School* (Garden City: Doubleday, Page, 1923), 3, 2; Flexner, *Universities: American, English, German* (New York: Oxford University Press, 1930). On this strand of curricular reform, see Reuben, *The Making of the Modern University*, 203–210; and Thomas Neville Bonner, *Iconoclast: Abraham Flexner and a Life in Learning* (Baltimore: Johns Hopkins University Press, 2002).

Unlike the scientific democrats, however, Flexner did not see science as the ethical source of such a culture. He described "high scholarly attainment" and "serious intellectual effort" as the central values of scientific study, while relegating "character, service, good fellowship, or whatnot" to the humanities. Flexner proposed a two-tiered system of higher education. In the top tier, specialized preprofessional courses would give a foundation of "professional competency" to students headed for advanced study in fields such as engineering, law, teaching, the ministry, and business. Meanwhile, those "nondescripts" who still managed to make it into the lower-tier colleges after what Flexner hoped would be a substantial tightening of secondary-school requirements would choose from two prescribed tracks, "one humanistic, the other scientific and humanistic." Flexner described the sciences as the technical core of a modern culture, and the humanities as the source of its ethical framework.[9] In other words, he joined the New Humanists in holding that the humanities disciplines could take the place of the Protestant churches, providing the values needed to keep a technologically sophisticated polity on track. This vision of intellectual life and curricular organization, which provided no ethical role for the human sciences as such, would become widespread after World War II.

The model of knowledge production that emerged under the auspices of the Social Science Research Council comported well, in practice if not in theory, with a view of the university as a provider of specialized research and researchers. Charles E. Merriam and other SSRC leaders disagreed with Flexner's narrow reading of science and believed that the proliferation of scientific knowledge would not only improve society by solving concrete problems but also infuse public culture with a new conception of social ethics. Still, these figures worried comparatively little about reforming collegiate education. Engaging primarily with citizens and decision-making bodies, they assumed that the professor's main role was to produce more knowledge and train more graduate students, not to instill ethical values in undergraduates.

But such tensions remained largely latent in the 1920s. With the institutional stakes still vague, a widely shared commitment to building up the knowledge base and professional authority of the scientific disciplines cut across lines of difference regarding the nature of the ideal polity. Thus, scientific democrats of various stripes collaborated easily with one another and with other scientific thinkers in building up the survey and orientation courses that sprouted from the interwar general education movement. These thinkers hoped that such courses would produce citizens steeped in the scientific attitude toward ethical and social questions. It is often said that the interwar general education movement, like its post-World War II successor, took its shape from a humanistic Great Books approach and the postulation of a coherent "Western civilization" stretching across the centuries. Certainly courses of this variety can be found between the wars, but the dominant impulse toward general education in the 1920s and 1930s actually came from scientific democrats who sought to make

[9] Flexner, *A Modern College and a Modern School*, 54, 27, 58.

the ethical lessons of the modern human sciences available to students in read-ily digestible form.[10]

Alexander Meiklejohn launched the first full-fledged survey course in the human sciences, which premiered at Amherst in 1915. Meiklejohn was a sharp critic of Dewey's pragmatism who viewed Kantian idealism as the proper foun-dation for a modern, secular system of ethics. He insisted that educators, rather than assuming the prior existence of a genuinely democratic public, needed to train students to become intellectually free, all the while preserving the illusion that the students had chosen their own paths. Tapped in 1912 as the first non-clergyman to occupy Amherst's presidency, Meiklejohn set about preaching from Aristotle's writings and progressive political journals at morning chapel services. Meanwhile, he drew up an ambitious, four-year core curriculum of tightly linked courses centered on the social sciences and philosophy. Although Meiklejohn dialed back the program to a pair of first-year electives after much of the faculty balked, he still hoped that "Social and Economic Institutions," led by the institutional economist Walton H. Hamilton, would orient students productively to the "moral, social, and economic scheme" emerging all around them.[11]

With America's entry into World War I, however, Amherst's renowned course gave way to the practical, nationalistic "War Issues," sponsored by the Student Army Training Corps. Meiklejohn, a fierce opponent of military preparation, protested in vain, but the war had wreaked havoc on the nation's universities; by June of 1918, only 6 of Amherst's 125 seniors remained on campus. After the war's end, Meiklejohn sought all the more energetically to alter American culture, so that such a cataclysm could not recur. He concluded that big busi-ness represented the biggest threat to intellectual freedom and became a lead-ing proponent of workers' education, contributing to his ouster by Amherst's alarmed trustees in 1923.[12]

Elsewhere, the path to curricular reform led through the War Issues experi-ence rather than around it. Most scientific democrats, seeing in the war effort a salutary commitment to national service, seized on the opportunity to estab-lish a peacetime curriculum addressing the everyday problems of an indus-trial democracy. To be sure, much of the War Issues program had introduced students – dressed in military uniform and subject to military discipline – to such mundane topics as ordnance and the reading of maps. But at Columbia,

[10] Reuben, *The Making of the Modern University*, 163–167, 201–210, 228.

[11] Adam Nelson, *Education and Democracy: The Meaning of Alexander Meiklejohn, 1872–1964* (Madison: University of Wisconsin Press, 2001), 71–73. Illustrating the permeable boundary between Christianity and social science at the time, Meiklejohn apparently considered both Edward A. Ross and the Social Gospel theologian Walter Rauschenbusch as course heads before settling on Hamilton (73).

[12] Nelson, *Education and Democracy*, 91–100. On human scientists' enlistment in the war effort, see Carol S. Gruber, *Mars and Minerva: World War I and the Uses of the Higher Learning in America* (Baton Rouge: Louisiana State University Press, 1975); and Paul B. Cook, *Academicians in Government from Roosevelt to Roosevelt* (New York: Garland, 1982).

President Nicholas Murray Butler, dean of the college Herbert E. Hawkes, and graduate dean Frederick J. E. Woodbridge seized on another War Issues course that had defined the war as a struggle between competing forms of civilization, drawing its material from the social sciences, history, and philosophy. Science-minded philosophers had played a large role in the administration of the War Issues program, with the pragmatists James H. Tufts and George Herbert Mead serving as regional directors. Many scientific thinkers, including administrators at Columbia, reasoned after 1918 that the cultural confusions of peacetime called for a course that, without any specific vocational slant, gave students "a common starting point and a single point of vantage from which to study, to understand and to appreciate the world of nature and of men."[13]

Launched to great fanfare in the fall of 1919, Columbia's required Contemporary Civilization course (known as CC) featured an array of idealistic young instructors, primarily from philosophy and history but also sociology, economics, and political science. They employed small-group methods to introduce freshmen to "the controlling ideas of the world of today." These, as the course head explained, included a "view of nature as subject to man's control through science, and of man himself as perfectible by natural means; the change in production methods from home to factory and the great social changes which have accompanied this economic revolution; [and] the abandonment of monarchical forms of political control for democratic, nationalistic rule." Students then discussed how such changes bore on tasks such as harmonizing political stability with flexibility, achieving national efficiency, designing a socially integrative and personally liberating system of education, and "produc[ing] many and cheap goods without sacrificing human nature."[14]

Under the nose of the hidebound nationalist Butler, CC instructors advanced the democratic socialism of the Progressive movement's center-left wing. Most were either students or admirers of John Dewey and followed fairly closely his ethical and political tenets and his instrumental conceptions of social institutions and knowledge claims. For example, in the first textbook for the course, the young philosopher Irwin Edman equated reason with the criticism of existing institutions from the standpoint of a particular ethical ideal: "a way of life in which no natural impulse shall be frustrated." Edman combined this claim with the concept of a social self, as did his colleague John Storck in the second textbook. "The feeling that we are distinct individuals, bound only by external ties to other persons and to our group," according to Storck, dated back to

[13] Nelson, *Education and Democracy*, 91; Gilbert Allardyce, "The Rise and Fall of the Western Civilization Course," *American Historical Review* 87, no. 3 (2005), 706, 708 (quoted); Frank Aydelotte, *Final Report of the War Issues Course of the Students' Army Training Corps* (Washington: War Department, 1919), 15–16; James Campbell, *A Thoughtful Profession: The Early Years of the American Philosophical Association* (Chicago: Open Court, 2006), 210–211. See also Timothy P. Cross, *An Oasis of Order: The Core Curriculum at Columbia College* (New York: Columbia College, 1995).

[14] John J. Coss, "The New Freshman Course in Columbia College," 247–249. Like Meiklejohn, many of the early CC instructors hoped to reform the entire undergraduate curriculum: Harry J. Carman, "Reminisces of Thirty Years," *Journal of Higher Education* 22 (1951), 117.

the social science of the eighteenth century. That body of knowledge, with its foundational postulate of individual autonomy, had originally emerged as an ideological weapon during the liberation of "the rising industrial and commercial bourgeoisie from mercantilistic governmental interference with business." Storck thus described the concept of a discrete, atomistic individual as simply another human product, one that failed when judged against the ethical standard of genuine self-realization. In *The Making of the Modern Mind*, originally written for the CC course, John Herman Randall Jr. likewise traced classical individualism to the political ascendancy of the commercial middle class. These figures presented a shift from competition to cooperation as a necessary step in the full democratization of the United States.[15]

Despite these substantive claims, many of CC's architects insisted that the course freed students from dogmatic convictions and taught them to think for themselves. Hawkes described the course's central goal as leading students to the insight "that there are two or more sides to most questions, and that those who support each side are equally honest." Even better would be the further recognition that "no one can have an intelligent opinion on any controversial subject until he has tried to look at it through the eyes of the man with whom he does not agree." Hawkes saw in the course's discussion-based approach a discursive outlook holding "that when any man … has a real conviction it is treated with respect." In short, he saw CC as a vector for the scientific attitude.[16]

Columbia's primarily historical and philosophical introduction to the modern scientific outlook served as a model for many similar offerings elsewhere. In fact, Rutgers adopted CC entirely unmodified: syllabus, readings, and all. Other 1920s orientation courses approached questions of modern culture and institutions through material drawn from economics and political science. Often titled "Problems of Citizenship," as at Stanford, these courses told students that the "scientific attitude," as a general method of problem solving, could defuse all kinds of social problems. According to one course leader, producing an "alert, intelligent, and efficient citizenry" entailed giving students not only knowledge of "functions and theories of government" but also information regarding the basic characteristics of the human person: his "relation to his environment, his capacity to feel and to think, his powers of discrimination, his dependence upon public opinion, [and] his intelligence" – in short, "his capacities as a thinking man."[17]

[15] Irwin Edman, *Human Traits and Their Social Significance* (Boston: Houghton Mifflin, 1920), 431; John Storck, *Man and Civilization: An Inquiry Into the Bases of Contemporary Life* (New York: Harcourt, Brace and Company, 1927), 64. On Randall's intention to write a CC textbook, see the early edition, *The Western Mind: Its Origins and Development* (n.p., 1924); and Jacques Barzun, "Foreword," in Randall, *The Making of the Modern Mind: A Survey of the Intellectual Background of the Present Age* (New York: Columbia University Press, 1976 [1926]).

[16] Herbert E. Hawkes, "On the College Frontier III: Experimenting at Columbia," *Nation* 131 (October 15, 1930), 399.

[17] Edgar Eugene Robinson, "Citizenship in a Democratic World" (1928), printed in Carnochan, *The Battleground of the Curriculum*, 134; Reuben, *The Making of the Modern University*,

Instructors were hardly shy about the ethical and political values that they identified with modern science. The University of Utah's team told students that, because "every individual in a civilized community today profits beyond calculation by the labors of his fellow-men, past and present," all citizens bore a debt of "service to present and future generations." Enlisting student initiative wherever possible, the instructors moved through various examples of "human conservation," from "the democratic assumption" that individuals should be fully equal with respect to "the use of natural resources" to thorny questions about the "ethical grounds" and "legal and ethical limitations" of "the right to hold private property." At that point, according to course head Milton Bennion, the students freely settled on a broad, universal principle of justice: "Each individual has the moral right and the duty to develop to the utmost his socially valuable native capacities, and to give in social service in proportion to his abilities and his opportunities; he should also receive in proportion to his needs, fairly and objectively determined." In subsequent weeks, students systematically compared this conception of justice to those "stated or implied in the Mosaic law, Plato's *Republic*, the Sermon on the Mount, and Herbert Spencer's philosophy." As Bennion summarized, the Utah course gave students "an empirical basis for moral and social obligation."[18]

The University of Wisconsin's Experimental College, created by Alexander Meiklejohn a few years after his acrimonious firing from Amherst, employed a historical approach similar to that of CC, though in the service of a view of politics more republican than liberal. With enthusiastic support from Wisconsin president Glenn Frank, Meiklejohn built a small learning community, comprised of carefully selected freshmen and sophomores, within the larger university. He sought to turn students into "free and responsible human beings" by encouraging "a certain mode of living – that of seeking intelligence by means of books." In fact, Meiklejohn famously defined a college as simply "a group of people, all of whom are reading the same books."[19]

Meiklejohn's definition notwithstanding, the Experimental College curriculum bore only a passing resemblance to the Great Books approach developed by John Erskine at Columbia and taken up by Robert M. Hutchins and Mortimer Adler at Chicago. Meiklejohn's stress on book learning merely reflected his belief in the importance of a common body of knowledge shared by all citizens, coupled with the practical observation that books and their authors currently provided the only reliable source of "the values of liberal understanding" in

163–164. For the dominance of philosophers and historians in the early years of CC, see Charles H. Russell, "The Required Programs of General Education in the Social Sciences at Columbia College, the College of the University of Chicago, and Harvard College" (PhD dissertation, Columbia University, 1961), 46–47. CC directly replaced mandatory freshman courses in history and philosophy: "Introductory Note," in *Introduction to Contemporary Civilization: A Syllabus, Second Edition* (New York: Columbia University Press, 1920).

[18] Milton Bennion, "Orientation through Social Ethics," *Journal of Higher Education* 2 (1931), 255–257.

[19] Alexander Meiklejohn, *The Experimental College* (New York: Harper, 1932), 22, 122, 40.

a commercial society deeply hostile to those values. Launched in 1927, the Experimental College curriculum offered a year on ancient Athens, viewed through the lens of Plato's *Republic* and assorted other readings, followed by a second year on modern America, with the pessimistic Brahmin Henry Adams as the students' primary guide. This comparison, Meiklejohn explained, showed the enrollees "what a civilization is" and spurred them to remake their own nation in that mold. Dewey, though a frequent target of Meiklejohn's philosophical critiques, nevertheless found the Experimental College quite congenial. Both men believed that, at least in the intellectual sphere, "the art of democracy is the art of thinking independently together." But the wider Madison community seemed not to agree. Meiklejohn's experiment began to fall apart within a few years because of its mismatch with the social views of parents and local residents.[20]

The interwar turn toward orientation courses lagged perceptibly in the natural sciences, except in teacher-training programs. Where natural science surveys did emerge, moreover, they tended to be abbreviated summaries of multiple discrete courses, featuring individual units that imparted basic knowledge of various fields and were taught by an array of disciplinary experts in succession.[21] This may be because of the centrality of the human sciences, rather than the natural sciences, to the broad project of scientific democracy. In fact, by the early 1930s, a group of social psychologists developed measures for the scientific attitude and concluded that courses in the natural sciences – especially high school courses – did not effectively teach it.[22]

Nevertheless, a number of higher educators did explicitly draw ethical lessons from the natural sciences in the 1920s. These figures typically portrayed scientific knowledge, not as a set of humanly constructed tools, but rather as a body of observer-independent facts and laws to which human beings needed to conform. Dartmouth's compulsory freshman course in evolution offers a clear example of this approach, which identified scientific laws as fully objective and therefore normative for human behavior. Developed by the reform-minded zoologist William Patten, the course drew sweeping conclusions about social behavior and political organization from its descriptions of biological processes. Patten, as we have seen, viewed evolution as a progressive, integrative affair – and science as a potent source of cultural authority. He urged each listener to "voluntarily

---

[20] *Ibid.*, 39, 138, 68–69, 107; Nelson, *Education and Democracy*, 225; John Dewey, "The Meiklejohn Experiment," *New Republic* 72 (August 17, 1932): 23–24.

[21] Donald R. Watson, "A Comparison of the Growth of Survey Courses in Physical Science in High Schools and in Colleges," *Science Education* 24, no. 1 (January 1940): 14–20; Peter S. Buck and Barbara Gutmann Rosenkrantz, "The Worm in the Core: Science and General Education," in Everett Mendelsohn, ed., *Transformation and Tradition in the Sciences: Essays in Honor of I. Bernard Cohen* (Cambridge: Cambridge University Press, 1984): 371–394.

[22] Victor H. Noll, "The Habit of Scientific Thinking," *Teachers College Record* 35 (1933): 1–9; Noll, "Measuring the Scientific Attitude," *Journal of Abnormal and Social Psychology* 30 (1935): 145–154; Elliot R. Downing, "Does Science Teach Scientific Thinking?" *Science Education* 17 (1933): 87–89.

submit to the exaction which modern science, in an ever increasing degree, must impose upon his conduct in commerce, agriculture, public health and other phases of social life." If students hewed to the "pre-established natural laws which stand over and above man-made laws," Patten promised, nature would shower its gifts upon them. In fact, Patten found such an expansion of both social sympathy and social discipline inevitable in the end, for nature itself pointed toward "the upbuilding of a great world-state, with the aid, and for the benefit, of all its constituents." In Patten's portrayal, the realm of ethical choice was as narrow as the demands of ethics were broad.[23]

An active engagement with religious questions characterized the scientific offerings at Oberlin, which reorganized its entire curriculum around the concept of evolution in 1930.[24] The economist Harvey A. Wooster explained that the evolutionary dynamic had "come to be basic, not only to all science, both natural and social, but to philosophy, ethics, religion, and history, as well, including the special histories of literature, education, and religion." Its influence had even been felt in the fine arts, he noted. Oberlin's new curriculum took the form of an ascending hierarchy of knowledge: physical science; biological science; psychology; social science; language, mathematics, and logic; literature and the arts; and finally philosophy and religion at the top of the pyramid. It presented students with what Wooster called a "frankly anthropocentric" view, designed to replace both the "theocentrism of the medieval Church" and the "cosmogonistic slant" that natural scientists had long offered in its place. The individual, in Oberlin's evolutionary narrative, was at once "one of many characters in the great drama of ceaseless change, a modifier of the consequences of natural law, and a subject of evolutionary change himself." Here, unlike at Dartmouth, human beings were charged with choosing and promoting their ends in accordance with the inexorable laws of nature, rather than simply obeying those laws.[25]

The more influential experiments at Chicago and Northwestern in the 1920s reveal the same division on the relative weight of human choice in history. Much less integrated than their counterparts at Dartmouth and Oberlin, each of these amounted an all-star survey course in which a series of specialists introduced students to recent developments in their fields, including certain areas of social science. The mastermind of Chicago's "The Nature of the World and of Man," a two-quarter course launched in 1924, was the geneticist Horatio Hackett Newman, who wrote for popular audiences on eugenics and

[23] Quoted in Gregg Mitman, "Evolution as Gospel: William Patten, the Language of Democracy, and the Great War," *Isis* 81, no. 3 (September 1990), 460; Patten, *Growth: An Introduction to the Study of Evolution* (Hanover: Dartmouth Press, 1929 [1923]), 9, 12, 11.

[24] Bruce A. Kimball, *Orators and Philosophers: A History of the Idea of Liberal Education* (New York: Teachers College, 1986), 188.

[25] Harvey A. Wooster, "To Unify the Liberal-Arts Curriculum," *Journal of Higher Education* 3 (1932), 375, 377; Ernest H. Wilkins, "The Revision at Oberlin College: An Interpretation of the Seven-fold Plan as the Basic Philosophy of the New System," *Journal of Higher Education* 2 (1931), 66–67.

other subjects. According to Newman and his collaborators, science's great lesson for public and private behavior alike was its revelation of "the universal orderliness of nature." Other contributors to the course text – characteristically, a series of separate chapters on individual sciences – likewise stressed the unifying theme of natural order and defined human progress in terms of the gradual ascertainment of that order. Thus, the astronomer Forest Ray Moulton saw in the man's discovery of the "majestic succession of the celestial phenomena" a source of the supreme "hope that we shall be able to understand not only the exterior world but also our own bodies and our own minds." Rollin T. Chamberlin took a similar lesson from the geological image of history as a "great drama" composed of repeating cycles in a "beautifully regulated system of balance." According to Chamberlin, such orderliness proved that "there will be many acts and scenes in a long future to come," and that "the human race will have almost unlimited opportunity to develop along its chosen lines." For these authors, as for Patten, science's social promise lay in its ability to identify unchanging natural laws to which human beings could conform their behavior and reap untold rewards. The story of science, Moulton summarized, was "the story of the efforts of the human race to conquer the physical world, to learn how to live happily with itself, and to direct its own improvement."[26]

At Northwestern, meanwhile, the philosopher Baker Brownell's "Problems of Contemporary Thought" offered a richer vision of human creativity. Unlike the Chicago faculty, Brownell embraced a highly spiritualized form of naturalism grounded in the work of Santayana, Royce, and James. He sharply attacked both specialization and materialism as social dangers, insisting that abstract science, for all its value, missed both the details of experienced reality and the underlying unity behind them. To provide a foundation for modern civilization, he insisted, science needed to "develop a metaphysical base and understanding far more mature than the innocent positivism of its past." Brownell treated his students to a parade of leading scholars, writers, and political figures, including scientific scholars such as Richard T. Ely, C. Judson Herrick, Edward Sapir, and the British philosopher Bertrand Russell as well as Clarence Darrow, the novelist Jean Toomer, and the reformer Charlotte Perkins Gilman. In fact, "Problems of Contemporary Thought" was ultimately less a course than a lecture series. But Brownell's initiative, much more than the Chicago course – the tone of which was set by physical scientists – presented the claims of the Progressive social theorists and their post-Progressive successors in the interwar cultural sciences. Like Brownell's 1926 book *The New Universe* and his New Deal-era advocacy of small, decentralized communities, his Northwestern course attacked all attempts to divorce science from

---

[26] Mary Morrice Bogin, "Newman, Horatio Hackett," *Dictionary of Scientific Biography* (New York: Scribner, 1987), 668; Reuben Frodin, "Very Simple, But Thoroughgoing," in *The Idea and Practice of General Education: An Account of the College of the University of Chicago* (Chicago: University of Chicago Press, 1950), 43; H. H. Newman, ed., *The Nature of the World and of Man* (Chicago: University of Chicago Press, 1926), 162, 30, 55, 1.

values and traced the future survival of democracy to "the importation of good will into the deep structure of industrial society" by thinkers who recognized that nature itself – the "commandment in things" – transcended all attempts to force it into conceptual boxes and supported a fundamentally humanistic, democratic form of culture.[27]

Through these survey courses and the wider curricular initiatives from which they sprang, scientific democrats of various stripes sought to tear down students' preconceptions, train them in more scientific modes of thought, and thereby reshape American public culture. They drew on new conceptions of biological and social reality to challenge traditional sources of cultural authority, and in many cases even the authority of science itself, at least in its more mechanistic forms. But these initiatives revealed a deep tension between a definition of science as a body of authoritative, expert-created knowledge of fixed natural laws and a competing image of science as an embodiment of the lesson that knowledge was a human construction, perhaps conditioned by external structures of some kind but ultimately validated by its contributions to human well-being. This division, as we have seen, also emerged quite clearly in the theoretical productions of the 1920s. It did not, however, become a source of internal dissension among scientific democrats until the 1930s, when a new set of political conditions forced epistemological and methodological issues to the fore.

ENGAGING THE PUBLIC

Had it taken place in New York, Baker Brownell's Northwestern course could easily have graced the stage of the Labor Temple, the People's Institute, or one of the many other institutions that sprang up on the revitalized lecture circuit of the 1920s. Developments in the cultural sciences, especially psychology, intersected with immense popular interest in easily digested summaries of academic knowledge after the war. This "outline boom," kicked off by the enthusiastic response to the British futurist H. G. Wells' *The Outline of History* (1920) and the equally striking success of the Dutch émigré Hendrik Willem Van Loon's *The Story of Mankind* (1921), gave scientific democrats a golden opportunity to speak directly to the American public, or at least a wider portion of it than they could expect to find in their undergraduate classes. Titles such as George A. Dorsey's *Why We Behave Like Human Beings* (1925), Harry A. Overstreet's *About Ourselves: Psychology for Normal People* (1927), and Raymond B. Fosdick's *The Old Savage in the New Civilization* (1929) offered sweeping observations on the past and the present of the human enterprise, informed both by recent developments in the cultural sciences and by the wider project of scientific democracy. In fact, one of the greatest successes of the era, *The Story of Philosophy* (1926), by the Columbia PhD Will Durant, hewed

[27] Baker Brownell, *The New Universe: An Outline of the Worlds in Which We Live* (New York: Van Nostrand, 1926), 5, 344, 347–348, 268; Brownell, ed., *The World Man Lives In* (New York: Van Nostrand, 1929), frontispiece.

fairly closely to Dewey's version of that program. Durant and other scientific democrats also took to the podiums at the Labor Temple and elsewhere, often while keeping up demanding writing schedules on the side.[28]

But such attempts to harmonize modern science with accepted and emerging cultural meanings contended in the marketplace with the more sensational, deterministic claims of figures such as Darrow and John B. Watson. Scientific democrats also faced competition from science popularizers who sought merely to rev up the industrial machine and augment the authority of disciplinary professionals. Many leading natural scientists believed that a combination of technical knowledge and Protestant values would suffice to keep the nation on track. This view struck a chord with a public that thought of technological innovations, and particularly the products of the individual inventor, when it heard the term "science." Mass magazines in the 1920s presented figures whose fame rested on hard-headed technical achievements – Thomas Edison, the plant breeder Luther Burbank, even Henry Ford – as the public faces of science. Many natural scientists reinforced this implicit excision of ethical concerns from science's public engagements. Albert Einstein's immense renown gave physicists, in particular – many of whom rejected the tenets of scientific democracy – a considerable advantage in seeking to define the meaning of the contested term "science" in the 1920s. Thus, Robert A. Millikan, the best-known natural scientist after Einstein in those years and a vocal defender of evolution against its evangelical critics, found a wide audience for his claim that a liberal Christianity centered on self-sacrifice for the common good provided all the values an industrial society needed. Like many other physical scientists and engineers of the era, Millikan agreed with the former mining engineer Herbert Hoover that private enterprise, embedded in a Christian culture and perhaps augmented by a bit of coordination and information from the state, would dissolve the industrial crisis.[29]

[28] Joseph F. Kett, *The Pursuit of Knowledge Under Difficulties: From Self-Improvement to Adult Education in America, 1750–1990* (Stanford: Stanford University Press, 1994), 346–348, 350; George Cotkin, "Middle-Ground Pragmatists: The Popularization of Philosophy in American Culture," *Journal of the History of Ideas* 55 (1994): 283–302. Chicago's philosophers used Durant's book in their introductory undergraduate courses, illustrating the porosity of the boundaries between the university and other arenas of cultural production: James T. Kloppenberg, "The Place of Value in a Culture of Facts: Truth and Historicism," in *The Humanities and the Dynamics of Inclusion Since World War II*, ed. David A. Hollinger (Baltimore: Johns Hopkins University Press, 2006), 127. Durant had taken a PhD under Columbia's philosophers in 1913, writing a dissertation on the problem of harmonizing the social sciences with the natural sciences.
[29] John C. Burnham, *How Superstition Won and Science Lost: Popularizing Science and Health in the United States* (New Brunswick: Rutgers University Press, 1987); Marcel C. LaFollette, *Making Science Our Own: Public Images of Science, 1910–1955* (Chicago: University of Chicago Press, 1990); Daniel Patrick Thurs, *Science Talk: Changing Notions of Science in American Popular Culture* (New Brunswick: Rutgers University Press, 2007); Robert H. Kargon, *The Rise of Robert Millikan: Portrait of a Life in American Science* (Ithaca: Cornell University Press, 1982).

Physicists were not the only ones who believed that science's contributions to human flourishing were technical rather than ethical. Thus, the Dutch-American biologist Paul De Kruif's 1926 blockbuster *Microbe Hunters* focused exclusively on science's contribution to material wellbeing – in that case, its capacity to produce new medical technologies. Similarly, many of the popular works on psychology that sprouted like mushrooms in the Freud-crazy culture of the 1920s presented that discipline's primary contribution as facilitating individual adjustment to a new world of industrial capitalism. In the high schools, meanwhile, a growing body of curricular literature identified the "appreciation" of science as a primary aim, while the leaders of a sprawling constellation of science clubs joined De Kruif in presenting scientists as heroic votaries of Nature, driven by curiosity alone to make discoveries that extended the lives and raised the comfort level of their fellow citizens. Chemists made similar claims. A series of lectures by the Science Service's Edwin E. Slosson identified five major contributions of science to civilization in recent years: gasoline, refrigeration, photography, sugar, and coal-tar products. Although these figures often spoke of integrating science into a harmonious modern culture, they meant simply that citizens should learn to defer to scientists and welcome the technological advances they generated.[30]

Post-Progressive scientific democrats decried the relentless public emphasis on technology, efficiency, and individualism. But they often missed the theoretical chasm separating them from figures such as Millikan, De Kruif, and Slosson. Most scientific democrats believed that science pointed uniquely – perhaps even inexorably – toward their own ethical-political framework and, thus, that all efforts to increase the authority of science portended the political future they desired. This view led many of them to ignore difficult questions about epistemology and authority. So, too, did a grand historical narrative they shared with those touting science's industrial benefits. This narrative traced virtually all aspects of contemporary Western life, including capitalism and democracy, to the emergence of modern science and technology in the seventeenth century. Scientific democrats, of course, went on to argue that the great historical transformation that began with the Scientific Revolution – or perhaps the Reformation – would find completion in the widespread internalization of a scientifically grounded ethical framework by American citizens. Modern scholars, as specialists in the application of reason, would play a key role in harmonizing industrial production with the fulfillment of human needs,

[30] Paul De Kruif, *Microbe Hunters* (New York: Harcourt, Brace, 1926); Stephen Fried, *American Popular Psychology: An Interdisciplinary Research Guide* (New York: Garland, 1994); Paul Ammon Maxwell, *Cultural Natural Science for the Junior High School: Objectives and Procedures* (Baltimore: Williams & Wilkins, 1932); Hanor A. Webb, "Some First Hand Information Concerning Science Clubs," *School Science and Mathematics* 29 (1929): 273–276; A. Silverman, "The Influence of Modern Science on History and Civilization," *Science* n.s. 58, no. 1493 (August 10, 1923): 102–103; James Steel Smith, "The Day of the Popularizers: The 1920s," *South Atlantic Quarterly* 62 (1963): 297–309.

material and otherwise. But to many other 1920s Americans, industrial production appeared as a virtually automatic guarantee of social progress and
individual happiness in its own right. The overarching claim that whatever
was good in the modern world had resulted from the development of science
obscured this important line of division between scientific democrats and their
cultural competitors in the book market and on the lecture circuit.

The boundary between those scientific democrats who pursued careers as
lecturers or independent writers and those who plied their trade in academia
was equally fuzzy in the 1920s. A good example is the New Historian James
Harvey Robinson, who provided an influential theoretical framework for the
interwar surge of popularization in New York City and elsewhere. Robinson
seamlessly transitioned into a writing career shortly after World War I, resign-
ing his post at Columbia in 1919 and cutting all academic ties in 1922. He
followed his initial popular effort, *The Mind in the Making* (discussed in
Chapter 6), with *The Humanizing of Knowledge* (1923). There, he spelled out
the rationale driving many scientific democrats to wade into the public arena
in the 1920s. Robinson called for both the synthesis of modern science and its
expression in terms accessible to the layman. In a scientific age, he wrote, it was
no longer "safe ... to permit the great mass of mankind and their leaders and
teachers to continue to operate on the basis of presuppositions and prejudices"
formed in ages past and utterly out of tune with contemporary realities. Seeing
in modern psychology's emphasis on irrational motives merely a prod to couch
their appeals in emotional terms, Robinson and many other popularizers of
the era tended to work in the simple terms of intellectual freedom and democratic access. They assumed that the public needed little more than a key to the
vast storehouse of knowledge piled up by bustling researchers in the scientific
disciplines. Still, this was clearly a form of scientific democracy, going beyond
the claim that industrial growth and Christian faith alone could ground a democratic culture.[31]

Numerous scientific democrats shaped their careers in accordance with the
interwar stress on public engagement. During the New Deal, some would take
their cultural project to the highest reaches of government: former CC instructor
Rexford G. Tugwell as a member of the "Brain Trust," former Antioch College
president Arthur E. Morgan as director of the Tennessee Valley Authority (TVA),
and the University of Chicago philosopher T. V. Smith as a Democratic con-
gressman. In the 1920s, however, educational strategies for reform prevailed.
Scientific democrats retooled struggling colleges and launched new institutions
of higher education to better promote their cultural goals.

The boundary between such institutions and the wider public blurred as
a result. Extension courses and summer programs flourished as university

---

[31] Kett, *The Pursuit of Knowledge Under Difficulties*, 336–339; James Harvey Robinson, *The
Humanizing of Knowledge* (New York: G. H. Doran, 1923), 41. See also Rae H. Rohfeld,
"James Harvey Robinson: Historian as Adult Educator," *Adult Education Quarterly* 40 (1990):
219–228.

administrators seeking to capitalize on public interest in academic knowledge joined forces with reformist faculty members – often scientific democrats attracted by the opportunity to gain unprecedented access to the public mind. At Columbia, for example, the philosopher John J. Coss simultaneously oversaw the CC course and developed a thriving program of summer courses for adults. Many leaders of the "progressive colleges" created in the 1930s saw those institutions as filling the same cultural need through their everyday programs. At Bennington, Sarah Lawrence, Black Mountain, Goddard, Bard, and similar colleges, faculty sought to turn out leaders who could put modern knowledge to work in its social setting. Coss and other Columbia figures were trustees at Bennington, whose founding president, Robert Devore Leigh, had taught in CC during the 1920s.[32]

Other institutional innovations arose when individuals or groups concluded that existing universities could not fulfill the cultural tasks charged to them, even when equipped with new appendages such as the Experimental College. Particularly expansive ambitions characterized the reborn Antioch, where Arthur Morgan held court prior to his TVA appointment. Trained as a civil engineer, Morgan sought to counteract what he saw as the culturally fragmenting effects of modern specialization. He would give each student not only a practical vocation – Antioch gained renown for its "co-op" system, in which students worked at various industrial positions to gain real-world experience – but also the capacity to coordinate and harmonize any social situation in which they found themselves. Morgan wanted Antioch graduates to be philosopher-engineers like himself, trained in the "ability to gather together the various tangled threads of forces, conditions, and affairs, which make up the elements of any potential human accomplishment, and to weave them into a perfect fabric, showing the texture and design of a preconceived plan." As a key part of this plan, Morgan stressed the ability to think and act with a holistic understanding of the world. He worked so vigorously to make Antioch into the germ of a revolution in American moral character that he clashed repeatedly with his own faculty and students.[33]

The founders of New York's New School for Social Research, which opened its doors in 1919, likewise saw their hopes dashed against hard realities.

---

[32] Kett, *The Pursuit of Knowledge Under Difficulties*, 261; Irwin Edman, "John J. Coss," *Columbia University Quarterly* 32, no. 3 (October 1940), esp. 203; Frederick Rudolph, *Curriculum: A History of the Undergraduate Course of Study since 1636* (San Francisco: Jossey-Bass, 1977), 475–477; Louis T. Benezet, *General Education in the Progressive College* (New York: Teachers College, 1943).

[33] Roy Talbert Jr., *FDR's Utopian: Arthur Morgan of the TVA* (Jackson: University Press of Mississippi, 1987), 17, 48, 60–62, 64–66; Arthur E. Morgan, "What is College For?" *Atlantic Monthly* 86 (1922), 642–643, 649; Frederick Rudolph, *The American College and University: A History* (New York: Vintage, 1962), 474; Algo D. Henderson and Dorothy Hall, *Antioch College: Its Design for Liberal Education* (New York: Harper, 1946), 6. See also Judith Sealander, "'Forcing Them To Be Free': Antioch College and Progressive Education in the 1920s," *History of Higher Education Annual* 8 (1988): 59–78.

That institution grew out of a controversy at Columbia, where Robinson and Charles A. Beard resigned after becoming embroiled in an argument with President Butler over the limits of academic freedom during the war. Drawing on the intellectual and institutional resources of both Columbia and the *New Republic*, Beard and Robinson enlisted a remarkable group of figures, including Dewey, Herbert Croly, Wesley Mitchell, Thorstein Veblen, Walter Lippmann, Horace Kallen, and the legal scholars Learned Hand and Felix Frankfurter, and created an entirely new kind of institution centered on the application of the human sciences. Although the institution provided a congenial home for an array of leading scientific democrats in its first few years, it hit a rough patch in the early 1920s when Croly and Robinson clashed over the relative priority of research and teaching in the institution's mission. Many of the original faculty left, and the school soon became a provider of student-driven adult education classes, rather than the scholar-led instrument of public enlightenment envisaged by Robinson.[34]

Indeed, adult education emerged in the 1920s as a major outgrowth of the push toward scientific democracy – a "new means for liberals," as Eduard C. Lindeman put it in the *New Republic*. Reform-minded philosophers proved particularly eager to enlist, seeking to sidestep the influence of the prevailing school system and lead communities to develop new standards that they would then teach to the upcoming generation. Kallen put much of his energies into adult education during the 1930s, though he kept one foot at the New School and also contributed to the Depression-era "consumers' movement." Joining him in the adult education field was fellow cultural pluralist Alain Locke, an African-American graduate of Harvard who had likewise studied with William James. But the leading figure as the adult education movement took shape in the late 1920s and early 1930s was Lindeman. A pioneer in the field of social work who discovered the Danish "folk school" model in the early 1920s, Lindeman penned the bible of the new movement, *The Meaning of Adult Education*, in 1926.[35]

---

[34] Peter M. Rutkoff and William B. Scott, *New School: A History of the New School for Social Research* (New York: Free Press, 1986), 1–3, 10–11, 23, 28–31, 18, 36–38, 48; Harold W. Stubblefield, *Towards a History of Adult Education in America: The Search for a Unifying Principle* (New York: Croom Helm, 1988), 14. Also see James Harvey Robinson, "The New School," *School and Society* 11 (January 31, 1920): 129–131.

[35] Eduard C. Lindeman, "Adult Education: A New Means for Liberals," *New Republic* 54 (February 22, 1928): 26–29; Cotkin, "Middle-Ground Pragmatists"; Stubblefield, *Towards a History of Adult Education in America*, 130–134; Kett, *The Pursuit of Knowledge Under Difficulties*, 315–316; Horace M. Kallen, "Between the Dark and the Ivory Tower," *New Republic* 54 (February 22, 1928): 39–41; Amy D. Rose and Linda O'Neill, "Reconciling Claims for the Individual and the Community: Horace Kallen, Cultural Pluralism, and Persistent Tensions in Adult Education," *Adult Education Quarterly* 47, nos. 3–4 (Spring–Summer 1997): 138–152; Rudolph A. Cain, "Alain Leroy Locke: Crusader and Advocate for the Education of African American Adults," *Journal of Negro Education* 64, no. 1 (1995): 87–99; Lindeman, *The Meaning of Adult Education* (New York: New Republic, 1926). Some Americans sought to import the Danish model whole cloth during the 1920s and 1930s, producing such institutions as the Pocono

Innovations in adult education by scientific democrats hardly ceased there. As the Depression progressed, Lyman Bryson set up a "Readability Laboratory" at Columbia to determine standards for comprehensibility and to promote the adoption of a simpler prose style in non-fiction texts for the general public. By the 1930s, Alexander Meiklejohn had joined these figures in attempting to directly influence the general population. Enthused about adult education as early as 1924, he moved to Berkeley after the collapse of the Experimental College and helped establish the San Francisco School of Social Studies, which opened its doors in January 1934. Through that initiative, Meiklejohn hoped to promote what he saw as the core values of democracy and to demonstrate that these conflicted directly with the shallow materialism of the age.[36]

The Carnegie Corporation, led by former Columbia dean Frederick P. Keppel, bankrolled the rapid expansion of American adult education via the American Association for Adult Education (AAAE) in the 1920s and 1930s. Keppel hoped to steer adult learners away from vocational studies toward academic subjects such as history, literature, and science, both to inform the use of their leisure time and to save them from intellectual confusion in their daily lives. As part of this endeavor, Keppel and the AAAE sponsored a breakthrough psychological study that validated the enterprise of adult education as a whole. Much of the Progressive Era emphasis on primary and secondary education stemmed from the view – influentially promoted by William James in his classic textbook *The Principles of Psychology* (1890) – that "disinterested curiosity" waned in the individual after the age of twenty-five, meaning that individuals needed to develop healthy intellectual habits before their mind hardened. Assumptions about the mental inflexibility of adults continued to reverberate through the cultural sciences, as in Ogburn's theory of cultural lag and many Boasians' emphasis on the shaping power of culture. But in 1928, the psychologist Edward Thorndike and several Columbia colleagues published experimental evidence that learning ability did not decline until the age of 55, if at all. Adult education now had the firm imprimatur of modern science.[37]

Of course, scientific democrats already populated the nation's teachers' colleges and schools of education, which concerned themselves with the learning

People's College in Pennsylvania and the Highlander Folk School in Tennessee: Kett, *The Pursuit of Knowledge Under Difficulties*, 363–365; Stubblefield, *Towards a History of Adult Education in America*, 143, 145.

[36] Stubblefield, *Towards a History of Adult Education in America*, 42, 102–105; Kett, *The Pursuit of Knowledge Under Difficulties*, 375; Alexander Meiklejohn, "The Return to the Book" (1924), in C. Hartley Grattan, ed., *American Ideas About Adult Education, 1710–1951* (New York: Teachers College, 1959): 124–128.

[37] Stubblefield, *Towards a History of Adult Education in America*, 22, 24, 27, 29–30, 33; Ellen Condliffe Lagemann, *The Politics of Knowledge: The Carnegie Corporation, Philanthropy, and Public Policy* (Middletown: Wesleyan University Press, 1989), 100 (quoted); Alan Lawrence Jones, "Gaining Self-Consciousness While Losing the Movement: The American Association of Adult Education, 1926–1941" (PhD dissertation, University of Wisconsin-Madison, 1991), 62–63, 67, 69–70, 77–79, 119–120; Kett, *The Pursuit of Knowledge Under Difficulties*, 379–381.

processes of those much younger than Thorndike's subjects. At Columbia, it was a rare Teachers College professor – F. Ernest Johnson was one, and then only partially – who challenged the overall contours of Dewey's conception of education and vision of cultural change. Figures such as William H. Kilpatrick, George S. Counts, and John L. Childs developed various elements of Dewey's program, sometimes in collaboration with their mentor himself. The faculty there also ran an experimental "New College," oriented toward socioeconomic problems, for teachers in the 1930s. At Ohio State, meanwhile, the philosopher Boyd H. Bode offered broadly Deweyan conceptions of education and of science. In fact, teacher-training programs in Washington, Colorado, and elsewhere provided the sole venue in which survey courses in the natural sciences truly flourished between the wars. As an ever-widening array of critics recognized by the 1930s, the teachers' colleges represented the most effective – if not the most theoretically sophisticated – means of imbuing the public with new conceptions of ethics and the human person.[38]

THE CRITICS

Scientific democrats in the teachers' colleges and elsewhere believed that the conceptions of science and its cultural meaning that came to prevail in the popular mind would ultimately shape the contours of American life, even if college graduates wielded an outsized cultural influence. But these figures hardly stood in the same relation to the general public as they did to their students. Within the universities, the scientific democrats could influence many, though not all or even most, of the images of science and ethics available to students. In the wider public arena they faced much more powerful competitors, most of whom sought to keep science confined within narrow limits. As the 1920s gave way to the 1930s, a steadily expanding group of critics, both inside and outside the universities, worked to combat the cultural sway of the modern social sciences and philosophy.

   Adhering to a variety of religious and humanistic worldviews, these critics of "scientism" – the extension of scientific methods and concepts into the sphere of the human – represented a much more potent threat to the scientific democrats' program than had the New Humanists and anti-evolutionists of the 1920s. Believing that naturalistic philosophies shaped American culture most effectively through the social sciences, not biology, critics of scientism challenged the growing influence of what they took to be the mechanistic views of the person characteristic of those disciplines. They argued that social scientists claimed for themselves, and largely enjoyed, the exclusive right to shape social

[38] Ellen Condliffe Lagemann, "Prophecy or Profession? George S. Counts and the Social Study of Education," *American Journal of Education* 100, no. 2 (February 1992): 137–165; Herbert M. Kliebard, *The Struggle for the American Curriculum, 1893–1958* (New York: Routledge, 1995), 151–174; "New College: Addressing the Persistent Problems of Living," available online at http://www.tc.columbia.edu/news/article.htm?id=2911; Earl J. McGrath, *Science in General Education* (Dubuque, IA: William C. Brown, 1948), 195, 299.

values. Coupled with the assumption that social science was militantly materialistic, this meant that social scientists were directly imposing an amoral outlook on society. Such arguments took strength from, and also reinforced, the widespread popular identification of science with materialism.[39]

These critics of scientism disagreed on the question of science's neutrality. Most argued that social action should stem from a combination of the neutral, technical knowledge produced in the social sciences and normative principles derived from either literature and the arts or a religious tradition. A minority of the critics declared such neutrality impossible, insisting that all putatively detached knowledge actually embodied a hidden system of values, and that the social sciences should be infused with a more traditional normative frame.[40] But the two groups found common ground in echoing the claim of Noah Porter, James McCosh, Charles Eliot Norton, Irving Babbitt, and other earlier critics that the modern scientific enterprise could not in itself provide adequate guidance for social action.

Among these critics of scientism, Catholic neo-Thomists such as Fulton J. Sheen stood farthest from the centers of academic prestige. But they were highly organized and widely influential among a large segment of the public, especially in the big industrial cities. From the 1920s onward, American Catholic leaders worked to shed their image as backward-looking medievalists. Taking their cue from the writings of Thomas Aquinas, these modern scholastics identified the natural law – a set of moral prescriptions inscribed in the world and accessible through reason – as a reliable guide to action in any and all social contexts. In its emphasis on a divine system of moral government, neo-Thomism echoed key claims of nineteenth-century Protestant political thought that still resonated powerfully among American citizens.[41]

---

[39] On the era's broad tensions over cultural authority, see Edward A. Purcell Jr., *The Crisis of Democratic Theory: Scientific Naturalism and the Problem of Value* (Lexington: University Press of Kentucky, 1973), esp. 139–158; David A. Hollinger, "Science as a Weapon in *Kulturkämpfe* in the United States during and after World War II," in *Science, Jews, and Secular Culture: Studies in Mid-Twentieth-Century American Intellectual History* (Princeton: Princeton University Press, 1996): 155–174; and James Gilbert, "A World Without John Dewey," in *Redeeming Culture: American Religion in an Age of Science* (Chicago: University of Chicago Press, 1997): 63–93.

[40] The theologian Reinhold Niebuhr took the latter path, as had Porter earlier: Daniel F. Rice, "Niebuhr's Critique of Religion in America," in *Reinhold Niebuhr Revisited: Engagements with an American Original*, ed. Rice (Grand Rapids, MI: W. B. Eerdmans, 2009), 332–337; Daniel Breslau, "The American Spencerians: Theorizing a New Science," in *Sociology in America: A History*, ed. Craig Calhoun (Chicago: University of Chicago Press, 2007), 54. By contrast, the New Humanists adopted the neutrality line: Reuben, *The Making of the Modern University*, 217–218.

[41] William M. Halsey, *The Survival of American Innocence: Catholicism in an Era of Disillusionment, 1920–1940* (Notre Dame: University of Notre Dame Press, 1980), 138–168; Philip Gleason, *Contending with Modernity: Catholic Higher Education in the Twentieth Century* (New York: Oxford University Press, 1995), 105–136, esp. 117–119. See also Arnold Sparr, *To Promote, Defend, and Redeem: The Catholic Literary Revival and the Cultural Transformation of American Catholicism, 1920–1960* (New York: Greenport Press, 1990), 78–80. Gleason's account demonstrates that Catholics also echoed common-sense theories of knowledge.

Close students of political and legal theory, neo-Thomists targeted the scientific democrats' claim to possess a new cultural foundation for democracy. They argued that democracy rested on popular adherence to the natural law and could not survive in its absence. In fact, in a pattern common to all of the critics of scientism, neo-Thomists found John Dewey's politically and culturally engaged form of naturalism far more dangerous than the work of those natural scientists that simply carried out their research and eschewed attempts to guide public culture. American neo-Thomists portrayed Dewey's writings – especially his pedagogical prescriptions – as the entering wedge of a relativistic attack on the fundamental moral truths underpinning human action. In part because of their efforts, Dewey's instrumentalism became the symbol of everything materialistic and industrial among critics of scientism in the 1930s.[42]

Not that the other critics lacked reasons to target Dewey and his allies. The "neo-orthodox" Protestant theologian Reinhold Niebuhr shared many of the scientific democrats' political views, but he disagreed sharply with them on questions of cultural authority. He sought "a prophetic religion" that would "appropriat[e] what was true in the Age of Reason," rather than ceding the sphere of social knowledge to scientific investigators in the manner of Protestant modernism. Niebuhr thus adopted a kind of separate-spheres argument, claiming that the human sciences could not proceed without incorporating Christian insights, whereas the natural sciences could. Among political progressives outside the universities, Niebuhr's writings helped to fuel a growing reaction against naturalism in the human sciences.[43]

So, too, did the work of the independent cultural critic Lewis Mumford. Mumford did not speak in the voice of Christianity, however. Instead, he arrayed the authority of modern biology against scientism in the human sciences. Ascribing to Dewey and Charles Beard a "New Mechanism" that modeled social inquiry on physics, Mumford found a more satisfying basis for a naturalistic understanding of human behavior in emergent evolution, with its emphasis on individual freedom and moral choice. He saw there an "organic humanism" that would harmonize science with art and thus heal a cultural rift between facts and values that dated back to the disintegration of the medieval synthesis. Like the New Humanists, Mumford regarded literature and the arts,

---

[42] A classic statement of interwar neo-Thomism is Fulton J. Sheen's *God and Intelligence in Modern Philosophy* (New York: Longmans, Green, 1925). On Sheen, see Thomas C. Reeves, *America's Bishop: The Life and Times of Fulton J. Sheen* (San Francisco: Encounter Books, 2001) and Kathleen L. Riley, *Fulton J. Sheen: An American Catholic Response to the Twentieth Century* (Staten Island: St. Paul's/Alba House, 2004).

[43] Quoted in Daniel F. Rice, *Reinhold Niebuhr and John Dewey: An American Odyssey* (Albany: State University of New York Press, 1993), 87; R. Laurence Moore, "Secularization: Religion and the Social Sciences," in William R. Hutchison, ed., *Between the Times: The Travail of the Protestant Establishment in America, 1900–1960* (New York: Cambridge University Press, 1989), 245–248; Richard Wightman Fox, *Reinhold Niebuhr: A Biography* (Ithaca: Cornell University Press, 1996); Robert B. Westbrook, *John Dewey and American Democracy* (Ithaca: Cornell University Press, 1991), 523–532.

rather than science or religion, as the proper source of social ideals in a modern, industrial democracy.[44]

The New Humanism itself briefly flowered into a national movement in 1930. That year saw an intense exchange of articles, the appearance of edited volumes for and against the New Humanism, and a Carnegie Hall discussion featuring Irving Babbitt and the literary critics Henry Seidel Canby and Carl Van Doren that drew three thousand spectators. A loose alliance also began to form between New Humanists and religious traditionalists. Previously, the New Humanists had steered clear of orthodox forms of theism, but when faced with a choice, they feared scientism far more. "You must either be a naturalist or a supernaturalist," demanded the poet T. S. Eliot of his mentor Babbitt in 1929. Babbitt, conceding the need to take sides, chose supernaturalism. Only religion, he noted, acknowledged that even the most fractious modern possessed "an immortal essence presiding like a king over his appetites." Babbitt hoped to borrow the cultural authority of the Christian tradition for his secular version of psychological dualism.[45]

A turn to the Western heritage as a source of ethical guidance also characterized the Great Books approach, which flowered into a major cultural force in the 1930s. Not coincidentally, this pedagogical model first emerged at the two leading centers of social-scientific thought, Columbia and the University of Chicago. Columbia's John Erskine led the way in the 1920s, establishing a General Honors program wherein students read a series of classic works and discussed them in small groups, thus gaining unmediated access to the wisdom of the Western tradition. By the mid-1920s, then, there were two distinct forms of general education at Columbia as General Honors proceeded alongside Contemporary Civilization. The proponents of these two curricular models shared a desire to embed industrial production in an ethical matrix, but they implicitly disagreed on where the needed values and cultural authority lay.[46]

[44] Lewis Mumford, "Towards an Organic Humanism," in C. Hartley Grattan, ed., *The Critique of Humanism: A Symposium* (Port Washington: Kennikat Press, 1968 [1930]), 340, 359; Casey Nelson Blake, *Beloved Community: The Cultural Criticism of Randolph Bourne, Van Wyck Brooks, Waldo Frank, & Lewis Mumford* (Chapel Hill: University of North Carolina Press, 1990), 220; Robert B. Westbrook, "Lewis Mumford, John Dewey, and the 'Pragmatic Acquiescence,'" in *Lewis Mumford: Public Intellectual*, ed. Thomas P. Hughes and Agatha C. Hughes (New York: Oxford University Press, 1990); Westbrook, *John Dewey and American Democracy*, 380–387; Paul Forman, "How Lewis Mumford Saw Science, and Art, and Himself," *Historical Studies in the Physical and Biological Sciences* 37 (2007): 271–336; Shuxue Li, *Lewis Mumford: Critic of Culture and Civilization* (New York: Peter Lang, 2009).

[45] Hoeveler, *The New Humanism*, 25; Norman Foerster, ed., *Humanism and America: Essays on the Outlook of Modern Civilization* (New York: Farrar and Rinehart, 1930); Grattan, *The Critique of Humanism*; quoted in Sparr, *To Promote, Defend, and Redeem*, 82; Irving Babbitt, "Humanism: An Essay at Definition," in Foerster, *Humanism and America*, 39. Ronald Schuchard has challenged the common portrait of Eliot as an anti-Semite, noting his close friendship with Horace Kallen: "Burbank with a Baedeker, Eliot with a Cigar: American Intellectuals, Anti-Semitism, and the Idea of Culture," *Modernism-Modernity* 10, no. 1 (January 2003): 1–26.

[46] Alex Beam, *A Great Idea at the Time: The Rise, Fall, and Curious Afterlife of the Great Books* (New York: PublicAffairs, 2008), 17; Robert A. McCaughey, *Stand, Columbia: A History of*

The Depression brought such tensions to the fore, as a 1932 conference at New York University on "The Obligation of Universities to the Social Order" revealed. One speaker after another offered the now-familiar claim that academic scholars had created the modern world and bore a responsibility to address its ills. In the words of conference chairman Henry Pratt Fairchild, an NYU sociologist and sometime eugenicist, the natural sciences had "released economic and social impulsions for which no rational directions were indicated, and no adequate controls provided." But solutions were close at hand. Fairchild argued that the university's denizens now rejected "cloistered tranquility" and took their direction from "the needs of the community," understanding that "the organization and functioning of the university must be as dynamic and plastic as the social order itself." Like many of the other speakers, Fairchild insisted that modern universities should give their graduates a system of ethical values in addition to scientific knowledge.[47]

But the participants at the 1932 conference, like Columbia's advocates of general education, differed on the source of those values. NYU chancellor Elmer Ellsworth Brown recommended distilling "comprehensive social objectives" from research findings, in the form of a "higher synthesis" of the results of the physical and especially the social sciences. Walter Lippmann, rapidly souring on the social sciences, instead emphasized a set of universal intellectual virtues – "qualities of insight, imagination, and judgment which transcend the immediate and the momentary." For his part, Union Theological Seminary president Henry Sloane Coffin described each discipline itself as a "great tradition." Borrowing an argument from academic humanists, Coffin promised that students steeped in a disciplinary tradition would be "nonconformists to the vogue of the moment because they are conformed in some degree to the spirit of the master minds who influence the centuries." In the end, none of these outlooks found majority support. As Fairchild summarized, speakers of widely divergent backgrounds and views agreed on the "danger of the overdevelopment of the narrowly scientific outlook itself" and wanted the universities to provide "an essentially philosophical interpretation and guidance." But a key question remained unanswered, namely how to create the link between "spiritual values and specialized scientific expertness."[48]

Columbia University in the City of New York, 1754–2004 (New York: Columbia University Press, 2003), 285–299. Erskine's goals for the program appear in "General Honors at Columbia," *New Republic* 32 (October 25, 1922, Educational Supplement): 13, which built on Erskine's earlier essay collection *The Moral Obligation to be Intelligent, and Other Essays* (New York: Duffield, 1915).

47 Henry Pratt Fairchild, "Editorial Foreword" and "Retrospect and Prospect," in Fairchild, ed., *The Obligation of Universities to the Social Order* (New York: New York University Press, 1933), xv, 481, 483.

48 Elmer Ellsworth Brown, "Summary of the Conference"; Walter Lippmann, "The Permanent Elements in Social Change"; Henry Sloane Coffin, "Timeless Elements in Education"; and Fairchild, "Retrospect and Prospect," in *ibid.*, 449–450, 452, 261, 481, 483, 487, 491. David A. Hollinger assesses this conference in "Two NYUs and 'The Obligation of Universities to the Social Order' in the Great Depression," in *Science, Jews, and Secular Culture*, 60–79.

Open conflict over this issue broke out at the University of Chicago in the 1930s. New president Robert M. Hutchins had come to believe, as dean of Yale's law school, that science could not speak to value questions. He hired the philosopher Mortimer Adler to shake up the Chicago faculty, notorious for its support of formally nonpartisan and often quantitative approaches to scholarship. Adler, a sharp critic of Dewey's pragmatism during his student days at Columbia in the 1920s, insisted that all empirical inquiry rested on a foundation of metaphysical principles. He held that the supposedly neutral knowledge put forward by Chicago's social scientists was actually shot through with ethical and political commitments. Although he was a non-observant Jew, Adler sought to place the scientific enterprise on a foundation closely akin to the neo-Thomists' natural law.[49]

Significantly, both Hutchins and Adler developed their views about the relationship between science and values in direct opposition to the "legal realists" of the day. Whereas most other interwar human scientists sought to extricate themselves from the strict empiricism of the nineteenth century, advocates of a more scientific jurisprudence were pushing *toward* such a form of empiricism. Although cognizant of the complexities of modern epistemology, the legal realists were engaged in a struggle against the nineteenth-century vision of moral government that had been won decades earlier in most of the other disciplines. Popular and legislative understandings still built on the quasi-Platonic view that a nation's laws were simply pale imitations of a higher order of moral law. Additionally, the need to establish an authoritative base for courtroom decisions led them to speak more reverentially of scientific facts than did their peers in other fields. Latecomers to the rhetoric of science, legal realists often denied that hypotheses or generalizations of any kind – in other words, human constructions – had any place in their proposed "science of the law."[50] No wonder, then, that morally sensitive scholars such as Hutchins and Adler came to view social science as an inadequate basis for a modern philosophy of life. To both men, as to the legal scholars they criticized, "science" appeared in the guise of a narrow, mindless empiricism.

[49] Harry S. Ashmore, *Unseasonable Truths: The Life of Robert Maynard Hutchins* (Boston: Little, Brown, 1989); Mary Ann Dzuback, *Robert M. Hutchins: Portrait of an Educator* (Chicago: University of Chicago Press, 1991); William H. McNeill, *Hutchins' University: A Memoir of the University of Chicago, 1929–1950* (Chicago: University of Chicago Press, 1991), 34–38, 70. Outraged responses by scientific thinkers include Alfred North Whitehead, "Harvard: The Future," *Atlantic Monthly* 159 (1936): 260–270 and John Dewey, "Rationality in Education" (1936), in Jo Ann Boydston, ed., *The Later Works, 1925–1953, Volume 11* (Carbondale: Southern Illinois University Press, 1981), 391–396.

[50] Laura Kalman, *Legal Realism at Yale, 1927-1960* (Chapel Hill: University of North Carolina Press, 1986); Morton J. Horwitz, *The Transformation of American Law, 1870–1960: The Crisis of Legal Orthodoxy* (New York: Oxford University Press, 1992); John H. Schlegel, *American Legal Realism and Empirical Social Science* (Chapel Hill: University of North Carolina Press, 1995); Brian Z. Tamanaha, "Understanding Legal Realism," *Texas Law Review* 87, no. 4 (March 2009): 731–785.

To the extent that they managed to confine science within such conceptual limits, Hutchins and other critics of scientism dramatically narrowed the debate over science's relationship to culture in the 1930s. Hutchins, Adler, Lippmann, Mumford, Niebuhr, and a host of neo-Thomists framed the issue as the relationship between science – defined in advance as incapable of speaking to normative, evaluative questions – and a set of cultural commitments that took the form of answers to such questions. On the other side of the debate, meanwhile, the emergence of these new common enemies led scientific democrats either to close ranks, despite wide differences regarding science's epistemological and methodological orientations, or to fall on one another, seeking to establish once and for all which version of scientific democracy they should advocate against the likes of Hutchins and Adler. Although many of the institutional changes they favored would be attained in the 1930s and 1940s, the scientific democrats would never again regain the cultural initiative.

Like the tensions between scientific democrats themselves, this wider struggle between aspiring cultural leaders in the 1930s represented an internecine conflict among groups sharing a common enemy and broadly similar goals. All of these figures decried visions of the American future that subordinated "higher" values to the success of commercial enterprise. In a rare conciliatory moment, the Harvard New Humanist Louis J. A. Mercier acknowledged that his empirically minded opponents followed him in rejecting "mechanistic determinism, impressionism, the mood of disillusion and of escape, skepticism as a substitute for thought and effort, pseudo-realism, diffusiveness, formlessness, meaningless imagery, boob-baiting, and 'cackling-hen' criticism." From the other side of the divide, the philosopher Paul Arthur Schilpp listed Babbitt, More, and A. Lawrence Lowell alongside James, Dewey, and Meiklejohn as the voices of a new academic vision still awaiting realization. Ralph Barton Perry likewise held, in the relatively quiet days of the late 1920s, that the "humanism" of the literary scholars and the "humanitarianism" of the scientists represented, despite their different moods, part of a single movement of "modern secular progressivism," aimed at directing man's newfound power over nature to ethical ends and fighting naive materialism and ethical nihilism wherever they appeared.[51] These groups, however, never managed to make common cause. Instead, members of each aligned themselves with less culturally engaged – and in most instances less politically progressive – peers who shared their views on the superiority of scientific or religious ways of knowing. In doing so, they

[51] Louis J. A. Mercier, *The Challenge of Humanism: An Essay in Comparative Criticism* (New York: Oxford University Press, 1933), 256–257; Paul Arthur Schilpp, "The Most Critical Failure of the American College," in Schilpp, ed., *Higher Education Faces the Future* (New York: Horace Liveright, 1930), 217–218; Ralph Barton Perry, *General Theory of Value: Its Meaning and Basic Principles Construed in Terms of Interest* (New York: Longmans, Green and Co., 1926), 15. Cf. Hoeveler, *The New Humanism*, 27; David Hollinger, "The Knower and the Artificer, *with* Postscript 1993," in *Modernist Impulses in the Human Sciences, 1870–1930*, ed. Dorothy Ross (Baltimore: Johns Hopkins University Press, 1994), 42–43.

missed a historic opportunity to explore more nuanced conceptions of the scientific enterprise and to fight effectively against those who would subordinate intellectual life to the imperatives of capitalism.

Moreover, the scientific democrats did not fare as well as their humanistic and religious counterparts in the 1930s. In that decade's cultural battles, scientific thinkers faced unique obstacles in promoting their version of the shared ethical program, because their opponents managed to control the meaning of the term science in American cultural politics. In fact, by the late 1930s the critics of scientism would routinely insist that all attempts to bring a scientific outlook to bear on human values powerfully aided America's totalitarian enemies by destroying the intellectual basis for a normative commitment to democracy. To analyze any commitment scientifically, in this understanding, was to reduce it to a matter of cold, amoral facts – to eliminate that which made it a matter of human concern. Scientific democrats gave as good as they got, with Dewey and others charging that Hutchins and even Niebuhr joined the neo-Thomists in postulating "fixed and eternal authoritative principles as truths that are not to be questioned" and from which all behavior must be "deductively derived."[52] Here, insisted many scientific democrats, lay the true root of totalitarianism. But the deck was stacked against these figures, given that the prevailing public understanding of science rendered it incapable of serving as a basis for a democratic culture. To the extent that the scientific democrats managed to clear cultural space for science in subsequent decades, it would be science writ small, a matter of narrowly cognitive knowledge about physical and biological nature rather than an ethically robust conception of the human scene.

---

[52] John Dewey, "President Hutchins' Proposals to Remake Higher Education" and "Rationality in Education," in *The Later Works, Volume 11*, 398, 393.

PART III

SCIENCE AND POLITICS

"No philosopher in the world, however great, can help believing a million things on trust from others or assuming the truth of many things besides those that he has proved.... So somewhere and somehow authority is always bound to play a part in intellectual and moral life."

– Alexis de Tocqueville

The period from the start of the New Deal to the McCarthy years witnessed the spread of value-neutral conceptions of science and their growing entanglement with a managerial form of liberalism. Deweyan formulations of scientific democracy actually enjoyed their greatest support in the 1930s. But the political shifts of that decade – the launch of the New Deal and the rise of totalitarianism, followed by the coming of World War II – inaugurated profound changes in the epistemological claims and theoretical content of the human sciences, where scientific democracy had found its main home since the early twentieth century.

Superficially, both the natural and the human sciences benefited substantially from the New Deal, World War II, and the Cold War. These massive national endeavors brought greater financial support, opportunities to appeal to new audiences and patrons, and even direct avenues of employment. But the persistence of a crisis atmosphere for more than three decades silently ate away at the deliberative tendencies of scientific democracy, eliminating the perceived need for public discussion of the ends that the political system as a whole should pursue. This lack of normative open-endedness, sometimes erroneously credited to the impact of value-neutral theories of knowledge, was actually a pervasive feature of mid-twentieth-century American politics that profoundly shaped epistemological discourses and most other aspects of society. The era's immense and practically unchallengeable national projects tended to mute difficult questions about what the American public wanted and what the public good entailed. The answer always seemed clear: Create jobs. Win the war. Defeat communism. Deliberative understandings of democracy became attenuated as policy debate shrank to a set of choices between alternative means to predetermined ends.

Certain renderings of scientific inquiry fit relatively easily into the emerging structures of postwar American society, as did certain methods, findings, and applications. Others did not, and these faded into the background or disappeared entirely. The result was the image of science that most Americans carry around today. Science, in this understanding, means the disinterested, experimental study of physical and biological phenomena, undertaken within a set of massively capitalized and minutely coordinated institutions. However, this image did not represent either a timeless essence of science or a logical outcome of the campaign for scientific democracy. Nor was the mid-century turn toward value-neutrality either an interwar phenomenon that paved the way for the postwar military-industrial complex or a straightforward response by postwar scholars to the emergence of that complex. All of these interpretations miss the salience of conflicts between groups with competing cultural programs and the capacity of the nonscientific combatants to shape public perceptions of science's social meaning. Political outcomes arise through complex interactions between multiple groups of actors, and politically relevant conceptual changes are no exception to that rule.

The postwar understanding of science as a politically and morally detached form of inquiry represented a joint product of several groups of thinkers: physical scientists and engineers seeking to distinguish their work from the politically controversial social sciences; Congressional leaders and other political figures who saw social-scientific research as thinly veiled advocacy for socialism and atheism; the growing number of human scientists who, for various reasons, gravitated toward the value-neutral stance of the natural sciences; and humanistic scholars and religious leaders – including political liberals such as Reinhold Niebuhr as well as conservative evangelicals and Catholics – who portrayed themselves as the sole authorities on questions of public morality. Despite their profound differences, each of these groups had a strong interest in establishing today's prevailing view that science cannot, by definition, speak to normative questions. Their efforts catalyzed a profound transformation in the political meanings of American science.

Amid the era's political shifts, the post-Progressive version of scientific democracy that dominated the human sciences between the wars reached a high-water mark in the 1930s and generated remarkable new theories of knowledge and language. But it simultaneously began to metamorphose into something quite different from what the Progressive theorists could have imagined, something less concerned with public culture and less attentive to the psychological and sociological dimensions of behavior. The most radical decade of the twentieth century also inaugurated a long retreat from the lofty cultural ambitions of most interwar scientific democrats. In itself, the New Deal probably would not have caused this transformation of scientific democracy. But the conservative backlash against Roosevelt's initiatives, followed by ideological changes associated with the military campaigns against fascism and communism, led human scientists to adopt a far less interventionist stance toward American public culture. From the late 1930s onward, human scientists increasingly spoke of protecting core institutions from erosion, rather than breaking them down and building them back up. With the forces of conservatism aroused and the institutional victories of the New Deal era under their belts – or so they thought – postwar scientific democrats sought to consolidate their political gains.

To be sure, the New Deal did not conform particularly well to the ideals of many scientific democrats. Its extensive, if piecemeal, regulatory efforts seemed legitimate to them only to the extent that these rested on a foundation of public commitment to a humanitarian framework of social ethics. Many of the social scientists employed by federal agencies tried to direct popular support for economic relief into cooperative and deliberative channels. In fact, leaders of the Departments of Education and Agriculture helped spearhead a national discussion movement that had ordinary citizens vigorously debating policy questions across the nation. Department of Agriculture administrators also sought to turn that agency itself into an avenue for decision-making by citizens.[1] But powerful business interests inevitably put their stamp on Roosevelt's reforms, producing something like the corporate liberalism that Walter Weyl had feared. Worried about the New Deal's centralization of power, technocratic tendencies, and friendliness toward business, John Dewey and many others favored alternative approaches based on the "self-directed activities of autonomous groups performing necessarily social functions." To these skeptics, Roosevelt's experts and bureaucracies seemed positively threatening, absent a public sensitized to the common good and empowered to enforce that good by its possession of effective power and a scientific attitude.[2]

This was all the more true because economists, and to a lesser extent sociologists and political scientists, became major players in Washington during the 1930s. Roosevelt took on social scientists as top administrators and advisers, aligning his presidency with the academic disciplines to a degree unimaginable

---

[1] David Goodman, "Democracy and Public Discussion in the Progressive and New Deal Eras: From Civic Competence to the Expression of Opinion," *Studies in American Political Development* 18 (2004): 81–111; Robert Kunzman and David Tyack, "Educational Forums of the 1930s: An Experiment in Adult Civic Education," *American Journal of Education* 111, no. 3 (May 2005): 320–340; Jess Gilbert, "Democratic Planning in Agricultural Policy: The Federal-County Land-Use Planning Program, 1938–1942," *Agricultural History* 70 (1996): 233–250; Andrew Jewett, "Philosophy, the Social Sciences, and the Cultural Turn in the 1930s USDA," *Journal of the History of the Behavioral Sciences* (forthcoming). Cf. James T. Kloppenberg, "Deliberative Democracy and the Problem of Poverty in America," in *The Virtues of Liberalism* (New York: Oxford University Press, 1998), 100–123.

[2] Dewey voted four times for the Socialist Party's Norman Thomas over Roosevelt: Robert B. Westbrook, *John Dewey and American Democracy* (Ithaca: Cornell University Press, 1991), 440–452 (quote on 452), 458, 461. Other critics included W. C. Allee, Gordon Allport, L. L. Bernard, Charles Ellwood, and Howard Jensen: Gregg Mitman, *The State of Nature: Ecology, Community, and American Social Thought, 1900–1950* (Chicago: University of Chicago Press, 1992), 55; Ian A. M. Nicholson, *Inventing Personality: Gordon Allport and the Science of Selfhood* (Washington: American Psychological Association, 2003), 193; Robert C. Bannister, "Principle, Politics, Profession: American Sociologists and Fascism, 1930–1950," in *Sociology Responds to Fascism*, ed. Stephen P. Turner and Dirk Käsler (New York: Routledge, 1992), 184.

228   *Science and Politics*

just a few years earlier. Institutional economists, including Columbia's Rexford G. Tugwell and Wesley C. Mitchell, figured prominently in the administration's decision-making process and public profile. Tugwell, who had taught for many years in the CC course, became a leading member of the fabled Brain Trust that offered policy guidance to the president. Meanwhile, Mitchell directed a pair of massive empirical ventures, *Recent Social Trends in the United States* (1933) and *Technological Trends and National Policy* (1937). Many other social scientists populated Roosevelt's "alphabet agencies," often bringing the progressive economic views of the institutionalists Thorstein Veblen and John R. Commons into the very heart of American governance.[3]

Social scientists' engagement with the state accelerated a set of vituperative debates over epistemology and methodology, especially among sociologists. Seeking to appear reliable and noncontroversial, many of the government-employed social scientists carried forward the nonpartisan rhetorical stance characteristic of earlier work for foundations and the SSRC. Mitchell, for example, declared his reports purely empirical and technical, with no policy implications.[4] In the disciplines, however, a group led by the sociologists Charles A. Ellwood, Robert M. MacIver, Pitirim A. Sorokin, and Robert S. Lynd faced off against William F. Ogburn and other advocates of strict value-neutrality. This was not a merely two-sided dispute, as the participants' epistemological claims lined up in varying ways with their views on state activism. But epistemology became the front along which human scientists mobilized in their battles of the 1930s. Despite their profound differences, Ellwood, MacIver, Sorokin, and Lynd stood together against a detached, nonnormative conception of social science. In fact, a surprisingly large number of scholars during the New Deal years rejected the idea that they sought neutral knowledge applicable to any and all ends. To many, that conception sounded suspiciously like the moral relativism of the Nazis.

The era's political conflicts also called attention to the fact that scientific reasoning, like all forms of thought, took place in specific social contexts. While Mitchell and Ogburn stressed value-neutrality, other theorists described science as shot through with values and inextricably intertwined with cultural and economic forces. Such claims proliferated in the early days of the emerging fields of philosophy of science, sociology of science, history of science, intellectual history, and linguistics, although even the most context-sensitive theorists stopped short of full-blown cognitive relativism and argued that science offered unique benefits unavailable from other cultural practices.

Within a few years, however, contextual approaches became intellectually and culturally marginal in the United States. Those scientific democrats

[3] Mark C. Smith, *Social Science in the Crucible: The American Debate Over Objectivity and Purpose, 1918–1941* (Durham: Duke University Press, 1994), 49–83; Theodore Rosenof, *Economics in the Long Run: New Deal Theorists and Their Legacies, 1933–1993* (Chapel Hill: University of North Carolina Press, 1997), 28–43; Michael A. Bernstein, *A Perilous Progress: Economists and Public Purpose in Twentieth-Century America* (Princeton: Princeton University Press, 2001), 73–90.
[4] Smith, *Social Science in the Crucible*, 74–78.

advocating contextual views failed to establish a mainstream, publicly recognized conception of science that highlighted its cultural embeddedness and included interpretive or hermeneutic forms of analysis. In each of the aforementioned professionalizing disciplines, the founding paradigms instead reinforced the emerging image of an autonomous, rationalistic scientific enterprise. Meanwhile, in the arena of what might be called "popular epistemology," contextualists found their efforts scotched by the overlapping efforts of multiple groups who deemed science thoroughly value-neutral. Each of these culturally influential groups defined science as – for better or for worse – a politically and morally detached practice.

The struggle against totalitarianism made it particularly easy for religious and humanistic thinkers to press the case against scientific forms of social knowledge. Niebuhr, Lewis Mumford, the neo-Thomists, and other long-standing critics charged that scholars and citizens who were in thrall to science's moral relativism possessed no basis on which to choose democracy over its terrifying, inhuman rivals. More prominent figures joined the chorus too, including the poet Archibald MacLeish, then serving as Librarian of Congress, and Walter Lippmann, America's most prominent journalist and an erstwhile ally of the cultural scientists who fiercely criticized the New Deal and science-centered reform efforts after 1935.[5] During the war, these highly visible critics helped to ensure that ordinary citizens viewed science as ethically and politically sterile. The postwar years brought an additional surge of public criticism by religious leaders claiming that democracy rested on Christian or Judeo-Christian foundations. Meanwhile, right-wing anti-communists, who often combined theistic and political critiques, became a major force in national politics and targeted even profoundly anti-communist human scientists, equating their naturalistic vision with atheistic Marxism.

It was not institutional religion alone that gained from this mid-century contest over moral authority. After World War II, the humanities became more visible, more influential, and friendlier to modern industrial society. Their practitioners – the Great Books and Book-of-the-Month advocates, the extollers of Western civilization, the middlebrow defenders of modernism for the masses – stepped into the cultural role that leading human scientists had long claimed for themselves. These humanists, who stood at a much safer distance from Marxian materialism and determinism than did the human scientists, took up the task of sustaining American democracy by shaping the culture at large. They claimed the ability to advance the cause of freedom by introducing Americans to great artistic movements and traditions of humanistic learning. Meanwhile, they repeatedly denied that science could speak to the normative

[5] Lewis Mumford, "The Corruption of Liberalism" (1940), reprinted in *Values for Survival: Essays, Addresses, and Letters on Politics and Education* (New York: Harcourt Brace, 1946): 20–44; Archibald MacLeish, *The Irresponsibles: A Declaration* (New York: Duell, Sloan and Pearce, 1940); Walter Lippmann, *An Inquiry Into the Principles of the Good Society* (Boston: Little, Brown, 1937).

questions at stake in postwar politics. Not all humanists adopted this project, by any means. Lionel Trilling and many other disaffected champions of modernist literature had become deeply suspicious of ordinary citizens by the 1950s. Yet in the universities, the torch of democratic culture-building clearly passed from human scientists to humanists. The arts and literature, and the humanists who interpreted them, emerged as major sources of authority in a society whose citizens were self-consciously assuming cultural leadership of the West.

As the human scientists contended against their many critics, a new political economy of science, centered on the nexus of the physical sciences, engineering fields, high-tech industrial firms, and the military, coalesced around them. This development, too, generated powerful images of a morally and politically detached science. Top physical scientists and engineers had long stood beyond the sway of the post-Progressive project. They saw free markets and industrial innovation, perhaps augmented by liberal Christianity and the voluntary sharing of technical advances between firms, as the solution to modern social problems. Although Vannevar Bush and other science administrators gradually set aside their suspicion of the state, reconciling themselves to federal funding for wartime research and later for its peacetime equivalent, they insisted that the scientific community could function only if it stood apart from the rest of society.

In this, as in many other respects, leading physical scientists believed that the social sciences had failed utterly. They deemed the social sciences insufficiently value-neutral, citing either their practitioners' political leanings or the intrinsic slipperiness of the subject matter. Many physical scientists, disturbed by the cultural initiatives and ethical claims of their counterparts, joined conservative politicians in seeing the social sciences as hotbeds of socialism and materialism. More broadly, postwar physicists and even some chemists rejected the claim of social relevance central to many versions of scientific democracy. These "atomic scientists" came out of the war with more cultural authority than they had ever expected – provided they did not direct it against the nascent arms race, as they quickly learned. Fearing that the public and policymakers now expected them to single-handedly cure domestic ills and defeat communism abroad, many powerful researchers and administrators hedged against future disappointments – and present political entanglements – by downplaying science's relevance to practical problems. The Rockefeller Foundation's Warren Weaver and others wanted citizens and politicians to see science as "a noble intellectual and artistic pursuit," not a source of "gadgets, rockets, antibiotics, [and] plastics."[6] Long inclined to trumpet science's industrial contributions, prominent physical scientists now worked to ratchet down expectations for its real-world impact in an age of atomic fear and technological enthusiasm.

As they worked to break the association in the public mind between science and industrial production, many of these figures sought to forge a new link

---

[6] Quoted in John Rudolph, *Scientists in the Classroom: The Cold War Reconstruction of American Science Education* (New York: Palgrave, 2002), 170.

between science and art. Science administrators such as James B. Conant and J. Robert Oppenheimer described science as an exercise in pure, creative freedom. These physical scientists put science's products on a par with the most magnificent works of art and traced them to the same salutary rejection of social conformity. Hardly a source of bombs and refrigerators – let alone reconstructed social institutions – science was instead a grand achievement of the human spirit. Conant, Oppenheimer, and others sought to integrate science into an essentially humanistic "Western tradition" by stressing its affinities with artistic creation. Both science and art, they argued, took their shape from individual acts of creative imagination, and both offered society, not practical results, but rather an essentially aesthetic pleasure coupled with shining exemplars of resistance to the prevailing cultural winds. Like art, science provided democratic citizens – at least those with the proper tastes – a model of individual freedom coupled with materials for reflection on the human condition.

Postwar physical scientists thus aligned themselves with professional humanists in and beyond the universities, rather than with social scientists. Humanists, like their scientific counterparts, identified the absolute freedom of the individual as the very heart and soul of democracy, as well as the source of all creative achievement. And they, too, held that public appreciation for the practical benefits of their work threatened to reduce it to an instrument of prevailing social norms, thereby eliminating its capacity to serve as a model of democratic freedom. The postwar convergence of physical scientists and humanists on a shared politics of individual autonomy found expression in a picture of the intellectual universe as divided between the natural sciences and the humanities, a view famously captured by the British physicist-novelist C. P. Snow in *The Two Cultures and the Scientific Revolution* (1959).[7] But while Snow saw a major disjuncture between the scientific and humanistic modes of thought, most American advocates of the two-cultures image instead deemed them harmonious, and in crucial respects even identical. Still, neither version of this two-cultures analysis left a place for the social sciences as distinct, authoritative bodies of knowledge. Science's most visible postwar advocates, including Conant and Oppenheimer, simply erased the social sciences from their portraits of modern thought. Meanwhile, postwar humanists filled much of the cultural space hitherto claimed by scientific democrats in the human sciences, increasingly winning the mantle of cultural leadership that the crankier New Humanists had failed to capture.

The image of a socially detached, creative individual, shared by postwar physical scientists and humanists, deviated sharply from the understandings of human psychology and social relations that underwrote many post-Progressive versions of scientific democracy. It ignored the modern human sciences' characteristic emphasis on the operation of irrational motives stemming from material needs, childhood experiences, and cultural environments. In fact, postwar

---

[7] C. P. Snow, *The Two Cultures and the Scientific Revolution* (New York: Oxford University Press, 1959).

physical scientists and humanists tended to eradicate the entire social dimension of human existence, or at least to confine it to the very local level in small professional communities. They described science and art as the products of individual minds grappling with a vast and potentially alienating cosmos.

Facing a new set of cultural and institutional configurations after the war, scientific democrats in the human sciences struggled to make sense of their public roles. The prevailing image of science as autonomous and value-neutral seemingly forced human scientists to choose between two avenues to public relevance and cultural authority. They could adopt the prevailing understanding of science and portray themselves as either neutral technologists of human affairs or pure theorists in the mold of Conant and Oppenheimer's physicists. Alternatively, they could abandon their self-understanding as scientific thinkers, cede the marker "science," and reinvent themselves as humanists, perhaps by taking up the normative defense of Western civilization.

However, a third option presented itself in the form of "consensus liberalism." This understanding of American society and history had taken shape during the early years of World War II. The impending conflict had provoked many human scientists to ardently defend their nation's culture and institutions, rather than challenging them from the standpoint of a rigorous ethical ideal. But such ideological initiatives cut against the ideal of political neutrality, which the fierce attacks of anti-New Deal critics made increasingly attractive. Consensus liberalism resolved that tension, and produced many other convenient effects, by positing that the American people, from the colonial period forward, had embraced a set of universally shared and broadly liberal values. Progressive and post-Progressive theorists had argued that the human sciences revealed the existence and function of an interpersonal sphere of communication, while simultaneously infusing it with science's technical and ethical insights. But according to those who developed consensus liberalism in the 1940s and 1950s, social investigation showed that, in the United States, this interpersonal sphere was already steeped in humanitarian ethical principles and needed only technical knowledge from the sciences to make those values effective. Many scientific democrats' long-standing emphasis on the psychological roots of social order now became a much more specific claim about the content of American "political culture" – namely, that it reliably imbued citizens with the ethical orientation that scientific democrats associated with science. Human scientists could produce a fully scientific society simply by reminding citizens of their own deeply held values and revealing how technical insights from the natural and human sciences intersected with them. In short, scientific scholars could serve as ethical guides without engaging in ethical innovation. Quite the opposite, in fact: the task at hand was ethical conservation.

Consensus liberalism also resolved many of the epistemological difficulties racking fields such as sociology by authorizing scholarly action on behalf of progressive values without requiring a particular stand on the vexed question of value-neutrality. The new outlook portrayed the liberal consensus as both an empirical fact and a normative imperative, at least for Americans. Thus, scholars who were actively engaged in implementing the public's liberal values could describe that task as either a neutral technical intervention or a normative enterprise. They could speak authoritatively on behalf of the public good without resolving their interminable disputes over epistemology. In fact, in the new understanding, social scientists were allowed, and perhaps even obliged, to challenge the conscious beliefs of the many Americans who claimed to reject liberal values. Conservatives, racists, and fundamentalists could all be seen as the true outsiders, either selfishly elevating their interests over the public will or lacking the moral self-consciousness and scientific acumen to make their core commitments manifest. Human scientists did not even need to breach the sacred obligation of value-neutrality in order to press liberal principles in the face of such errors. After all, they were simply channeling the public's values, not seeking to change them.

Consensus liberalism had important political advantages as well. Although these benefits helped make consensus liberalism a centerpiece of social-scientific thought in the early Cold War years, it originally emerged from the confluence of the struggle against totalitarianism with the domestic political battles of the late New Deal years. Facing stiff challenges from the right, many scientific democrats began to portray the egalitarian, cooperative principles they saw behind the New Deal – rather than competitive individualism – as the centerpiece of the American political tradition. Claiming the immense inertia of culture for their side, rather than blaming it for the strength of the opposition, the early advocates of consensus liberalism portrayed *laissez-faire* as a misguided interpretation of universally held liberal values rather than the defining commitment of a large and fundamentally conservative majority.

Putting the point too reductively, these theorists invented a tradition to support Roosevelt's departures from American political precedent – and, many hoped, to point beyond the New Deal toward a more extensive reform program. Like so many photographers, folklorists, folk musicians, documentary filmmakers, novelists, and cultural historians in the 1930s, they looked around them and discovered a usable American past. These scientific democrats contended that the nascent welfare state possessed the full weight of conformity – that it was backed by all of the irrational and emotional elements that caused people to fit themselves to the prevailing cultural mold. Consensus liberalism meant that disagreeing with the underlying principles of the New Deal meant disagreeing with the tenets of American democracy itself, which were embodied in both the nation's founding documents and a deep cultural consensus that long predated the Revolution. On this view, the liberal tradition, not the *laissez-faire* argument arrayed against it by misguided conservatives, was the essential element in American culture, and in the long run the most causally powerful element as well. Consensus liberalism framed political change as a matter of purification rather than transformation, of getting back to age-old American values rather than creating new ones.

It should come as no surprise that consensus liberalism took hold quickly, given the remarkably wide range of psychological and strategic advantages it offered. Rhetorically, it enabled scientific democrats to appeal to the better angels of the American public's nature, rather than telling citizens they had gotten everything wrong. Politically, it fostered unity among the many schools of reformist thought, which could adopt different interpretations of the liberal consensus and different interpretations of the institutional configurations it implied without calling attention to those variations. As anti-communism loomed larger in American politics, consensus liberalism also allowed scientific democrats to frame their critiques of American culture as celebrations of it, claiming loyalty to the nation (and denying it to their opponents) even as they challenged many of its policies and practices. Consensus liberalism also explained away the frustrating gap between prevailing beliefs and what scientific democrats believed to be the public good, by distinguishing between authentic and inauthentic public opinion. No longer did scholars need to view themselves as swimming against the stream in an individualistic, materialistic culture. Finally, and crucially, consensus liberalism reconciled the ideal of popular sovereignty with the fact of expert administration by lodging ultimate authority in the people without actually requiring the submission of key questions to public debate. Experts could speak for a liberal consensus that floated somewhere above or below the existing body of public opinion. Consensus liberalism solved all manner of problems for mid-twentieth-century human scientists by deeming them the authoritative spokesmen for American values.

Accordingly, cultural initiatives became less central to the human sciences after World War II, while the twin tasks of defeating communism and ensuring smooth economic growth took up more and more scholarly energy. To the extent that scientific democracy survived the war, it did so in a highly attenuated form. The leading postwar calls for a scientific culture had little of the economically critical edge of interwar versions, and the proximity of these proposals to the military-industrial complex and other instantiations of scientific expertise made them look to challengers like little more than rationalizations of the status quo. Although a handful of human scientists, most notably the anthropologist Margaret Mead and the sociologists David Riesman and C. Wright Mills, served as public critics in the 1950s, most of their contemporaries remade themselves as "behavioral scientists" and conceded to humanists the task of direct, normative engagement with citizens. These scholars described themselves as analyzing, in a purely disinterested manner, patterns of human behavior that stemmed from normative commitments. They emphasized the ubiquitous and constitutive role of psychological and cultural commitments in the behavior of ordinary individuals, but they denied that such commitments influenced their own behavior during the research process. This bifurcated explanation of behavior, ascribing sociologically purer motives to scientists than to laypeople, would come under sharp attack in the 1960s.

8

Science and Its Contexts

As Chapter 9 shows, the sociologist William F. Ogburn and other "neo-positivists" built on the post-World War I turn toward quantification and advocated a strictly value-neutral conception of social science in the 1930s. They proposed that investigators should bracket their own convictions, even when studying value-driven human behavior. But other theorists of the era viewed science as a real-world practice inextricable from normative concerns. This chapter reveals that the 1930s represented a high point of concern with a theme also ascendant in recent decades: the interpenetration of scientific inquiry with its social, cultural, and political contexts. The New Deal years witnessed a tremendous burst of theoretical innovation in this area, driven by the basic claim that human constructs – words, concepts, theories, presuppositions, frameworks – underlay all conscious mental activity, including scientific research.

Closely connected to scientific democracy, the contextual understandings of science that emerged in the 1930s took their shape from a desire to augment science's cultural influence rather than to challenge it, as is typical today. Trusting that the scientific community reflected the most progressive possibilities in the surrounding society, many advocates felt comfortable rejecting strong claims of disinterestedness and declaring science a culturally embedded practice. Thus, the interwar push to use scientific knowledge to reshape American culture fueled a parallel tendency to interpret the practice of science as part of an integrated cultural matrix. Bringing science closer, conceptually, to public culture seemed to open possibilities for fruitful interchange in both directions. The 1930s theorists described in this chapter saw their explorations into the nature of language, symbolism, theory construction, and other intellectual processes as the entering wedge for cultural change.

The first three sections of the chapter explore the fate of such analyses in the "science studies" subdisciplines – philosophy of science, sociology of science, history of science, and the closely related field of intellectual history – that took shape in the late 1930s. In each of these areas, strongly contextual understandings of science gave way in the World War II years to images of the scientific community as an autonomous, self-regulating mechanism, its activities shaping the surrounding society and culture but not reflecting them. A similar transition occurred in the scientific study of language, the subject of the chapter's final section. Interwar research on language raised even more radical possibilities regarding science's social entanglements and revealed the line beyond which scientific democrats could not go without abandoning their core commitment to scientifically driven cultural change.

A number of conceptually distinct theories of science developed side by side in the 1930s. Neo-positivists in Ogburn's vein portrayed human ends as private and subjective, untouched by the empirical findings of science. In this portrait, emotionally tinged commitments could only hinder the rational analysis of empirical data and thus needed to be excluded from the scientific process at all costs. That process, argued neo-positivists, eventuated in purely technical, value-neutral knowledge that was applicable to any imaginable ends.

A similar isolation of science from normative values characterized most versions of operationalism, which became the model of choice for many cutting-edge theorists in the New Deal years. Whereas Ogburn described the body of scientific knowledge as a set of photographic representations of reality, operationalists saw in it a set of tools for manipulating experience. In fact, they reduced the meaning of scientific concepts to nothing more than the specific operations employed in measuring them. Codified by the Harvard physicist Percy W. Bridgman in 1927 and picked up by leading psychologists in the 1930s, operationalism, like neo-positivism, comported well with the interwar turn to quantitative measurement as a means of eliminating bias.[1]

But these two theories of science hardly swept the field, as most of our histories would lead us to believe. A surprising number of American scholars in the 1930s agreed with John Dewey that emotional commitments and biological habits could hardly be isolated from cognitive beliefs within the common matrix of organismic behavior. In this view, empirical knowledge of causal relations profoundly shaped what human beings sought to achieve in the world, while such normative pursuits led them to develop empirical knowledge in the first place. Not all exponents of this conception of science learned it at the foot of Dewey, or even agreed with his specific formulations of it. But he served as a hero to so many of them that it is entirely appropriate to call this view "Deweyan." It was close in spirit to operationalism, insofar as it portrayed science as a tool for manipulating reality. But the Deweyans meant social reality, not the material reality of instruments and objects in the library. They drew on a non-reductive form of social psychology that described knowledge and purposive action as inseparable in practice.[2]

---

[1] Joel Isaac, *Working Knowledge: Making the Human Sciences from Parsons to Kuhn* (Cambridge, MA: Harvard University Press, 2012), 92–124.

[2] I call this position "Deweyan" rather than "pragmatist" because its adherents often self-identified as naturalists, or even behaviorists, rather than pragmatists. Moreover, Dewey himself eschewed the latter term between the wars, preferring "instrumentalism" or "naturalism." Revealingly, Walton Hamilton called the neo-positivists "scholastics" and their Deweyan critics "protestants": "A Deweyesque Mosaic," in *The Philosopher of the Common Man: Essays in Honor of John*

A fourth, "perspectivalist" conception of science appeared only sporadically in the 1930s, although the sociologist Talcott Parsons' writings would give it greater cache in the postwar years. This view, which we saw in the work of J. E. Creighton, defined the sciences as the products of divergent perceptual filters applied to a single, if highly complex, object of knowledge. Here, as for the neo-positivists, science remained a matter of representation rather than manipulation, although the entities represented could be actions or even purposes rather than simply objects moving in physical space. Like operationalism and the Deweyan theory, however, perspectivalism called attention to the indispensable role of symbols and concepts in the knowing process.[3] These three theories identified scientific knowledge as a product of human activity – at the very least, an experimental manipulation or a choice between competing frames of reference – rather than a mere imprint of reality on the passive mind of an observer. The conjunction of these new philosophies of knowledge in the 1930s fueled tremendous interest in the roles of presuppositions, concepts, models, heuristic devices, and frames of reference in scientific inquiry.[4]

Amid the ferment of these multiple, competing understandings of science, the initially intertwined subfields of philosophy of science, sociology of science, history of science, and intellectual history began to differentiate themselves in the late 1930s and early 1940s. Before World War II, Deweyan thinkers mingled freely with neo-positivists, operationalists, and others in a wide-open, interdisciplinary dialogue about the character of science. But as scientific democrats muted their criticism of American institutions and culture during the war years, and as physical scientists vocally asserted their own, context-free visions of science, Deweyan theories largely vanished from the professionalizing science studies fields. Meanwhile, leading operationalists forged points of connection with advocates of logical empiricism, a new philosophy of Central European origin, and with the neo-positivists who touted the objective, unbiased observer. The confluence of neo-positivism, operationalism, and logical empiricism enforced a narrow definition of science that excluded normative concerns – and the Deweyans who placed them at the heart of the knowing process.[5]

PHILOSOPHY OF SCIENCE

Although speculation on the deep meaning of scientific inquiry was coextensive with philosophy itself, the philosophy of science arose as a professional field in the United States only in the late 1930s. This development largely stemmed

Dewey to Celebrate His Eightieth Birthday (New York: Greenwood Press, 1968 [1940]), 148–149, 171.

[3] Perspectivalism exhibited strong affinities with German neo-Kantianism, and many of its adherents, especially at Harvard, followed German developments closely: Isaac, *Working Knowledge*, 16–27. However, I have chosen a term that implies no claim of direct influence.

[4] Peter Novick, *That Noble Dream: The "Objectivity Question" and the American Historical Profession* (New York: Cambridge University Press, 1988), 133–167.

[5] Cf. Novick, *That Noble Dream*, 293–301.

from the arrival of a group of Central European émigrés known collectively as logical positivists or, as they preferred in their later years, logical empiricists. Their provocative questions and sophisticated analyses brought a new focus – though hardly a consensus – to the broad, heterogeneous American discourse on the nature of scientific knowledge. That heterogeneity began to vanish as the philosophy of science took on the form of a professional discipline and Deweyan perspectives fell out of the field.[6]

As late as 1930, Dewey and Morris R. Cohen were two of the very few Americans who could be described as philosophers of science. The vast majority of those writing on the topic, including Ogburn, Bridgman, and Boas, worked in the natural or human sciences and explored methodological issues on the side. Still, their efforts proved sufficiently vigorous and numerous to help generate a journal, *Philosophy of Science*, in 1934. Meanwhile, Deweyan thinkers figured prominently in the *Journal of Philosophy*, run by the Columbia naturalists. Originally titled the *Journal of Philosophy, Psychology and Scientific Methods*, that journal provided a forum for what it called "scientific philosophy," a term later captured by the logical empiricists. Contributors in the 1930s skewed toward pragmatism, naturalism, and realism, although those of other stripes also published in what had become the nation's leading philosophical organ. Deweyan perspectives also appeared in the pages of the rapidly multiplying social science journals, where methodologically self-aware investigators sought to articulate the relationship between their work and broader intellectual and cultural concerns.[7]

By the late 1930s, in fact, a Deweyan conception of science had emerged as the leading option for scholars dissatisfied with the view of the investigator as an ivory-tower specialist. In academic circles, the move toward Dewey's approach was particularly pronounced among applied social scientists; among philosophically minded biologists; in human sciences that were not readily amenable to quantification, such as anthropology, philosophy, and history; and in public universities, especially those of the Midwest and upper South. All of these institutional locations featured scholars who either emphasized the practical results of their research or, conversely, worried about its lack of obvious

[6] Alan Richardson, "Philosophy of Science in America," in *The Oxford Handbook of American Philosophy*, ed. Cheryl J. Misak (New York: Oxford University Press, 2008): 339–374; George A. Reisch, *How the Cold War Transformed Philosophy of Science: To the Icy Slopes of Logic* (New York: Cambridge University Press, 2005); Bruce Kuklick, *A History of Philosophy in America, 1720–2000* (New York: Oxford University Press, 2001), 232–237.

[7] Reisch, *How the Cold War Transformed Philosophy of Science*, 96–117; Daniel J. Wilson, *Science, Community, and the Transformation of American Philosophy, 1860–1930* (Chicago: University of Chicago Press, 1990), 54; Kuklick, *A History of Philosophy in America*, 190; Cornelis de Waal, "Introduction," in *American New Realism, 1910–1920*, ed. de Waal (Bristol: Thoemmes, 2001), xiv; Alan Richardson, "Scientific Philosophy as a Topic for History of Science," *Isis* 99, no. 1 (2008), 92. After 1936, the *Journal of Symbolic Logic* also published certain kinds of inquiries in the field: Curt J. Ducasse and Haskell B. Curry, "Early History of the Association for Symbolic Logic," *Journal of Symbolic Logic* 27, no. 3 (1962): 255–258.

applications and stressed their work's relevance to the task of changing public attitudes.[8]

Many such scholars took inspiration from Dewey's claim that scientists were not "men and women who have so broken the bonds of habits that pure reason and emotion ... speak through them," but simply individuals who had adopted the "specialized infrequent habit" of empirical inquiry in order to address specific real-world problems. Dewey insisted that this habit had proven, over the centuries, to generate the most reliable predictions of causal relationships in the world of human experience. These predictions, he cautioned, could not be understood as pictures of the world – or rather, one could never know whether they were faithful pictures or not, because one could not see around or through these humanly constructed concepts to the underlying reality itself. Yet it was clear from the past progress of science, said Dewey, that empirical inquiry produced instrumentally useful guidelines for manipulating reality in the service of human values.[9]

Few Deweyans said more than this about the nature of scientific practice. Seeking primarily to change the public's mind about the meaning and utility of science, they traded heavily in criticism of classical images of objectivity. Rather than detailed descriptions of the scientific enterprise, they instead tended to offer sweeping denunciations of what the Ohio State philosopher and educational theorist Boyd H. Bode called "our whole Graeco-Christian tradition of transcendentalism." Bode and other Deweyans argued that any definition of truth as "a matter of conforming to an immediate structure of reality" was "extra-scientific, or even unscientific."[10]

Their critique of objectivity rested on a presumption that thinking, as a mode of activity, could not be isolated from other forms of action. Although practical action, moral valuation, and cognitive belief might be conceptually distinguishable, they were always intertwined in practice, rendering untenable any sharp separation between facts and values. There could be no "knowledge

[8] These are impressions based on many years of research; much work remains to be done in pinning down the extent and nature of Dewey's influence in the interwar cultural sciences. At least one physicist, Edward U. Condon, cited Dewey in adopting a functionalist view of knowledge as a tool for predicting future experience: Michael Day, "E. U. Condon: Science, Religion, and the Politics of World Peace," *Physics in Perspective* 10, no. 1 (March 27, 2008), 15.

[9] John Dewey, *The Public and Its Problems* (1927), in Jo Ann Boydston, ed., *The Later Works, 1925–1953, Volume 2* (Carbondale: Southern Illinois University Press, 1981), 334–336; Dewey, *Freedom and Culture* (1939), in *The Later Works, Volume 13*, 165. Helpful summaries appear in Robert B. Westbrook, *John Dewey and American Democracy* (Ithaca: Cornell University Press, 1991), 140–147; and William R. Caspary, "'One and the Same Method': John Dewey's Thesis of Unity of Method in Ethics and Science," *Transactions of the Charles S. Peirce Society* 39, no. 3 (2003): 445–468.

[10] Boyd H. Bode to Richard H. Shryock, February 29, 1944, in Shryock papers, American Philosophical Society, folder "Bode, Boyd H." One exception is Dewey student Joseph Ratner's lengthy introduction, personally approved by Dewey, to *Intelligence in the Modern World: John Dewey's Philosophy* (New York: Modern Library, 1939). But Dewey himself, in *How We Think* (1910) and elsewhere, offered a highly generalized account of the problem-solving process, not a precise description of the creation of scientific knowledge.

of objective facts," declared R. Freeman Butts of Columbia's Teachers College, because facts "depend upon the point of view of the one who does the discovering or the formulating." In turn, the observer's point of view took its shape from "the social situation, interests, biases, values, loyalties, and the whole 'scheme of reference' in which that knowledge is created." Still, Butts and other Deweyans rejected the charge of relativism. Butts hastened to deny that evidence could be "willfully overlooked or mutilated" in scientific inquiry: "The point of view ... must be as fair to other points of view and must be as broad and inclusive as the materials used will allow." Deweyans identified good communicative practices, rather than precise mechanical or quantitative techniques aimed at eliminating biases, as the ultimate source of reliable knowledge.[11]

That reliability could only be approximate, however. Deweyans rejected strict determinism in the realm of human action, tending to argue instead that the "laws" of society were mere statistical generalizations, abstractions from concrete cases that might prove highly trustworthy but were never infallible. Many Deweyans explicitly applied this analysis to natural laws as well as social ones, rejecting "fatalistic predeterminism" across all domains of science. Franz Boas suggested in 1928 that strict determinism, with its denial of purpose and contingency, was on the wane throughout the intellectual world. Max C. Otto stressed that no future outcome, whether natural or social, could be fully predicted by even the best science, because no two occurrences were exactly alike. He denied that so-called natural laws exercised coercive force, calling them instead "interpretations of the way specific things behave, arrived at by ingeniously relating particular occurrences to other occurrences not obviously of the same type." In a characteristic Deweyan sentiment, Otto described such generalizations as simultaneously human constructions and elements of an external, experienced world. The Deweyan conception of human ideals as desired future states of affairs required the existence of both free will and reasonably comprehensive generalizations linking actions to consequences.[12]

All of these commitments shaped the Deweyans' negative response to logical empiricism when its exponents arrived in the 1930s. They had long contended against a "separate spheres" strategy that erected a high wall between factual claims and normative ends and relegated the latter to some nonempirical sphere. Seeing this ethical transcendentalism as the common ground between apparently opposed movements such as neo-positivism and Christian traditionalism, the Deweyans argued that it disabled the project of science-based cultural reform by ruling off limits to empirical criticism the foundational normative commitments of the existing culture. When several of the leading logical empiricists emigrated from Austria and Germany to the United States, the

---

[11] R. Freeman Butts, *The College Charts Its Course: Historical Conceptions and Current Proposals* (New York: McGraw-Hill, 1939), 338, vii.

[12] Edwin Grant Conklin, *The Direction of Human Evolution* (New York: Charles Scribner's Sons, 1922), 188; Franz Boas, *Anthropology and Modern Life* (New York: W. W. Norton, 1928), 154; Max C. Otto, *Natural Laws and Human Hopes* (New York: H. Holt, 1926), 80–81.

Deweyans simply slotted them into their well-worn critique. Many philosophical idealists did the same. Having long since rejected the New Realists' version of the separate spheres approach, most American pragmatists, naturalists, idealists, and critical realists viewed logical empiricism as a major step backward in both the philosophy of knowledge and the project of scientific reform.[13]

Despite their deeply held – mostly socialist – political commitments, the logical empiricists gained a reputation for political passivity after arriving in the United States. On the whole, American philosophers both mistrusted and misconstrued the movement's core philosophical positions, wrongly identifying it as a version of neo-positivism. For example, American interpreters tended to fixate on the logical empiricists' early verificationism, which defined the meaning of a scientific statement solely in terms of its fit with lower-level, empirical sentences. In its original formulation, verificationism held that sentences lacking empirically testable referents should be banned from both scientific and philosophical discourse, being literally meaningless. But the leading logical empiricists dropped this theory by the mid-1930s and came to define meaning solely in terms of the internal coherence of a theoretical system. Few Americans noticed the shift. Similarly, most American critics misread the "physicalism" championed by many logical empiricists – the postulate that all of the valid sentences in any science could be translated into the spatio-temporal language of physics – as a crude reductionism or materialism that reduced all human communication to artificial, colorless terms.[14]

But most of the criticism focused on "emotivism," an ethical theory holding that value judgments fell into the category of meaningless – unscientific, non-referential – sentences. Few of the logical empiricists ever hewed cleanly to emotivism, and none of them meant by it that value judgments were unimportant or irrelevant to social action. But the movement's leading intellectual spokesman in the United States, the reticent Austrian Rudolf Carnap, came close enough to this position in his public pronouncements to cement logical empiricism's reputation as utterly dismissive of the ethical concerns shaping daily life. More than anything else, this reading of emotivism explained the widespread perception that logical empiricism was politically quiescent, countenancing a retreat into ivory-tower speculation and fruitless logic-chopping. Carnap and

[13] The following events are discussed in Andrew Jewett, "Canonizing Dewey: Columbia Naturalism, Logical Empiricism, and the Idea of American Philosophy," *Modern Intellectual History* 8, no. 1 (April 2011): 91–125. See Brand Blanshard, *The Nature of Thought* (London: G. Allen & Unwin, 1939), 405–455 for an idealist critique.

[14] Reisch, *How the Cold War Transformed Philosophy of Science*, esp. 27–56; Thomas Uebel, "Political Philosophy of Science in Logical Empiricism: The Left Vienna Circle," *Studies in History and Philosophy of Science* 36 (2005): 754–773; A. W. Carus, *Carnap and Twentieth-Century Thought: Explication as Enlightenment* (New York: Cambridge University Press, 2007), esp. 36; Cheryl J. Misak, *Verificationism: Its History and Prospects* (New York: Routledge, 1995), 55–79; Uebel, *Empiricism at the Crossroads: The Vienna Circle's Protocol-Sentence Debate* (Chicago: Open Court, 2007). Also see Gary Hardcastle and Alan Richardson, eds., *Logical Empiricism in North America* (Minneapolis: University of Minnesota Press, 2003).

his counterparts engaged with public questions in a very different way than did the Deweyans and many other American thinkers.[15]

This is not to say that Americans universally rejected the movement. Many human scientists committed to strict forms of behaviorism found logical empiricism congenial, though it appears that few of them fully understood it. For example, the psychologists Clark L. Hull and S. S. Stevens picked up many of the logical empiricists' terms, if not necessarily their underlying meanings. More consequentially, the young Harvard philosopher W. V. O. Quine saw in the work of Carnap and his compatriots a refreshing antidote to the American philosophical mainstream's morally inflected forms of naturalism, realism, and idealism. Like Hull and Stevens, Quine worked to reshape his discipline from within, pushing it away from the cultural orientation of the Deweyans.[16]

In contrast to the logical empiricists' philosophical claims, their Unity of Science Movement struck many Americans as a salutary effort at cultural reform. The Fifth International Congress for the Unity of Science, held at Harvard in September 1939, attracted many science-minded scholars. From the host institution came Bridgman, Talcott Parsons, the historian of science George Sarton, and the physiologist-methodologist Lawrence J. Henderson. Another sociologist, Louis B. Wirth, made the trip from Chicago, and Horace M. Kallen came from New York. Edward C. Tolman and a young B. F. Skinner, then a colleague and friend of the logical empiricist Herbert Feigl at Minnesota, also found much of value in the Unity of Science Movement. Even Dewey contributed, albeit reluctantly, to the first volume of the movement's massive *International Encyclopedia of Unified Science* project. Many of those who firmly rejected the logical empiricists' philosophical claims could nevertheless applaud and aid their attempts to raise science's public profile.[17]

But it was philosophical tenets, rather than a cultural program, that survived into the postwar years and shaped the professionalizing field of philosophy of science. In the interim, Chicago's Charles W. Morris, a student of

---

[15] The British philosopher A. J. Ayer's *Language, Truth, and Logic* (London: V. Gollancz, 1936) played a key role in identifying the movement with emotivism, as did C. L. Stevenson's "The Emotive Meaning of Ethical Terms," *Mind* n.s. 46, no. 181 (January 1937): 14–31 and *Ethics and Language* (New Haven: Yale University Press, 1944). A characteristic statement by Carnap is "Logic," in *Factors Determining Human Behavior* (Cambridge, MA: Harvard University Press, 1937): 107–118; cf. Reisch, *How the Cold War Transformed Philosophy of Science*, 47–53.

[16] Laurence D. Smith, *Behaviorism and Logical Positivism: A Reassessment of the Alliance* (Stanford: Stanford University Press, 1986); Kuklick, *A History of Philosophy in America*, 252–257; Joel Isaac, "W. V. Quine and the Origins of Analytic Philosophy in the United States," *Modern Intellectual History* 2 (2005): 205–234.

[17] Joergen Joergensen, "The Development of Logical Empiricism," in *International Encyclopedia of Unified Science, Volume 2, no. 9*, ed. Otto Neurath, Rudolf Carnap, and Charles W. Morris (Chicago: University of Chicago Press, 1951), 47–48; Westbrook, *John Dewey and American Democracy*, 403–404; Reisch, *How the Cold War Transformed Philosophy of Science*, 83–95. Dewey made the case for an empirical approach to values in his *Theory of Valuation* (1939), published as Volume 2, Number 4 of the *Encyclopedia*.

George Herbert Mead, and Ernest Nagel, who studied with Cohen and then Dewey before becoming the latter's colleague at Columbia, attempted to knit together logical empiricism with indigenous currents of pragmatism and naturalism. Whereas most American philosophers identified emotivism as the heart of logical empiricism, Morris and Nagel thought it was an unnecessary excrescence on a school that implicitly embodied the ethical engagements characteristic of Dewey's work. But Morris, Nagel, and others failed to make a Deweyan emphasis on the inseparability of empirical inquiry and valuation central to the new field. By the late 1940s, Morris had moved on to other subjects and Nagel, a temperamentally cautious civil libertarian, marked the outer edge of ethical commitment in a discipline dominated by the logical empiricists and their younger American successors. The latter group, led by Quine, rejected key tenets of logical empiricism but retained its highly analytical style, which they combined with a profound distaste for political commitment on the part of scholars. In the hands of Quine and his allies, postwar philosophy of science focused closely on causation, imputation, and other elements of scientific reasoning, saying little about the social and cultural settings in which that reasoning took place.[18]

SOCIOLOGY OF SCIENCE

The sociology of science would seem an obvious candidate to have addressed the embeddedness of the scientific enterprise in the postwar years. To a limited extent, that held true. But American sociologists of science had come to rely on a highly attenuated picture of the social context for research. Led by Columbia's Robert K. Merton, postwar sociologists of science focused on explicating the internal dynamics of a freestanding "scientific community," which they portrayed as set off entirely from the surrounding society. The field-defining sociological accounts of the World War II and early Cold War years stressed science's institutional autonomy, not its cultural matrix. In sociology of science, as in philosophy of science, a relatively narrow professional discipline emerged from the strikingly heterogeneous discourse of the 1930s.

Deweyans were less prominent in that interwar sociological discourse than in its philosophical counterpart. They concerned themselves with science's broad historical and cultural contexts, not the specifics of knowledge creation in local, disciplinary settings. Moreover, they devoted comparatively little attention to the natural sciences, locating the promise of science largely in its extension to

---

[18] Charles W. Morris, "The Unity of Science Movement and the United States," *Synthese* 3, no. 12 (November 1938): 25–29; Morris, *Foundations of the Theory of Signs, International Encyclopedia of Unified Science, Volume 1, no. 2,* ed. Otto Neurath, Rudolf Carnap, and Morris (Chicago: University of Chicago Press, 1938); Ernest Nagel, "The Fight for Clarity: Logical Empiricism," *American Scholar* 8 (1939), 47; Nagel, "The Eighth International Congress of Philosophy," *Journal of Philosophy* 31 (1934), 591–592; Kuklick, *A History of Philosophy in America,* 243–258; Isaac, "W. V. Quine and the Origins of Analytic Philosophy in the United States."

social affairs. The American field of sociology of science, which came to focus squarely on knowledge production in the natural sciences, emerged primarily from debates about the arguments of Soviet Marxists, of radical British scientists, and of the Hungarian-German sociologist Karl Mannheim.

At the 1931 International Congress of the History of Science and Technology in London, a group of Soviet scholars electrified the English-speaking participants by giving them their first close look at Marxist conceptions of science. The physicist Boris Hessen made a particular splash by arguing that Isaac Newton's natural philosophy took its shape from the economic agenda of the emerging bourgeoisie. The "formation of ideas," he insisted, must be "explained by reference to material practice," especially "productive forces and productive relationships." To use a contemporary distinction, Hessen saw commercial needs behind both "problem choice" and "theory choice" in Newton's work. First, he argued that practical issues related to navigation, water pumps, and cannons drew Newton to problems in mechanics. Then, at the level of theory choice, Hessen asserted that Newton's isolated, mechanical atoms in motion reflected the emerging capitalist mode of production, with its focus on individual preferences. In place of this "conservative nature," Hessen endorsed Friedrich Engels' "historical view of nature," which featured "the self-movement of matter" and a strong developmental progression over time.[19]

A group of British scientists, many of whom attended the Congress, responded enthusiastically to Hessen's approach. But their self-proclaimed Marxism differed markedly from that of the Soviets. The crystallographer J. D. Bernal and other British leftists rejected strict social determinism in problem choice and exempted theory choice altogether from the shaping force of society. In fact, the Bernalists tended to reverse the direction of influence between science and society, arguing that the fruits of empirical research would utterly transform social institutions once science had been freed from the constraints placed on it by competitive capitalism. End "the frustration of science," the Bernalists declared in a 1935 collaborative volume, and the whole population would enjoy the material well-being latent in science-based technology, preventing an otherwise inexorable slide toward fascism.[20]

---

[19] Boris Hessen, "The Social and Economic Roots of Newton's 'Principia,'" in *Science at the Cross Roads: Papers Presented to the International Congress of the History of Science and Technology* (London: Kniga, 1931), 4, 53, 39, 38. The decidedly heterodox Hessen may not have endorsed his own claims: Loren R. Graham, "The Socio-political Roots of Boris Hessen: Soviet Marxism and the History of Science," *Social Studies of Science* 15, no. 4 (November 1985): 705–722.

[20] Frederick Soddy et al., *The Frustration of Science* (London: George Allen & Unwin, 1935); Gary Werskey, *The Visible College: The Collective Biography of British Scientific Socialists of the 1930s* (New York: Holt, Rinehart, and Winston, 1978); William McGucken, *Scientists, Society, and State: The Social Relations of Science Movement in Great Britain, 1931–1947* (Columbus: Ohio State University Press, 1984); Werskey, "The Marxist Critique of Capitalist Science: A History in Three Movements?" *Science as Culture* 16, no. 4 (December 2007): 397–461. The Bernalists assumed that even upper-class scientists (as many of them were) could bracket their class interests at the level of problem choice.

Bernal and the biologist Julian Huxley (Thomas Henry's grandson) introduced this program to the readers of *Harper's* in the winter of 1934–1935. They found another American outlet for their writings in *Science & Society*, a Marxist journal whose foreign editors included Bernal and his compatriots Lancelot Hogben and Joseph Needham. Launched in 1936, *Science & Society* sold up to 10,000 copies in its early years. It featured a broad spectrum of socialist thought on scientific matters, keeping its distance from political parties and showing little hesitation to take on the Soviet Union despite several editors' membership in the Communist Party. Contributors in the late 1930s not only debated the scientific merits of Freudian psychoanalysis and attacked anti-materialist physicists such as Arthur Eddington and James Jeans but also published some of the earliest exposés of the abuses of Soviet genetics under Trofim Lysenko. For the *Science & Society* circle, the cause of scientific radicalism trumped that of socialism in one country.[21]

Meanwhile, Mannheim's *Ideology and Utopia*, informed by the fractious clashes of 1920s Central Europe and translated into English in 1936, provided a more politically mainstream focal point for the brewing debate over science's social relations. As Merton noted approvingly, Mannheim sought to turn various forms of ideology critique – claims that intellectual opponents were merely couching their self-interest in intellectual terms – into a generalized theoretical account of the interaction between ideas and their social matrices. American thinkers who were accustomed to seeing science as one social practice among many could and did find this approach amenable. But Mannheim, who taught at the London School of Economics after 1933, made two additional moves that riled many American scientific democrats. First, he seemed to exempt intellectuals as a group from the shaping influence of class interests. Updating the orthodox Marxists' portrait of the proletariat as a universal class capable of puncturing economically interested ideologies, Mannheim argued that a "socially unattached intelligentsia," drawn from socially diverse backgrounds and guided by noneconomic motives, would be structurally immunized against partisanship. Although members of the intelligentsia could not access reality in an unmediated fashion, they could produce reliable representations of it by reconciling and integrating the partial accounts offered by those from other, interested social groups. Second, Mannheim categorically exempted natural science ("formal mechanistic knowledge") from social influence, limiting the scope of his sociological analysis to other forms of cultural production.[22]

Most American reviewers of Mannheim's book found his sociological account of ideas old hat – "hardly more than an elaborate German statement of

---

[21] David Goldway, "Fifty Years of *Science & Society*," *Science & Society* 50, no. 3 (Fall 1986): 260–279; "The First Half Century: Reminisces and Reflections," *Science & Society* 50, no. 3 (Fall 1986): 321–341.

[22] Karl Mannheim, *Ideology and Utopia: An Introduction to the Sociology of Knowledge* (New York: Harcourt Brace, 1936); David Kettler and Volker Meja, *Karl Mannheim and the Crisis of Liberalism: The Secret of These New Times* (New Brunswick: Transaction, 1995).

what American philosophers and sociologists have together already made clear to themselves," wrote the philosopher T. V. Smith tartly. "Though individual thinkers do the knowing," Smith summarized, "society provides the individual with his mind…, furnishing him both its content and its configurations." George Herbert Mead had already said as much decades earlier, Smith declared. Plus, many reviewers added, Mannheim's American counterparts admitted that they spoke as members of social groups themselves, even when they used the voice of science. These reviewers took for granted that thinkers who recognized "the perspectival character of every specific bit of social knowledge" could produce "*relatively* more adequate knowledge." But for this very reason, they balked at Mannheim's apparent exemptions of natural science and the intelligentsia from social influences. "Wherein does this differ from the Marxian assertion that the 'proletarian vanguard' foresees the day of salvation?" complained Howard Becker, a University of Wisconsin sociologist.[23]

A heterogeneous group of Americans had already produced their own sociological accounts of knowledge production by the late 1930s and early 1940s. The literary theorist Kenneth Burke joined Mannheim in tracing modern thoughtways to social disharmony and questioning the viability of asymmetrical debunking strategies, although he hardly viewed the intelligentsia as socially detached. S. C. Gilfillan developed a sociological account of technological innovation that posited a far greater role for broad social needs than for individual genius. The Chicago School sociologist Florian Znaniecki undertook a detailed analysis of academic roles and institutions, as did Tulane's Logan Wilson. And the radical critic C. Wright Mills, at Columbia, traced what he saw as the complacent, moralizing stance of mainstream sociology to the small-town, rural roots of its producers.[24]

[23] T. V. Smith, review of Karl Mannheim, *Ideology and Utopia, International Journal of Ethics* 48 (1937), 120–121; Eduard C. Lindeman, review of Karl Mannheim, *Ideology and Utopia, Sociometry* 1 (1937), 263; Paul Arthur Schilpp, review of Karl Mannheim, *Ideology and Utopia, Philosophical Review* 49 (1940), 268; Howard Becker, review of Karl Mannheim, *Ideology and Utopia, American Sociological Review* 3 (1938), 261. Cf. C. Wright Mills, "Methodological Consequences of the Sociology of Knowledge," *American Journal of Sociology* 46, no. 3 (November 1940), 319. The University of Chicago's Louis Wirth framed the point more positively: "Preface," in Mannheim, *Ideology and Utopia*, xix-xx, xxx. By contrast, another group of reviewers who had received their intellectual training in Europe blasted the entire idea of a sociology of knowledge, denying that psychological and sociological forces could affect the validity of scholarly knowledge: Robert M. MacIver, review of Mannheim, *Ideology and Utopia, American Historical Review* 43, no. 4 (July 1938): 814–816; Hans Speier, review of Karl Mannheim, *Ideology and Utopia, American Journal of Sociology* 43 (1937): 159–161; Alexander von Schelting, review of Karl Mannheim, *Ideology and Utopia, American Sociological Review* 1 (1936): 674.
[24] Kenneth Burke, *Permanence and Change: An Anatomy of Purpose* (New York: New Republic, 1935), 68, 74; S. C. Gilfillan, *The Sociology of Invention* (Chicago: Follett, 1935); Florian Znaniecki, *The Social Role of the Man of Knowledge* (New York: Columbia University Press, 1940); Logan Wilson, *The Academic Man: A Study in the Sociology of a Profession* (New York: Oxford University Press, 1942); C. Wright Mills, "The Professional Ideology of Social Pathologists," *American Journal of Sociology* 49, no. 2 (September 1943): 165–180.

By deftly weaving together these strands of analysis and adding new elements of his own, Merton developed the American field of sociology of science. Trained at Harvard by the historian of science George Sarton and the historically minded sociologist Pitirim A. Sorokin, Merton wrote his dissertation and several early papers on the history of science and technology, viewed from a sociological perspective. He engaged directly and critically with the theories of scientific development put forth by Hessen, the Bernalists, and Mannheim, while also drawing on the work of Burke, Gilfillan, and many other Americans.[25] But Merton worried about the political effects of these studies, noting with alarm that "totalitarian theorists have adopted the radical relativistic doctrines of *Wissenssoziologie* [sociology of knowledge] as a political expedient for discrediting 'liberal' or 'bourgeois' or 'non-Aryan' science." For Merton, any stress on science's connection to culture represented an existential threat to it, rather than a necessary first step toward building a new society. Eschewing the Deweyan image of a social organism remaking its collective habits through empirical inquiry, Merton described the "scientific community" as one element in a pluralistic matrix of social subsystems, each featuring its own distinctive "ethos." Although these institutions followed different and often competing norms, Merton explained, their relative autonomy enabled them to coexist more or less harmoniously. In his view, only science's near-total isolation from other social institutions – above all, from the political system – ensured its success.[26]

It did so, Merton famously explained, by allowing scientists to operate solely in accordance with four behavioral norms comprising the "ethos of science," namely disinterestedness, organized skepticism, universalism, and communism, or the collective ownership of intellectual products. Merton's pluralistic model of society helped him to square recent insights about the potent influence over cognition of cultural presuppositions, social interests, and sheer irrationality with his belief that science produced reliable, universally applicable truths. He

[25] Almost every item in Merton's bibliography before 1942 cited one or more of these figures, culminating in "A Note on Science and Democracy," *Journal of Legal and Political Sociology* 1 (1942): 115–126. On this celebrated text, see David A. Hollinger, "The Defense of Democracy and Robert K. Merton's Formulation of the Scientific Ethos," in *Science, Jews, and Secular Culture: Studies in Mid-Twentieth-Century American Intellectual History* (Princeton: Princeton University Press, 1996): 80–96; Everett Mendelsohn, "Robert K. Merton: The Celebration and Defense of Science," *Science in Context* 3 (1989): 269–289; David Kaiser, "A Mannheim for All Seasons: Bloor, Merton, and the Roots of the Sociology of Scientific Knowledge," *Science in Context* 11, no. 1 (1998): 51–87; and Lawrence T. Nichols, "Merton as Harvard Sociologist: Engagement, Thematic Continuities, and Institutional Linkages," *Journal of the History of the Behavioral Sciences* 46, no. 1 (2010): 72–95.

[26] Robert K. Merton, "Science and the Social Order," *Philosophy of Science* 5, no. 3 (July 1938), 328; Hollinger, "The Defense of Democracy." It is relevant here that Merton was the son of poor Jewish immigrants: Merton, "A Life of Learning," in *Sociological Visions*, ed. Kai Erikson (Lanham: Rowman & Littlefield, 1997): 275–295. Unlike the culturally Protestant liberals and humanists who led the charge for scientific democracy until the 1930s, Merton could not simply presume that most Americans shared his basic commitments.

argued that the distinctive local culture of the autonomous scientific community – the ethos of science – turned the deeply fallible, irrational beings of
modern psychology and sociology into the selfless, disinterested investigators
of scientific lore. By channeling behavior into narrowly defined social roles,
Merton explained, the ethos of science transmuted each individual's desire for
acceptance and status into a collective pursuit of truth. Merton's scientific community amounted to a social machine for generating reliable, depersonalized
claims, an institutionalized subculture whose judgments approximated those
that would be reached by all of humanity reasoning together. He warned that
science would wither and die if scientists upset this delicate balance by adopting another institution's nonscientific norms, as the Nazis had forced them to
do and the Marxists, Bernalists, and Deweyans – and the capitalists – now
urged them to do voluntarily.[27]

Merton's philosophy of science deviated sharply from neo-positivism, despite
the shared emphasis on the dangers of extrinsic motivations. He outlined what
we would now call a theory of "underdetermination," centered on the claim
that many different generalizations can account for any set of empirical results.
From the group of possible laws exhibiting "truth," or "correspondence to the
facts," said Merton, scientists chose one on the basis of its "meaning," the
"intellectual satisfaction" it gave. To "'prove' a theory," he explained, a scientist needed to win over other specialists in the field, drawing not only on
the theory's truth – its empirical adequacy – but also on its meaning. And the
meaning, and thus the persuasive success, of a theory derived "largely from the
cultural scheme of orientation," said Merton. Here the scientific ethos came
in. It provided the cultural norms that scientists used to judge the meaning
of a proposed theory, and thereby ensured that, despite the theoretical indeterminacy of every body of empirical evidence, scientists would construe that
evidence in a manner comporting with the interpretive needs of humanity as a
whole. Merton had found a way to read the era's contextually sensitive views
of science as evidence that it should be isolated from the rest of society in order
to protect its distinctive subculture.[28]

Thus, while Merton followed Dewey in seeing the intellectual freedom characteristic of the scientific attitude as the result of a certain set of cultural norms
and habits – a "free culture," in Dewey's words – he had a profoundly different
sense of the near-term prospects for such a culture. Whereas Dewey held out
hope of transforming American culture as a whole, Merton believed that the
prevailing culture was far too strong and hostile for such a direct approach.

[27] Merton, "A Note on Science and Democracy." Merton adopted the "ethos" concept from
Sumner's *Folkways* and Hans Speier's "The Social Determination of Ideas," *Social Research* 5
(1938): 182–205. He also cited the Marxist conception of "institutional compulsives": "A Note
on Science and Democracy," 124.
[28] Robert K. Merton, "Science, Population and Society," *Scientific Monthly* 44, no. 2 (February
1937), 169. Also see Merton's review of George Herbert Mead, *Mind, Self and Society, Isis* 24
(1935): 189–191, where he argued that even science's core "assumption of an order of nature"
was "socially founded."

Although Merton took his own advice and rarely engaged political issues directly, he appears to have disliked the emerging political culture of the New Deal era almost as much as the expiring one. Rejecting the millenarian optimism of so many American human scientists, Merton saw the 1930s as a time of profound, even unprecedented, danger to the scientific enterprise. Many progressive thinkers thought capitalism was giving way to some form of socialism. By contrast, Merton's articles from the late 1930s and early 1940s reveal a persistent undercurrent of suspicion about corporate influences on the New Deal, along with mordant observations about the limits of social planning and the shortcomings of bureaucracies as tools for promoting human values.[29]

Indeed, Merton seems to have viewed the American polity's decentralized character as its sole redeeming feature. But it was a highly important one. Only in the United States were scientists not automatically subordinated to a powerful state or party. Merton seems to have hoped that if American science could be pried from the clutches of the capitalists, then it could perform its appointed historical task by gradually and indirectly – the more discreetly the better, so as not to awaken active opposition – spreading universalistic values throughout society. Merton shared with those fleeing Hitler's Europe a strong sense that science was fragile, that it was not widely endorsed by the general population, that it was not on the verge of ousting its cultural rivals, and indeed that it could be extinguished in a heartbeat if it were not fiercely and strategically defended. In the scientific community, carefully walled off from corrupting political entanglements, he saw the West's only remaining hope.

Merton's "ethos of science" would become a staple of postwar arguments about the proper institutionalization of scientific research. By that time, he would appear even less politically engaged than in his early writings, which referenced the wartime context by identifying a pluralistic, liberal democracy as the optimal setting for scientific research. After the war, with the Nazi threat neutralized and capitalism even more powerful at home, he banished the entire question of science's relationship to democracy from his writing, along with any discussion of competing answers to that question. Merton also added a significant element of self-interest – by way of eponymy, the honor of having one's name attached to a law or other scientific object – to his picture of the scientist's motives. Fine-tuning the balance between extrinsic rewards and internalized habits in his model, he brought it more closely into line with the self-seeking tone of postwar American life. Far earlier, however, Merton had prefigured the postwar discourse of scientific autonomy and detachment by domesticating the modern human sciences' seemingly disruptive emphasis on the impurity of human motives.[30]

[29] See especially Robert K. Merton, "Role of the Intellectual in Public Bureaucracy," *Social Forces* 23 (1945), 408; Merton, "The Unintended Consequences of Purposive Social Action," *American Sociological Review* 1 (1936), 898–899; and Merton, "Bureaucratic Structure and Personality," *Social Forces* 18 (1939), 565.

[30] Hollinger, "The Defense of Democracy," 82; Mendelsohn, "Robert K. Merton," 287; Merton, *Social Theory and Social Structure* (Glencoe: Free Press, 1957).

SCIENTIFIC HISTORIES

As philosophers and sociologists of science began to write for smaller, more specialized audiences in the 1940s, historians continued to paint influential portraits of science for a wider public. But the content of those portraits changed substantially. As elsewhere, the closely related fields of intellectual history and history of science each shed the contextually sensitive theories that had flourished in the preprofessional days of the 1920s and 1930s and coalesced around a non-Deweyan, institution-building "founder." Indeed, these fields soon came to be remembered as having been single-handedly created by Arthur O. Lovejoy and George Sarton, respectively. Important earlier contributions, particularly by figures drawn to Dewey's conception of science, have largely fallen out of the historical record.

The Deweyans saw the relationship between science and society as a two-way street, a view closely tied to their cultural-political project. In fact, many Progressive Era formulations of scientific democracy had directly implied such a view of science's social embeddedness. Progressive theorists frequently ascribed the influence of existing scientific ideas – particularly those in the human sciences – to the cultural power of conservative social groups. Thus, Dewey and other critics of classical political economy, empiricist epistemology, and ethical transcendentalism often traced these intellectual tendencies back to some combination of traditional Christianity, monarchy, and capitalism. As an antidote, these Progressive thinkers often called, not for purely neutral truths, but rather for forms of knowledge that explicitly embodied the normative foundations of what they took to be an emerging industrial democracy.[31]

Dewey's classic books of the 1920s, which proved a remarkably fertile decade for him, followed many of his earlier works by weaving holistic narratives of modern intellectual history. He sought to identify the common intellectual flaws shared by contending philosophical concepts or theories – usually a pair of alleged opposites that turned out to be closely aligned – and to demonstrate the pernicious historical effects produced by the application of these inadequate interpretations to social action. Dewey followed the recent trajectories of the core elements of modern thought, including concepts of knowledge, of nature, of human nature, of the state, and of individual freedom. He found the roots of each faulty theory in a bygone, undemocratic form of collective life and ascribed the present-day remnants of that social-political form to the lingering influence of the associated ideas among scholars and the general public. Taken together, Dewey's writings of the 1920s added up to a broad-gauged, philosophically committed intellectual history of the West.[32]

[31] Westbrook, *John Dewey and American Democracy*, 148, 347–361. For another example, see the analysis of shifting conceptions of the absolute offered by Isaac Woodbridge Riley, a Yale-trained philosopher influenced by William James: "The Fifth Meeting of the American Philosophical Association," *Journal of Philosophy, Psychology and Scientific Methods* 3 (February 1, 1906), 72.

[32] These were *Reconstruction in Philosophy* (1919), *Human Nature and Conduct* (1922), *Experience and Nature* (1925), *The Public and Its Problems* (1927), *The Quest for Certainty*

At Columbia, the overlapping influence of James Harvey Robinson's version of the New History and Dewey's naturalism generated a substantial body of interwar work that ranged freely across what we today would call intellectual history, history of philosophy, and history of science. In the philosophy department, many of Dewey's students filled out his broad sketches with detailed historical-philosophical studies of their own. John Herman Randall Jr. and Herbert W. Schneider led the way in developing Dewey's emphasis on the practical and social character of thought into a highly contextual, theoretically self-conscious mode of historical analysis. Columbia's philosophers jointly issued volumes of *Studies in the History of Ideas* in 1918, 1925, and 1935. Having taken undergraduate courses from Robinson and his student Carlton J. H. Hayes before moving on to graduate study with Dewey and his colleagues, these second-generation Columbia naturalists insisted that all philosophers needed a firm grasp on the history of ideas. The most famous product of this milieu was Randall's *The Making of the Modern Mind*, discussed in Chapter 6, which inspired a Peter Arno cover for the *New Yorker* and remained in print and on undergraduate syllabi for decades. While Randall continued his historical explorations and eventually produced the two-volume blockbuster *The Career of Philosophy* in the 1960s, Schneider wrote the definitive history of American philosophy in 1946 and helped found the leading journal in that field.[33]

Randall articulated the theoretical underpinnings of this work most clearly, adopting a view of the social relations of thought that his younger colleague Ernest Nagel dubbed "objective relativism" or "contextualistic naturalism" and deemed uniquely suited to democracy. Randall argued in 1939 that objectivity "means always being *objective for* something, just as necessity means always being *necessary for* something: there can be no objectivity without an objective." Like so many interwar scientific democrats, Randall viewed the context for science and philosophy in economic and political terms, linking nineteenth-century scientific materialism to the political power of the commercial middle class and looking beyond it to new conceptions of natural and social reality.

(1929), and *Individualism, Old and New* (1930). Hamilton argued that Dewey's thought would find its fullest development in "a genetic account of orthodox social theory": "A Deweyesque Mosaic," 161.

[33] Columbia University, Department of Philosophy, *Studies in the History of Ideas* (New York: Columbia University Press, 1918/1925/1935); John Herman Randall Jr., "On Understanding the History of Philosophy," *Journal of Philosophy* 36 (August 17, 1939): 460–474; Randall, *The Making of the Modern Mind* (Boston: Houghton Mifflin, 1926); Randall, *The Career of Philosophy* (New York: Columbia University Press, 1962–1965); Herbert W. Schneider, *A History of American Philosophy* (New York: Columbia University Press, 1946); Richard H. Popkin, "Founding Editor's Note," *Journal of the History of Philosophy* 40, no. 4 (2002): 417–419. On these figures' historical work, see Victorino Tejera, *American Modern, The Path Not Taken: Aesthetics, Metaphysics, and Intellectual History in Classic American Philosophy* (Lanham: Rowman & Littlefield, 1996), 176–194; Robert Piercey, "Doing Philosophy Historically," *Review of Metaphysics* 56, no. 4 (2003): 779–800; and Joseph L. Blau, "The Philosopher as Historian of Philosophy: Herbert Wallace Schneider," *Journal of the History of Philosophy* 10, no. 2 (April 1972): 212–215.

He thus described the mechanistic ideals of Bacon and Descartes as "natural to a society bent on mastering nature by mechanical means" and linked subsequent intellectual developments to shifts in the fortunes of the commercial middle class. Hessen's claim that modern science emerged from the commercial revolution was a staple of interwar Columbia naturalism, as was the further claim that scientists' more recent focus on the social realm mirrored the inclusion of the working class in the democratic process. With the partial exception of Nagel, the Columbia naturalists insisted on the mutual influence of social changes and intellectual developments such as the rise of science.[34]

Randall and Nagel also published a number of detailed, empirical articles on the history of science after the late 1930s. Indeed, Randall played a small but important role in that field's postwar solidification by demonstrating important continuities between medieval Aristotelianism and modern science. But Robinson's students made even more direct contributions to the field, extending his twin emphases on the power of ideas in history and the centrality of science in modern thought. Robinson trained several important historians of science, including Dorothy Stimson and Lynn Thorndike. Stimson, who taught at Goucher College in Baltimore, served for many years as president of the History of Science Society, while Thorndike's 1905 dissertation on *The Place of Magic in the Intellectual History of Europe* has been called the most important American contribution to the history of science prior to World War I. Another Robinson student, Cornell's Preserved Smith, featured science prominently in *A History of Modern Culture* (1930/1934), an important early contribution to intellectual history. Robinson's own writings focused broadly on the scientific attitude and its contemporary applications, but his students pushed on to the study of specific ideas, institutions, and figures.[35]

As the 1930s began, history of science was the best established of the American science studies fields, the History of Science Society (HSS) having emerged in 1924. Sustained work in the field dated back to the Progressive Era, and in some cases even earlier. Already, in 1915, Stanford's Frederick E. Brasch

---

[34] Ernest Nagel, "Philosophy and the American Temper" (1947), in *Sovereign Reason* (Glencoe, Ill., 1954), 51–53, 55, 57; Randall, "On Understanding the History of Philosophy," 472; quoted in Nathan Reingold, "Uniformity as Hidden Diversity: History of Science in the United States, 1920–1940," in *Science, American Style* (New Brunswick: Rutgers University Press, 1991), 369. Cf. Thomas Robischon, "What is Objective Relativism?" *Journal of Philosophy* 55, no. 26 (1958): 1117–1132.

[35] Ernest Nagel, "'Impossible Numbers': A Chapter in the History of Logic," in *Studies in the History of Ideas, Volume III*: 429–74; Nagel, "The Formation of Modern Conceptions of Formal Logic in the Development of Geometry," *Osiris* 7 (1939): 142–222; John Herman Randall Jr., "The Development of Scientific Method in the School of Padua," *Journal of the History of Ideas* 1, no. 2 (April 1940): 177–206; Randall, "The Place of Leonardo Da Vinci in the Emergence of Modern Science," *Journal of the History of Ideas* 14, no. 2 (April 1953): 191–202; Reingold, "Uniformity as Hidden Diversity," 367, 369; I. Bernard Cohen, "Eloge: Dorothy Stimson, 10 October 1890–19 September 1988," *Isis* 81, no. 2 (June 1990): 277–278; Marshall Clagett, "Eloge: Lynn Thorndike (1882–1965)," *Isis*, 57, no. 1 (Spring 1966): 85–89. Of course, Randall's *The Making of the Modern Mind* also centered on the growth of modern science.

had noted a burst of curricular growth in the history of science, amounting to 176 courses at 113 colleges and universities. He traced this development to the reaction against disciplinary specialization and free election. However, Brasch's data revealed multiple, conflicting rationales for such courses. Chicago's Charles R. Mann and Brasch's Stanford colleague Harold Chapman Brown offered a Deweyan portrait, arguing that scientists had actively and intentionally contributed to the rise of modern civilization. By contrast, John F. Woodhull, who worked alongside Dewey at Columbia's Teachers College, nevertheless held that scientists toiled solely for "the love of knowledge," not for social improvement. According to Woodhull, teaching the history of science would pull students away from their obsession with the practical and the relevant by honing their "esthetic faculties" and inculcating "lofty ideals, elevated conceptions and noble thoughts."[36]

This Victorian emphasis on refined character and high ideals found its most influential advocate in George Sarton. Sarton brought over his journal *Isis* from Belgium during World War I and subsequently shuttled between Harvard and the Carnegie Institution of Washington, authoring numerous influential books and devoting his prodigious energy and talent to building a professional discipline. Like Woodhull, he viewed science as one of the humanities, its selfless practitioners standing apart from the grubby details of the practical world. Sarton's writings featured paeans to "the love of truth, and the disinterested search for it, irrespective of desires and consequences." He favored this nobility of purpose in the historical study of science as well, framing the discipline as a pure research field centered on what he took to be the highest sciences, namely those formulated in mathematical terms. Indeed, Sarton saw the field as the opening wedge for a "new humanism" that would unite the sciences with the arts. "The history of science," he wrote in the first issue of *Isis*, "is the history of mankind's unity, of its sublime purpose, of its gradual redemption." Sarton's narrative of spiritual progress ignored the social sciences and associated reform efforts, along with the material contexts in which natural scientists plied their trade.[37]

Most Progressive Era contributions to the history of science had stemmed from either this Victorian idealism or simple antiquarian curiosity on the part of scientific practitioners. But between the wars, as we have seen, more closely contextual portraits of science emerged at a number of sites. One, of course, was Columbia, which featured a wide range of projects in the field. Columbia's

---

[36] Arnold Thackray and Robert K. Merton, "On Discipline Building: The Paradoxes of George Sarton," *Isis* 63 (1972), 483; Frederick E. Brasch, "The Teaching of the History of Science," *Science* 42 (November 26, 1915), 746–748.

[37] Thackray and Merton, "On Discipline Building," 479, 480 (quoted); George Sarton, "An Institute for the History of Science and Civilization (Third Article)," *Isis* 28 (February 1938), 13; Reingold, "Uniformity as Hidden Diversity," 363–365. See also Eugene Garfield, "The Life and Career of George Sarton: The Father of the History of Science," *Journal of the History of the Behavioral Sciences* 21 (April 1985): 107–117 and the articles in *Isis* 100, no. 1 (March 2009): 60–117.

official historian of science was the physical chemist Frederick Barry, the first
full-time professor of the subject in the United States. (Sarton taught only part-
time at Harvard.) Barry managed to combine a Sartonian emphasis on disinterested motives with a pragmatic account of truth akin to that of William
James. A music enthusiast, he portrayed the scientist as "a creative artist ...
the romanticist's simple child of nature," exuberantly traipsing through the
fields of pure experience. Barry sought a new form of humanistic education
centered on the "scientific habit of thought," the powerful combination of
tough-mindedness and secular idealism that he saw remaking modern societies.
Despite their epistemological differences, Barry found strong support from the
Deweyan naturalists in the philosophy department, who implored the administration to hire a replacement and sustain the field after Barry died in 1943.
On the sociological side, both Robert K. Merton and the committed Marxist
Bernhard J. Stern also contributed to Columbia's flourishing conversation on
the history of science.[38]

Meanwhile, the universities of the Midwest proved extremely hospitable
to Deweyan modes of historical analysis. Unlike most of their Eastern counterparts, Midwestern scholars tended to view science in thoroughly practical
terms. Thus, they embraced the idea that scientists took cues from the social
context, at least at the level of problem choice. The Iowa State historian Earle
D. Ross and other Midwesterners saw their roles in very different terms than
those in traditional liberal-arts institutions, given that they trained "future
agricultural leaders, public utility executives, labor organizers, [and] experts
in child care, public health, and dietetics," as well as numerous "governmental administrators and specialists, farmers, professional men and women, and
homemakers – responsible citizens all." Dewey's theory of science, with its
emphasis on the interplay between academic knowledge and public culture,
made implicit sense to many professors in the land-grant colleges.[39]

This curricular emphasis gave rise to numerous courses in the history of
science, which flourished as early as 1915 at Wisconsin, Illinois, Michigan,
Indiana, and Iowa State. Chicago's private universities, Northwestern and the
University of Chicago, also played important roles in developing the field. Like
Columbia, these metropolitan universities in a bustling commercial center
combined ample resources with a commitment to public service and an egalitarian sensibility. At Northwestern, Baker Brownell's interwar "Contemporary

[38] Frederick Barry, *The Scientific Habit of Thought: An Informal Discussion of the Source and
Character of Dependable Knowledge* (New York: Columbia University Press, 1927), 276, vii;
Lynn Thorndike, "Frederick Barry (1876–1943)," *Isis* 34, no. 4 (Spring 1943): 339–340; Ernest
Nagel to Frank Fackenthal, April 22, 1944, Ernest Nagel papers, Rare Book and Manuscript
Library, Columbia University, Box 1, folder "Nagel, Ernest"; Samuel W. Bloom, *The Word as
Scalpel: A History of Medical Sociology* (New York: Oxford University Press, 2002), 84, 96.
Stern was a co-founder and editor of *Science & Society* whose 1927 dissertation prefigured
Merton's (and predated Hessen's) analysis of the material context for the Scientific Revolution.
[39] Earle D. Ross, "History in the Land-Grant College," *Mississippi Valley Historical Review* 32
(March 1946), 579, 581.

Thought" program expanded on Progressive Era efforts by the historian of biology W. A. Locy. Also before World War I, the Chicago faculty tried to institute a coordinated set of courses on the history of the sciences. Although that effort collapsed, George Herbert Mead joined Charles R. Mann and others in teaching the history of ancient and modern science. After Mead died in 1931, his student Charles W. Morris took up the role of missionary for the history and philosophy of science. In 1939, Morris called on the land-grant schools to create departments in "the logic and the history and the sociology of science" and thereby fuel cultural regeneration in the heartland.[40]

By that point, Chicago's "Logic of Science Discussion Group" sponsored up to seventy lectures a year, and Wisconsin, Minnesota, Illinois, and Iowa State featured similar groups. At Iowa State, president Charles E. Friley personally led a faculty-wide effort to integrate new theories of science into the entire undergraduate curriculum. Some sixty to eighty professors met biweekly to discuss the process, and scholars in fields from economics to English incorporated insights from the history, sociology, and philosophy of science – especially those regarding the "mutual interaction between intellectual activity and the social environment" – into their writings and pedagogy.[41]

The University of Wisconsin, where work in the history of science dated back to the 1890s, emerged as an even more important base for the field. In 1941, six years after Edward A. Ross and fifteen colleagues established a local chapter of the HSS, Wisconsin created the first American history of science department. It housed a required sophomore course on the "History and Significance of Science," which was taught for the first time – and only time, as it turned out – by the Bernalist Lancelot Hogben, author of the bestsellers *Mathematics for the Millions* (1936) and *Science for the Citizen* (1938). Although Hogben's lectures were far too complex for the students, the faculty flocked to hear his socially grounded account.[42]

---

[40] Brasch, "The Teaching of the History of Science," 749–750, 753–756; George Herbert Mead, "The Teaching of Science in College," *Science* 24 (September 28, 1906): 390–397; Charles R. Mann, "The History of Science – An Interpretation," *Popular Science Monthly* 78 (1908): 313–322; quoted in Carl F. Taeusch's memo to members of the American Philosophical Association, Ernest Nagel papers, Box 26, folder "Agriculture, Department of."

[41] Victor L. Hilts, "History of Science at the University of Wisconsin," *Isis* 75 (March 1984), 66–67; Carl F. Taeusch memo; William H. Nicholls, "Social Biases and Recent Theories of Competition," *Quarterly Journal of Economics* 58 (November 1943), 1. See also E. W. Lindstrom, "Teaching to Think in a Field Rather than About It," *Iowa Academy of Science Proceedings* 49 (1942): 461–465; I. E. Melhus and G. C. Kent, "Application of the Group Conference Method of Teaching to Beginning Classes in Plant Pathology," *Iowa Academy of Science Proceedings* 50 (1943): 309–331; and Pearl Hogrefe, "Our Opportunities for Democracy Today," *College English* 1 (April 1940): 595–604. Iowa State was a major player in American graduate education between the wars, far outstripping other separate land-grant institutions in total enrollment and ranking thirteenth nationally in PhDs awarded. Earle D. Ross, *A History of Iowa State College* (Ames: Iowa State College Press, 1942), 356.

[42] Hilts, "History of Science at the University of Wisconsin," 63–64, 66–70; Curtis P. Nettels to Richard H. Shryock, February 5, 1941, Shryock papers, folder "History of Science Society #5."

But these Midwesterners remained peripheral to the HSS, whose powerful Eastern leaders tended to be less sensitive to science's social context and less attuned to the social sciences. Moreover, their social concerns did not carry through to the Cold War years. Sarton's 1930s work – his 1936 methodological statements *The Study of the History of Science* and *The Study of the History of Mathematics* and his application of those principles in the massive *Introduction to the History of Science* (1927–1948) – set the tone for the field after World War II. To be sure, few postwar historians echoed Sarton's expansive scheme for spiritualizing Western civilization. But they retained his sense that a sheer desire for knowledge, rather than social concerns, drove scientific inquiry. Scientists appeared in the leading postwar accounts as Sarton's disinterested heroes, not as members of cultures and social groups engaged with real-world problems.[43]

Harvard's president, the chemist James B. Conant, helped to build the postwar field and to define its less Victorian – and less Deweyan – tone. Although the hard-boiled Conant disliked much of Sarton's program, they both saw the history of science as central to a modern curriculum. Harvard initiated a PhD program in history of science three years after Conant took office in 1933. By the late 1930s, three professors taught the subject full-time alongside the polymathic L. J. Henderson. After the war, Conant's curricular reforms and semipopular books strengthened a growing sense that history of science courses best familiarized students with the scientific mode of thought. By hiring Thomas S. Kuhn and other young scholars, Conant helped to create a market for PhDs and to establish the field as a professional pursuit. But whereas Kuhn famously departed from the Sartonian model, most postwar practitioners, including Sarton's student Henry Guerlac at Cornell, portrayed science as a matter of disinterested, curious individuals probing the nonhuman world.[44]

[43] George Sarton, *The Study of the History of Science* (Cambridge, MA: Harvard University Press, 1936); Sarton, *The Study of the History of Mathematics* (Cambridge, MA: Harvard University Press, 1936); Sarton, *Introduction to the History of Science*, 3 vols. (Baltimore: Williams & Wilkins, 1927–1941). Of course, more radical forms of history of science could be found on the Eastern seaboard, where the Johns Hopkins historian of medicine Henry E. Sigerist and his Columbia counterpart Bernhard J. Stern endorsed Soviet-style economic planning. Meanwhile, HSS leaders attempted to redress the group's regional imbalance, though with limited success: Richard Shryock to Henry R. Viets, November 7, 1939; Shryock to Arno B. Luckhardt, February 23, 1942; Dana B. Durand to William J. Carson, May 6, 1942; and Shryock to Chauncey Leake, September 23, 1942; in Shryock papers, folders "History of Science Society #1," "History of Science Society #23," "History of Science Society #28," and "History of Science Society #31." But the intensely practical tenor of the Midwestern conversation never became mainstream in the national HSS.

[44] James B. Conant, "George Sarton and Harvard University," *Isis* 48 (September 1957), 304; I. Bernard Cohen, "George Sarton," *Isis* 48 (September 1957), 296; Cohen, "A Harvard Education," *Isis* 75 (March 1984), 16; Joy Harvey, "History of Science, History and Science, and Natural Sciences: Undergraduate Teaching of the History of Science at Harvard, 1938–1970," *Isis* 90 (1999): S270–S294; Thackray and Merton, "On Discipline Building," 494; Thomas S. Kuhn, "Professionalization Recollected in Tranquility," *Isis* 75 (March 1984), 29; Cohen, "The 'Isis' Crises and the Coming of Age of the History of Science Society: With Notes on the Early Days of

Arthur O. Lovejoy, who initiated a History of Ideas Club at Johns Hopkins in 1923, became Sarton's counterpart in intellectual history. He diligently developed the discipline, framing it as the study of purely cognitive worldviews and their component "unit-ideas." As we have seen, Lovejoy advocated logical and evidential reasoning in all spheres. His historical work traced long-term shifts in conceptions of the natural world, focusing on the domain in which natural scientists contended with philosophers and theologians. As early as his 1917 APA address, Lovejoy tasked philosophers with deconstructing the "highly accidental and highly unstable compound[s]" of unit-ideas comprising existing philosophical systems and putting those elemental building blocks into their logically entailed relations. According to Lovejoy's biographer, this historical methodology, as exemplified in *The Great Chain of Being* (1936) and defined in the introduction to that volume, aimed at establishing the existence of a transcendental order of "objective, verifiable, and clearly communicable truths" that enabled human beings to cognitively grasp their world.[45]

Contextual accounts of knowledge, including those written by the Columbia naturalists, appeared in the *Journal of the History of Ideas* (founded in 1940) during Lovejoy's editorship. But his own goal was to de-historicize human thoughtways by shaking the underlying, transcendental truths free of their historically arbitrary encrustations. By the 1950s, Lovejoy's rationalistic approach had become the dominant mode of intellectual history in the United States. Here, as in the history, sociology, and philosophy of science, the World War II years brought a substantial narrowing of an intellectual conversation in which Deweyan conceptions of science had jostled against more rationalistic, less contextual understandings. This transition accompanied a shift of research initiative and professional leadership away from the teaching-oriented figures

the Harvard Program in History of Science," *Isis* 90 (1999): S37–S42. By contrast, Columbia's efforts languished until 1961, when Randall, Nagel, and other supporters convinced their former student Jacques Barzun, now provost and dean of faculties, to establish a history of science program: John Herman Randall Jr. to Jacques Barzun, July 5, 1961 and Barzun to Randall, July 7, 1961, John Herman Randall Jr. papers, Rare Book and Manuscript Library, Columbia University, Box 1, folder "Barzun, Jacques." Cf. Reingold, "Uniformity as Hidden Diversity," 367.

[45] Daniel J. Wilson, *Arthur O. Lovejoy and the Quest for Intelligibility* (Chapel Hill: University of North Carolina Press, 1980), 139–156, 188; Arthur O. Lovejoy, "On Some Conditions of Progress in Philosophical Inquiry," *Philosophical Review* 26 (1917), 160, 163; Lovejoy, *The Revolt Against Dualism: An Inquiry Concerning the Existence of Ideas* (Chicago: Open Court, 1930). See also Wilson, "Lovejoy's *The Great Chain of Being* after Fifty Years," *Journal of the History of Ideas* 48, no. 2 (1987): 187–206 and Leo Catana, "Lovejoy's Readings of Bruno: Or How Nineteenth-Century History of Philosophy Was 'Transformed' Into the History of Ideas," *Journal of the History of Ideas* 71, no. 1 (2009): 91–112. John C. Higham's classic article on the field's development, which mentioned Robinson but ignored the philosophers, may have helped to cement Lovejoy's perceived status as its sole founder: "The Rise of American Intellectual History," *American Historical Review* 56 (1951): 453–471. Anthony Grafton briefly notes the field's early diversity: "The History of Ideas: Precept and Practice, 1950–2000 and Beyond," *Journal of the History of Ideas* 67, no. 1 (2006), 2.

at Columbia and the Midwestern schools toward the more professionalized programs at other Eastern universities.[46]

SCIENCE AND LANGUAGE

If interwar interpreters of modern science had the option of viewing science as a contextually grounded practice, those who sought to apply scientific techniques to the study of language were virtually compelled to do so. But this body of work also exhibited the theoretical boundaries of interwar contextualism, which almost always accompanied a desire to heighten science's cultural influence. The task of developing a scientific account of language pressed insistently on scientific democrats because language provided the medium of public culture, the ultimate target of their reform efforts. Various behaviorists, Boasians, and Deweyans joined hands in developing a more contextual, contingent understanding of language – the "tool of tools," in Dewey's words.[47]

Language had not been a major topic of scientific study in the Progressive Era, when most reform-minded scholars seem to have viewed it as a more or less transparent medium for transmitting beliefs and values. That changed drastically in the 1920s, as scholars in fields ranging from literature and the modern languages to anthropology, psychology, mathematics, and philosophy directed their attention to the scientific and political uses of language, and of symbolic systems more generally. By the end of that decade, claims about language's historicity, contingency, and influence over thought had become common coin in the cultural sciences. This theoretical shift gained some of its impetus from the political failure of Progressivism, which directed reformers' attention to the terms in which they had framed their proposals. It received another boost in the 1930s, when the growing use of propaganda techniques and fierce political battles over economic regulation and totalitarianism highlighted the essentially contested nature of terms such as equality, justice, and freedom. Theoretical work undertaken in this fertile interwar context birthed the American discipline of linguistics and substantially reshaped philosophy as well. Outside the academy, meanwhile, language became a major topic of discussion among the educated middle class by the late 1930s. By the start of World War II, linguistic insights had become central to the campaign to make America scientific.[48]

But scientific democrats repeatedly found ways of exempting scientific discourse from their observations about the linguistically conditioned character

---

[46] Wilson, *Arthur O. Lovejoy and the Quest for Intelligibility*, esp. 156. Andrew Abbott and James T. Sparrow find this eastward flow of prestige in the sociology discipline: "Hot War, Cold War: The Structures of Sociological Action, 1940–1955," in *Sociology in America: A History*, ed. Craig Calhoun (Chicago: University of Chicago Press, 2007), 294.

[47] Brigitte Nerlich, "The 1930s – At the Birth of a Pragmatic Conception of Language," *Historiographia Linguistica* 22, no. 3 (1995): 311–334; quoted in Robert B. Westbrook, *John Dewey and American Democracy* (Ithaca: Cornell University Press, 1991), 336.

[48] Stuart Chase stressed the connection to Progressivism: *The Tyranny of Words* (New York: Harcourt, 1938), 3.

of human thought. They treated science as a uniquely powerful tool for the reconstruction of language, rather than as one of many competing linguistic frames for interpreting reality. Even Edward Sapir and Benjamin Lee Whorf, whose names are indelibly linked to the concept of "linguistic relativity" – the idea that our thought is profoundly shaped by culturally conditioned linguistic filters through which we perceive the world – treated science as a privileged form of linguistic practice, unique in its capacity to link human needs to prevailing conditions. The underlying framework of scientific democracy made it impossible for these figures to treat science as simply contingent and culturally relative.

The pattern is clearly illustrated in the writings of Franz Boas, whose 1911 introduction to the Bureau of American Ethnology's *Handbook of American Indian Languages* laid important groundwork for interwar linguistics and pointed toward the idea of linguistic relativity. Unlike many of his successors, however, Boas compared highly divergent culture groups in order to sift out universal cultural traits. He deemed language invaluable for that comparative task: linguistic categories were far more pristine than other cultural phenomena, because their users almost never reflected on them consciously. Language patterns thus laid bare the "fundamental ethnic ideas" of a culture group, Boas wrote.[49]

Boas built on a psychological foundation much like that of William James. He argued that the external world appeared to human actors as an infinitely variable flux of perceptions, no two of which were actually identical, rather than a series of stable, discrete objects and actions. Linguistic concepts, in this view, embodied the categories of classification by which individuals grouped together, on the basis of formal similarities or common elements, aspects of the flux in order to communicate experiences to one another and reflect on them in their own minds. Gathering together a range of related but hardly identical experiences, each of these mental constructs represented – like a scientific theory – a fallible hypothesis about a world seen only dimly, through a dense fog of linguistic mediation. Boas held that language users inherited the vast majority of these hypotheses from their cultural surroundings, rather than inventing them individually. Thus, a language reified the categories of the prevailing culture by projecting them onto reality itself. It portrayed the group's favored abstractions as concrete entities in an external world rather than mere heuristic devices.[50]

Boas found this process of abstraction both necessary and dangerous. On the one hand, language use gave rise to society itself, by allowing individuals to

[49] Franz Boas, "Introduction," in *Handbook of American Indian Languages*, Bulletin 40, Part 1, Bureau of American Ethnology (Washington, DC: Government Printing Office, 1911); Murray B. Emeneau, "Franz Boas as a Linguist," in Thomas A. Sebeok, ed., *Portraits of Linguists: A Biographical Source Book for the History of Western Linguistics, 1746–1963*, vol. 2 (Bloomington: Indiana University Press, 1966), 123, 127; Stephen O. Murray, *Theory Groups and the Study of Language in North America: A Social History* (Philadelphia: John Benjamins, 1994), 47–65.
[50] Boas, "Introduction."

share elements of their experience and thus to act in concert. Trained in physics and geography, Boas viewed language, like all phases of culture, from a practical perspective. He held that language enabled social groups to turn natural resources to maximum advantage, allowing them to coordinate economic processes and to sustain group integrity by preventing wasteful internal conflicts. The inertial quality of language, however, also made cultures highly inflexible. Internalizing the cognitive filters of a language forced the individual's very perceptions of reality into predetermined forms, and thus embedded the culture's assumptions in the unconscious mind. Boas' insistence that even individual thought depended on linguistic categories pointed toward the alarming prospect of a thoroughgoing cognitive relativism.[51]

Boas, however, did not adopt such a view. He believed that the flux of experience posed significant resistance to certain patterns of categorization. Specifically, Boas held that a group's cultural patterns were not random, but instead referred to its material conditions. Boas always assumed that culture groups existed for the benefit of their individual members, and his 1911 text repeatedly invoked the metaphor of a mutually beneficial social contract forged deep in the past. He identified the "chief interests" of the group – its characteristic mode of ensuring biological survival – as the source of its distinctive perceptual filters, its tendency to fix on certain similarities between immediate experiences while ignoring others. In Boas' famous example, the Eskimo needed many words for "snow" in the context of their daily activities, whereas groups in the tropics had little use for such fine-grained distinctions.[52]

What happened, though, when the conditions of life changed, as they had with the West's move into the industrial "environment"? How could patterns of thought evolve if they were formed by fundamental linguistic categories? According to Boas, multiple sources of linguistic change, and thus of cultural flexibility, came into play in such instances of cultural lag. One was cultural diffusion. Individuals from different cultures who came into contact and compared their values and languages could borrow elements from one another and integrate them into their own cultures. In keeping with his broadly materialist perspective, Boas assumed that concepts and words, like all other tools, gained adherents in proportion to their usefulness. Thus, he believed that the contact of cultures would make them increasingly alike, as each converged on the best available resources for dealing with a given set of material circumstances.[53]

But scientific inquiry offered a second, and potentially more powerful and rapid, source of linguistic change. Boas identified comparative analysis as the

[51] *Ibid.*

[52] *Ibid.* Cf. William Graham Sumner, *Folkways: A Study of the Sociological Importance of Usages, Manners, Customs, Mores, and Morals* (Boston: Ginn & Co., 1907), 134, 140. Boas' illustration was apparently misinformed: Laura Martin, "Eskimo Words for Snow: A Case Study in the Genesis and Decay of an Anthropological Example," *American Anthropologist* n.s. 88, no. 2 (1986): 418–423.

[53] Regna Darnell, *Invisible Genealogies: A History of Americanist Anthropology* (Lincoln: University of Nebraska Press, 2001). For contract images, see Boas, "Introduction," 12, 19–20, 23, 31.

essence of science. He held that physics and many other sciences sharpened languages by carefully examining how well their terms fit with experienced causal dynamics in the material world. Anthropology, however, offered additional benefits through a second-level form of comparison and clarification. In that field, comparative analysis simulated the process of direct contact with other cultures, enabling much more rapid and effective forms of borrowing. Crucially, it also minimized the friction introduced into scientific communication by the culturally conditioned character of language. This was a major source of error, according to Boas. Two scientists could use the same term differently, or the term could be too broad or too narrow to express an important finding. Like many of his contemporaries, Boas worried especially about the incursion of philosophical abstractions such as "essence" and "existence" into scientific discourse. But cross-cultural comparison, he contended, helped scientists root out the biases in their theories and concepts by calling attention to their reification of cultural categories. By temporarily adopting other cultural lenses, scientists could become aware of their own, and eventually assimilate into their perceptual habits the fact that linguistic categories stood in a logically arbitrary relation to experience. Moreover, cultures informed by such a comparative science would be able to assess, without preconceptions, the functional utility of the cultural tools that other groups brought to the table. Boas viewed scientific analysis as both a powerful aid to cultural diffusion and a direct source of cultural flexibility in its own right.[54]

Boas' student Edward Sapir and Sapir's student Benjamin Lee Whorf found new ways of protecting science from the dangerous implications of their "Sapir-Whorf hypothesis" of linguistic relativity. As with Boas, the term "relativism" fits only insofar as Sapir and Whorf argued, first, that a given experiential phenomenon did not uniquely determine the linguistic terms employed to represent it, and second, that Western languages were not superior by definition. Sapir and Whorf assumed that some languages enabled the fulfillment of human needs better than others and that science offered unique contributions toward that goal.[55]

Sapir, who took charge of the linguistic side of the Boasian program after World War I, replaced Boas' emphasis on individual terms with the emerging "structuralist" view of a language as a holistic system, the essential meaning

---

[54] Boas, "Introduction."

[55] Discussions of this theory include John E. Joseph, "The Immediate Sources of the 'Sapir-Whorf Hypothesis,'" *Historiographia Linguistica* 23 (1996): 365–404; and Stephen O. Murray, *American Sociolinguistics: Theorists and Theory Groups* (Philadelphia: John Benjamins, 1998), 19–25. Many other human scientists noted the power of language over thought between the wars: e.g., Robert M. MacIver: *Society: Its Structure and Changes* (New York: R. Long & R. R. Smith, 1931), 5; Scott L. Pratt, "'A Sailor in a Storm': Dewey on the Meaning of Language," *Transactions of the Charles S. Peirce Society* 33, no. 4 (1997): 839–862. Nor was the insight new in the twentieth century: John Leavitt, "Linguistic Relativities," in *Language, Culture, and Society: Key Topics in Linguistic Anthropology*, eds. Christine Jourdan and Kevin Tuite (New York: Cambridge University Press, 2006): 47–81.

of which resided in its deep grammatical categories. His pioneering 1921 textbook described language as the very "mold of thought" – a closed, all-encompassing, structurally integrated "symbolic system" that subjected speakers to its "crushing historical precedents." Elsewhere, alluding to mathematics, Sapir called each language a "frame of reference" to which all of a speaker's concepts referred and from which there was "no escape." Sapir's reputation as a thoroughgoing relativist rests on such passages.[56]

But Sapir also wrote that "it is always the individual that really thinks and acts and dreams and revolts," despite the enormous influence of linguistic and cultural categories on even the most creative, nonconforming minds. Like many of the Deweyans, Sapir viewed social-scientific laws, including his own proposal that language profoundly shaped thought, as statistical generalizations with isolated exceptions, rather than fully deterministic causal relations of the type found in classical physics. Sapir always sought to maximize the space for individual freedom of interpretation, in his theory of science as in his ideal society. He described society itself as "reanimated or creatively reaffirmed from day to day by particular acts of a communicative nature." According to Sapir, social "reality" amounted to an endless series of choices in which individuals either followed prevailing practices or flouted them. This vision of society as a psychological, discursive entity left open the possibility of more or less conscious deviations from the norm, and thus cultural change. Like water eroding sandstone, small acts of resistance to prevailing forms altered the cultural landscape over time.[57]

In fact, Sapir took individual creativity as his standard in distinguishing between "genuine" and "spurious" cultures. The former enabled individuals to gradually create new cultural forms, whereas the latter repressed individual creativity. Departing from Boas again, Sapir saw a language as more akin to a work of art than a scientific representation of the world. Largely ignoring the considerations of practical utility stressed by his teacher, he instead emphasized the ability of language and other culture elements to elicit, and respond to, creative acts by individuals. Although the Boasians typically rejected "great man" theories of history, Sapir stressed the contribution of "striking and influential personalities" such as "Aristotle, Jesus, Mahomet, Shakespeare, Goethe,

[56] Edward Sapir, *Language: An Introduction to the Study of Speech* (New York: Harcourt, Brace, 1921), 21–22, 17; Sapir, "The Grammarian and His Personality," in David G. Mandelbaum, ed., *Selected Writings of Edward Sapir in Language, Culture, and Personality* (Berkeley: University of California Press, 1963), 153.

[57] Edward Sapir, "Do We Need a 'Superorganic'?" *American Anthropologist* n.s. 19, no. 3 (July–September 1917), 442, 447; Sapir, "Communication," in Mandelbaum, ed., *Selected Writings of Edward Sapir*, 104. Regna Darnell discusses Sapir's perspectivalism in *Invisible Genealogies*, 107–117. See also Darnell, *Edward Sapir: Linguist, Anthropologist, Humanist* (Berkeley and Los Angeles: University of California Press, 1990); Murray, *Theory Groups and the Study of Language in North America*, 77–111; Darnell, "Camelot at Yale: The Construction and Dismantling of the Sapirian Synthesis, 1931–1939," *American Anthropologist* 100 (1998): 361–372; and E. F. K. Koerner, ed., *Edward Sapir: Critical Assessments of Leading Linguists*, 3 vols. (New York: Routledge, 2007).

[and] Beethoven" to social progress. At the same time, however, he believed that a comparative science of language could reveal the essentially creative, socially constitutive character of language and thereby ensure that its users would maximize their opportunities for individual self-expression.[58]

In his seminal articles of the late 1930s and early 1940s, Whorf turned back to Boas' emphasis on the representational character of language, combining it with attention to Einstein's theory of relativity and the unfolding crisis of capitalism. Like Sapir, he focused on the deep structures – in Whorf's case, conceptions of time and space – underlying each language or "thought world." These, he wrote, embodied the inertial force of "the mass mind" and thus exhibited far greater rigidity than other culture elements. Yet while Whorf pressed the relativist claims that language structured thought and that all languages represented "equally logical, provisional analyses" of reality, he also praised certain languages for being closer to "naïve" experience than others.[59]

An enthusiast of theosophy whose metaphysical views resembled Alfred North Whitehead's, Whorf found Hopi far more suited to emerging understandings of the social and natural world than the language complex he called "Standard Average European" (SAE). Venturing into the sociology of knowledge, he traced the objectified, quantitative, spatialized thought world of SAE to "a commercial structure based on time-prorata values" such as wages, rent, and credit. "Newtonian space, time, and matter," Whorf declared, "are recepts from culture and language. That is where Newton got them." The result was a view of time – as divided into separate, discrete moments – that contradicted "the subjective experience of duration." Whorf thought scientists had recently begun to shake off the sterile intellectual categories of a capitalist society and move toward the more intuitively satisfying Hopi view, which focused on dynamic, multifarious processes of change and described each moment as carrying "impresses, both obvious and occult," of future goals and movements. Yet, he argued, the very structure of SAE languages, through which commercial values still molded "the 'common sense' of the Western average man," thwarted a non-objectifying, post-Newtonian science. Whorf urged linguistic "inventors and innovators" to target the temporal and spatial assumptions underlying SAE and allow the revolutionary worldview of modern science to play its appointed cultural role by creating a harmonious, non-materialistic, post-capitalist society. Although he stressed a different normative goal than Boas or Sapir – an embrace of the underlying unity of experience, rather than material fulfillment or creative freedom – Whorf agreed that the scientific study

---

[58] Sapir, "Do We Need a 'Superorganic'?", 443.

[59] Benjamin Lee Whorf, *Language, Thought, and Reality: Selected Writings of Benjamin Lee Whorf*, ed. John B. Carroll (Cambridge, MA: Technology Press, 1956), 147, 156, 244, 221. Specifically, Whorf argued that the world consisted of innumerable existents, each of which exhibited a characteristic "mode of duration," or means of growth. Some existents, he specified, "become later and later" by "growing like plants, some by diffusing and vanishing, some by a procession of metamorphoses, some by enduring in one shape till affected by violent forces." *Language, Thought, and Reality*, 147.

of language could help counter the immense cultural influence of outdated linguistic forms.[60]

So, too, did the many scientific democrats who understood the relation between science and language in much simpler terms. Language, complained James Harvey Robinson in 1921, "is not primarily a vehicle of ideas and information, but an emotional outlet, corresponding to various cooings, growlings, snarls, crowings, and brayings." Many other theorists shared Robinson's view of everyday language as a roiling sea of meaningless utterances that should give way to a thoroughly scientific mode of communication. Joining the interwar burst of theorizing on language and symbolism, a host of behaviorists and semanticists distinguished sharply between objective, scientific analysis and emotionally biased, cognitively empty gibberish.[61]

This view found powerful expression in *The Meaning of Meaning* (1923), by two British scholars, the psychologist C. K. Ogden and the literary theorist I. A. Richards. Widely read in the United States, Ogden and Richards' book codified the dichotomy between "emotive" and "referential" ("symbolic," in their terminology) uses of language that Robinson and many other frustrated scientific democrats invoked in response to Progressivism's defeat and the rise of new propaganda techniques.[62]

Ogden and Richards argued that some of the signs human beings invented to represent the world matched its structure better than others. Exhibiting a structure isomorphic to reality, these signs made possible the manipulation of observed experience. The authors argued that a word, when used in this "symbolic" fashion, stood in for the act of pointing to an object, a referential gesture in physical space. However, words – in many cases, the same words – also generated emotional responses. Ogden and Richards urged readers to distinguish carefully between the descriptive and hortatory aspects of language. Such linguistic prophylaxis, they argued, would instantly dissolve all manner of social and intellectual disagreements, not least the "fictitious" disputes between "Vitalism and Mechanism, Materialism and Idealism, Religion and Science, etc."[63]

---

[60] Whorf, *Language, Thought, and Reality*, 153, 139, 148, 156, 152. See also Penny Lee, *The Whorf Theory Complex: A Critical Reconstruction* (Philadelphia: John Benjamins, 1996); Darnell, *Invisible Genealogies*, 173–191; and Darnell, "Benjamin Lee Whorf and the Boasian Foundations of Contemporary Ethnolinguistics," in *Language, Culture, and Society*, eds. Jourdan and Tuite, 82–95. The work of the Vassar anthropologist Dorothy Demetracopoulou Lee further illustrated theorists' interest in languages that drew distinctions fundamental to scientific thought: "Conceptual Implications of an Indian Language," *Philosophy of Science* 5, no. 1 (January 1938), 102; "A Primitive System of Values," *Philosophy of Science* 7, no. 3 (July 1940): 355–378.

[61] James Harvey Robinson, *The Mind in the Making: The Relation of Intelligence to Social Reform* (New York: Harper, 1921), 224.

[62] C. K. Ogden and I. A. Richards, *The Meaning of Meaning* (New York: Harcourt, Brace & Co., 1923); W. Terrence Gordon, *C. K. Ogden: A Bio-Bibliographic Study* (Metuchen, NJ: Scarecrow Press, 1990).

[63] Ogden and Richards, *The Meaning of Meaning*, viii.

The linguist Leonard Bloomfield, who moved from Ohio State to the University of Chicago in 1927, developed the distinction between scientific and nonscientific utterances in a slightly different manner. A committed behaviorist, Bloomfield found logical empiricism attractive in the late 1930s and worked to inject insights from linguistic analysis into that framework. A few years earlier, in his classic *Language* (1933), Bloomfield framed this linguistic project in relentlessly populist terms. The book he called "my high-school text" was hardly clear enough to reach that audience, but it did purport to make available to ordinary citizens a set of analytical tools, hitherto monopolized by specialists, that they could use to demystify claims of cultural authority and break down social hierarchies. Linguistic study, Bloomfield told his readers, could restore self-government by proving that existing categories did not reflect "universal forms of human thought" or a "cosmic order," and also by challenging "verbal response habits" that failed to address real human needs. Bloomfield saw his field as a "mental or, as they used to say, moral science."[64]

Language portrayed society as a complex network of linguistic interchanges that could be tuned for maximum efficiency. Bloomfield followed his former Ohio State colleague Albert Paul Weiss in revising John B. Watson's behaviorism by describing the environment that shaped behavior as fundamentally social. Within that environment, he explained, language offered individuals a powerful tool for meeting their needs. Each utterance acted as a stimulus to those within earshot, ideally provoking a favorable "handling response" such as providing food. These exchanges linked together individual nervous systems, knitting them into what was literally a "social organism" that held the potential to respond quickly and effectively to the needs of each of its individual components.[65]

Yet, Bloomfield complained, the prevailing linguistic categories routinely blocked such responses by obscuring reality with layer upon layer of verbal confusion. He traced these misguided forms back to a group of eighteenth-century grammarians who had built their class interests into the language itself by condensing the "speculations of ancient and medieval philosophers" into an authoritarian grammatical system. However, Bloomfield continued, that system gained its authority solely from the choice of each individual to conform to it at any given moment. His own theory, he promised, would empower

[64] Leonard Bloomfield, review of Arthur F. Bentley, *Linguistic Analysis of Mathematics* and *Behavior, Knowledge, Fact*, *Language* 12 (1936), 139; Bloomfield, *Language* (New York: Rinehart and Winston, 1933), 5, 507; quoted in Bernard Bloch, "Leonard Bloomfield," in Sebeok, ed., *Portraits of Linguists*, 515; Bloomfield, "Why a Linguistic Society?" *Language* 1, no. 1 (March 1925), 1. See also Robert A. Hall Jr., *A Life for Language: A Biographical Memoir of Leonard Bloomfield* (Philadelphia: John Benjamins, 1990); Murray, *Theory Groups and the Study of Language in North America*, 113–135; and Charles F. Hockett, "Leonard Bloomfield: After Fifty Years," *Historiographia Linguistica* 26, no. 3 (1999): 295–311.
[65] Leonard Bloomfield, *Linguistic Aspects of Science*, *International Encyclopedia of Unified Science*, Volume 1, no. 4, ed. Otto Neurath, Rudolf Carnap, and Charles W. Morris (Chicago: University of Chicago Press, 1939), 15.

citizens to throw off their linguistic chains and align everyday language with their actual needs.[66]

Specifically, Bloomfield hoped his behaviorist approach would allow readers to scrap the scientifically dubious categories of contemporary language and view the world solely through the lens of factual reference. He looked to the schools as a particularly important locus of linguistic reform. The schools, he asserted, could perform a crucial service by orienting children toward the "actual environment" rather than wasting their time on the purely verbal phenomena of languages and literature. Bloomfield declared science and technology utterly devoid of "linguistic features," explaining that the terms of these languages referred directly to experienced reality rather than to verbal representations of it. Thus, the schools could arm students against linguistic phantoms by versing them in scientific, "non-linguistic" fields. Teachers could train students to employ direct behavioral referents and continually remind them that language use was simply one among many forms of behavior – "the noise you make with your face," as Bloomfield put it elsewhere.[67]

The basic distinction between referential and merely emotive uses of language gained numerous adherents in interwar scholarship and the wider culture. Often this distinction came packaged with the further claim that using language in an emotive manner flouted human needs and represented a drain on society. Richards offered a particularly striking formulation, given his employment as a literary critic. Rather than reveling in the richness and specificity of each language, he called the fact that words carried multiple connotations "wickedly wasteful of mental energy, the most valuable commodity in the world." Of course, many scientific democrats declined to call the study of literature and other such pursuits intrinsically antisocial. Still, when these figures addressed questions of language, most sought, in the words of the NYU philosopher Sidney Hook, to discover "under what conditions it is possible for discourse to render what is not discourse with a minimum of ambiguity and indeterminacy." Like Boas, they assumed that language use, insofar as it appeared as a subject for scientific analysis, aimed at "the communication of

---

[66] Bloomfield, *Language*, 3.

[67] Leonard Bloomfield, review of Bentley, 139; Bloch, "Leonard Bloomfield," 514; Bloomfield, *Language*, 499, 506–507. Not all advocates of a behavioristic approach to linguistic study sought merely to strip existing languages of their non-referential content, however. Operating outside Bloomfield's circle, George K. Zipf joined many of his Harvard colleagues in worrying about the problem of social stability and finding solace in the concept of a self-correcting social mechanism, in this case language. Undertaking a series of complex statistical analyses of word usage patterns, he sought to prove that speech acted as a kind of interpersonal parliament, naturally restoring "harmony and equilibrium" whenever social conflict disturbed "the great complex of individual-environment": *The Psycho-Biology of Language: An Introduction to Dynamic Philology* (Boston: Houghton Mifflin, 1935), 294–295, 264. On Zipf's thought, see Jennifer Platt, *A History of Sociological Research Methods in America: 1920–1960* (New York: Cambridge University Press, 1996), 87–90, 212–223; and Regina Pustet, "Zipf and His Heirs," *Language Sciences* 26, no. 1 (January 2004): 1–25.

ideas." In short, scientific democrats typically focused on what scholars now call the "pragmatics" of language.[68]

Moreover, even scholars attuned to the poetic and figurative dimensions of language, such as Sapir, shared important points of commonality with those suspicious of such uses. Bloomfield thus worked surprisingly closely with Sapir, despite their divergent psychological frameworks. He and his students would later dominate the post-World War II discipline, after Sapir died in 1939 and the federal government based its wartime language preparation materials on Bloomfield's work. But between the wars, Sapir and Bloomfield served as the twin leaders of American linguistics, joining forces in the Linguistic Society of America and other forums. In fact, Sapir found considerable inspiration in Ogden and Richards' work as well. Although their reform strategies and theoretical frameworks differed substantially, all of these theorists sought to create a scientific culture by using powerful new linguistic tools to empirically analyze the character of cultural phenomena.[69]

This approach proved particularly attractive in the 1930s, when ideological clashes and advances in the use of propaganda forced questions of language and meaning to the fore in American academic life. The involvement of interwar scientific democrats with three concrete projects says a great deal about the perceived promise of linguistic study in general and the appeal of the more specific claim that science could cut through merely emotive uses of language. The first of these projects was the Institute for Propaganda Analysis (IPA). Founded in 1937, that organization operated under the leadership of a pioneering public opinion researcher, Hadley Cantril of Princeton. Its *Propaganda Analysis* newsletter promised to help everyday Americans see through specious claims in the media to a "clear understanding of conditions and what to do about them." The IPA's board included a host of other scientific democrats, including the geologist Kirtley F. Mather, the historian Charles Beard, the social worker and adult educator Eduard Lindeman, and the sociologist Robert S. Lynd. The organization's membership overlapped substantially with the list of contributors to the *Public Opinion Quarterly*, which was launched in the same year and likewise featured numerous calls for the public to adopt a more referential outlook. Contributions to *Propaganda Analysis* and the *Public Opinion Quarterly* identified propaganda and science as diametrically opposed modes

---

[68] I. A. Richards, *Basic English and its Uses* (New York: W. W. Norton, 1943), 36; John Paul Russo, *I. A. Richards: His Life and Work* (Baltimore: Johns Hopkins University Press, 1989); Sidney Hook, "The Nature of Discourse," *Saturday Review of Literature* 10 (March 10, 1934), 547; Boas, "Introduction," 23. Richards also offered social and political commentary in popular magazines: e.g., "Changing American Mind," *Harper's* 154 (1927): 239–245; "Our Lost Leaders," *Saturday Review of Literature* 9 (1933): 509–510; "Psychopolitics," *Fortune* 26 (1942): 108–109.

[69] John E. Joseph, *From Whitney to Chomsky: Essays in the History of American Linguistics* (Philadelphia: John Benjamins, 2002), 158; Stephen O. Murray, "The First Quarter Century of the Linguistic Society of America, 1924–1949," *Historiographia Linguistica* 18 (1991): 1–48; Joseph, "The Immediate Sources of the 'Sapir-Whorf' Hypothesis"; Darnell, *Invisible Genealogies*, 188–191; Nerlich, "The 1930s – At the Birth of a Pragmatic Conception of Language."

of speech, one resting on emotional appeals and the other gaining clear sight by eschewing emotionality. A similar distinction undergirded a massive program in communication research inaugurated by the Rockefeller Foundation in the late 1930s.[70]

Many scientific democrats also found highly appealing the idea of an international "auxiliary language" that could aid cultural exchange, prevent misunderstandings, and point the way to world peace by enabling free and fluid communication among the peoples of the world. Recognizing the futility of attempts to simply replace existing languages with Esperanto or another substitute, many interwar thinkers reasoned that a simple, flexible second language could provide many of the same benefits. Ogden and Richards stumped for "Basic English," which Ogden created by taking a language already associated with the impersonal exchanges of trade and reducing it to a mere 850 words, each referring to an agent or operation in the shared, public world of space-time. By contrast, Edward Sapir, one of the most vigorous American advocates of an auxiliary language, enlisted Boas and Bloomfield in opposition to the use of English as its basis. Although these figures questioned the uncritical use of Western forms, however, they joined the call for a language based on the "structural 'lowest terms.'" They, too, hoped to facilitate the sharing of common experiences by creating a language that was devoid of emotive trappings and featured purely functional terms suitable for "scientific, business or other practical work." Like the activities of the IPA, those of the New York-based International Auxiliary Language Association, founded in 1924, demonstrated the profound tension in interwar formulations of scientific democracy between a commitment to accommodating diverse beliefs and an assumption of the superiority of science as a way of knowing. This tension bridged the implicit divide between those scientific democrats who thought that science united individuals because it directly accessed reality and those who viewed it in more discursive terms as a human construct.[71]

70 Advertisement in *Forum and Century* 99 (January 1938): iii; Floyd H. Allport, "Toward a Science of Public Opinion," *Public Opinion Quarterly* 1, no. 1 (January 1937): 7–23; "Public Opinion in a Democracy," *Public Opinion Quarterly* 2, no. 1, supp. (January 1938); Brett Gary, "Communication Research, the Rockefeller Foundation, and Mobilization for the War on Words, 1938–1944," *Journal of Communication* 46 (1996): 124–147. A statement of the IPA's aims appears in the October 1937 issue of *Propaganda Analysis*. On opinion polling, see Sarah E. Igo, *The Averaged American: Surveys, Citizens, and the Making of a Mass Public* (Cambridge, MA: Harvard University Press, 2007).
71 I. A. Richards, "Basic English," *Fortune* 23 (1941), 111–112; C. K. Ogden, *Debabelization* (London: Kegan Paul, Trench, Trubner & Co., 1931), 15; Andrew Large, *The Artificial Language Movement* (New York: Basil Blackwell, 1985), 162–173; W. Terrence Gordon, "Undoing Babel: C. K. Ogden's Basic English," *ETC: A Review of General Semantics* 45 (1988): 337–340; Edward Sapir et al., "Memorandum on the Problem of an International Auxiliary Language," *Romanic Review* (1925), 244, 251, 245–246, 248–251; Darnell, *Edward Sapir*, 272–276; International Auxiliary Language Association [IALA], *Annual Report for 1938* (New York: IALA, 1938), 2–4, 5–7; IALA, *General Report for 1945* (New York: IALA, 1945), 13–15, 20. MIT president Karl T. Compton, the psychologist Edward L. Thorndike, and the educator Arthur E. Morgan joined Sapir on the board of the IALA. As early as 1920, Columbia's CC syllabus touted the benefits

However, only a narrower spectrum of scientific democrats could support a project that arrayed science against philosophy and even the universities, rather than literary study. Such was the general semantics movement, led by the Polish-born Alfred ("Count") Korzybski. Picking up on Tolman's neobehaviorism and certain aspects of Ogden and Richards' work, Korzybski developed the referential/emotive distinction into a form of linguistic therapy that appealed to many middle-class professionals. Originally trained as a chemical engineer, he argued that the linguistic disorders of the modern age extended from the public sphere of communication into the brains of individuals. Developing various techniques of neurological retraining, Korzybski lectured widely and found a growing audience among natural scientists and members of the helping professions. In 1938, a bestseller called *The Tyranny of Words* by the inveterate popularizer Stuart Chase propelled general semantics into the national spotlight. Accredited courses in general semantics appeared in university curricula by 1941, and *Time* reported that many scientists thought Korzybski's theories would soon "rank in historic importance with the work of Aristotle and Einstein." But it was primarily natural scientists and those in professional fields that found general semantics attractive. Although they often disagreed with elements of Korzybski's system, they took seriously his attempt to launch what amounted to a clinical tradition in the philosophy of science and language. By contrast, Korzybski appealed much less strongly to scientific democrats in other fields, likely because of his insistence on the mathematical character of all genuine knowledge and his charge that modern scholarship was hopelessly corrupted by "Aristotelianism," a tendency to reify mental constructs rather than recognize their purely contingent and functional character.[72]

of an international language: *Introduction to Contemporary Civilization: A Syllabus, Second Edition* (New York: Columbia University, 1920), 65.

[72] Ross Evans Paulson, *Language, Science, and Action: Korzybski's General Semantics: A Study in Comparative Intellectual History* (Westport, CT: Greenwood, 1983); Robert P. Pula, "Alfred Korzybski, 1879–1950: A Bio-Methodological Sketch," *Polish American Studies* 53, no. 2 (Autumn 1996): 57–105; Neil Postman, "Alfred Korzybski," *ETC: A Review of General Semantics* 60, no. 4 (Winter 2003): 354–361; Chase, *The Tyranny of Words*; M. Kendig, "Introduction," in *Papers from the Second American Congress on General Semantics*, ed. Kendig (Chicago: Institute of General Semantics, 1943); "General Semantics," *Time* 32 (November 21, 1938), 35; R. Alan Lawson, *The Failure of Independent Liberalism, 1930–1941* (New York: Putnam, 1971), 226–231. A long list of prominent scholars, including most of the biologists discussed in Chapter 5, wrote promotional blurbs for Korzysbki's 1933 book *Science and Sanity*. Honorary trustees of the Institute of General Semantics included George E. Coghill, the legal scholars Thurman Arnold and Roscoe Pound, the psychoanalyst Abraham A. Brill, the physical anthropologist Earnest A. Hooton, the psychiatrists Smith Ely Jelliffe and Adolf Meyer, the mental hygienist Stewart Paton, and Walter L. Treadway, the medical director of the Public Health Service. Ruth Benedict invoked Korzybski's term "time-binding" as a widely understood referent in *Patterns of Culture* (Boston: Houghton Mifflin, 1934), 231, as did Harvey A. Wooster in "To Unify the Liberal-Arts Curriculum," *Journal of Higher Education* 3 (1932), 375 and T. V. Smith in "Opportunism," *International Journal of Ethics* 45, no. 2 (January 1935), 237. For additional uses of Korzybski's terminology by scientific thinkers, see also William Morton Wheeler, "Social Life Among the Insects," *Scientific Monthly* 14, no. 6 (June 1922), 506; William E. Ritter, "Why Aristotle Invented

The scientific study of language thus revealed a conflict between the recognition of science's cultural embeddedness and contingency by so many interwar theorists and an underlying faith that science offered resources unavailable through any other cultural practice. Few, however, felt the tension, given that a belief in science's unique character was constitutive of scientific democracy itself. Interwar students of language consistently found strategies for insulating science from the general practice of unmasking linguistic terms and categories as human constructs. Even Sapir and Whorf, known as thoroughgoing relativists, drew the line at calling science a mere cultural product.

Meanwhile, strongly contextual understandings of science, which also raised the prospect of full-blown cognitive relativism that loomed in linguistics, faded rapidly from the professionalizing science studies fields during the 1940s. Those fields became a major lost opportunity for scientific democrats who sought to widen the definition of science and foster normative public engagement by scholars. In particular, Dewey's brand of contextualism, with its emphasis on the intrinsically public, critical character of science, disappeared almost entirely from the science studies disciplines. His approach largely ignored what the logical empiricist Hans Reichenbach famously dubbed the "context of discovery" and the "context of justification" and instead focused on what might be called the "context of internalization" – that part of the scientific enterprise wherein new knowledge lodged itself in the motivational structures of ordinary citizens. In the 1920s and 1930s, virtually every scientific thinker could find something of value in Dewey's work. His use of everyday language and his identification of science as part of culture created bridges across the disciplines, across much of the political spectrum, and across the gap between experts and citizens. But the 1940s brought dramatic changes in the organization of American science that undermined Dewey's appeal. The campaign against totalitarianism and the emergence of the military-industrial complex not only altered research priorities in the disciplines but also transformed philosophical, methodological, and political descriptions of the scientific enterprise.[73]

As Chapters 10 and 11 show, there was no academic constituency for Dewey's theory in the new world of big science and federal funding. Every group of scholars to whom Dewey's theory of science might have appealed needed something other than what it provided. Working researchers of that era

the Word Entelecheia (Continued)," *Quarterly Review of Biology* 9, no. 1 (March 1934), 11; and Arthur F. Bentley, cited in Harold Thayer Davis, "The Ninth Annual Meeting of the Indiana Section," *American Mathematical Monthly* 39, no. 8 (October 1932), 444. Philosophers, not surprisingly, proved most critical: e.g., Ernest Nagel, review of Korzybski, *Science and Sanity*, *New Republic* 79 (August 1, 1934): 327; "Mr. Nagel Answers," *New Republic* 81 (December 26, 1934): 195; and Sidney Hook, "The Nature of Discourse."

[73] Hans Reichenbach, *Experience and Prediction: An Analysis of the Foundations and the Structure of Knowledge* (Chicago: University of Chicago Press, 1938). Philip Mirowski uses Dewey and Reichenbach as symbols for interwar and postwar philosophies of science: "The Scientific Dimensions of Social Knowledge and their Distant Echoes in 20th-Century American Philosophy of Science," *Studies in History and Philosophy of Science* 35A, no. 2 (June 2004): 283–326.

placed a premium on objectivity, rhetorically distancing themselves from the political fray and often from practical applications as well. Intellectual historians turned away from the sciences in seeking the deep roots of American political culture. And those in the professionalizing science studies fields pursued highly detailed accounts of theoretical changes in the natural sciences, focusing primarily on the context of justification and writing the context of internalization out of the definition of science. These self-appointed spokespersons for science found the prospect of a two-way traffic between the scientific disciplines and the wider culture profoundly alarming in the age of totalitarianism. To the extent that their work registered in the intellectual commons, it tended to reinforce a rationalistic image of science as the discovery of politically neutral truths about the natural world through the exercise of disinterested reason. By 1950, resources in the science studies fields flowed disproportionately to interpreters who portrayed the scientist as a heroic individual motivated by curiosity and other personal traits, not by explicit social goals. At just the moment when the military-industrial complex took shape, the science studies fields portrayed science in highly individualistic, decontextualized terms.[74]

For a time in the 1940s, Dewey's broad political concerns and emphasis on the human sciences found a new home in the emerging interdisciplinary field of American studies. Under the pressure of the mobilization for war, the cross-fertilization of work by Deweyans such as Herbert W. Schneider and the Wisconsin historian Merle Curti with the socially grounded literary history of the University of Washington's Vernon L. Parrington and Harvard's Howard Mumford Jones put the study of American thought and culture on firm professional ground. But postwar practitioners of American studies moved away from viewing ideas as products of their contexts and toward viewing ideas themselves as the contexts for, and the causes of, action. In fact, as Chapter 11 shows, many posited a static American "consensus" that dated back to the Founders or even the Puritans. Meanwhile, intellectual history on Lovejoy's model explored the grand sweep of Western history, tracing tectonic changes over centuries through the writings of canonical figures, especially European philosophers and political theorists. It no longer served to enlist the American public in what leading interwar scientific democrats viewed as an epochal transition to a new form of collective life and a corresponding, contextually sensitive, self-reflective form of intellectual work.[75]

---

[74] David A. Hollinger has noted this irony: "Free Enterprise and Free Inquiry: The Emergence of Laissez-Faire Communitarianism in the Ideology of Science in the United States," in *Science, Jews, and Secular Culture*, 97–120.

[75] Philip Gleason, "World War II and the Development of American Studies," *American Quarterly* 36, no. 3 (1984): 343–358; Gleason, "The Study of American Culture," in *Speaking of Diversity: Language and Ethnicity in Twentieth-Century America* (Baltimore: Johns Hopkins University Press, 1992), 188–206; Leila Zenderland, "Constructing American Studies: Culture, Identity, and the Expansion of the Humanities," in *The Humanities and the Dynamics of Inclusion Since World War II*, ed. David A. Hollinger (Baltimore: Johns Hopkins University Press, 2006), esp. 273–286.

9

The Problem of Values

The critics of scientism who became such a powerful presence in New Deal-era public discourse defined science as a set of neutral tools to be used by the inhabitants of an existing culture, not as a resource for remaking that culture. In the 1920s, this view also prevailed among physical scientists and gained adherents in the human sciences as well. The travails and opportunities of the 1930s provided new incentives for economists, political scientists, and sociologists to adopt a value-neutral stance. To be sure, Franklin D. Roosevelt embraced a more historicist view of social knowledge when he declared in his 1932 acceptance speech that "economic laws are not made by nature. They are made by human beings." Most experts employed by New Deal agencies agreed. But the fierce ideological contests of the era also encouraged them to portray themselves as technicians implementing the public will rather than ethical guides on the path to a cooperative future. Political nonpartisanship, and perhaps even absolute value-neutrality, was the order of the day. Like the SSRC funding of the 1920s, but on a far larger scale, Roosevelt's embrace of social-scientific research favored a rhetoric of objectivity, in the public sphere if not always among administrators themselves.[1]

The champions of objectivity insisted that a modern society could not function without the technical knowledge provided by scientific scholars. But they rejected the idea that science could also provide normative guidance. By contrast, leading scientific democrats denied that science could be separated so easily from considerations of value. A running argument over science and values among American sociologists clearly revealed these increasingly polarized

[1] Quoted in Terry A. Cooney, *Balancing Acts: American Thought and Culture in the 1930s* (New York: Twayne, 1995), 42; Charles Camic, "On Edge: Sociology During the Depression and the New Deal," in *Sociology in America: A History*, ed. Craig Calhoun (Chicago: University of Chicago Press, 2007), 249. The major exception to the new orientation was the U.S. Department of Agriculture: Olaf F. Larson and Julie N. Zimmerman, *Sociology in Government: The Galpin-Taylor Years in the U.S. Department of Agriculture, 1919–1953* (University Park: Penn State University Press, 2003); Andrew Jewett, "Philosophy, the Social Sciences, and the Cultural Turn in the 1930s USDA," *Journal of the History of the Behavioral Sciences* (forthcoming).

positions. Deeply engaged in the political innovations of the New Deal era, yet lacking even the tenuous cultural authority of the economists, sociologists struggled to understand their discipline's relation to public concerns. William F. Ogburn pushed for a sharp divide between a value-neutral sociology, on the one hand, and applied fields such as social work, on the other. His critics, such as the liberal Protestant Charles A. Ellwood and the European émigrés Robert M. MacIver and Pitirim A. Sorokin, stressed the intimate connections between scientific knowledge and human purposes. Columbia's Robert S. Lynd capped off the debate by declaring in his widely read *Knowledge for What?* (1939) that the responsible social scientist should work to create public support for economic planning in the name of human welfare. By that time, however, the emergence of consensus liberalism had introduced new resources for navigating tricky epistemological questions, and the outbreak of war soon thereafter would lead sociologists to circle the wagons again, defending democracy against totalitarianism – and science against those who deemed it a threat to democracy.

THE SOCIOLOGISTS DIVIDED

On the academic front, at least, the very idea that science could be defined in relation to a single set of entities collectively described as "values" was an innovation of the early twentieth century, when the philosophical field of value theory emerged as a secular successor to parts of nineteenth-century moral philosophy. That field took shape in the context of philosophers' bitter arguments over the nature of science and its relation to the wider culture.[2] The idealists Wilbur Marshall Urban and Hugo Munsterberg both proposed a general field of "value theory" in 1909. Value, as these two conceived it, was like the "nature" of the emergentists, in that it was monistic in substance but pluralistic in form – in other words, general and universal in application but experiential and particular in its concrete instances. By the 1920s, American idealists, realists, and pragmatists had all adopted Urban and Munsterberg's view that value could be discussed in unitary terms, as a single, general concept about which universal statements could be made and whose proper definition promised to unite the disparate realms of ethics, aesthetics, religion, economics, politics, and social science, "whether in federation or empire." Ralph Barton Perry's 1926 *General Theory of Value*, which we encountered in Chapter 6,

---

[2] It is suggestive that three of the four members of the APA committee which chose to center the 1913 meeting on value – Walter B. Pitkin of Columbia, Harry A. Overstreet of the City College of New York, and Edward G. Spaulding of Princeton – would later spend the bulk of their careers writing and lecturing to the general public rather than speaking primarily to fellow specialists: "Subject of Discussion for the Next Meeting of the American Philosophical Association," *Journal of Philosophy, Psychology and Scientific Methods* 10, no 6 (March 13, 1913), 167–168. Not surprisingly, Dewey led the pragmatists in the discussion of value, though he did not attend the 1913 conference. See Dewey, "The Problem of Values," *Journal of Philosophy, Psychology and Scientific Methods* 10, no. 10 (May 8, 1913), 268–269.

represented the most prominent contribution to the emerging field, though his purely descriptive definition of a value as "any object of any interest" fueled continuing controversy. Despite their differences, however, Perry and his interlocutors generally agreed that the concept of value provided a universal currency with which to relate and compare the full range of ethical, aesthetic, economic, and religious impulses.[3]

Unlike the older term "metaphysics," the unitary category of "values" allowed latter-day positivists to ban a range of phenomena from science without also denigrating them linguistically. Nonetheless, the term trickled into the human sciences only slowly and took hold across the disciplines only in the mid-1930s, when human scientists vigorously debated their relationship to political concerns.[4] But the unprecedented uses of social science – for decidedly non-metaphysical purposes – by the federal government in the 1930s prepared the way for this generalized concept to come into play as a shorthand descriptor for all the realms into which true scientists refused to venture.

Although he had not yet picked up the language of "values" in its systematic, philosophical meaning, William F. Ogburn gave the new form of social-scientific neo-positivism its most visible and controversial statement in "The Folk-Ways of a Scientific Sociology," his 1929 presidential address to the American Sociological Association. Not coincidentally, Ogburn had been a major recipient of SSRC funding in the late 1920s. He subsequently did more than any other sociologist to bring the resources of his field to bear on the economic crisis, insisting all the while that sociology offered only technical resources, not normative guidance.[5]

Ogburn's work for the federal government predated Roosevelt's election. When President Hoover, a strong advocate of statistical research and rational coordination in industry, appointed a group of experts to study recent economic changes in the wake of the 1929 stock market crash, Ogburn quickly took charge of the study. Appearing just before Roosevelt's inauguration, *Recent Social Trends in the United States* ran to 1,568 densely packed pages, accompanied by innumerable charts and tables. It followed Ogburn's earlier argument that all sectors of society – "agriculture, labor, industry, government, education, religion, and science" – should advance in tandem, through national

---

[3] Abraham Edel, "The Concept of Value and Its Travels in Twentieth-Century America," in *Values and Value Theory in Twentieth-Century America: Essays in Honor of Elizabeth Flower*, ed. Murray G. Murphey and Ivar Berg (Philadelphia: Temple University Press, 1988), 12–15 (quote on 12).

[4] Neither "value-neutral" nor "value-neutrality" appears in the online JSTOR database of professional journals until 1937, and the next appearance after that is not until 1944. "Science and values" generates a small stream of hits after the psychologist Edward Thorndike's presidential address of that title to the American Association for the Advancement of Science in late 1935. This usage overlaps directly with the onset of the "first scientists' movement" described in Chapter 10.

[5] William Fielding Ogburn, "The Folk-Ways of a Scientific Sociology," *Scientific Monthly* 30 (April 1930): 300–306; Barbara Laslett, "Biography as Historical Sociology: The Case of William Fielding Ogburn," *Theory and Society* 20 (1991): 511–538.

planning aimed at minimizing cultural lag. However, the report, written during the Hoover years, expressed little confidence that the haphazard, *laissez-faire* approach to institutional change could be overcome. Roosevelt's election changed all that, though planning enthusiasts like Ogburn continued to chafe at the New Deal's limitations. In fact, the developments of the 1930s only strengthened Ogburn's commitment to technological determinism, the view that technological changes drove everything in their wake and social institutions, including governmental structures, inevitably changed in response.[6]

Ogburn believed that science served to minimize cultural lag by helping citizens see that they needed to change their institutions in keeping with technological shifts. Yet it offered only purely descriptive knowledge, he insisted. Ogburn famously declared that sociology, as a science, "is not interested in making the world a better place in which to live, in encouraging beliefs, in spreading information, in dispensing news, in setting forth impressions of life, in leading the multitudes or in guiding the ship of state." A scientific sociology had but a single goal, he insisted: "discovering new knowledge." Ogburn argued that sociologists would make their political mark indirectly, by giving information to "some sterling executive who will appear to do the actual guiding." Noting that no one could actually "guide the course of evolution" as a whole, Ogburn nevertheless predicted "fair success in using approximate knowledge," presumably to foster social adjustment to technological changes.[7]

Ogburn did not simply define away the normative dimension of scientific knowledge. He also relied on a fundamentally different conception of the knowing process than did the advocates of a normatively committed science. Edward C. Tolman, for example, distinguished immediate experience from its expression in logical, scientific terms. From this standpoint, scientific knowledge represented an abstraction from experience – its codification in a form suited to intersubjective exchange. But Ogburn distinguished instead between scientific thought and the mental processes associated with action. On that basis, he erected a high wall between science and fields of action such as politics. Action, Ogburn explained, "tends to follow directly out of emotion," while science demanded the rigorous suspension of judgment and all other "associations that disturb the closeness of the connection between the thinking

[6] Quoted in Arlene Inouye and Charles Susskind, "'Technological Trends and National Policy,' 1937: The First Modern Technology Assessment," *Technology and Culture* 18 (October 1977), 597–598; William F. Ogburn, "Technology and Governmental Change," *The Journal of Business of the University of Chicago* 9 (January 1936), 13; Subcommittee on Technology to the National Resources Committee, *Technological Trends and National Policy* (Washington: Government Printing Office, 1937). On the 1933 report, see also Mark C. Smith, *Social Science in the Crucible: The American Debate Over Objectivity and Purpose, 1918–1941* (Durham: Duke University Press, 1994), 72–75; John M. Jordan, *Machine-Age Ideology: Social Engineering and American Liberalism, 1911–1939* (Chapel Hill: University of North Carolina Press, 1994), 179–184; and William A. Tobin, "Studying Society: The Making of 'Recent Social Trends in the United States, 1929–1933,'" *Theory and Society* 24, no. 4 (August 1995): 537–565.
[7] Ogburn, "The Folk-Ways of a Scientific Sociology," 300–301, 304–305.

and the data." The scientist needed to temporarily eliminate the feelings and value judgments that caused irrelevant mental associations to appear, allowing reliable knowledge to emerge. Thus, whereas Tolman viewed science as the communication of experience through its expression in abstract, systematic form, Ogburn identified it with a certain solidity, permanence, or protection from criticism, achieved through the rigorous suppression of emotions.[8]

Still, Ogburn's line between science and other realms did not divide scientists from other people, but instead divided the scientist as such from the scientist as a full human being. In his view, engaging in scientific research meant playing a role, undertaking a special form of behavior because it generated a socially useful product: a "photographic record" of reality. In their capacities as citizens, he stressed, individual sociologists would of course "want to seek for knowledge that will be of benefit to mankind," and this would and should shape their choice of research questions, though not their answers to those questions. The "excellent social activities" of "education, propaganda, ethics, journalism, literature, religion and ... executive leadership," along with popularization, would continue unabated in Ogburn's ideal world, and any scientist could participate in such activities "in the capacity of another self." Ogburn fully expected dramatic social change to take place, partly as a result of his studies. He trained and strongly supported the Marxist sociologist of medicine Bernhard J. Stern, and his own take on technology's social implications was too radical for the Civilian Conservation Corps, which blocked the distribution of a pamphlet he wrote for the agency in 1934.[9]

It is also worth noting that Ogburn's ban on emotional engagement applied only to theory choice and scholarly writing. He predicted that sociological publications would soon be "wholly colorless," read only by specialists and featuring only verified results. They would omit both hypotheses and philosophical speculation, except in the form of "syntheses of broad researches." But social philosophy would flourish in other venues, Ogburn noted. Although he believed that philosophical speculation primarily represented "a rationalization of wishes" derived from the "*Zeitgeist*," he recognized that it also generated the ideas that scientists tested. Ogburn stressed that the maturation of a science meant, not the diminution of theoretical and philosophical work in the field, but rather the exclusion of such activities from its published forms – the withholding of the honorific term "science" from the untested products of speculative, associational thought. As always, Ogburn drew a functional

[8] *Ibid.*, 304, 302, 306. Cf. Ogburn, "Bias, Psychoanalysis, and the Subjective in Relation to the Social Sciences" (1922), reprinted in *On Culture and Social Change: Selected Papers* (Chicago: University of Chicago Press, 1964), 293.

[9] William F. Ogburn, "Studies in Prediction and the Distortion of Reality," *Social Forces* 13 (December 1934), 224; Ogburn, "The Folk-Ways of a Scientific Sociology," 306, 300–301; Samuel W. Bloom, *The Word as Scalpel: A History of Medical Sociology* (New York: Oxford University Press, 2002), 84, 96; Arlene L. Barry, "Censorship during the Depression: The Banning of 'You and Machines,'" *OAH Magazine of History* 16, no. 1 (Fall 2001): 56–61. Ogburn's vision of scientific politics strongly echoed that of the British Bernalists described in Chapter 8.

distinction between the scientist and the philosopher (or the social worker). A given individual could play multiple roles, as long as the production and publication of empirical results remained thoroughly untouched by emotion.[10]

Nor did Ogburn demand that sociologists limit themselves to quantitative, statistical methods. Despite his well-known proclivity for those approaches, Ogburn also listed "historical" and "descriptive" methods as legitimate substitutes for the outmoded method of theoretical speculation. In fact, he estimated that less than half of sociology's subject matter would prove susceptible to numerical analysis. A few years earlier, Ogburn and the Boasian anthropologist Alexander Goldenweiser had called man "a subjective and capricious creature," adding, "[b]oth man and history ... are relatively impervious to the concept of law and but partly subject to control."[11]

Still, Ogburn's conception of knowledge was anathema to most scientific democrats in sociology and elsewhere. In his own field, figures such as Duke's Charles A. Ellwood read Ogburn as trying to run the advocates of non-quantitative methodologies out of the discipline altogether. In a published response to Ogburn's address, Ellwood discerned in his writings and those of other sociological "hyper-scientists" a campaign to officially establish within the discipline a single methodology and an associated view of reality, namely a "practical monis[m]" that interpreted social reality on the model of physics. Ellwood firmly resisted the imposition of such methodological and philosophical tests for admission to the sociological guild. Moreover, he blasted Ogburn's approach as insufficiently empirical. The true scientist, Ellwood wrote, worked with "any evidence which experience seems to afford," rather than distorting reality to fit the narrow method of quantification. Aligning himself with "the cultural conception of human society," which stressed human freedom and intentionality, Ellwood declared quantitative methods unsuited to the study of social reality.[12]

Ellwood was an unreconstructed Social Gospeler who titled a 1940 book *The World's Need of Christ*. Whereas Ogburn worried about the corrupting

---

[10] Ogburn, "The Folk-Ways of a Scientific Sociology," 301–305; Ogburn, "Trends in Social Science," *Science* 79 (March 23, 1934), 260–262.

[11] Ogburn, "Trends in Social Science," 261–262; Ogburn and Alexander Goldenweiser, eds., *The Social Sciences and Their Interrelations* (Boston: Houghton Mifflin, 1927), 9.

[12] Charles A. Ellwood, "Scientific Method in Sociology," *Social Forces* 10 (October 1931), 15–16; Ellwood, "The Uses and Limitations of the Statistical Method in the Social Sciences," *Scientific Monthly* 37, no. 4 (October 1933), 354. The best account of this print debate between Ogburn, Ellwood, and others is Robert C. Bannister, *Sociology and Scientism: The American Quest for Objectivity, 1880–1940* (Chapel Hill: University of North Carolina Press, 1987), 188–230. See also Stephen Turner and Jonathan H. Turner, *The Impossible Science: An Institutional Analysis of American Sociology* (Newbury Park: Sage Publications, 1990), 65–69; Smith, *Social Science in the Crucible*, esp. 72–75, 142–145; Elżbieta Hałas, "How Robert M. MacIver was Forgotten: Columbia and American Sociology in a New Light, 1929–1950," *Journal of the History of the Behavioral Sciences* 37, no. 1 (2001), 35–39; and Stephen Turner, "A Life in the First Half-Century of Sociology: Charles Ellwood and the Division of Sociology," in *Sociology in America: A History*, ed. Craig Calhoun (Chicago: University of Chicago Press, 2007): 115–154.

influence of personal emotions in the sciences, Ellwood sought to combat irrationality in the broader, sociological form of group oppression and blind adherence to tradition, including *laissez-faire* principles and the amoral materialism of a consumer society. Against these forces, he arrayed a scientifically grounded system of ethics. Ellwood actually agreed with Ogburn that sociologists should reject "ultimate value-judgments" of the type announced by traditional Christians, for these were based solely on personal whims. But no scientist or other modern thinker had believed in such transcendentals since the nineteenth century, he averred. All now focused on conditional, if-then value judgments that stated the means of achieving human ends. These judgments, Ellwood insisted, fell squarely in the domain of sociology. He deemed it crucial to affirm, in his professional role as a sociologist, the existence of "a social ethics made up of these relative value-judgments." In his view, social ethics stood shoulder to shoulder with "the other social sciences" in the task of providing "a new and scientific social philosophy" to guide human behavior in the modern world.[13]

In 1933, Ellwood sharpened his critique of "Emasculated Sociologies" and codified his alternative conception of the field in *Methods in Sociology*. Linking epistemology firmly to politics, he charged that figures like Ogburn adopted a pose of *faux* neutrality to convince "traditionalists and vested interests within the Church or the State" that "the sociological movement is quite harmless." Yet this "separatist" or "fractional" strategy would simply reify existing values and "destroy any chance of sociology becoming the scientific guide of human society," Ellwood protested. Deeming the United States the most backward of all nations in its collective self-understanding, Ellwood traced the problem directly to the state of its scholarship. He argued that scientists, hemmed in by vested economic and religious interests, had failed to stand up and serve the people by offering "rational, intelligent criticism" of existing social forms. Rather than taking on "the responsibilities of leadership in civilization," they had retreated to the shelters of "pure" knowledge and quantification, while artificially narrowing the goal of education to mere "individual development." Given that all education was "necessarily a propaganda for something," Ellwood insisted that the schools and universities should seek to produce "the completely social man," though without engaging in direct social indoctrination. In both his educational prescriptions and his conception of knowledge, the avowed Christian Ellwood echoed the naturalist John Dewey. Both of these scientific democrats rejected Ogburn's rigorous separation of knowing from emotional processes, insisting that rationality was instead a matter of choosing and reinforcing emotional habits with an eye to their social consequences.[14]

[13] Charles A. Ellwood, *The World's Need of Christ* (New York: Abingdon-Cokesbury, 1940); Ellwood, "Emasculated Sociologies," *Sociology and Social Research* 17 (1933), 224–225; Ellwood, "Scientific Method in Sociology," 20.

[14] Ellwood, "Emasculated Sociologies," 228–229, 221–222; Ellwood, "Social Education in the United States," in Paul A. Schilpp, ed., *Higher Education Faces the Future* (New York: Horace Liveright, 1930), 262–263, 259–260. Ellwood set great store in the method of direct confrontation

Similar critiques emerged from many precincts of sociology during the 1930s. Thinkers who disagreed on many other matters could agree that science needed to be understood quite broadly, because social behavior was less regular and predictable than natural phenomena and its causes were essentially different. If the task of assessing social phenomena turned out to be beyond the capacity of existing scientific methods, then so much the worse for those methods; they would need to be modified.

In his classic textbook *Society* (1931), the Scottish-born Columbia sociologist Robert M. MacIver challenged Ogburn's emphasis on emotional detachment from a different angle than Ellwood. He declared that the holistic nature of the social whole meant that the investigator, firmly embedded in that whole, could neither attain an objective view of it nor approach it through quantitative tools. "Society is a game of social relationships," he wrote, "and we know these relationships in the same manner that we enter into them, directly, qualitatively." MacIver added that the scientist could never capture the entire social whole even directly and qualitatively, as it was both inordinately complicated and a moving target. Thus, the student of society needed to employ artistry as well as empirical analysis, selecting carefully from among the immense range of phenomena to explore. Not that such work amounted to a mere tracing of fixed contours. Like the Progressive sociologists, MacIver held that the fluid, intersubjective character of social phenomena rendered scholarly analysis a powerful tool for social change. He offered the familiar portrait of society as an organic, psychological complex, within which the spread of a new understanding of the relationships between the whole and its parts would lead inexorably to institutional changes. In particular, MacIver hoped that the sociologists of his day would work to replace the prevailing view of society as a mechanical object, ruled by material forces, with a more adequate portrait of an organic whole constituted by consciously held meanings. Why, he asked, "should the misunderstood name of science limit the sociologist to the arid schematism of figures and tables and classifications, so that the student often finds a clearer illumination of the working of society in the fragmentary revelations of the social novelist, dramatist, and essayist?" MacIver worried that the social sciences stood in danger of losing out to the arts as a source of social guidance if they confined themselves to the mechanistic mode of explanation characteristic of the natural sciences.[15]

Divergent though they were, the methodological prescriptions offered by Ogburn, Ellwood, and MacIver hardly accounted for the full range of conceptions of science put forth by American cultural scientists in the 1930s. The

as a means of changing minds; see, for example, his letter to the editor in *American Journal of Sociology* 40 (July 1934), 140.

[15] Robert M. MacIver, *Society: Its Structure and Changes* (New York: R. Long & R. R. Smith, 1931), x, 527–528, 546, vii, ix. On MacIver, see Robert Bierstedt, *American Sociological Theory: A Critical History* (New York: Academic Press, 1981), 243–297; and Halas, "How Robert M. MacIver was Forgotten."

sprawling collaborative volume *Methods in Social Science* (1931), produced by an SSRC committee that was chaired by MacIver and included scientific democrats such as Edward Sapir, captured both the methodological diversity and the surprisingly broad challenge to mechanistic methods that character-ized the cultural sciences in those years. Editor Stuart A. Rice, a statistically inclined sociologist from the University of Pennsylvania, adopted an ecumeni-cal approach to the vexed concept of "scientific method," noting that the book could do no better than to survey the landscape of existing methodological approaches. The volume offered a series of case studies in which prominent scholars dissected the techniques employed by other notable scholars. The sub-jects of these studies ranged from Voltaire and Comte forward, including fig-ures from history, social geography, anthropology, and linguistics. Contributors both studied and employed an enormous range of techniques, both quantita-tive and qualitative.[16]

Rice, however, declared one conception of the scientific method off limits: the classic empiricist portrait in which scientists proceeded inductively from facts to generalized laws. He charged that those who employed this model were "accepting without criticism generalized symbols of experience, already interpreted in such a way as to fit within the modes of a given ideational sys-tem." One needed to define problems to study, Rice pointed out, and this choice necessarily involved the elaboration of some theoretical constructs prior to investigation. According to Rice, even physical scientists were "becoming less and less certain of the meaning of 'facts,' and more and more interested in the underlying concepts which have the power to make facts appear and disap-pear." Facts could be true only in relation to the theoretical systems they were invoked to support or demolish, he insisted. Moreover, the "facts of human history" were particularly "dependent ... upon the presuppositions of those who record and interpret them," though all facts exhibited this dependence in some degree. And, he added, "the flow of events in their entirety escapes the perception of the most gifted." According to Rice, every viable concept of sci-entific method presented theoretical presuppositions as "*instruments* as well as *frameworks* of investigation."[17]

Within the broad boundaries set by this shared rejection of classical empir-icism, the sociologists' debate over methodology and epistemology simmered through the 1930s and into the war years, occasionally flaring up around a par-ticularly controversial text. New figures also weighed in on the question of value-neutrality, prominently including Harvard's Pitirim A. Sorokin. Hitherto known as a solid researcher for his 1920s work in the field of rural sociology, the fiery Russian émigré joined the philosophical fray with a 1931 article on "Sociology

[16] Stuart A. Rice, "Introduction," in Rice, ed., *Methods in Social Science* (Chicago: University of Chicago Press, 1931), 4. Cf. Jennifer Platt, *A History of Sociological Research Methods in America: 1920–1960* (New York: Cambridge University Press, 1996), 20.

[17] Rice, "Introduction," 8–10. Also surprisingly diverse, especially given its editor's proclivities, was Luther Lee Bernard's *The Fields and Methods of Sociology* (New York: R. Long & R. R. Smith, 1934).

as a Science" and put his theory into action with the four-volume *Social and Cultural Dynamics* (1937–1941). Sorokin dismissed "sociological preachers" and neo-positivists such as Ogburn alike. In fact, he challenged virtually everyone who occupied the American discipline of sociology. For example, although Sorokin joined MacIver in stressing the unity and holistic character of the social complex, he believed that scholars could reveal its contours and principles of change by painstakingly reducing enormous masses of quantitative data to mathematical laws. *Social and Cultural Dynamics* argued that Western nations had cycled steadily through three types of culture, which Sorokin dubbed ideational, idealistic, and sensate. Each cultural mode, Sorokin tried to demonstrate, gave its color to all prevailing artistic and intellectual expressions, as well as to social relationships. Sorokin lambasted the sensate culture of the contemporary era and sought to make his critique of it available to the masses through a semipopular summary, *The Crisis of Our Age* (1941).[18]

But the most polarizing contribution to the sociologists' debate in its later years was *Knowledge for What?* (1939), a brief for systematic economic planning by Columbia's Robert S. Lynd. The co-author of *Middletown* attacked his fellow sociologists for their detachment from the political struggles of the day. He argued that, unless they chose their problems and interpreted their results in the light of the universal moral standard Lynd identified with democracy – the imperative to satisfy the "basic cravings of human personality" across the entire population – then they would allow "the National Association of Manufacturers, the American Federation of Labor, the advertising man, the American Legion, and so on" to set the terms for American culture instead. Like many contemporaries, Lynd cautioned that values could never be allowed to "bias one's analysis or the interpretation of the meanings inherent in one's data" but insisted that these were central at the levels of both problem choice, where investigators' values directed them to truly important public problems, and social application, where they connected research findings back to those generating problems. Lynd framed his conclusions in radical terms, writing that capitalism "probably cannot ... assure the amount of general welfare to which the present stage of our technological skills and our intelligence entitle us." His ideal society would "see that its citizens from birth to death had as little chance as possible to invest their savings ignorantly, to purchase sub-standard commodities, to marry disastrously, to have unwanted children 'accidentally,' to postpone needed operations, to go into blind-alley jobs, and so on."[19]

---

[18] Pitirim A. Sorokin, *Social and Cultural Dynamics*, 4 vols. (New York: American Book Company, 1937–1941); Sorokin, *The Crisis of Our Age: The Social and Cultural Outlook* (New York: Dutton, 1941); Barry V. Johnston, *Pitirim A. Sorokin: An Intellectual Biography* (Lawrence: University Press of Kansas, 1995), 49–51. On Sorokin, see also Bierstedt, *American Sociological Theory*, 299–347; Gary Dean Jaworski, "Pitirim A. Sorokin's Sociological Anarchism," *History of the Human Sciences* 6, no. 3 (1993): 61–77; and Lawrence T. Nichols, "Science, Politics, and Moral Activism: Sorokin's Integralism Reconsidered," *Journal of the History of the Behavioral Sciences* 35, no. 2 (Spring 1999): 139–155.

[19] Robert S. Lynd, *Knowledge for What? The Place of Social Science in American Culture* (Princeton: Princeton University Press, 1939), 205, 177, 182–183, 226, 235. The best treatment

Though widely hailed – then and now – as a progressive blow for democracy against an inherently conservative positivism, Lynd's book actually reveals that a commitment to normative engagement by scholars did not necessarily entail a lofty view of the average citizen's cognitive capacities or the potential of public deliberation as a political force. When Lynd called for a "more active" means of "structuring rank-and-file participation in, and responsibility for, authority," he meant by this that the public should elevate social scientists to pride of place in the polity and then defer to their authority. In an article published shortly after the 1939 book, he described the key to the contemporary crisis of democracy as "persuad[ing] our fellow citizens to take their hands off the *details* of intricate public matters and to recognize the need of delegating decisions ... to expert surrogates." Lynd insisted that scholars needed both to produce empirical research and to pronounce authoritatively on its normative, political implications. No longer could Americans innocently rely on "the informal pressures of neighborly life to curb unsocial expressions of personal freedom," he wrote. Here, as elsewhere, they needed to establish conscious, centralized control over individual behavior, in keeping with the findings of specialized experts.[20]

CULTURE AND PERSONALITY

Lynd's (and Ellwood's) emphasis on keeping empirical research and normative guidance fused in a single discipline actually claimed greater cultural authority for the scientist than did Ogburn's approach. Lynd wrote that "social science must be prepared to tell democracy what functions may and may not wisely – that is, in the public interest – be left to various types of democratic action, and under which types of leadership," while also indicating "what types of learning need to be the subject of constant public propaganda" in order to sustain the system as a whole. Social scientists would set the boundaries of democratic deliberation and form citizens who would not seek to go beyond those boundaries.[21]

By the start of World War II, however, the emergence of consensus liberalism provided many scholars with a way out of this manipulative stance by suggesting that scientists could openly apply normative arguments to public problems without imposing their own values, by simply voicing the preferences of an existing, thoroughly liberal public. Consensus liberalism entailed a new twist

of the book and the ensuing controversy, which made the pages of *Time*, is Smith, *Social Science in the Crucible*, 120–158; cf. "Knowledge for What?" *Time* (April 17, 1939): 70–71.

[20] Lynd, *Knowledge for What?*, 212, 237–238; Lynd, "Democracy in Reverse," *Public Opinion Quarterly* 4, no. 2 (June 1940): 218–220.

[21] Lynd, *Knowledge for What?*, 237–238. Crane Brinton noted Lynd's assertion of authority in a review: "What's the Matter With Sociology?" *Saturday Review of Literature* (May 6, 1939): 3. Cf. Christopher Shannon, *Conspicuous Criticism: Tradition, the Individual, and Culture in American Social Thought, from Veblen to Mills* (Baltimore: Johns Hopkins University Press, 1996), 105–132.

on a claim shared by scientific democrats of the early twentieth century: that, as Lynd put it, social order rested on "a central core of emotionally resonant loyalties widely shared by the mass of the people."[22] Any science of society worth its salt would obviously take these shared values as part of its empirical data, the argument continued. But what were those values at present, and what role for the scholar did they imply? Lynd argued that scientists needed to create the normative foundation of a new culture, in an era when traditional religious faiths were vanishing. By contrast, advocates of consensus liberalism declared that the United States already featured a core of fundamentally liberal values that, under current circumstances, dictated the conscious coordination of economic behavior. Consensus liberalism identified the New Deal as entirely continuous with American values, not as a sharp departure from prevailing sentiment. American history, in this reckoning, no longer featured a sharp cultural divide brought on by industrialization, but rather a temporary maladjustment between Americans' fundamental commitments and the political structures designed to reflect them.

The culture and personality school provided the matrix within which consensus liberalism began to take shape as a recognizable intellectual orientation. A new, transdisciplinary formation of the 1930s, the culture and personality approach was less a single set of tenets than a running argument among anthropologists, psychologists, and psychiatrists over the concrete meaning of a simple insight: that neither the wider culture nor individual personality formation could be explained without reference to the other.[23] Culture and personality theorists identified a strong causal link between personality disorders and social disorganization. Yet latent ambiguities and outright disagreements regarding the nature and scope of individual freedom ran through their work.

One thing was clear, as the postwar application of this insight to racial segregation in *Brown v. Board of Education* would demonstrate: The culture

---

[22] Lynd, *Knowledge for What?*, 238–239. On related developments in polling, see Jean M. Converse, *Survey Research in the United States: Roots and Emergence 1890–1960* (Berkeley: University of California Press, 1987); and Sarah E. Igo, *The Averaged American: Surveys, Citizens, and the Making of a Mass Public* (Cambridge, MA: Harvard University Press, 2007).

[23] Recent studies of culture and personality include Dennis Bryson, "Personality and Culture, the Social Science Research Council, and Liberal Social Engineering: The Advisory Committee on Personality and Culture, 1930–1934," *Journal of the History of the Behavioral Sciences* 45, no. 4 (Fall 2009): 355–386; and Joanne Meyerowitz, "'How Common Culture Shapes the Separate Lives': Sexuality, Race, and Mid-Twentieth-Century Social Constructionist Thought," *Journal of American History* 96, no. 4 (March 2010): 1057–1084. The movement's roots are unearthed in Regna Darnell, *Edward Sapir: Linguist, Anthropologist, Humanist* (Berkeley: University of California Press, 1990), 288–308; Katherine Pandora, *Rebels Within the Ranks: Psychologists' Critiques of Scientific Authority and Democratic Realities in New Deal America* (New York: Cambridge University Press, 1997), 6; and Andrew Abbott and James T. Sparrow, "Hot War, Cold War: The Structures of Sociological Action, 1940–1955," in *Sociology in America: A History*, ed. Craig Calhoun (Chicago: University of Chicago Press, 2007), 305. Some historians also picked up on the culture and personality approach: e.g., Caroline Ware, ed., *The Cultural Approach to History* (New York: Columbia University Press, 1940).

and personality approach offered a potent tool for political argumentation. It declared that the adoption of misguided policies could disfigure the very psyches of citizens. By identifying freedom from neurosis as a primary element of political freedom, and thus a proper site for state intervention, the culture and personality school and its successors radically reshaped postwar politics.[24]

But the direction of causation between individual and social problems remained ambiguous, and solutions could be proposed at either end of the chain. Did a given maladjustment between individual personalities and social institutions call for political changes, or just therapeutic interventions to create better-adjusted personalities? Other questions emerged as well: To what extent should scholars adopt an activist stance toward cultural and political change? If so, what change should they seek? What features of modern American life accounted for personality problems?

The competing approaches of the Boasian anthropologist Ruth Benedict and the psychiatrically inclined foundation executive Lawrence K. Frank reveal one axis of disagreement: the relationship between psychological health and individuals' ability to choose their own cultural identities. Benedict traced the bulk of psychological disorders in modern America to the rigid molds imposed on a diverse set of individuals. In *Patterns of Culture* (1934), a foundational text of the culture and personality school, she argued that a culture functioned by imposing a set of predetermined social roles on the relatively plastic stuff of humanity. Daily confronted with society's disapproval of her sexual preferences as well as her progressive political views, Benedict emphasized the crushing weight of the cultural frame. But she also held out hope that "sane and scientific guidance" of cultural change could move Americans toward a society that provided greater latitude for divergent forms of self-expression. Benedict hoped to reinscribe individualism – now understood in terms of respect for diversity, not market competition – as the underlying norm of American life. Describing cultures as filters that favored specific personality types, Benedict sought to create a new type of American personality that was more open to individual differences.[25]

---

[24] Ellen Herman, *The Romance of American Psychology: Political Culture in the Age of Experts* (Berkeley: University of California Press, 1995), esp. 174–207, 241–257; Daryl Michael Scott, *Contempt and Pity: Social Policy and the Image of the Damaged Black Psyche, 1880–1996* (Chapel Hill: University of North Carolina Press, 1997); John E. Jackson Jr., "Creating a Consensus: Psychologists, the Supreme Court, and School Desegregation, 1952–1955," *Journal of Social Issues* 54, no. 1 (Spring 1998): 143–177. For longer vistas on the political meaning of modern psychology, see Kurt Danziger, *Constructing the Subject: Historical Origins of Psychological Research* (New York: Cambridge University Press, 1990); Nikolas Rose, *Inventing Ourselves: Psychology, Power, and Personhood* (New York: Cambridge University Press, 1996); and Jeffrey P. Sklansky, *The Soul's Economy: Market Society and Selfhood in American Thought, 1820–1920* (Chapel Hill: University of North Carolina Press, 2002).

[25] Ruth Benedict, *Patterns of Culture* (Boston: Houghton Mifflin, 1934). Recent treatments include Marc Manganaro, *Culture, 1922: The Emergence of a Concept* (Princeton: Princeton University Press, 2002), 151–174; and Howard Brick, *Transcending Capitalism: Visions of a New Society*

By contrast, Frank, who bankrolled the development of the culture and personality school from his position as director of the Laura Spelman Rockefeller Memorial, argued that Americans had too many choices, rather than too few. Like Benedict, he called for "new ego-ideals that look toward non-pecuniary goals." Against Benedict's emphasis on diversity, however, Frank contended that the characteristic ills of modern society stemmed from its utter lack of accepted standards of personal behavior. Westerners, he declared, had lacked a shared culture for some 300 years, ever since natural science had begun to supersede traditional faiths. Accelerating immigration worsened the problem considerably, Frank added. Adopting a remarkably manipulative tone, he updated the age-old image of society as a biological organism by portraying the social scientist as a doctor working to remedy cultural diseases. In fact, Frank sought to replace the entire post-Renaissance edifice of individualism: "The individual, instead of seeking his own personal salvation and security, must recognize his almost complete dependence upon the group life and see his only hope in and through cultural reorganization." Whereas Benedict advocated pluralism and wanted to open up cultural space for individuals opposed to competitive ideals, Frank sought to create a society that featured a single type of individual: one committed to social service. He did not see the imposition of a homogeneous cultural standard on the growing child as a source of psychological tension, but rather as a process of psychological release – the attainment of a culturally derived "superego that is integrated and wholesome."[26]

An even wider range of political emphases characterized the intertwined fields of personality psychology and social psychology, which powerfully fed into the culture and personality school in the 1930s. Gordon Allport, the leading figure in personality psychology, and Gardner Murphy, who helped develop social psychology with his wife Lois Barclay Murphy, shared many theoretical concepts and a powerful antipathy to reductionist forms of behaviorism. But they offered very different interpretations of the social conditions

in Modern American Thought (Ithaca: Cornell University Press, 2006), 95–98. On Benedict, see also Barbara A. Babcock, "'Not in the Absolute Singular': Re-Reading Ruth Benedict," *Frontiers* 12 (1992): 39–77; Regna Darnell, *Invisible Genealogies: A History of Americanist Anthropology* (Lincoln: University of Nebraska Press, 2001), 191–200; Daniel Rosenblatt, "An Anthropology Made Safe for Culture: Patterns of Practice and the Politics of Difference in Ruth Benedict," *American Anthropologist* 106, no. 3 (2004): 459–472; and Richard Handler, *Critics Against Culture: Anthropological Observers of Mass Society* (Madison: University of Wisconsin Press, 2005), 96–140. A recent biography, with important lecture notes appended, is Virginia H. Young, *Ruth Benedict: Beyond Relativity, Beyond Pattern* (Lincoln: University of Nebraska Press, 2005).

[26] Lawrence K. Frank, "Social Planning and Individual Ideals," *International Journal of Ethics* 45 (October 1934), 85; Frank, "Society as the Patient," *American Journal of Sociology* 42 (November 1936), 339–343. See also Stephen John Cross, "Designs for Living: Lawrence K. Frank and the Progressive Legacy in American Social Science" (PhD dissertation, Johns Hopkins University, 1994); Dennis Raymond Bryson, *Socializing the Young: The Role of Foundations, 1923–1941* (Westport: Bergin & Garvey, 2002); and Pier Francesco Asso and Luca Fiorito, "Lawrence Kelso Frank's Proto-Ayresian Dichotomy," *History of Political Economy* 36, no. 3 (Fall 2004): 557–578.

for psychological health. The divergence roughly mapped onto that between Benedict and Frank: whereas Allport stressed the importance of individual self-determination, Murphy called for extensive social engineering. Allport and Murphy also disagreed on the capacity of bureaucratic organizations to advance human needs. Finally, the two adopted divergent approaches to reforming their discipline and the wider culture, with Murphy favoring a mode of direct confrontation that Allport eschewed.[27]

Not that Allport had any kind words for behaviorism. He complained privately of the "crude and arrogant" tone of his field, its "excessive empiricism, grotesque nativism, traffic in boggled ethics, superficiality, and undue abstractness." Allport hoped to replace the "generalized mind" of the prevailing schools with the richer, more empirically adequate data of individual minds. He also joined Tolman in identifying "complex autonomous units of motivation," or "traits," rather than the orthodox behaviorists' narrowly physiological stimulus-response circuits, as the immediate source of action. Allport believed that emphasizing the variable and culturally conditioned character of motives explained the emergence of "socialized and civilized behavior" from the "entirely wolfish and piggish" responses of children, while still protecting the all-important foundation of individual freedom by describing individual personalities as something more than mere reflexes of the wider culture. "Nature, as Goethe said, seems to have planned everything with a view to individuality," Allport stated at the beginning of his influential 1937 textbook. He closed the book by writing that the individual "struggles on even under oppression, always hoping and planning for a more perfect democracy where the dignity and growth of each personality will be prized above all else."[28]

In keeping with his emphasis on individual freedom and self-determination, Allport pursued his psychological agenda indirectly. He called for intellectual pluralism, in the form of a broad conception of psychology that could accommodate both the behaviorists' experiments with rats and his own method of receptive apprehension – "to put a specimen before one's eyes and look at it repeatedly until its essential features sink indelibly into one's mind," as he later wrote. Murphy, on the other hand, understood social difference in terms of group competition. He, too, steered a course between cultural determinism and biological determinism, while allowing for substantial variation between individuals. "If man is to be moulded to society," he wrote, "society must also be moulded to man." Unlike Allport, however, Murphy viewed individual variation as a product of membership in social subgroups. He also believed that

---

[27] Allport's Harvard colleague Henry A. Murray followed Murphy in this regard: Ian A. M. Nicholson, *Inventing Personality: Gordon Allport and the Science of Selfhood* (Washington, DC: American Psychological Association, 2003), 184.

[28] Quoted in Pandora, *Rebels Within the Ranks*, 3; Allport, "The Functional Autonomy of Motives," *American Journal of Psychology* 50 (1937), 142, 151, 156; Allport, *Personality: A Psychological Interpretation* (New York: Holt, 1937), 3, 566. See also Nicole B. Barenbaum, "How Social Was Personality? The Allports' 'Connection' of Social and Personality Psychology," *Journal of the History of the Behavioral Sciences* 36, no. 4 (Fall 2000): 471–487.

fierce opposition by conservatives to new forms of social organization would force psychologists to take an active role in the political struggles of the day.[29]

Specifically, Murphy sought to guide Americans between the poles of conservatism and Marxism, each of which, he believed, defined social health in economic rather than psychological terms. To his mind, this political task called for a psychology that could, among other things, explain "resistance to social change" and reveal "ways of making people aware of their own needs and interests." Murphy's sense of the political centrality of psychology at a dangerous moment of transition guided his vigorous institutional work in the 1930s, in forums ranging from his editorship of *Sociometry* to his participation in the Society for the Psychological Study of Social Issues (SPSSI). Like Frank, and unlike Benedict and Allport, he described the society of the future as one characterized, not by substantive individual freedom *per se*, but instead by the universal sway of a form of selfhood oriented toward social service.[30]

The temperamental and political differences between Murphy and Allport shaped their conceptions of science itself. Both sought to establish the scientific credentials of their versions of psychology. Murphy, however, joined the behaviorists in seeking a set of consensually held, experimentally verified, and universally applicable laws of psychic behavior. He took issue only with how the behaviorists sought such laws. Murphy advocated an ascending hierarchy of research practices that culminated in, rather than began with, laboratory work. First, he wrote, should come "the description of social interactions as given by historian, economist, sociologist or ethnologist," then "the tentative outline of psychological hypotheses regarding these interactions," then the testing of some of these hypotheses through a comparative analysis of cultures, then "systematic observation of the [individual] in his natural environment under different culture conditions," and only then "systematic experimentation" on groups of individuals. Given that the preliminary investigations had not yet been completed, Murphy argued, the existing body of experimental findings merely revealed the contours of behavior and personality in certain Western contexts. These were not yet the absolute, universal laws that psychology ultimately promised, but Murphy expected such laws to emerge. He rejected what he took to be the increasingly influential position that "'alternative modes of

[29] Quoted in Pandora, *Rebels Within the Ranks*, 73; Gardner Murphy, "The Research Task of Social Psychology," *Journal of Social Psychology* 10 (1939), 111. Allport equated his method of receptive apprehension with Sorokin's "logico-meaningful" standard for integration. In fact, he declared this the natural orientation of the human mind, arguing that holistic thinking ("sensitivity to form") did not need to be taught, whereas the reductive procedures of analytic thinking did. *Personality*, 360, 546–547.

[30] Gardner Murphy, "The Research Task of Social Psychology," *Journal of Social Psychology* 10 (1939), 119; Pandora, *Rebels Within the Ranks*. See also Lois Barclay Murphy, *Gardner Murphy: Integrating, Expanding, and Humanizing Psychology* (Jefferson: McFarland, 1990); and Ian A. M. Nicholson, "The Politics of Scientific Social Reform, 1936–1960: Goodwin Watson and the Society for the Psychological Study of Social Issues," *Journal of the History of the Behavioral Sciences* 33 (1997): 39–60.

conceptual analysis,' or 'mutually exclusive sets of presuppositions,'" served as legitimate bases for multiple scientific schools. Down the line, Murphy expected psychologists, and indeed all social scientists, to achieve unity on the grounds of hard experimental evidence.[31]

By contrast, Allport heartily endorsed the perspectivalism and interpretive pluralism that Murphy rejected. He viewed the diversity of theories in modern psychology as a positive good, not a sign of the field's infancy. The mind "may be surveyed and divided in an infinite number of ways," he wrote, and the only possible standard of judgment for any analysis was its fit with the investigator's stated purpose. Each "map-maker" would naturally have distinct goals in mind, said Allport. Problems would arise only if those surveying the terrain for one purpose presented their findings as the means to different interpretive ends. "The multiplication of points of view is not evil so long as they are allowed to act upon one another," he insisted. "The more ways we have the better. None that can stand the tests of fidelity to fact and self-consistency should be excluded." Eschewing Murphy's talk of a definitive, empirical resolution, Allport proposed instead a *modus vivendi* wherein all schools of psychology could proceed unmolested, if they granted the validity of the others and passed the tests of empirical adequacy and logical coherence.[32]

One of the leading advocates of this perspectival approach in the late 1930s was Harvard's Talcott Parsons, who helped to integrate sociology with the culture and personality orientation. Unlike Allport, however – and like the Kantian J. E. Creighton earlier – Parsons employed perspectivalism in the service of shoring up his own discipline's claim to autonomous status within a broader field of sciences, not to carve out room for multiple theoretical perspectives within a single discipline. Parsons portrayed sociology as a special science within a broader, overarching "theory of action." His integrative ambitions for the social sciences directly paralleled his sense that a set of shared values bound together even a highly differentiated modern society. Parsons portrayed the normative core of a pluralistic, democratic society as an intricate hierarchy of values, differentiated toward the bottom but held together at the top by a few core commitments such as individualism and freedom. This distinctive combination of synthesis and pluralism made Parsons' writings central both to the behavioral sciences and to consensus liberalism in the 1950s.[33]

[31] Gardner Murphy, Lois Barclay Murphy, and Theodore M. Newcomb, *Experimental Social Psychology: An Interpretation of Research Upon the Socialization of the Individual* (New York: Harper, 1937), 11, 6–7, 22–23; Murphy and Friedrich Jensen, *Approaches to Personality* (New York: Coward-McCann, 1932), vi.

[32] Allport, *Personality*, 564–565.

[33] Brick, *Transcending Capitalism*, esp. 127–128. Edward Shils dubbed this view "consensual pluralism." Quoted in Thomas Bender, "Politics, Intellect, and the American University, 1945–1995," *Daedalus* 126, no. 1 (Winter 1997), 35. See also Charles Camic's introduction to Talcott Parsons, *The Early Essays* (Chicago: University of Chicago Press, 1991); Uta Gerhardt, *Talcott Parsons: An Intellectual Biography* (New York: Cambridge University Press, 2002): ix–lxiv; and Camic, "On Edge: Sociology During the Depression and the New Deal," in *Sociology in America: A History*, ed. Craig Calhoun (Chicago: University of Chicago Press, 2007), 276–280.

The 1937 classic *The Structure of Social Action* set forth a vision of interpretive pluralism that Parsons developed as part of the "Pareto Circle," a group of scholars that the physiologist and historian of science Lawrence J. Henderson led in vigorous discussions of the philosophy of knowledge during the late 1930s. Parsons argued that each of the sciences addressing human action represented the application of a specialized frame of reference to that phenomenon. More generally, he explained, every scientific field took its shape from practitioners' adoption of a transcendental, *a priori* first principle. Physicists, for example, adopted the postulate of a "space-time framework." Without such a conceptual apparatus, Parsons declared, one could not even have facts, because "a fact is not itself a phenomenon at all, but a proposition *about* one or more phenomena." When scientists engaged in the process they typically called "description of the facts," they were actually imposing a preexisting conceptual framework on raw sensory data.[34]

On the basis of this perspectival conception of knowledge, Parsons proposed a division of intellectual labor in which each of the behavioral sciences would focus solely on one of the overlapping motives that shaped behavior: the biological motive, the economic motive, and so forth. Sociologists, he argued, should analyze the shaping force of values in human action, filtering out all other motives and concentrating on how human beings strove to realize normative ideals. Parsons described sociologists' distinctive *a priori* postulate as "the voluntaristic theory of action": the view of human beings as agents motivated by value judgments rather than material forces. This postulate was not an inductive generalization from evidence, he stressed, but rather a working presupposition, chosen for its heuristic capacity – its ability to explain certain phenomena or relationships. In Parsons' proposed system of disciplines, sociology would be the special science of the value-oriented dimension of behavior.[35]

Defined in this manner, however, sociology had a political importance that the other special sciences lacked. It distinguished democracy from both totalitarianism and *laissez-faire* capitalism, both of which Parsons took to deny the operation of normative ideals in human relations by offering forms of economic determinism. Deeply engaged in the ideological struggle against totalitarianism, yet also implicitly critical of modern capitalism, Parsons believed that the American public possessed a deep reservoir of moral idealism that could be tapped in the cause of institutional reform. Yet, he lamented, social scientists constantly sowed cynicism among the public by portraying partial, monocausal

---

[34] Talcott Parsons, *The Structure of Social Action* (New York: McGraw-Hill, 1937), 41–42, 28. A cogent analysis of Parsons' "analytical realism" appears in Joel Isaac, "Theories of Knowledge and the American Human Sciences, 1920–1960" (PhD dissertation, University of Cambridge, 2005), 62–70.

[35] Cf. C. Lloyd Morgan, "A Philosophy of Evolution," in J. H. Muirhead, ed., *Contemporary British Philosophy: Personal Statements, First and Second Series* (New York: Macmillan, 1953 [1924]), 305. For Parsons, as for Creighton, perspectivalism served, like emergent evolution, as a naturalistic justification for disciplinary specialization.

explanations of human behavior – economic determinism, biological determinism, and so on – as the whole truth. Identifying the voluntaristic theory as the needed complement to these forms of determinism, Parsons attempted to sketch a synthetic picture of behavior as a whole. After the war, he would integrate anthropology and psychology into his model of a composite behavioral science, further advancing his long quest to downgrade economics to a secondary and derivative form of social analysis, and thereby to elevate normative concerns over pecuniary interests as a guiding force in American life.[36]

CONSENSUS LIBERALISM

Parsons' postwar thought represented one of the era's most influential expressions of consensus liberalism. That new conception of American culture and politics was strongly in the ascendant by the early 1940s, as the outbreak of war in Europe and America's deepening involvement in the conflict, among other factors, led scientific democrats to adopt a more celebratory, less critical stance toward the prevailing institutions and values. They typically believed that the war offered an opportunity to expand their program of cultural reconstruction around the world. With the partial exception of philosophers and historians, scientific democrats numbered among the earliest and most vocal supporters of America's entry into the war. They joined the neo-Thomists, Niebuhr, Mumford, and other aspiring cultural leaders in declaring that Americans' commitment to democratic ideals required them to fight fascism abroad, both to safeguard self-rule at home and to extend its reach across the globe. Lyman Bryson put the question directly to the public in 1940: "Do We Love Freedom for Ourselves Alone?"[37]

Scientific democrats in the universities, acting individually or through groups such as Ralph Barton Perry's Universities Committee on Post-War International Problems (UCPIP), generated thousands of books, pamphlets, and articles designed to counteract public skepticism about the war by explaining and promoting the American cause. In an extended replay of the World War I experience, they declared that the Allies fought selflessly for a new world order, one that would bring political freedom and material comfort to the global masses and put an end to armed conflict once and for all. Scientific democrats used Hitler's emergence from the wreckage of World War I as evidence that this time around, Americans would need to win the peace by building a strong world organization. This war, they insisted, did not pit nations against one another, but rather civilization against barbarism, with the fate of the whole world at stake. Believing that political ideologies, not national interests, lay behind the

---

[36] Brick, *Transcending Capitalism*, 135–151.
[37] Frank A. Warren, *Noble Abstractions: American Liberal Intellectuals and World War II* (Columbus: Ohio State University Press, 1999); Lyman Bryson, "Do We Love Freedom for Ourselves Alone?" *Journal of Adult Education* 12 (1940): 363–365.

conflict, scientific democrats offered a compelling image of the postwar order that they hoped would gird Americans for the battle against fascism.[38]

Along with the critics of scientism, these interventionist scientific democrats tended to be early and vociferous users of the new language of "totalitarianism." This analytical framework equated the fascism of Hitler and Mussolini with the communism of Stalin, holding that these regimes attempted to retain legitimacy by controlling the minds of their subjects. Scientific democrats argued that fascism and communism, like *laissez-faire* and traditional religion at home, violated the deepest needs of the human person. Advocates of these systems retained power by using fear and other emotional appeals to colonize the inner sphere of belief, rather than by giving citizens the security and freedoms that they actually needed. More than a struggle for political influence or institutional control, the conflict between democracy and totalitarianism appeared to these scientific democrats as a direct, zero-sum competition for cultural authority, in which militant, transcendentalist ideologies contended with science to fill the cultural space left open by the decline of theism and free-market conservatism.[39]

Interventionist scientific democrats typically recoiled against the prospect of preserving American economic superiority, even as they advocated a massive exportation of ideals and knowledge through a concerted cultural offensive. MacIver and many others called for "equal access to the natural resources of the world" and the elimination of the profit motive as a spur to economic development. Likewise, Perry rejected the idea that "backward societies should remain backward in order to be picturesque, and thus afford entertainment to spectators who live in advanced hotels." Yet, in keeping with their emphasis on ideas and values, commentators believed that the United States would need to make only minimal economic sacrifices to bring material plenty to the whole globe. The Harvard geologist Kirtley F. Mather predicted that Americans would briefly engage in direct economic assistance after the war, but would then primarily export "knowledge and techniques … ideas and ideals." This approach, said Mather, would end the exploitation of poor peoples and enable them to enjoy the fruits of their own natural resources. Leading scientific democrats hoped for the worldwide spread of the true industrial democracy they thought they saw emerging in 1930s America.[40]

[38] Ralph Barton Perry, *Final Report on the Work of the Committee, 1942–1945* (Boston: Universities Committee on Post-War International Problems, 1945). Ruth Nanda Anshen's edited volumes *Freedom: Its Meaning* (New York: Harcourt, Brace, 1940), *Science and Man* (New York: Harcourt, Brace, 1942), and *Beyond Victory* (New York: Harcourt, Brace, 1943) provide useful overviews of this wartime discourse.

[39] Benjamin L. Alpers, *Dictators, Democracy, and American Public Culture: Envisioning the Totalitarian Enemy, 1920s-1950s* (Chapel Hill: University of North Carolina Press, 2003).

[40] Robert I. MacIver, *Towards an Abiding Peace* (New York: Macmillan, 1943), 79; Ralph Barton Perry, *One World in the Making* (New York: Current Books, 1945), 191; Kirtley F. Mather, *Enough and to Spare: Mother Earth Can Nourish Every Man in Freedom* (New York: Harper & Brothers, 1944), 118–119.

At the same time, these figures believed a world state lay on the horizon. In holding forth the prospect of a global form of political organization, scientific democrats replayed familiar arguments about the potential of a shared, normative culture to harmonize unity with diversity and political coordination with individual freedom. Earlier, Progressive Era scientific democrats such as Charles Horton Cooley had argued that the mechanical interdependence of modern societies portended a more organic, idealistic form of social unity. During World War II, commentators applied the same argument to global affairs. Dewey, for example, called for "transforming physical interdependence into moral – into human – interdependence." Similarly, an early UCPIP planning memo declared that the world was in a state of crisis because it had "becom[e] physically and economically unified before its peoples are psychologically and spiritually prepared." In the earlier iteration of this argument, the scientific disciplines had appeared as the nucleus from which new forms of social relations would radiate through American society, bringing the culture into line with its new material foundations. Now, that remade culture itself could function as a new nucleus, according to leading scientific democrats. From it, scientific knowledge, ideals, and attitudes would spread outward to the whole world, allowing the inhabitants of other nations to respond constructively to the realities of global interdependence.[41]

Equipped with this understanding of the war's meaning and potential effects, scientific democrats, led by Parsons and other culture and personality theorists, jumped headfirst into the public arena. They worked feverishly to build support for the Allied cause and then took up various war-related posts when the United States joined the conflict. Human scientists participated in myriad aspects of the war effort, contributing to projects ranging from the management of Japanese internees to propaganda campaigns behind enemy lines in Europe and Asia. On this cultural-psychological front, a vast array of nongovernmental groups, funded by universities, foundations, the National Research Council, the Social Science Research Council, and the American Council of Learned Societies, assisted federal agencies such as the Office of Facts and Figures, the Office of Strategic Services, the Office of War Information, and the Allied Expeditionary Force's Psychological Warfare Division. The many thousands of anthropologists, sociologists, psychologists, and psychiatrists employed by this complex of institutions worked not only to win battles abroad, but also to shore up popular morale and minimize social conflict at home. The culture and personality school, and with it the consensus liberalism strategy, came to maturity in the context of World War II.[42]

---

[41] John Dewey, *Freedom and Culture* (1939), in Jo Ann Boydston, ed., *The Later Works, 1925– 1953, Volume 13* (Carbondale: Southern Illinois University Press, 1981), 181; Perry, *Final Report on the Work of the Committee*, 3.

[42] Brick, *Transcending Capitalism*, 139–145; Virginia Yans-McLoughlin, "Science, Democracy, and Ethics: Mobilizing Culture and Personality for World War II," in *Malinowski, Rivers, Benedict, and Others: Essays on Culture and Personality*, ed. George Stocking, Jr. (Madison: University of Wisconsin Press, 1986), esp. 197, 210; Herman, *The Romance of American Psychology*, 17–123

Culture and personality theorists reshaped their theories and public personas in keeping with their sense of the nation's needs during wartime. Today, we tend to remember this episode largely for its impact on American race relations. Arguing that group prejudice stemmed from the same emotional sources as the war itself, culture and personality theorists mobilized powerfully against racism, especially after a series of race riots in the summer of 1943. Participants in this antiracist campaign produced numerous arguments on behalf of public engagement by scholars. The anthropologist Gene Weltfish, who called the detached "robot scientist" far more dangerous than the "robot bomb," authored a kind of Hippocratic Oath for scientists: "I pledge that I will use my knowledge for the good of humanity and against the destructive forces of the world and the ruthless intent of men; and that I will work together with my fellow scientists of whatever country, creed, or color, for these, our common ends." Racism presented both an obvious target for progressive scholars and a potent obstacle to national morale.[43]

An equally striking product of the war years, however, was the concept of "national character," which powerfully reinforced the argumentative strategy of consensus liberalism. The Boasian anthropologist Margaret Mead joined two close associates, Ruth Benedict and the British theorist Geoffrey Gorer, in developing and promoting this theoretical innovation, which amounted to nothing less than the expansion of anthropology's scope to include ethnographic analyses of the developed nations of the Western world. After 1939, Mead and her husband, the British-born anthropologist Gregory Bateson, undertook a series of initiatives that the couple viewed as a prelude to the "restructuring of the culture of the world" in accordance with the democratic potentialities of scientific inquiry. They saw beneath the war a struggle between elites and the masses for control of the powerful tool of social science. Mead and Bateson began with a letter to Eleanor Roosevelt in August 1939, suggesting that knowledge of Hitler's psychological makeup might enable the president to turn him toward the peaceful reconstruction of Europe. Failing to stop

(see esp. 32–42, 68); Blair T. Johnson and Diana R. Nichols, "Social Psychologists' Expertise in the Public Interest: Civilian Morale Research During World War II," *Journal of Social Issues* 54, no. 1 (Spring 1998): 53–77; James H. Capshew, *Psychologists on the March: Science, Practice, and Professional Identity in America, 1929–1969* (New York: Cambridge University Press, 1999), 1–154; Lemov, *World as Laboratory*, 156–161, 170–187; Abbott and Sparrow, "Hot War, Cold War," esp. 286–290, 305–306; Mark Guglielmo, "The Contribution of Economists to Military Intelligence During World War II," *Journal of Economic History* 68, no. 1 (March 2008): 109–150; David H. Price, *Anthropological Intelligence: The Deployment and Neglect of American Anthropology in the Second World War* (Durham: Duke University Press, 2008). In psychology, war work – especially under the employ of the federal government – occupied one-quarter of those with graduate degrees in the field by the end of 1942: Herman, *The Romance of American Psychology*, 18.
43 Herman, *The Romance of American Psychology*, esp. 57–66; Gene Weltfish, "Science and Prejudice," *Scientific Monthly* 61, no. 3 (September 1945): 212. See also Ruth Benedict, *Race: Science and Politics* (New York: Modern Age Books, 1940); and Benedict and Weltfish, *The Races of Mankind*, Public Affairs Pamphlet 85 (New York: Public Affairs Committee, 1943).

the war, Mead then headed to Washington in late 1941 to run the Committee on Food Habits, organized under the National Research Council's Division of Anthropology and Psychology. There, she and the émigré psychologist Kurt Lewin brokered the government's use of anthropological experts. At every opportunity, Mead urged administration officials to take full account of the psychological makeup of the American people. She also sought to explicate that makeup herself in *And Keep Your Powder Dry* (1942), a seminal statement of the national character approach and a major contribution to the development of consensus liberalism.[44]

In alternately celebrating and chiding Americans for their characteristic modes of behavior, Mead's wartime book departed substantially from her earlier writings. In the late 1920s and early 1930s, a sharp undercurrent of anti-capitalism and, eventually, a strong technocratic impulse had repeatedly broken through the gentle, collective self-satire adorning her popular advice to parents. Under Benedict's influence, Mead had quickly abandoned the hope, expressed in her runaway bestseller *Coming of Age in Samoa* (1928), that changes in childrearing practices could generate systematic cultural change. By the time she published *Growing Up in New Guinea* (1930), Mead had adopted Benedict's claim that social institutions outside the schools forced children into preset cultural molds. Progressive educators, she wrote, instilled in children "an attitude which will find no institutionalised path for adult expression." Mead concluded that neither childrearing nor the current educational system could bear the weight of reform, given that the existing crop of parents and teachers took their cues from a culture of self-seeking individualism. She thus called for "emergency measures," in the form of a comprehensive, scientifically guided program of cultural retraining inspired by the "spectacular experiment in Russia." Mead proposed that a small intellectual elite should take control of the American school system, using new technologies such as radio and motion pictures to pipe their ideas into the nation's classrooms. The existing teachers, deeply infected with the virus of individualism, could serve only as "disciplinarians" and "record keepers" for the time being, passively overseeing students' assimilation of an expert-created curriculum until they had been adequately retrained as "the advance guard of civilisation."[45]

As the 1940s dawned, Mead continued to write critically of America's competitive, individualistic orientation. In October 1940, for example, Mead described Americans as culturally primed for manipulation by a demagogic leader. She flatly rejected the optimistic assumption "that because Americans

[44] Peter Mandler, "Margaret Mead Amongst the Natives of Great Britain," *Past and Present* no. 204 (August 2009): 195–233; Margaret Mead, *And Keep Your Powder Dry: An Anthropologist Looks at America* (New York: W. Morrow, 1942), 261; Yans-McLaughlin, "Science, Democracy, and Ethics," 209 (quoted), 194–195, 197; Herman, *The Romance of American Psychology*, 52. A recent study of Mead is Nancy Lutkehaus, *Margaret Mead: The Making of an American Icon* (Princeton: Princeton University Press, 2008).
[45] Margaret Mead, *From the South Seas: Studies of Adolescence and Sex in Primitive Societies* (New York: W. Morrow, 1939), 272, 275.

have always stood for democracy, or religious freedom, or some other ideal, they will always do so." Even in the aftermath of Pearl Harbor, she continued to describe the existing American culture as the primary obstacle to political reform and systematic cultural change as the solution.[46]

By contrast, *And Keep Your Powder Dry* offered something else entirely: a full-blown ethnographic analysis of the American national character framed in the comforting terms of Mead's popular articles. She purported to describe the prevailing culture, not to announce (and hasten) its imminent passing. Mead's book outlined the strengths and weaknesses of American character from the standpoint of the war effort and the forthcoming attempt to win the peace, rather than directly applying her earlier standard of critical judgment, the maximal promotion of human welfare. "We are the stuff with which this war is being fought," Mead declared. Of course, Mead, like other wartime scientific democrats, built the ideal of fulfilling human needs into her conception of the postwar order, which she hoped would be "a moral universe where success crowns the efforts of the efficiently good." Her rhetorical strategy, however, had clearly changed. In a revealing passage near the end of the book, she wrote that Americans would fight if, and only if, they believed "that they are fighting for a new and better world."[47]

Mead developed her newly appreciative portrait of American culture by conceptually isolating the "puritan tradition," with its "peculiar drive towards efficiency and success," from "the ruthless terms of big business" in which it had long been expressed. She held that the puritan tradition, freed from its *laissez-faire* fetters, offered "just the mechanism we need to build a new world." According to Mead, Americans' "need for success" and "genius for seeing themselves on the side of the good and right" produced, not merely an insufferable combination of materialism and moralism, but also a concrete drive, at times highly effective, to implement moral values through "purposeful thought and experimentation" – at least, so long as policymakers did not appear to undercut individual freedom by openly manipulating the political process from above. Overall, Mead's book described American culture as full of progressive potential, not as hopelessly corrupted by individualism. It suggested that, even if actual Americans routinely deviated from the national consensus, scholars and policymakers could call them back to the puritan tradition, which provided their core values and commitments.

Mead's text operated at a number of levels. As a tool for boosting morale, it aimed to make ordinary Americans "feel strong, not weak ... certain and proud and secure of the future." As a handbook for policymakers, it outlined the conditions under which wartime leaders could hope to make the most of American culture. Finally, as an ethnographic portrait of that culture, it gave

---

[46] Margaret Mead, "Social Change and Cultural Surrogates," *Journal of Educational Sociology* 14, no. 2 (October 1940), 107; "War Need Not Mar Our Children," *Journal of Educational Sociology* 16, no. 4 (December 1942 [1941]), 197.

[47] Mead, *And Keep Your Powder Dry*, 25, 261, 256, 254.

a major boost to consensus liberalism as a new basis for scientific democracy in the 1940s.[48]

Mead's change of argumentative strategy is all the more striking because she disavowed key tenets of consensus liberalism in speaking to fellow experts. Just before she published *And Keep Your Powder Dry*, Mead urged the intellectuals gathered for the second Conference on Science, Philosophy, and Religion in Their Relation to the Democratic Way of Life to "lead and change" a "chaotic, heterogeneous culture" toward "greater democracy." Suggesting that Americans' "intense emphasis upon the importance of competition" might prove "completely incompatible with democratic goals," Mead described the scholar's task as "the purposive cultivation of democratic values." Nine years later, speaking to the same group, she asserted that social scientists had made only limited progress toward that end.[49]

Many other scientific democrats also gravitated toward the affirmative, celebratory approach of consensus liberalism during World War II. To some degree, this move entailed the emergence of warmer feelings toward American institutions, as the war highlighted the existence of much greater evils than capitalism. Scientific democrats noted, often with considerable pride, that Americans had weathered the totalitarian storm and were now making grave sacrifices for the friends of democracy elsewhere. Unlike the economic situation at home, moreover, fascism and communism highlighted the positive effects of Americans' apparently endemic individualism. What leading scientific democrats had once considered a lamentable and dangerous obsession with individual freedom began to seem a source of stout resistance to dictatorship, if it could be channeled into support for widespread prosperity rather than *laissez-faire*. Meanwhile, scientific democrats viewed fascism as the political expression of pure philosophical irrationalism, the utter negation of science as well as democracy. Most extended this analysis to communism as well.[50] But the rise of consensus liberalism also reflected a new sense of the most effective

[48] *Ibid.*, 206–208, 199, 260. For analyses of Mead's book, see Philip Gleason, "The Study of American Culture," in *Speaking of Diversity: Language and Ethnicity in Twentieth-Century America* (Baltimore: Johns Hopkins University Press, 1992), 198–200; and Handler, *Critics Against Culture*, 141–153. In an unpublished manuscript written about this time, Benedict equated "the American dream" with her open, civil-libertarian social ideal: "Primitive Freedom" (1942), reprinted in Margaret Mead, ed., *An Anthropologist at Work: Writings of Ruth Benedict* (Boston: Houghton Mifflin, 1959), 391.

[49] Margaret Mead, "The Comparative Study of Culture and the Purposive Cultivation of Democratic Values," in Lyman Bryson and Louis Finkelstein, eds., *Science, Philosophy and Religion: Second Symposium* (New York: Conference on Science, Philosophy and Religion, 1942), 63–64; Mead, "The Comparative Study of Culture and the Purposive Cultivation of Democratic Values, 1941–1949," in Bryson, Finkelstein, and Robert M. MacIver, eds., *Perspectives on a Troubled Decade: Science, Philosophy, and Religion, 1939–1949: Tenth Symposium* (New York: Harper, 1950). In the earlier paper, Mead described research in the social-scientific disciplines as both "an instrument for undermining established beliefs" and the source of a more enlightened view to replace them (63).

[50] Alpers, *Dictators, Democracy, and American Public Culture*, esp. 129–156.

strategy for changing a national culture. Its advocates followed John Dewey in seeking to reformulate Americans' preexisting ideals, to give cherished linguistic markers new cultural import. Most saw the liberal consensus as an inbuilt, normative standard that scholars could mobilize to challenge its near-constant breaches in everyday practice.

This mode of cultural criticism had deep roots in Progressive thought. For one thing, it built on many Progressives' assumption that they could speak for the needs of the people against corporate interests – and, if necessary, even against prevailing public opinion. As we have seen, moreover, the anti-materialist form of social theory adopted by Progressive sociologists and anthropologists traced social order to a consensually held framework of beliefs and values. Following the prescriptions of Ross and Boas, scientific democrats had long sought to foster change in that psychological sphere by seeding it with new culture elements rooted in the findings of the social sciences. But whereas most Progressive critics took up this task by demonstrating the utter bankruptcy of dominant social values such as individualism, competition, and freedom, Dewey's treatment of the social self pointed toward an alternative reform strategy. Undertaking a subtle, internal critique of the existing culture, Dewey sought to infuse its terms and symbols with new meanings that supported the extension of democratic control into the economic sphere.[51]

Dewey targeted numerous items of American public discourse in his writings. Most obviously, he worked from the late 1880s forward to define "democracy" as the political expression of liberal Christianity – which Dewey later rechristened "naturalism" or "liberalism" – rather than the institutionalization of a negative form of liberty. In keeping with the general tendency of Progressive legal thought, Dewey also undertook to redefine freedom in positive terms as the concrete ability to realize one's goals rather than the mere lack of formal obstacles to action. Finally, the very discourse of the social self likewise involved a redefinition of a core American concept, that of self-interest. Dewey, who was steeped in the theory of learning – as a slow and incremental process, not a sudden conversion – articulated by nineteenth-century figures such as Horace Bushnell and William James, sought to guide Americans into a mental world more appropriate to the material world that they already inhabited. But he presented himself as, in his student John Herman Randall Jr.'s words, simply drawing out "the values immanent in our civilization."[52]

[51] Analogously, albeit more successfully, President Roosevelt adopted William Graham Sumner's image of a long-suffering "forgotten man" for the cause of his administration's economic progressivism: Amity Shlaes, *The Forgotten Man: A New History of the Great Depression* (New York: HarperCollins, 2007).

[52] Robert B. Westbrook, *John Dewey and American Democracy* (Ithaca: Cornell University Press, 1991), 42–51, 435–437; David M. Rabban, "Free Speech in Progressive Social Thought," *Texas Law Review* 74 (1996): 951–1038; John Herman Randall Jr., *Our Changing Civilization: How Science and the Machine are Reconstructing Modern Life* (New York, 1929), 354. Cf. Westbrook, *John Dewey and American Democracy*, 145–149, 192, 430–439, 444–445. A few years later, Dewey would likewise seek to recolonize "God" and "religious" with new meanings: *A Common Faith* (1934), in *The Later Works, Volume 9.*

Dewey's effort to recolonize the American political vocabulary found its clearest expression in a little book of 1930 entitled *Individualism, Old and New*. There, he followed the ethical economists' argument that the social organism had entered a new economic environment. Dewey allowed that *laissez-faire* policies had been appropriate under the wide-open frontier conditions of the early nineteenth century, but he insisted that industrialization had created new threats to individual freedom, requiring a new political instantiation of that core American value. In the wake of industrialization, Dewey argued, Americans would need to voluntarily abandon some of their economic latitude of action in order to attain the material security, cultural development, and opportunities for political participation that a more cooperative society would provide.

Viewed from the standpoint of the culture and personality discourse, Dewey's book directly paralleled Benedict's ironic claim that contemporary Americans embraced competitive individualism only as a result of unthinking conformity. These theorists portrayed "rugged individualists" as excessively conformist, not insufficiently social. They held that conservatives, rather than selfishly ignoring the functional requirements of a modern society, simply failed to think critically and thereby impeded the cause of self-realization – their own as well as that of others. Dewey, like figures ranging from Karl Marx through Émile Durkheim to Charles Taylor today, held that viewing oneself as a discrete, self-seeking individual was a habit instilled through powerful socialization processes, not a natural, pre-cultural fact.

Of course, many scientific democrats declared individualism unnatural and then proceeded to describe a life of mutual service as the truly natural one. But Dewey, like Benedict, defined all personality types as culturally contingent constructions, to be judged according to their impact on human flourishing under prevailing conditions rather than their fit with a putatively preexisting, universal human nature. Even critical thinking and cultural tolerance amounted to socially conditioned habits, Dewey and Benedict cautioned. Conceptually separating cultural affairs from economic matters, they called for greater tolerance of disagreements in the cultural sphere and greater cooperation in the economic realm. This move allowed Dewey and Benedict to urge the expansion, rather than the contraction, of individual freedom, redefined in terms of the remarkable diversity of individuals and the psychological as well as material nature of individual needs.[53]

During World War II, Dewey's students and admirers undertook, in a highly revealing – and not terribly successful, in the long run – expression of consensus liberalism, to lodge his pragmatism at the heart of the American political tradition. These figures declared Dewey a worthy successor to "Jefferson, Emerson, Whitman or Lincoln" as "a symbol to the people, a molder of tradition, [and]

---

[53] Brick writes of interwar social liberals "fashioning new spaces for individuality beyond the cultural straitjacket of economic individualism": *Transcending Capitalism*, 16.

an architect of the future." Horace Kallen, adding Thomas Paine and William James to the list, declared Dewey far more important culturally than "the Fords and the Edisons," with their "lucky inventions and financial good fortunes." Dewey's eightieth birthday in 1939 generated a volume declaring him *The Philosopher of the Common Man*. James' centenary three years later produced similar paeans to his thoroughly American outlook, but the homespun Dewey proved easier to integrate into the pantheon. A selection from Dewey's writings capped off the seminal documents of the democratic tradition, reaching back to the Hebrew prophets, that Irwin Edman anthologized in *Fountainheads of Freedom* (1941). This celebration of Dewey's uniquely American viewpoint continued into the 1950s, and postwar intellectual historians such as Merle Curti, Henry Steele Commager, Morton White, and Henry May contributed to the cause by making a central place for pragmatism in their accounts of American thought.[54]

But the wartime struggle against racism provided the impetus for the most influential expression of consensus liberalism in those years: the Swedish sociologist Gunnar Myrdal's 1944 study *An America Dilemma*. Myrdal explained that the racism pervading American society stood in fundamental contradiction to the nation's deepest cultural commitments, a set of values that Myrdal termed "the American Creed." He wrote that "Americans of all national origins, classes, regions, creeds, and colors" shared this fundamental "social *ethos*," the "cement in the structure of this great and disparate nation." Myrdal emphasized that the American Creed not only served as "the implicit background of the nation's political and judicial order," but also enlisted the active support of all Americans as those "principles which *ought* to rule." In short, the American Creed amounted to "the national conscience." But Myrdal, noting that "the political creed of America is not very satisfactorily effectuated in actual social life," proceeded to devote almost 1,500 pages to the failures of application. Still, Myrdal, like other advocates of consensus liberalism, expected to foster change by calling on Americans to live up to their values, not to change their values. He concluded that "the conquering of color caste ... is America's own innermost desire," because "[t]he main trend in [American] history is the gradual realization of the

---

[54] Joseph Ratner, "Foreword," and Horace Kallen, "Freedom and Education," in *The Philosopher of the Common Man: Essays in Honor of John Dewey to Celebrate His Eightieth Birthday* (New York: Greenwood Press, 1968 [1940]), 7, 16; Brand Blanshard and Herbert W. Schneider, eds., *In Commemoration of William James, 1842–1942* (New York: Columbia University Press, 1942); Irwin Edman, *Fountainheads of Freedom: The Growth of the Democratic Idea* (New York: Reynal & Hitchcock, 1941); Sidney Hook, ed., *John Dewey: Philosopher of Science and Freedom* (New York: Dial, 1950); Edman, *John Dewey: His Contribution to the American Tradition* (Indianapolis: Bobbs-Merrill, 1955); James T. Kloppenberg, "Pragmatism and the Practice of History: From Turner and Du Bois to Today," *Metaphilosophy* 35, nos. 1–2 (January 2004), 209, 212–213. Cf. Justus Buchler, "The Philosopher, the Common Man, and William James," *American Scholar* 11 (August 1942): 416–426.

American Creed." In his final paragraph, Myrdal called it "the supreme task of social science" to "find the practical formulas for this never-ending reconstruction of society."[55]

As Myrdal's book showed, consensus liberalism suggested that the vast majority of Americans needed to assimilate new forms of knowledge in order to properly implement their deeply held values. In fact, expressions of consensus liberalism typically coupled a strategy of internal critique with a non-Marxist variant of the concept that Antonio Gramsci dubbed "hegemony." This side of consensus liberalism drew on a different legacy of Progressivism: namely, Progressive reformers' claim to speak for "the people" against "the interests," who were said to have clamped mental shackles on the *actual* people and caused them to misunderstand the public good. Scientific democrats since the days of Andrew Dickson White and Charles W. Eliot had distinguished the public interest, as represented by reformers and experts, from public opinion – the beliefs of the actual citizens. In the 1940s and 1950s, those employing consensus liberalism similarly argued that the small minority of cultural leaders who sought to change the values of their fellow citizens were the true spokesmen for American values, against which businessmen had turned the people themselves.

At the time Myrdal's book appeared, however, scientific democrats were turning their attention outward, to the rest of the world. They increasingly concerned themselves with the question of global reconstruction as the war ground toward its end. In 1945, Gardner Murphy gathered together some fifty psychologically informed thinkers to produce *Human Nature and Enduring Peace*, the third yearbook of the SPSSI. Finding "practically no remaining doubt" among "thinking people" regarding the necessity of a "world federation," Murphy and his collaborators – expecting political leaders to "gladly use" their results – called for intensive research into the kinds of selves appropriate to modern conditions. Meanwhile, Talcott Parsons took to the field directly at the war's end, helping to rebuild and reorient the German economy. Before then, he worked to prepare military officers for the postwar task of reconstruction. And Ruth Benedict capped off the body of wartime work on national character with *The Chrysanthemum and the Sword* (1946), her pioneering contribution to what Mead called "the study of culture at a distance." Assaying a systematic description of the underlying traits of Japanese culture that could guide the completion of the American war effort, Benedict spoke as clearly to American

---

[55] Gunnar Myrdal, *An American Dilemma: The Negro Problem and Modern Democracy* (New York: Harper & Row, 1944), 1, 3, 23, 1021, 1024. Also see Walter A. Jackson, *Gunnar Myrdal and America's Conscience: Social Engineering and Racial Liberalism, 1938–1987* (Chapel Hill: University of North Carolina Press, 1990); Herman, *The Romance of American Psychology*, 176–181; Wendy Wall, *Inventing the "American Way": The Politics of Consensus from the New Deal to the Civil Rights Movement* (New York: Oxford University Press, 2008), 95–100; and William J. Barber, *Gunnar Myrdal: An Intellectual Biography* (Basingstoke: Palgrave Macmillan, 2008), 64–85.

traits as to Japanese ones. In keeping with consensus liberalism, her analysis stressed the beneficent side of American culture.[56]

In fact, the postwar behavioral sciences as a whole took much of their shape from consensus liberalism, with its suggestion that Americans needed only to reactivate and reapply their core values in order to produce a world – and now, scholars really meant a world – of peace and plenty. The postwar integration and domestication of Freud's work by culture and personality theorists, working in tandem with refugees from Nazism such as the critical theorist Theodor Adorno, the psychologist Erich Fromm, and the psychiatrist Karen Horney, would provide consensus liberalism with a stable theoretical foundation for years to come. Throughout the 1950s and into the 1960s, a new generation of scholars would attempt, in the vein of consensus liberalism, to alter American political culture and social institutions by describing them anew. Frequently, however, such efforts ended up merely reinforcing more conservative readings of the same American ideals.

[56] Gardner Murphy, in Murphy, ed., *Human Nature and Enduring Peace* (Boston: Houghton Mifflin, 1945), v, 223; Uta Gerhardt, "Talcott Parsons and the Transformation of German Society at the End of World War II," *European Sociological Review* 12 (1996), 305, 310; Gerhardt, ed., *Talcott Parsons on National Socialism* (New York: De Gruyter, 1993). Critical accounts of Benedict's book include Christopher Shannon, "A World Made Safe for Differences: Ruth Benedict's *The Chrysanthemum and the Sword*," *American Quarterly* 47 (1995): 659–680; Adrian Pinnington, "Yoshimitsu, Benedict, Endō: Guilt, Shame and the Post-War Idea of Japan," *Japan Forum* 13, no. 1 (April 2001): 91–105; and Sonia Ryang, "Chrysanthemum's Strange Life: Ruth Benedict in Postwar Japan," *Asian Anthropology* 1 (2002): 87–116.

10

Two Cultures

All innovations in the human sciences notwithstanding, natural scientists, and especially physical scientists and engineers, would ultimately shape the definition of science at the national level. Beginning in the late 1930s, these figures powerfully staked their claim as spokespersons for science in general, but they did not share a single voice. In fact, during the tumultuous decade and a half bracketed by the wars in Spain and Korea, these figures divided sharply over questions regarding science's character and political implications. Disputants in the physical sciences and engineering, however, shared a common belief that scientific research took place in relative isolation from political commitments – and, in most cases, they also shared a common distrust of claims to scientific status on the part of human scientists. Even progressive advocates of "social responsibility" followed their more establishmentarian counterparts in ignoring and implicitly denigrating social scientists, even as they insisted on the importance of applying scientific techniques and findings to social problems. Although these critical scholars viewed political engagement as an integral part of the scientist's professional role, they still described the relationship between science and society in simple, one-way terms: scientific research drove progress. Neither their view nor the alternative, centrist vision made much room for the complexities of human motivation stressed by the human scientists of the interwar years.

Even these activists' relatively thin understanding of science's political ties became marginal, and then anathema, in American public discourse by the end of the 1940s. A small cadre of science administrators affiliated with the highly selective National Academy of Sciences (NAS) and Roosevelt's wartime Office of Scientific Research and Development (OSRD) – both of which excluded virtually all social scientists – became quasi-official spokesmen for the American scientific community. Led after 1940 by the electrical engineer Vannevar Bush, this group of physical scientists and engineers also included MIT president Karl T. Compton, Harvard president James B. Conant, and Bell Laboratories' Frank B. Jewett. They and their allies had worked since 1937 to convince political leaders to modernize the nation's armed forces by enlisting scientific research in the nation's defense. On the domestic front, they preferred Herbert Hoover's vision of voluntarily coordinated capitalism to the regulatory approach of the New Deal. Bush and his compatriots shared Hoover's belief that a combination of technological ingenuity, market freedoms, and government-sponsored coordination would suffice to bring science's material benefits to all. Thus, they hoped to insulate science from political influence, especially the decisions of would-be planners. Nevertheless, they had the ear of the president, who effectively ceded Bush responsibility for the nation's science policy during the war years. Needing all the help he could get in the uncharted field of federal science administration, Roosevelt deferred to these powerful figures in decisions related to the mobilization of the nation's scientific resources. Their vision for the administrative organization of science, if not their political-economic ideal, would take firm hold by the 1950s.[1]

Distrustful of state action in any form, these physical scientists and engineers established a relationship with the federal government during World War II that brought academic scientists generous financial support without day-to-day control. In the military branches, an emerging system of federal research contracts set the institutional precedent for the Cold War era's massive defense-related programs. Meanwhile, Bush and the NAS establishment powerfully reinforced the image of science as morally and politically neutral. Ironically, in fact, they helped to pave the way for the emergence of a profoundly militaristic national culture that would leave many of them out in the cold in the 1950s. Only a few years after they worked out a relationship with the federal government that suited their Hooverite fear of state control, they watched helplessly as anti-communists turned the whole apparatus of publicly funded science toward militaristic ends. The paradoxical spectacle of government-funded scientists insisting on their political neutrality while actively prosecuting the fight against communism was now in place, at least in the highly visible physical sciences.

THE PHYSICAL SCIENTISTS

Many historians have explored the struggle among physical scientists, engineers, and political leaders over the contours of a nascent, science-based political economy in the mid-twentieth century. Leading physical scientists undertook to systematically define their role in the American polity in the late 1930s, just as the science studies fields were shifting away from Deweyan

[1] Stuart W. Leslie, *The Cold War and American Science: The Military-Industrial-Academic Complex at MIT and Stanford* (New York: Columbia University Press, 1993); Daniel Lee Kleinman, *Politics on the Endless Frontier: Postwar Research Policy in the United States* (Durham: Duke University Press, 1995); Alfred K. Mann, *For Better or for Worse: The Marriage of Science and Government in the United States* (New York: Columbia University Press, 2000); Patrick J. McGrath, *Scientists, Business, and the State, 1890–1960* (Chapel Hill: University of North Carolina Press, 2002). For a good summary of Hoover's "associationalist" vision, see David M. Hart, *Forged Consensus: Science, Technology, and Economic Policy in the United States, 1921–1953* (Princeton: Princeton University Press, 1998), 18–20, 39–61.

perspectives and consensus liberalism was gaining a foothold in the social sciences. However, they contended bitterly with one another over the nature of that role. Those emphasizing the "social responsibility" of scientists insisted that they should actively promote social progress through their research, by allowing social values to guide them in choosing research problems. On this view, scientists bore an obligation as citizens to select research topics on the basis of their potential social benefits and to ensure that their findings were applied in accordance with the public good. Even in their professional capacity, scientists were still members of the American polity and needed to keep the needs of that polity constantly in mind. By contrast, defenders of scientific neutrality argued that research would only produce reliable outcomes if it were pursued in a thoroughly disinterested fashion, without any thought of real-world consequences.

Also well understood are the institutional structures and practices that emerged from the struggle, which the opponents of the social responsibility discourse won decisively. In the postwar system of "big science," money flowed freely from the National Science Foundation, the National Institutes of Health, and an array of military agencies to academic scientists who were engaged in research relevant to national interests. Coming with few strings attached, the new federal grants preserved scientists' subjective sense of freedom in choosing research goals, yet powerfully imprinted a distinctive pattern on the American research enterprise as a whole. Much to the chagrin of the social responsibility advocates, it was the NAS establishment that set the research priorities, rather than populists and progressives in Congress and the disciplines. Yet the critics rarely contested the Bush group's denigration of the social sciences, and often participated vigorously in the marginalization of those fields. As a new political economy of science emerged in the years around World War II, physical scientists and engineers of all ideological stripes consistently wrote the social sciences out of the national conversation about knowledge and politics, despite the substantial integration of social scientists into the postwar military-industrial complex.

In its American form, the discourse of social responsibility took shape in the context of the "first scientists' movement," a series of political initiatives lasting from roughly 1936 to 1941. During those tumultuous years, scientists mobilized for causes ranging from health care reform and the protection of civil liberties to anti-fascism and support for the Spanish Republic. In the run-up to the New York World's Fair of 1939, some contested the presentation of science as a magic fount of consumer goods lacking broader cultural and political implications. Others of a more radical bent sought to unionize bench scientists in laboratories. The push for social responsibility, which found a home in the American Association for the Advancement of Science (AAAS), produced strange political bedfellows, as the KGB spy William M. Malisoff worked alongside the fierce anti-New Dealer Robert A. Millikan in groups such as the Lincoln's Birthday Committee for Democracy and Intellectual Freedom. These politically disparate figures could agree, first, that public policies should be tuned to maximize the economic prosperity promised by science-based technologies, and second, that American citizens should mobilize against fascist governments, which would impose ideological controls on science.[2]

In part, this new attention to the social relations of science merely reflected a growing public concern about "technological unemployment" – that resulting from the replacement of human labor by machines. Many union leaders and politicians saw the hand of science behind the massive job losses of the early Depression years. In response, scientists such as Caltech's Millikan and the University of Chicago physicist Arthur Holly Compton declared that "Science Makes More Jobs," as the title of a 1934 symposium had it. These centrist defenders of science insisted that economic recovery awaited the maximal development of research rather than the regulation of industry. "The trouble with us today is not the overproduction of goods, but the underproduction of new ideas," declared General Electric's Charles P. Kettering.[3]

But the concern about science's social impact went far beyond concrete worries about technological unemployment. Progressives such as Secretary of Agriculture Henry A. Wallace focused on the broader phenomenon that William F. Ogburn had termed "cultural lag." At the 1933 AAAS meeting, Wallace chided natural scientists and engineers for having "turned loose upon the world new productive power without regard to the social implications." He urged them to focus on building up the "public consciousness" needed to support and operate the "Christian, cooperative, democratic state" that, in Wallace's view, lay implicit in modern science.[4]

Some commentators at the time called for an outright moratorium on scientific research, to provide time to address the social dislocations created by rapid technological development. Wallace instead proposed educational and regulatory solutions. So, too, did President Roosevelt, who registered the widespread concern with the social effects of technology at the highest level of government.

---

[2] Peter J. Kuznick, *Beyond the Laboratory: Scientists as Political Activists in 1930s America* (Chicago: University of Chicago Press, 1987), 181–182, 188, 227–252; Kuznick, "Losing the World of Tomorrow: The Battle over the Presentation of Science at the 1939 New York World's Fair," *American Quarterly* 46 (1994): 341–373. On Malisoff's Soviet connections, see George A. Reisch, *How the Cold War Transformed Philosophy of Science: To the Icy Slopes of Logic* (New York: Cambridge University Press, 2005), 105–107. Millikan identified a mix of liberal faith, technical knowledge, economic competition, and courses in the history of science as the cure for the world's ills. Science, he contended, revealed empirically that the leading stimuli to human action were wage differentials and the promise of equality and freedom: "Science and Social Justice: 'A Stupendous Amount of Woefully Crooked Thinking,'" *Vital Speeches of the Day* 5 (December 1, 1938), 98–101.

[3] Quoted in Kuznick, *Beyond the Laboratory*, 21; quoted in Karl T. Compton, "Engineering Research and National Welfare," in *Proceedings of the Association of Land-Grant Colleges and Universities, Fifty-Second Annual Convention* (New Haven: Quinnipiack Press, 1938), 39. See also Amy Sue Bix, *Inventing Ourselves Out of Jobs? America's Debate Over Technological Unemployment, 1929–1981* (Baltimore: Johns Hopkins University Press, 2000).

[4] Henry A. Wallace, "The Social Advantages and Disadvantages of the Engineering-Scientific Approach to Civilization," *Science* 79 (January 5, 1934), 2, 4.

In October 1936, Roosevelt assigned technology, and behind it the nation's educational system, much of the blame for the Depression. In an open letter, he urged MIT's Karl T. Compton and other leaders of engineering schools to cushion technology's negative effects by orienting their students toward questions of social relevance. The following year, in his second inaugural address, Roosevelt described New Deal agencies and policies as "moral controls" on science, aimed at rendering it "a useful servant instead of a ruthless master of mankind." The president argued that science was proceeding too freely and needed to be hemmed in, though not at the level of research itself.[5]

By contrast, advocates of social responsibility argued that science suffered from innumerable checks and hedges because of its entanglement in a capitalist economy. Like the British Bernalists, figures such as Edwin Grant Conklin, the educator Benjamin C. Gruenberg, and *New York Times* science editor Waldemar Kaempffert (a cousin of the logical empiricist Otto Neurath) decried the "frustration of science" under the system of profit-driven enterprise. As an antidote, Kaempffert and the Scottish journalist Ritchie Calder envisaged a worldwide organization of scientists who would orient their fields toward the public good. The liberal Protestant *Christian Century*, along with newspapers such as the *New York Times*, *Baltimore Sun*, and *Washington Post*, editorialized on behalf of the Kaempffert-Calder plan during the 1937 AAAS meeting in Indianapolis, which drew reporters from around the world. Meanwhile, Conklin, elected AAAS president the preceding year, frankly advocated "democratic socialism." Citing the Bernalists approvingly, he traced the ills of the day to "the conflict between altruistic science and acquisitive society." Conklin used his presidency to press his longstanding claim for the evolutionary superiority of altruism.[6]

But Conklin's vaguely defined Christian socialism coexisted with other conceptions of social responsibility, some more radical but many less so. In 1938, AAAS members signaled their concern for science's public role by awarding the presidency to a social scientist, the Columbia economist Wesley C. Mitchell. Mitchell's presidential address, however, called for a technocratic program of national planning, not a culture rooted in biological altruism. He argued that scientists should never be swayed by emotions and could serve the public solely by attaching themselves to public agencies as expert "technical advisers." Departing from the liberationist rhetoric of many activists, Mitchell cautioned

[5] Carroll Pursell, "'A Savage Struck by Lightning': The Idea of a Research Moratorium, 1927–37," *Lex et Scientia* 10 (1974): 146–161; Wallace, "The Social Advantages and Disadvantages of the Engineering-Scientific Approach to Civilization," 3, 5; Kuznick, *Beyond the Laboratory*, 64; Franklin D. Roosevelt, "The Responsibility of Engineering," with Compton's reply, *Science* n.s. 84 (October 30, 1936): 393; quoted in Daniel J. Kevles, *The Physicists: The History of a Scientific Community in Modern America, Revised Edition* (Cambridge, MA: Harvard University Press, 1995), 264. See also Christophe LeCuyer, "The Making of a Science Based Technological University: Karl Compton, James Killian, and the Reform of MIT, 1930–1957," *Historical Studies in the Physical and Biological Sciences* 23 (1992): 158–180.

[6] Kuznick, *Beyond the Laboratory*, 77–78, 72; Edwin Grant Conklin, "The American Association for the Advancement of Science and the Society of Sigma Xi," *Science* n.s. 83 (June 26, 1936), 609, 608.

against utopian expectations of science and the drawing of premature conclusions in socially relevant research fields, while urging scientists to instill
"respect for evidence" in the population. With centrists such as AAAS secretary
Forest Ray Moulton and his brother, Brookings Institution head Harold G.
Moulton, at the helm, the AAAS steered away from Conklin and Kaempffert's
vision of a worldwide campaign against pecuniary motives, despite continued
hopes for cultural transformation among prominent scientists such as Albert
Einstein and the chemist Harold C. Urey. As Europe drifted toward war, optimism still ran high among many scientific democrats that they would achieve
their cultural goals.[7]

The outbreak of war in Europe accelerated the processes that would squelch
those dreams. When, in 1940, Roosevelt installed Bush as chairman of what
would become the OSRD, it was not at all clear how the federal government
would mobilize the scientific and technological expertise it needed to prosecute
the war effort. But Bush vigorously countered efforts to have the government
employ scientists directly on the atomic bomb project and other military initiatives, and Roosevelt followed his lead. Rather than hiring its own scientific
staff or commandeering private enterprises, as it had with the railroads during
World War I, the administration simply drew on existing pools of academic
faculty and industrial researchers. To the greatest extent possible, it located
research on existing campuses. When secrecy required the construction of
dedicated facilities, as at Los Alamos, oversight fell to universities such as the
University of California-Berkeley, MIT, and the University of Chicago. Bush
developed the central administrative mechanism of this system: a contract
arrangement that compensated the universities at a level well beyond costs and
required little in the way of oversight. As the war unfolded and the federal government devoted ever-increasing sums to military technologies, Bush and other
science administrators strove to preserve a strict line of separation between
private bodies – universities and corporations – and the state.[8]

Meanwhile, the West Virginia senator Harley M. Kilgore, a sharp critic of
the Eastern science establishment, drafted the first of a long series of bills proposing federal funding of research and development through what became
known as the National Science Foundation (NSF). Well before the end of the
war, it was apparent both to populists in Kilgore's vein and to administrators
such as Vannevar Bush that such funding would eventually come. By 1945,

---

[7] Wesley C. Mitchell, "The Public Relations of Science," *Science* 90, no. 2348 (December 29,
1939), 604–605, 607; Kuznick, *Beyond the Laboratory*, 73–94. Mitchell echoed Ogburn's sharp
distinction between knowing and feeling, although he recognized how hard it was to achieve
that separation and how partial were most victories over subjectivity. His expert-centered vision
comes through more clearly in "Science and the State of Mind," *Science* 89, no. 2297 (January 6,
1939): 1–4; see also Mark C. Smith, *Social Science in the Crucible: The American Debate Over
Objectivity and Purpose, 1918–1941* (Durham: Duke University Press, 1994), 49–83.
[8] Rebecca S. Lowen, *Creating the Cold War University: The Transformation of Stanford* (Berkeley:
University of California Press, 1997), 58–66; Hart, *Forged Consensus*, 118, 122–128.

government agencies were footing a full 83 percent of the nation's research bill. The question was what form postwar research policy would take.[9]

In the ensuing debate, establishment scientists and research-oriented business leaders rejected all talk of science's social embeddedness, even at the level of problem choice. Abraham Flexner, now serving as head of the Institute for Advanced Study, insisted that the handful of geniuses responsible for scientific progress took their cues from sheer curiosity and could not be managed, administered, or directed. By contrast, the "clever inventor[s]" who turned science into practical devices were a dime a dozen. This "linear model" of technological development, as it is now called, framed technological innovation as mere "applied science," a pursuit in which entrepreneurs and government contractors capitalized on the "pure science" or "basic research" carried out by disinterested academic scientists.[10]

Bush codified this understanding of science and linked it to the American discourse of pioneering in *Science – The Endless Frontier*, his influential 1945 recommendation to President Truman regarding the structure of the proposed NSF. "Scientific progress on a broad front," he wrote, "results from the free play of free intellects, working on subjects of their own choice, in the manner dictated by their curiosity for exploration of the unknown." Although Bush's document appeared near the beginning of the bitter five-year debate that eventuated in the NSF's formation, the new agency essentially followed his prescription. What many commentators call the postwar "social contract" between science and the American state had been forged. With it came a powerful cultural understanding of science as a practice undertaken for morally and politically disinterested motives.[11]

The "second scientists' movement" that emerged in the wake of the bombing of Hiroshima and Nagasaki represented only an apparent deviation from this image of scientific neutrality. To be sure, the activists associated with

---

[9] Kilgore's proposal actually emerged from pork-barrel politics; he hoped that the new agency would serve as an engine of economic growth in West Virginia and other poor states. But Waldemar Kaempffert and many other backers vigorously denied the claim – shared by both Kilgore and Bush – that scientific knowledge could be turned to social advantage only through "the ordinary course of industrial activity." Kaempffert favored adding to profit-driven industrial research and curiosity-driven academic research a third system of government-funded research into problems central to the public welfare. Vannevar Bush, *Science – The Endless Frontier* (Washington: GPO, 1945), 68; Kaempffert, "The Case for Planned Research," *American Mercury* 57 (1943), 444–446. Cf. "A National Science Program," *New Republic* 113 (July 30, 1945): 116.

[10] Abraham Flexner, "The Usefulness of Useless Knowledge," *Harper's* 179 (1939), 545–548; Benoit Godin, "The Linear Model of Innovation: The Historical Construction of an Analytical Framework," *Science, Technology & Human Values* 31 (2006): 639–667. Cf. Frank B. Jewett, "The Challenges of Science," *Vital Speeches of the Day* 6 (1940), 575.

[11] Bush, *Science – The Endless Frontier*, 7. To be sure, the NSF budget remained minuscule until the Soviet launch of Sputnik in 1957. But the federal government had set the terms of its science policy, and the principles guiding the NSF also informed the vast expansion of research expenditures by other military agencies during the Korean War. David H. Guston and Kenneth Keniston, eds., *The Fragile Contract: University Science and the Federal Government* (Cambridge, MA: MIT Press, 1994).

the Federation of American Scientists (FAS) viewed democracy in thoroughly deliberative terms. They believed, as Albert Einstein put it, that the fate of the world rested on "decisions made in the village square." Thus, researchers from Los Alamos and other wartime laboratories sought to educate Americans about the policy implications of nuclear fission. These activists assumed that a public familiar with the ins and outs of nuclear technology would institute the information-sharing and international control that they recommended, leading to a harmonious world order rooted in the peaceful use of science. Through publications such as the *Bulletin of the Atomic Scientists* and the 1946 book *One World or None*, as well as radio addresses, interviews, speeches, and numerous other means of publicity, FAS members and other concerned scientists worked to mobilize public support for their proposals. In fact, these efforts spawned a minor revival of the 1930s discussion movement, as local groups gathered to explore the question of nuclear proliferation.[12]

By 1950, however, secrecy rather than publicity was the name of the game in American science and scholars could question Cold War goals only at risk to their livelihoods. As the goal of international control slipped from activists' grasp in 1947, Truman's establishment of security checks for an ever-growing circle of federal employees – soon to include academic recipients of federal research grants – reflected a burgeoning fear of domestic subversion. Many progressive physical scientists underwent a change of heart about the nation's international priorities, as did FAS stalwart Harold C. Urey when he concluded in 1948 that communism was more dangerous than nuclear proliferation. The new anti-communism squeezed out not only Henry A. Wallace and other progressives but also antistatists such as Bush. By the time Bush wrote *Modern Arms and Free Men* in 1949, his call to protect science from politics by maintaining civilian control over military agencies sounded positively subversive. In the fiercely anti-communist political order of the 1950s, only those scientists who favored American military supremacy in all spheres and at all costs, such as the ardent cold warrior Edward Teller, could make their voices heard on policy questions.[13]

[12] Jessica Wang, "Scientists and the Problem of the Public in Cold War America, 1945–1960," *Osiris* n.s. 17 (2002), 328, 330–331 (quote on 330); Paul S. Boyer, *By the Bomb's Early Light: American Thought and Culture at the Dawn of the Atomic Age* (New York: Pantheon, 1985), esp. 59–64; Megan Barnhart, "Selling the International Control of Atomic Energy: The Scientists' Movement, the Advertising Council, and the Problem of the Public," in *The Atomic Bomb and American Society: New Perspectives*, ed. Rosemary B. Mariner and G. Kurt Piehler (Knoxville: University of Tennessee Press, 2009), 103–119; Lawrence S. Wittner, *Confronting the Bomb: A Short History of the World Nuclear Disarmament Movement* (Stanford: Stanford University Press, 2009).

[13] Richard Beyler, Alexei Kojevnikov, and Jessica Wang, "Purges in Comparative Perspective: Rules for Exclusion and Inclusion in the Scientific Community under Political Pressure," *Osiris* 20 (2005), 40–41; Wang, *American Science in an Age of Anxiety: Scientists, Anticommunism, and the Cold War* (Chapel Hill: University of North Carolina Press, 1999), 55–58; G. Pascal Zachary, *Endless Frontier: Vannevar Bush, Engineer of the American Century* (New York: Free Press, 1997), 306–310; McGrath, *Scientists, Business, and the State*, 128–193; Charles Thorpe, *Oppenheimer: The Tragic Intellect* (Chicago: University of Chicago Press, 2006), 200–242.

NEW ALLIANCES

If the final outcome of the fifteen-year political mobilization by physical scientists dashed Bush's hopes, the entire phenomenon proved an unmitigated disaster for scientific democrats in the human sciences. Leading physical scientists denigrated or ignored the social sciences, while at the same time piggybacking on the cultural authority of literature and the arts. The outcome was a version of the "two-cultures" framework put forward by C. P. Snow in his 1959 book. In the American rendering of that framework, science and the humanities represented alternative but complementary ways of knowing. Whereas Snow identified a contextually sensitive form of social history as a practice that could combine, and thus harmonize, the two cultures, physical scientists in the United States tended to denigrate such hybrid pursuits and to present a division of labor between science and the humanities as permanent and salutary. By science, they meant the natural sciences, and especially the physical sciences. American commentators presented the natural sciences and the humanities as constitutive elements of the Western heritage, while the social sciences either vanished from their accounts of modern culture or appeared as failed attempts to apply the scientific method in an unsuitable domain.

Unlike most interwar scientific democrats, Bush and the other physical scientists who powerfully shaped the national discourse on science in the middle years of the twentieth century drew a sharp line between science and society. They portrayed science as utterly deaf to human concerns and sought to insulate the research process from all political pressures, advocating what one historian has called "laissez-faire communitarianism." This argument, which echoed Robert K. Merton's writings, described the scientific community as a self-contained, self-regulating entity that would inevitably spin off technologies of use to the wider society, if it were given substantial resources and left alone. Advocates of this view joined Merton in portraying scientific knowledge as the product of an autonomous community of researchers, jealously guarding a space of knowledge that stood outside, beyond, and above politics – in other words, a space untouchable by both the state and the horizontal communication between citizens on which the Progressives and post-Progressives had rested their political hopes.[14]

Moreover, even the progressive activists of the first and second scientists' movements adopted certain tenets of laissez-faire communitarianism. Taken for granted across the divide between social responsibility advocates and their neutralist critics, these presumptions operated silently but effectively to erode the cultural commitments and conceptions of science characteristic of interwar scientific democracy. At first glance, the social responsibility discourse of

[14] David A. Hollinger, "Free Enterprise and Free Inquiry: The Emergence of Laissez-Faire Communitarianism in the Ideology of Science in the United States," in *Science, Jews, and Secular Culture: Studies in Mid-Twentieth-Century American Intellectual History* (Princeton: Princeton University Press, 1996): 97–120.

the 1930s would seem to have matched up neatly with the prevailing expressions of scientific democracy. After all, activists in that vein joined hands, politically, with Franz Boas and other culture-minded progressives in the human sciences. Yet most of them followed Ogburn in viewing those fields as, at most, devices for enabling the population to adjust to technological progress. Outside biology, at least, most social responsibility advocates – like the Bernalists – believed that the public merely needed access to the technological fruits of scientific research, not a new culture or new values *per se*. Questions of culture, political participation, and the like took a back seat to economic production and distribution in the potent but highly ambiguous language of social responsibility. Its users rarely shared the emphasis of a Boas or a John Dewey on the psychological dimensions and social embeddedness of human action and flourishing.[15]

Still less did activists in the physical sciences view scientific research through such lenses. Here, too, they agreed with both the Bernalists and the neutralists that scientists could not allow social values into the realm of theory choice. The social responsibility framework held that social values properly – indeed, necessarily – shaped problem choice and the application of results, but left untouched the thoroughly disinterested process of testing theories against evidence that connected these earlier and later stages of scientific inquiry. All of the physical scientists, whatever their political views, assumed that scientists, during the research process itself – in other words, when operating in Reichenbach's context of justification – could and should bracket their emotions and normative commitments. In short, they took for granted something like William F. Ogburn's description of the psychology of scientific work, with its sharp separation of emotionally resonant action from purely cognitive observation.

In addition to these subtle conceptual departures, most of the physical scientists who forayed into the political sphere either explicitly or implicitly denigrated the social sciences as they did so. To be sure, they called attention to questions about science's social relations that fell into the existing domain of the social sciences. At the same time, however, most physical scientists took it for granted that they alone could understand and solve such problems regarding science and society. This assumption devalued existing approaches to social problems and suggested that the truth would continue to elude social scientists in the future as well.

Open criticism of the social sciences flourished among the neutralists, particularly during the NSF debate. That struggle highlighted the political marginality of scientific democrats in the social sciences, who were forced to adopt a stricter rendering of science's neutrality than many might have preferred. Meanwhile, the NSF controversy gave physical scientists a highly visible and consequential public forum in which to distance themselves from

[15] Doug Russell offers a similar critique of the parallel discourse of the Bernalists in "Popularization and the Challenge to Science-Centrism in the 1930s," in *The Literature of Science: Perspectives on Popular Scientific Writing* (Athens: University of Georgia Press, 1993), 37–53.

other forms of work claiming the label of "science." In 1945, President Truman urged the inclusion of the social sciences in the proposed agency. So, too, did the leaders of the SSRC, who quickly overcame their fear that federal patronage would impose political constraints on progressives and came to see being left out as the greater danger. Talcott Parsons joined more orthodox advocates of value-neutrality, such as Ogburn, Mitchell, and the psychologist Robert M. Yerkes, in seeking to shore up the scientific credentials of the social sciences and make them palatable to the establishment physical scientists and conservative Congressional leaders who would set the terms for the agency. But their efforts proved fruitless. The NSF's founding mandate allowed but did not require it to fund research in the social sciences. In its final form, the agency built on an understanding of science as a thoroughly value-neutral enterprise encompassing only the phenomena of the natural world.[16]

The physical scientists were not initially unified in rejecting the inclusion of the social sciences. Both Harold Urey and the Harvard astronomer Harlow Shapley favored their inclusion in 1945. But most of the physical scientists in the NSF debate dismissed the social sciences as either ineffectual or politically subversive. Even when writing on other topics, Conant made a point of lambasting social scientists for their dalliances with philosophy and with social applications. Several other members of Bush's circle, including Karl Compton, the physicist I. I. Rabi, and Johns Hopkins president Isaiah Bowman, testified against including the social sciences in Congressional hearings in late 1945. They argued that investigators in these fields had not attained, and probably could never attain, the rigor and objectivity required of publicly funded research. Behind the scenes, more pragmatic considerations also operated, given that controversies over the political bent of the social sciences threatened to halt the entire NSF campaign in its tracks. Kilgore, Truman, and the Bush group proved unwilling to endanger the whole prospect of federal funding by stoking the fires of Robert A. Taft and other Congressional conservatives, who equated social science with socialism – as did many of the physical scientists themselves. Using the language of laissez-faire communitarianism, leading physical scientists insisted that social scientists had proven unable to establish the requisite distance from cultural and political forces. Their own invocations

[16] Mark Solovey, "Riding Natural Scientists' Coattails Onto the Endless Frontier: The SSRC and the Quest for Scientific Legitimacy," *Journal of the History of the Behavioral Sciences* 40, no. 4 (Fall 2004): 393–422; David Paul Haney, *The Americanization of Social Science: Intellectuals and Public Responsibility in the Postwar United States* (Philadelphia: Temple University Press, 2008), 29–38. Talcott Parsons, paid handsomely by the SSRC to produce a counterstatement to Bush's report, toiled for years on a massively nuanced document that satisfied no one and stood no chance of achieving its goal in the wider public forum, where in any case it never appeared. Although the SSRC got better results from Stuart Chase, who published a breathless overview titled *The Proper Study of Mankind* (New York: Harper, 1948), Congress simply ignored social scientists' input after the 1945 hearings: Solovey, "Riding Natural Scientists' Coattails Onto the Endless Frontier," 414–415. When federal funding for the social sciences did come in 1954, half of it went to economists: Michael A. Bernstein, *A Perilous Progress: Economists and Public Purpose in Twentieth-Century America* (Princeton: Princeton University Press, 2001), 101.

of political autonomy in the late 1940s and early 1950s went hand in hand with their sharp criticism of the social sciences.[17]

In contrast to the centrist celebrators of neutrality, the social responsibility advocates tended to leave their criticism of the social sciences largely unstated. Still, from the very start, most ignored or denigrated existing work in those fields, even as they called attention to certain questions associated with them. For example, in his December 1934 presidential address to the History of Science Society, the Yale neurosurgeon Harvey Cushing called for a moratorium on research in the natural sciences so that Americans could address the effects of technological development. He imagined a future historian writing of the 1930s that, "under a quickly spreading Religion of Humanity, there began a new era – one in which scientists took a commanding position in a rapidly changing world" and created a "new and rational science of society" through "well-planned and executed experiments." But Cushing's moratorium proposal portrayed the existing crop of social scientists as ineffectual, even as it asserted the intellectual authority of science in general. His point was that the social problems associated with modern technology could be addressed only when natural scientists shifted their full energy to the characteristic questions of the social sciences, whose practitioners presumably would get nowhere unless the natural scientists stepped in.[18]

The advocates of social responsibility in the 1930s and early 1940s shared with many human scientists the assumption that scholars could speak authoritatively about the public's genuine needs, even when public opinion diverged markedly from the actual public good. Yet these activists primarily emphasized the physical sciences and their contributions to material prosperity. Moreover, many of them joined Robert S. Lynd in hoping to replace popular participation with decision-making by experts. Far more than their counterparts in the human sciences, politically engaged physical scientists tended to favor managerial solutions to the problem of economic injustice, rather than looking to public culture. Most would give the public little to do but sit back and watch as experts set the terms for implementing scientific advances and technological innovations. When these figures called for the conscious direction of science's application in accordance with social needs, they typically meant its direction by scientists themselves.

The FAS leaders of the late 1940s gave the discourse of social responsibility a strong deliberative turn. Most eschewed the managerial, expertise-centered conceptions of political action espoused by their earlier counterparts. Yet they also jettisoned the idea that political engagement was part of their professional work as scientists. Instead, the atomic scientists drew Ogburn's line between scientific and political roles, arguing that their research was politically relevant

---

[17] Solovey, "Riding Natural Scientists' Coattails Onto the Endless Frontier," 407, 403–404; James B. Conant, "The Scientific Education of the Layman," *Yale Review* 36 (1946), 33, 32, 35.
[18] Harvey Cushing, "The Humanizing of Science," *Science* 82 (February 8, 1935): 137–143 (quote at 143). Cf. Millikan, "Science and Social Justice," 98–99.

only to the extent that it had given them a lengthy head start on their fellow citizens in thinking about a single policy issue: the prospect of a nuclear world. Viewing their political activism as the result of an unusual and highly contingent imbalance in the information available to them and to other citizens, these activists did not claim that scientists, as such, possessed either a unique capacity or a moral obligation to speak to social and political questions. Thus, even as they turned toward the public, they backed away from claims about the inevitably social character of problem choice and technological applications. Meanwhile, their writings continued to marginalize the social sciences as resources for building the peaceful world that FAS activists sought.[19]

As leading physical scientists continued to push away from the social sciences in the late 1940s, they began to actively align themselves with literature and the arts. The concept of creativity anchored a postwar epistemological discourse that identified scientific knowledge as a product of the individual imagination, which to many commentators represented the common ground between scientific research and artistic creation. For centuries, the leading lights of Western science had proven thoroughly allergic to the human imagination. During the early Cold War period, however, Conant, Oppenheimer, Philipp Frank, and many others deemed the imagination the defining characteristic of humanity, the essentially human faculty. Portraying themselves as creators and originators rather than passive recipients of natural truths, these physical scientists identified scientific creativity and artistic creativity as the twin pillars of Western civilization, each simultaneously expressing and advancing the cause of imaginative freedom.[20]

In part, this humanistic turn among physical scientists reflected the fact that theoretical developments since the 1920s seemed to have pushed their fields decisively beyond common-sense understandings, and perhaps even beyond the public's cognitive capacity. Moreover, it was hardly clear, even to specialists, what general relativity or quantum theory indicated for the conduct of everyday affairs. Physical scientists had witnessed with disgust a flurry of wanton analogizing from the concept of relativity between the wars, and they wanted no part of such airy speculation. Rejecting a long-standing hope that they could teach citizens the truth and suggest its practical implications, many concluded that the barriers to public comprehension were simply too high: scientists would always occupy a lonely and isolated – if profoundly beautiful – intellectual world.[21]

But it took the events of the late 1940s to turn this assumption into a full-fledged theory of scientific practice. Emerging from their laboratories and

[19] Wang, "Scientists and the Problem of the Public in Cold War America," 332.
[20] Lorraine Daston, "Fear and Loathing of the Imagination in Science," *Daedalus* 127, no. 1 (Winter 1998): 73–95; Jamie Cohen-Cole, "The Creative American: Cold War Salons, Social Science, and the Cure for Modern Society," *Isis* 100, no. 2 (June 2009): 219–262.
[21] Ronald C. Tobey, *The American Ideology of National Science, 1919–1930* (Pittsburgh: University of Pittsburgh Press, 1971).

offices, wartime science administrators such as Conant and Oppenheimer suddenly found themselves expected to comment authoritatively on all manner of issues, and soon to single-handedly win the brewing Cold War as well. In the unprecedented position of enjoying too much cultural authority rather than too little – unless they questioned national military policy – these figures developed new conceptions of science as a form of communal practice in which researchers combined local knowledge and craft skills confined to a few specialists with an independence and flexibility of mind that all citizens could profitably emulate. Conant, Oppenheimer, and like-minded commentators decisively altered the long-standing conception of the scientific community as a model for the wider democratic polity.

Not all physical scientists gave up the traditional, engineering-centered view of science as an agent of industrial progress and material comfort, produced through diligence and self-sacrifice rather than fertile acts of imagination. Vannevar Bush embodied that image perfectly. Although he had not spent significant time in industry, he had several notable inventions to his credit and had co-founded what became the Raytheon Company. In the Cold War years, rank-and-file engineers, who were much more likely to be committed theists than were scientists in the liberal-arts disciplines, tended to play up the practical applications of their work. They described themselves as promoting the "American way of life" by producing consumer goods and weapons technologies. But newer conceptions of science coexisted with these invocations of practicality. In the midst of the national panic about scientific manpower spawned by the Soviet launch of Sputnik in 1957, even Bush himself worried publicly that the nation was focusing too heavily on technical achievements and forgetting the educational ideal of the "gentleman of culture," the "full man" who could resist popular fads and speak for the true public interest.[22]

Whereas Bush portrayed the scientist as a consummate republican leader, selflessly committed to the good of others, Conant crafted a thoroughly humanistic conception of science in a series of widely read books and articles. As president of Harvard and an integral part of the wartime bomb team, he quickly became an influential commentator on educational policy and the cultural meaning of science. A relentless meritocrat, Conant viewed full mobility throughout the occupational hierarchy as the key to social health. Although he also favored general education in the sciences and occasionally spoke of creating "a unified, coherent culture" suited to an "age of machines and experts," he sought primarily to familiarize American decision-makers and leaders of opinion with science's distinctive capacities – and its limitations. The science-centered curriculum Conant proposed would teach future bureaucrats and

[22] Quoted in Thorpe, *Oppenheimer*, 263. On science's popular image after the war, see John C. Burnham, *How Superstition Won and Science Lost: Popularizing Science and Health in the United States* (New Brunswick: Rutgers University Press, 1987) and Marcel C. LaFollette, *Making Science Our Own: Public Images of Science, 1910–1955* (Chicago: University of Chicago Press, 1990).

business managers how to assess whether a capital-intensive project proposed by an overly breathless researcher merited funding. As part of this project of training knowledgeable patrons and minimizing their expectations of practical results, he described science as an imaginative, essentially artistic pursuit restricted to a handful of specialists.[23]

Conant promoted his educational proposal in a series of semipopular books on the history and practice of science. Like Merton before him, he adapted interwar insights about the profoundly human character of scientific inquiry to the task of defending its autonomy from political ideologues and those who expected immediate practical results. Rather than Merton's theory of underdetermination, however, Conant touted the new understandings – he had in mind operationalism of his Harvard colleague Percy W. Bridgman – of a scientific concept in terms of a "fruitful guide to action" rather than a map to part of the world and science as a dynamic rather than representational practice. And like Charles S. Peirce before him, Conant defined the fruitfulness of scientific concepts in terms of their ability to generate future research, not their capacity to guide everyday action or drive technological progress. He called science "a very human adventure of quarrelsome individuals," not a product of "secular saints." Yet in his portrait, these quarrelsome individuals sought merely to know the world – to manipulate it, yes, but in pursuit of conceptual gains, not practical results. Genuine scientists, according to Conant, worked to keep the research process going, not to attain immediately applicable results. In fact, Conant called science "nothing but a game" and insisted that policymakers should never peg funding levels to short-term, practical outcomes.[24]

---

[23] James B. Conant, *On Understanding Science: An Historical Approach* (New Haven: Yale University Press, 1947), 4, 3; Conant, *Science and Common Sense* (New Haven: Yale University Press, 1951), 2; Steve Fuller, *Thomas Kuhn: A Philosophical History for Our Time* (Chicago: University of Chicago Press, 2000), 9; Joel Isaac, *Working Knowledge: Making the Human Sciences from Parsons to Kuhn* (Cambridge, MA: Harvard University Press, 2012), 203–210. Conant worried especially about the funding process at the Department of Defense, where the modern military's "almost fanatic enthusiasm for research and development" intersected with scientists' inevitable tendency to mystify their pet theories by using technical terms: *Modern Science and Modern Man* (New York: Columbia University Press, 1952), 67–68. In a series of NSF-sponsored textbooks, biologists joined physicists in seeking to create more informed leaders and describing science as a creative, artistic process. However, they also stressed its practical relevance: John Rudolph, *Scientists in the Classroom: The Cold War Reconstruction of American Science Education* (New York: Palgrave, 2002), esp. 162, 136.

[24] Conant, *Modern Science and Modern Man*, 57, 54; Conant, *The Citadel of Learning* (New Haven: Yale University Press, 1956), 8–10. Over time, Conant modified and honed his model, synthesizing a range of contributions to the philosophy of science. He combined Peirce's sense that scientific theories "worked" when they pushed forward the research frontier and Merton's emphasis on the ability of the scientific community to draw depersonalized knowledge out of a motley collection of individuals with various resources from Harvard scholars: Bridgman's operationalism, the physiologist Lawrence J. Henderson's rendering of science as a craft tradition rooted in the elaboration of "conceptual schemes," and later the philosopher W. V. O. Quine's conception of science as "a search for warranted belief" along the "edges of the man-made fabric of ideas." Conant insisted that the small cadres of scientific experts succeeded because they were steeped in communal traditions of interpretation akin to what his young protégé, the historian

Still, Conant argued that society needed to fund scientific research, and to do so handsomely. Rather than trumpeting the indirect, long-term, technological payoff of science, Conant emphasized two other contributions. One was its diffusive influence on the dense fabric of common-sense understandings that, in Conant's view, guided all human action. Initially serving as "guides to the action of scientists," he wrote, scientific concepts eventually shaped "common-sense ideas about the material universe." Yet Conant leaned much more heavily on science's aesthetic value as a product of the creative imagination. In his view, modern science represented a body of work no less sublime than the Western musical canon, and its foundational theories paralleled "the Parthenon and the cathedrals of the Middle Ages." Its creators needed "imaginative vision" above all, and they were driven by "the pure joy of its creativeness." Meanwhile, science's patrons and audiences needed to view it as a "triumph of the creative spirit, one manifestation of those vast potentialities of men and women that make us all proud to be members of the human race." In Conant's widely read books, science's technological impact vanished and its contributions to common sense paled next to its sheer beauty. He insisted that science should be judged by the non-material standards appropriate to all creative endeavors, as a contribution to the spiritual side of modern civilization.[25]

But Conant saw no symphonies, monuments, or cathedrals in the social sciences, which were vitiated by a relentless search for practical results. From the start of his presidency in 1933, Conant was notoriously skeptical of those fields. Privately, he believed that social science was the modern equivalent of theology, lacking empirical grounding but forced on undergraduates anyway as a kind of surrogate religion. As a university president, however, Conant needed to go easy on social scientists in public. He often simply identified as contentious and important the question of whether scientific methods could be extended to human affairs, while gently suggesting that the combatants would quickly reach agreement if only they understood what the term science truly meant. At times, however, he directly targeted social scientists. Sharply distinguishing between science, philosophy, and invention, Conant called most investigators in the field mere "social *philosophers*" and deemed their alleged practical contributions the equivalent of new "manufacturing processes and methods of transportation," created by canny inventors rather than disinterested scientists. In his view, almost all contributions to the so-called social sciences were merely trial-and-error tinkering or mindless data collection, rather than the high-flying acts of conceptual insight that constituted true science. Casting about in 1947 for instances of truly scientific reasoning in those disciplines, he found a few in certain areas of psychology – in "vision and audition," though not learning or personality differences – but hardly anything of note in other fields. Conant

of science Thomas S. Kuhn, would later call "paradigms." Conant, *On Understanding Science*, 7, 24, 138; Conant, *The Citadel of Learning*, 11–12; Conant, *Modern Science and Modern Man*, 66. On these epistemological exchanges at Harvard, see Isaac, *Working Knowledge*.
[25] Conant, *Modern Science and Modern Man*, 97, 111, 58–59; Fuller, *Thomas Kuhn*, 215–216.

claimed to recognize the long-term value of social science, but his portrait of it was hardly complimentary.[26]

In fact, Conant denigrated the social sciences even when he fashioned himself their defender. In his most extensive discussion, in the 1952 book *Modern Science and Modern Man*, Conant urged readers not to be discouraged by the mass of errors to which those fields currently amounted, because medicine and other genuine sciences had gone through similar periods of infancy a century or two earlier. Like biologists in the early days, he explained, social scientists were haltingly turning "common-sense fuzzy ideas about consciousness, love, or the zest for power" into fruitful guides to research, and ordinary citizens could hardly be expected to assimilate these early results. In fact, the social sciences were full of charlatans, he wrote. But leaders should support the truly "uncommitted" practitioners anyway, without expecting practical results, because there was no way to distinguish charlatans from pioneers except with the advantage of long hindsight. Society would generate the technical improvements on its own, as it had always done in industry. Here as elsewhere, the most important thing was to recognize that "planned attack and exhortation" – the prevailing approaches in the social sciences, according to Conant – could not possibly generate either new insights or immediately applicable results.[27]

While Conant theorized the physical scientists' humanistic turn, the more urbane J. Robert Oppenheimer literally embodied it. Cultured, well connected, and wealthy – his house featured a van Gogh painting he had inherited – Oppenheimer merged the sciences and the humanities in his very person and added a dash of Eastern religion to the mix. He agreed with Conant that science was an artisanal practice embodied in a guild-like structure, that its practitioners should take their inspiration solely from a desire to know the world, and that its "sober, modest attempt to penetrate the unknown" was entirely distinct from technology, the "frantic exploitation of the known." But Oppenheimer found Conant's proposal for a science-based general education far too populist. He held that only a few practicing researchers could ever have a deep understanding of scientific inquiry, and that a vast cultural gulf would always separate them from the laity. Science could not, as Conant hoped, reshape common sense over time.[28]

According to Oppenheimer, science offered only two contributions to the wider polity. First, practical results would emerge, albeit slowly and indirectly, from the research process, because social needs inevitably shaped the minds of individual investigators via "complex mechanisms of education, taste, and

[26] James G. Hershberg, *James B. Conant: Harvard to Hiroshima and the Making of the Nuclear Age* (Stanford: Stanford University Press, 1993), 94; Conant, *On Understanding Science*, 5, 27–28, 23, 26; Isaac, *Working Knowledge*, 58.

[27] Conant, *Modern Science and Modern Man*, 77–79. Conant now saw a bit more progress "in pedagogy, in handling some types of abnormal psychology, in a few restricted areas of economics, perhaps in certain kinds of human relations" (79).

[28] Thorpe, *Oppenheimer* 11, 255, 190, 174 (quoted); J. Robert Oppenheimer, *Atom and Void: Essays on Science and Community* (Princeton: Princeton University Press, 1989), 72.

value." Second, the scientific community, with its array of smaller disciplinary and sub-disciplinary communities, could serve as a model for human interaction in a society that Oppenheimer described as a loosely linked congeries of specialized professions. In the modern age, he wrote mordantly, massive bureaucratic structures endangered "true human community" – "the integrity of the intimate, the detailed, the true art, the integrity of craftsmanship and the preservation of the familiar, of the humorous and the beautiful." Yet, Oppenheimer averred, scientists illustrated how free, creative individuals could sustain communal relations by combining an unshakeable commitment to their tiny guilds with a cosmopolitan embrace of the vast diversity of human practices and thoughtways.[29]

Throughout his postwar writings, Oppenheimer analogized from the problem of knowledge in an age of specialization to the problem of social order in an age of centrifugal social forces. In each case, he argued, the key was to recognize that the "ideal of brotherhood" was utopian in its usual, absolute formulation. Yet it could be approximated in small human communities, which were "not ideal, not universal, imperfect, impermanent, as different from the ideal and as reminiscent of it as are the ramified branches of science from the ideal of a unitary, all-encompassing science of the eighteenth century." Science could teach moderns to temper their expectations of social unity and adopt the humility needed to coexist in a world of immense diversity and even fragmentation – in an "open society" akin to the realm of modern science.[30]

Such portraits of the scientific community as a guild of artists or craftsmen left the status of the social sciences unclear. Oppenheimer's institutional position, unlike Conant's, did not require him to comment on those fields. And he did not, except to briefly mention the anthropological lesson of cultural relativism – a favorite theme of postwar physical scientists who felt compelled to speak about the social sciences. Moreover, even those physical scientists who worked closely with social scientists often upheld the two-cultures image of an intellectual world divided between (natural) scientists and humanists. The émigré physicist and logical empiricist Philipp Frank, who had landed at Harvard in 1939, sought to revive the Unity of Science movement after 1946. He developed the outlines of his plan while participating in numerous interdisciplinary discussion groups with the likes of Talcott Parsons, the anthropologist Clyde Kluckhohn, and numerous psychologists. But Frank's books include virtually no discussion of the social sciences, even where it would seem eminently warranted.[31]

For instance, one might expect Frank's *Philosophy of Science* (1957) to have addressed the social sciences, given its lament of "the deep rift between our

---

[29] Thorpe, *Oppenheimer*, 171 (quoted), 257, 260 (quoted).
[30] Oppenheimer, *Atom and Void*, 70, 73.
[31] *Ibid.*, 65–66; Peter Galison, "The Americanization of Unity," *Daedalus* 127 (1998): 45–71; Gerald Holton, "Philipp Frank at Harvard University: His Work and Influence," *Synthese* 153, no. 2 (November 2006): 297–311.

rapid advance in science and our failure in the understanding of human problems." But Frank quickly restated this problem as "the rift between science
and the humanities." To fill that gap, Frank looked, not to the social sciences,
but rather to the philosophy of science, "a coherent system of concepts and
laws" encompassing "the natural sciences, ... philosophy and the humanities."
Only after 360 pages on modern physics did Frank finally mention "the sciences of human behavior," in the book's last sentence. He likewise ignored
those fields in all twenty-five of his contributions to the annual Conference on
Science, Philosophy and Religion in Their Relation to the Democratic Way of
Life (CSPR), at which Frank spoke as a representative of the scientific community throughout the 1940s. Behind the scenes, Frank believed that social scientists could do important work in helping natural scientists root out socially
conditioned biases in their theories. In print, however, he joined Conant and
Oppenheimer in simply erasing the social sciences from his portrait of modern
thought.[32]

Identifying the imagination as the shared root of science and the humanities
offered these physical scientists a number of important benefits, although one
must not simply read back from effects to motives. First, the analogy between
science and art isolated scientific research from practical applications – a major
concern for national leaders, such as Conant and Oppenheimer, who worried
about the potential for scientific research to be skewed by the immediate needs
of military agencies. Second, the discourse of scientific creativity enabled physical scientists to distance themselves from the social sciences, which many
politicians and other commentators saw as breeding grounds for materialism
and socialism. The writings of physical scientists highlighted the emphasis on
individual freedom that anchored postwar political culture, while detaching
this thread from its usual entanglement with business acumen and practical
success. Portraying scientists as *bona fide* artists, these scientists updated the
individualistic images of heroic pioneers and brave explorers that had underpinned Bush's rhetoric in *Science – The Endless Frontier*. The post-World War
II discourse of scientific creativity focused on a small handful of generative
individuals, not a nationwide network of social selves.

SCIENCE AND VALUES AGAIN

These writings also spoke to, and reflected, the spiritual anxieties that plagued
so many Americans after World War II. Conant openly addressed religious
questions in *Modern Science and Modern Man*, targeting the assumption of
"the usual naturalistic moralist" that "value judgments based on science" could
replace "those now accepted as part of our Judaic-Christian tradition." Conant
charged that such a view not only amplified society's expectations of practical

[32] Philipp Frank, *Philosophy of Science: The Link Between Science and Philosophy* (Englewood
Cliffs: Prentice-Hall, 1957), xii–xv; Frank, *Relativity: A Richer Truth* (Boston: Beacon, 1950);
Reisch, *How the Cold War Transformed Philosophy of Science*, esp. 299.

results but also obscured the "altruistic and idealistic" impulses manifest in the nonscientific "facts of human history." He further declared that such a misguided emphasis on the power of human reason led directly to the Marxists' dialectical materialism. Drawing heavily on the Book of Job's portrait of an essentially incomprehensible universe, Conant aligned himself with the "heretical Christians" of his day, adopting the critique of "scientism" offered by neo-orthodox theologians such as Reinhold Niebuhr. Even when science had been "fully assimilated into our cultural stream," he insisted, "all our common-sense ideas about the universe and human behavior, all our ethical principles, and our moral convictions" would remain firmly in place. Conant blasted those who defined ethics in purely social terms, leaving "no place for any theistic interpretation." Oppenheimer, too, held that scientific exploration fostered personal insight into the spiritual meaning of human existence, as well as sheer aesthetic delight.[33]

The idea that science, as such, could not speak to values – that a purely naturalistic ethics embodied a category error, confusing descriptive truths with normative obligations – reinforced broader humanistic and religious critiques of science as a depersonalizing force that needed to be powerfully disciplined by a framework of values derived from other intellectual sources. The interwar search for a naturalistic ethics ran afoul of both the physical scientists' campaign for autonomy and the critique of putatively value-neutral social science levied by rafts of humanistic and Christian critics in the postwar years. All of these groups denied that science could provide normative guidance on political matters. In fact, a figure like Dewey, who insisted that science validated democratic values, challenged the cultural projects of humanists and religious leaders in a way that the likes of Ogburn, who declared science value-neutral and left ethical concerns to artists, literary scholars, and church leaders, did not. Most humanistic critics of scientific democracy, and virtually all Christian critics, insisted that science could not ground – indeed, implicitly threatened – democracy because it failed to legitimate a solid foundation of universal, normative commitments. In short, these critics used the view of science articulated by Ogburn against Dewey's cultural and political program. Dewey and his allies became the target of multiple, overlapping arguments against scientism in the 1940s and 1950s.[34]

The critics, however, were not necessarily fussy about epistemological questions. Whatever epistemological package surrounded it, the claim that scientific

[33] Conant, *Modern Science and Modern Man*, 97–99, 90–92; Conant, *On Understanding Science*, 2. Oppenheimer found particularly distasteful the concept of "universal knowledge," which he identified with a "monistic view of the world in which a few great central truths determine in all its wonderful and amazing proliferation everything else that is true." *Atom and Void*, 69.
[34] Reinhold Niebuhr, *The Irony of American History* (New York: Charles Scribner's Sons, 1952), 80; Haney, *The Americanization of Social Science*, 172–202. Even a report sponsored by the National Education Association held that democratic government had "a moral and spiritual basis" and found its highest purpose in serving "God's man." Quoted in "Two Statements on Religion in American Society," *America* 44, no. 10 (December 3, 1955): 275.

inquiry could speak to human affairs rankled many physical scientists, humanists, and religious thinkers. On the religious side, a postwar surge of public
piety went hand in hand with deep suspicion of scientific approaches to human
affairs. Its leaders generally accepted the natural sciences but drew a line at
the human sciences, with their secular conceptions of the person. Religious
spokesmen and activists from across the theological and political spectrums
could argue that the social sciences improperly ignored God's role in shaping
behavior. Meanwhile, many scholarly humanists held that the arts and literature, rather than scientific analysis, revealed the deep springs of human action
and indicated the means and direction of its improvement. Physical scientists,
for their part, could adopt either of these conceptions of the source of human
values. To the degree that any of these critics deemed science valid, they meant
primarily the physical sciences, perhaps with a depoliticized biology as a junior
partner. Thus, the conceptions of science's disciplinary scope and cultural role
put forward by its most influential postwar advocates meshed well with the
arguments of its sharpest critics.[35]

This epistemological overlap between the physical scientists and the religious and humanistic critics of scientism was clearly on display during the war
years at the meetings of the Conference on Science, Philosophy, and Religion.
Conceived in late 1939 by a group of religious leaders in conjunction with
the physicist Arthur Holly Compton, the astronomer Harlow Shapley, and the
political scientist Harold D. Lasswell, the CSPR was designed to serve as an
"organized conscience for American civilization," creating a "pluralistic but
well regulated universe of thought" in which proponents of various intellectual
traditions could hash out differences and discover points of commonality. The
goal was explicitly political: CSPR members, including the five Nobel laureates
in the physical sciences that eventually joined the founding group, traced the
conflicts of the modern era to the fact that failures of communication and fault
lines of disagreement had prevented intellectuals from giving shape to the more
egalitarian culture latent in extant social conditions.[36]

From the start, however, those scientific democrats who identified scientific
inquiry as a source of values to guide social behavior were marginal to the
CSPR. This was true despite their representation at the highest organizational
level. The group's board of directors featured several prominent scientific
democrats, including Lasswell, Lawrence K. Frank, Alain Locke, and Robert
M. MacIver. Lyman Bryson participated in the Conference Circle, which was
largely responsible for the direction of the conference, and edited several of the
conference volumes. But most of the group's nonscientific members, and even
some of the natural scientists, attributed the troubles of the modern era directly
to the views espoused by these figures. No one ever topped Mortimer Adler's

[35] E.g., Fulton J. Sheen, *Communism and the Conscience of the West* (Indianapolis: Bobbs-Merrill,
    1948), 52; Joseph Wood Krutch, *The Measure of Man: On Freedom, Human Values, Survival,
    and the Modern Temper* (New York: Grosset and Dunlap, 1953).
[36] Fred W. Beuttler, "Organizing an American Conscience: The Conference on Science, Philosophy
    and Religion, 1940–1968" (PhD dissertation, University of Chicago, 1995), xiv, 6–7.

blast at the first meeting, where he famously argued that "the positivism of the professors" represented a greater danger to democracy than "the nihilism of Hitler." But attacks on naturalism, coupled with explicitly religious articulations of democracy, punctuated the yearly programs. The religious leaders involved in the early planning – Louis Finkelstein, F. Ernest Johnson, George N. Shuster, and John LaFarge – all insisted, with varying degrees of stridency, that democracy took its shape from religious principles. In fact, the CSPR's statement of purpose traced American democracy to "the religious principle of the Fatherhood of God and the worth and dignity of Man when regarded as a child of God."[37]

Some of the natural scientists attending the early CSPR convocations raised their voices in protest. By the end of the first day of the first meeting, the Clark University biologist Mark Graubard and others already feared that the conference had become an attack by metaphysicians on empirical science. Only one of the theologians that spoke had questioned the claim that democracy took its shape from Judeo-Christian principles. The next day, Finkelstein apologized profusely to the public on behalf of the CSPR after Albert Einstein opined that belief in a personal God prevented cooperation between scientists and religious thinkers. At the second meeting, meanwhile, the defenses of traditional theism grew so thick that a Muslim participant stood up and asked whether he could be a democrat at all without converting to Christianity.[38]

But many of the physical scientists in the group ignored or condoned such attacks on naturalism and granted church leaders pride of place in the articulation of ethical values. On the scientific side, the Nobel Prize-winning chemist Harold Urey, a key figure in both the first and second scientists' movements, answered a challenge by a Hindu scholar by insisting that Christianity was the "mother of democracy." Arthur Compton, a CSPR founder and a prominent voice for science in the wider public arena, likewise argued that democracy required a combination of scientific know-how with Christian ethics. Most of the CSPR participants, including these prominent physical scientists, shared with their critics a separate spheres view of science and religion in which the sciences merely implemented ethical values drawn from theistic sources. As elsewhere, they tended to rhetorically erase the social sciences and to portray social change as a joint project of scientists, humanists, and religious thinkers.[39]

[37] *Ibid.*, 101, 121; James Gilbert, *Redeeming Culture: American Religion in an Age of Science* (Chicago: University of Chicago Press, 1997), 80, 83, 91–92; Mortimer Adler, "God and the Professors," in *Science, Philosophy, and Religion: A Symposium* (New York: Conference on Science, Philosophy and Religion in Their Relation to the Democratic Way of Life, 1941), 128; quoted in *Science, Philosophy, and Religion*, 8.

[38] Beuttler, "Organizing an American Conscience," 173, 150, 231. Tensions between naturalists and theists already ran high by the time of the group's launch, as a result of a bitter controversy over the famed atheist Bertrand Russell's proposed appointment to the City College of New York faculty: Thom Weidlich, *Appointment Denied: The Inquisition of Bertrand Russell* (Amherst: Prometheus Books, 2000).

[39] Quoted in Beuttler, "Organizing an American Conscience," 199; Arthur H. Compton, "For Good or Ill?" *American Magazine* 129 (April 1940): 172. Cf. Compton, "What Science Requires of the

The politics of anti-communism also worked powerfully against interwar versions of scientific democracy, intertwining with the increasingly vocal and influential critiques of scientism. As far back as the late 1930s, anti-communism had begun to pit scientific democrats against one another. During the Spanish Civil War, Franz Boas launched groups such as the American Committee for Democracy and Intellectual Freedom (ACDIF) to mobilize scientific thinkers against the fascist threat. However, John Dewey and Sidney Hook's Committee for Cultural Freedom (CCF), organized in May 1939, attacked the ACDIF for refusing to identify the Soviet Union as a totalitarian state and for allowing communists on its roster. After the Nazi-Soviet pact of August 1939 effectively cemented the equation of Soviet communism with totalitarianism in the public mind, the ACDIF rapidly lost influence. Anti-communists from both the left and the right also targeted Goodwin Watson's Society for the Psychological Study of Social Issues, which sought to foster social change by distributing psychological expertise among the masses.[40]

To be sure, those scientific democrats who took up the cause of anti-communism did so on different grounds than their conservative counterparts. Four different features of the Soviet system – the Stalinist regime's official atheism, its abridgement of economic liberties taken for granted in the West, its deterministic theory of history, and its attempts to control the expression of dissenting opinions – shaped the competing versions of American anti-communism that jostled for supremacy by the 1950s. Conservatives worried most immediately about Soviet atheism and collectivism, in varying mixtures. By contrast, liberal anti-communists, though concerned about overly strict controls over economic behavior, focused more centrally on determinism and especially civil liberties – specifically, freedom of public expression, understood as a proxy for a broader freedom of thought. Scientific democrats in this vein argued that, whatever the virtues of regulating economic exchange, the Soviet regime's refusal to provide open space for political dissent rendered it profoundly dangerous. They

New World," *Science* 99 (January 14, 1944): 23–28; Compton, "Science and the Supernatural," *Scientific Monthly* 63 (December 1946): 441–446; Robert A. Millikan, "Knowledge is Power – Not Wisdom," *Rotarian* (January 1945), 7; and Igor Sikorsky, quoted in Robert M. Bartlett, *They Work for Tomorrow* (New York: International Committee of Young Men's Christian Associations, 1943), 18–19. These writings reveal the ability of a value-neutral conception of science to unite Bush's linear model of technological growth with free-market policies and a belief that the churches, not scientists, should provide moral and social guidance in the modern world. Both Compton and Millikan, in fact, deemed the global spread of Christianity inevitable, given that new military technologies made utter destruction the only alternative. By contrast, biologists attending CSPR meetings often argued that evolutionary theory proved the validity of Western values: e.g., Edwin Grant Conklin, "Science and Ethics," in Ruth Nanda Anshen, ed., *Science and Man* (New York: Harcourt, Brace, 1942), 436–452; Walter B. Cannon, "The Body Physiologic and the Body Politic," in *ibid.*, 287–308; and Edmund W. Sinnott, "The Biological Basis of Democracy," *Yale Review* 35 (1945): 61–73.
40 Kuznick, *Beyond the Laboratory*, 208–226; Ian A. M. Nicholson, "The Politics of Scientific Social Reform, 1936–1960: Goodwin Watson and the Society for the Psychological Study of Social Issues," *Journal of the History of the Behavioral Sciences* 33 (1997), 39, 41, 43–47.

believed that the totalitarian Soviet state went far beyond the blunt force of traditional despotism, rendering its subjects docile by controlling their minds and not just their behavior.[41]

Liberal anti-communists such as Dewey, Hook, T. V. Smith, Lyman Bryson, and Arthur O. Lovejoy linked their political critique closely to their conception of science, equating democratic processes with scientific inquiries. "If every man can say what he pleases, we have a fair chance of getting at the truth," Bryson explained. By contrast, the political "truth" could not emerge where the state shackled the expression of opinions. This epistemological reading of civil liberties implied a direct and fundamental conflict between scientific democracy and totalitarianism, and it seemed to justify strict domestic security measures. Because the Stalinists retained their hold by controlling the minds of adherents, Communist Party members in the United States could not be trusted to uphold democracy. Since no one would freely choose Soviet communism, the argument ran, party members and fellow travelers must be in thrall to a foreign power.[42]

Yet in the end, the anti-communist initiatives of these scientific democrats came back to haunt them. Postwar events enabled Joseph McCarthy, J. Edgar Hoover, and their allies to paint progressive intellectuals themselves with the red brush. In the wider public culture, atheism and challenges to capitalism remained the most salient features of the communist menace. McCarthy and Hoover portrayed academia as a central pillar of the American fifth column and the nation's professors, many of whom advocated controls on free enterprise and a secular understanding of the human person, as the entering wedge for communism. After all, the Soviets, too, opposed both unfettered capitalism and traditional theism. A few prominent scholars had run in the same circles as Communist Party operators during the Popular Front era of the late 1930s, giving the claim all the plausibility it needed to convince a jittery public.[43]

One of those erstwhile fellow travelers was Oppenheimer, the former head of the Manhattan Project and the highest-profile scientific casualty of McCarthy's campaign. In 1954, Oppenheimer was stripped of his security clearance after his consistent opposition to the development of the hydrogen "super-bomb" led Edward Teller and other critics to charge that he remained a threat to the

[41] E.g., T. V. Smith, "The Democratic Process," *Public Opinion Quarterly* 2 (1938), 17; Lyman Bryson, *Which Way America? Communism – Fascism – Democracy* (New York: Macmillan, 1939), 81.

[42] Bryson, *Which Way America?*, 7; Leo P. Ribuffo, *The Old Christian Right: The Protestant Far Right from the Great Depression to the Cold War* (Philadelphia: Temple University Press, 1983); Andrew Feffer, "The Presence of Democracy: Deweyan Exceptionalism and Communist Teachers in the 1930s," *Journal of the History of Ideas* 66, no. 1 (January 2005): 79–97; Robert B. Westbrook, *John Dewey and American Democracy* (Ithaca: Cornell University Press, 1991), 491–494.

[43] On the conservative tendencies of postwar politics, see Elizabeth A. Fones-Wolf, *Selling Free Enterprise: The Business Assault on Labor and Liberalism, 1945–60* (Urbana: University of Illinois Press, 1994); and Jonathan Bell, *The Liberal State On Trial: The Cold War and American Politics in the Truman Years* (New York: Columbia University Press, 2004).

nation. Yet while physical scientists bore much of the brunt of McCarthyism, a remarkably wide swath of thought drew the attention of right-wing critics. J. B. Matthews, a Methodist priest and prominent red-hunter, lumped John Dewey with two of his intellectual enemies, Rudolf Carnap and Robert M. Hutchins. Other scientific democrats on Matthews' list of subversives included Kirtley F. Mather, Harlow Shapley, Horace M. Kallen, and Edward C. Tolman. Meanwhile, Hoover's FBI kept tabs not only on the antiracist anthropologists Gene Weltfish and Margaret Mead, but also on sociologists from all sides of the 1930s epistemological dispute, including Ogburn, Talcott Parsons, Robert S. Lynd, and Pitirim A. Sorokin. The epistemological and metaphysical divisions between these intellectual combatants meant little to Matthews, Hoover, and other right-wing anti-communists. Virtually any deviation from the political status quo could bring out the hounds of McCarthyism, no matter how it was framed theoretically.[44]

Postwar anti-communism meshed easily with the massive, heterogeneous cultural campaign against intellectuals committed to naturalistic conceptions of selfhood. Here, social scientists and philosophers, rather than natural scientists, became the main targets. Pressing the familiar claims against scientism, figures such as Fulton J. Sheen argued that these scholars treated living, breathing, freely willing persons as mere animals or machines, subject to the iron-clad laws of the subhuman world. To these critics, science, when applied to human affairs, entailed a reductive mode of analysis that eliminated human freedom and ideals as explanatory factors.[45]

Free-market thinkers increasingly joined religious traditionalists in targeting the social sciences on religious as well as economic grounds. By the mid-1950s, the circle of conservatives around William F. Buckley Jr.'s *National Review* had begun to weld free-enterprisers and religious thinkers into a loose coalition of those who saw the American way – God-fearing and free – under assault from the Soviets and the intellectuals alike. The Supreme Court's *Everson* (1947) and *McCollum* (1948) decisions, affirming Thomas Jefferson's metaphor of a "wall of separation" between church and state, heightened already intense fears of the secularization of American culture at the hands of its professors.[46]

---

44 Thorpe, *Oppenheimer*, 200–242; Wang, *American Science in an Age of Anxiety*, 80; David Kaiser, "The Atomic Secret in Red Hands? American Suspicions of Theoretical Physicists During the Early Cold War," *Representations*, no. 90 (Spring 2005): 28–60; Lawrence Badash, "Science and McCarthyism," *Minerva* 38, no. 1 (March 2000), 63–64; Beyler, Kojevnikov, and Wang, "Purges in Comparative Perspective," 40–41; Reisch, *How the Cold War Transformed Philosophy of Science*, 188–189; Mike Forrest Keen, *Stalking Sociologists: J. Edgar Hoover's FBI Surveillance of American Sociology* (New Brunswick: Transaction Publishers, 2004); David H. Price, *Threatening Anthropology: McCarthyism and the FBI's Surveillance of Activist Anthropologists* (Durham: Duke University Press, 2004).

45 Philip Gleason, *Contending with Modernity: Catholic Higher Education in the Twentieth Century* (New York: Oxford University Press, 1995), 264–268; John T. McGreevy, *Catholicism and American Freedom: A History* (New York: W. W. Norton, 2003), 192–193; Haney, *The Americanization of Social Science*, 172–202.

46 Kevin Mattson, *Rebels All! A Short History of the Conservative Mind in Postwar America* (New Brunswick: Rutgers University Press, 2008); Patrick Allitt, *The Conservatives: Ideas*

On the other hand, some of the most vigorous critics of scientism leaned left. Sheen, for example, combined neo-Thomism with calls for social justice, after the fashion of many twentieth-century Catholics. More mainstream liberals likewise participated energetically in the postwar resurgence of faith. Well before President Eisenhower's famous invocations of God, Harry Truman traced American democracy to the "unchanging truths of the Christian religion." Yet even as religiosity flourished, critics such as Reinhold Niebuhr declared that the United States labored under a scientific, and thus technocratic, culture that ignored both the essential freedom and the irrational tendencies of the human person. Equating Dewey with the radical behaviorist B. F. Skinner, Niebuhr argued in his bestselling *The Irony of American History* (1952) that social science meant social determinism – the faulty attribution of fixed, causal relationships to the free actions of individuals. Many other Christian and Jewish liberals likewise discerned a strong secularizing tendency in postwar America and pinned the blame on Dewey and other scientific democrats.[47]

Humanistic scholars often employed essentially the same critique of scientism as their theistic counterparts, embracing the natural sciences but drawing a hard line at social scientists' conceptions of the human person. They held, however, that the needed ethical matrix could be found in the West's best-known textual and artistic objects, as spoken for by humanists. Although science could never move from an "is" to an "ought," they argued, the poet and the novelist stepped in where the sociologist dared not go. "[W]hat a Shakespeare has to say about human nature and human conduct," wrote the literary critic Joseph Wood Krutch, "is likely to be as true as, and rather more important than, what the summarizer of ten thousand questionnaires can tell us." The Harvard literature professor Howard Mumford Jones likewise insisted that the humanities complemented the nation's "passion for expertise," preserving democracy by rendering each citizen "a decider whose life has been enhanced by an experience of what the best and happiest minds can tell him."[48]

and Personalities Throughout American History (New Haven: Yale University Press, 2009), 158–190.

[47] Quoted in "Truman's Simple Truths," *America* 83, no. 18 (August 5, 1950): 460; "The Root of Our Moral Crisis," *Christian Century* (March 8, 1952): 294; William Inboden, *Religion and American Foreign Policy, 1945–1960: The Soul of Containment* (New York: Cambridge University Press, 2008); Niebuhr, *The Irony of American History*, 8, 147, 84–85, 81, 106. Niebuhr drew on both Gardner Murphy's critique of empiricism and Karl Mannheim's insights into the ideological character of social thought in his attacks (9, 164).

[48] Quoted in Alexander Miller, review of Krutch, *The Measure of Man*, *Christian Scholar* 38, no. 1 (March 1955), 70; quoted in Joan Shelley Rubin, "The Scholar and the World: Academic Humanists and General Readers in Postwar America," in *The Humanities and the Dynamics of Inclusion Since World War II*, ed. David A. Hollinger (Baltimore: Johns Hopkins University Press, 2006), 96–97 (italics removed). The social scientists' claim to speak for a liberal consensus actually comported tolerably well with the humanists' claim to speak for a Western tradition. But few humanists saw a practical need to look to the explorations of the social scientists, believing that they already had the West's canonical texts before them. Many went further, portraying social scientists as militant determinists and identifying their own work as the needed ethical complement to the scientists' technical achievements.

But much had changed since the days of Babbitt and More – not to mention the Nashville Agrarians and other critics of an industrial order. In the late 1930s, a new generation of urbane, forward-looking humanists had set aside the war on industrial modernity and taken up the campaign against totalitarianism. They saw their task as giving expression to Western values and defending individual autonomy within the context of modern, bureaucratic societies. Their political views ranged from the relatively sunny, forward-looking stance of the French émigré Jacques Barzun to the dark proto-conservatism of Barzun's Columbia colleague Lionel Trilling. But these postwar humanists shared a new sense of responsibility to existing American institutions and a new interpretation of the cultural needs of the citizenry, whom they took to have inherited from Europe the solemn task of upholding the core values of the West. They framed the ideal of individual cultivation in democratic rather than aristocratic terms and portrayed the Western tradition as a necessary complement to science rather than an antidote to it. This new stance, coupled with geopolitical developments, caused the stock of academic humanists to rise dramatically after World War II. Having guided the world's most powerful societies for more than two millennia, the Western tradition could surely redeem the industrial age.[49]

For these postwar humanists, the idea of a coherent "Western civilization," comprising a relatively static collection of broad values such as individual freedom and dignity, played much the same role that the liberal consensus did for social scientists: it enabled them to believe that the American public ultimately shared their views. Thus, it facilitated their adoption of a less militant, less totalizing mode of cultural criticism. With the Western civilization framework in hand, humanists could position themselves as mouthpieces for universally shared values rather than lone voices against a powerful, oppressive culture.

Of course, this impulse found its highest expression in the famed "Western Civ" courses that became a staple of postwar general education programs. Such courses tended to present the sciences and the humanities as the two halves of the Western tradition, the humanities conserving and expressing that tradition's core values and the sciences demonstrating how to implement them under prevailing conditions. If nothing else, the relative age of the natural sciences and humanities, as compared with the upstart social sciences, gave them pride of place in the narrative. Dating back to the Greeks, rather than to Comte or Spencer, the natural sciences and humanities represented the twin poles of Western thought in postwar courses, whereas the social sciences appeared only at the end of the story, if at all.[50]

[49] Rubin, "The Scholar and the World."
[50] General overviews include Gilbert Allardyce, "The Rise and Fall of the Western Civilization Course," *American Historical Review* 87, no. 3 (June 1982): 695–725 and Daniel A. Segal, "'Western Civ' and the Staging of History in American Higher Education," *American Historical Review* 105, no. 3 (June 2000): 770–805.

More importantly, the dominant version of the Western Civ framework did not grant the social sciences a normative role in guiding human behavior. Instead, it described human action in the West – including the practice of modern science itself – as one long attempt to realize a set of core values inherited from ancient sources. Although advocates recognized that these ethical and political commitments had so far appeared only in the West, and could thus have concluded that they were historically contingent, they instead presented such values as extraordinarily stable and often deemed them universally true. In this usage, Western values could serve as the functional equivalent of the previous century's transcendental truths: scholars could take Western values for granted while developing technical tools for implementing them. As in consensus liberalism, science appeared in the Western Civ framework as the faithful servant of a preexisting, non-controversial ethical tradition. In this manner, Western Civ powerfully reinforced the emerging image of science as a purely technical, value-neutral way of knowing.

But those who rejected such a view of scientific knowledge could still get on board with the concept of a coherent, age-old Western tradition. Western Civ, like the parallel framework of consensus liberalism, papered over the epistemological divide between advocates and critics of value-neutrality. More broadly, some interpreters deemed Western civilization deeply scientific, while others rooted it in literature and still others thought it was fundamentally religious. But each camp agreed that there was an ethically charged Western tradition that needed to be upheld against communism. Thus, the Western Civ framework steered many in the human sciences toward a new defense of modern liberalism, one that rooted values such as equality and tolerance in the Western tradition rather than finding their source in contemporary social science. Tracing liberal values to the distant past was also far safer, politically, than attempting to draw values out of contemporary social experience or wading into contentious debates about the character of the American liberal tradition.[51]

As with so much else in the late 1940s and 1950s, the basic components of the Western Civ framework dated back to the ideological offensive against fascism. For example, the leading textbooks used in Western Civ courses had mostly appeared between 1938 and 1941. Prior to that, the now-familiar meaning of "the West" had remained rather obscure to popular audiences. But interventionist historians had painted a portrait of shared cultural roots under direct fire from a new set of barbarians. A slightly different defense of democracy traced it deep into the past in order to emphasize its staying power, rather than the dangers to it. Irwin Edman, the most aesthetically inclined of the Columbia naturalists, sought to reassure readers by stressing "how old and

---

[51] As far back as the 1920s, Columbia's trustees had looked into the CC course, fearing that its focus on contemporary social problems and the human sciences would lead students into the arms of the communists. Charles H. Russell, "The Required Programs of General Education in the Social Sciences at Columbia College, the College of the University of Chicago, and Harvard College" (PhD dissertation, Columbia University, 1961), 168.

persistent is the democratic idea, how sturdily and tenaciously it reasserts itself, in reform, reconstruction, and revolution against governments and men that violate its ancient tenets of liberty and justice for all." Although Edman, as a student and ally of John Dewey, rejected the possibility of a value-free science, his historical narrative described American democracy as the product of an ancient ethical tradition, not a modern scientific culture.[52]

Under Conant, Harvard became the leading academic sponsor of the Western Civ approach, although it famously failed to heed its own curricular recommendations. While Conant pushed for a science-centered general education for nonscience majors, he simultaneously hoped to give future scientists and engineers a solid ethical foundation by steering them toward humanities courses. In this context, Conant described science as a technical adjunct to the ethical core of Western civilization, a means through which its deepest values found their realization. As he put it, science contributed to progress by "strengthening the hand of the Good Samaritan." This argument found powerful expression in Harvard's famous "Redbook," a 1945 faculty report officially titled *General Education in a Free Society*. The Redbook's image of science as a means to the ends given by the Western ethical tradition set the tone for the postwar conversation on general education, much as Columbia's human science-oriented CC had provided a model for the previous round of innovation in that area. The Harvard report largely cut the social sciences out of the curricular picture, leaving the humanities as the sole source of ethical guidance for students of all varieties.[53]

Conant shared with many other postwar administrators, including his MIT counterparts Karl T. Compton and James R. Killian Jr., a sense that science and engineering majors needed a strong dose of the humanities to orient them toward ethical values. That approach gradually grew out of an older model that highlighted the social sciences as a source of moral guidance. During the early

---

[52] Segal, "'Western Civ' and the Staging of History in American Higher Education," 781; Irwin Edman, *Fountainheads of Freedom: The Growth of the Democratic Idea* (New York: Reynal & Hitchcock, 1941), vii, 3. John Herman Randall Jr.'s publisher urged him to use *The Making of the Modern Mind*, not *The Western Mind*, fearing the latter title would lead "a very large portion of the general trade" to think the book addressed "the intelligence of California or something of that sort." W. E. Spaulding to Randall, February 17, 1925, Randall papers, Rare Book and Manuscript Library, Columbia University, Box 4, folder "Correspondence, 1920–1929."

[53] Conant, *Modern Science and Modern Man*, 86; Paul H. Buck et al., *General Education in a Free Society* (Cambridge: Harvard University Press, 1945). Jamie Cohen-Cole analyzes other aspects of the report in *The Open Mind: Cold War Politics and the Sciences of Human Nature* (Chicago: University of Chicago Press, forthcoming). It would be too simple to argue that Columbia's curricular philosophy (and broader academic leadership) gave way to that of Harvard in the postwar years. Yet there is something to the generalization. Harvard led the charge for the Western Civ model, with its more literary orientation and value-neutral conception of science. Although World War II consolidated a turn to consensus liberalism in CC, the course retained a social-scientific flavor, in keeping with the distinctive proclivities of Columbia's philosophers and historians. Russell, "The Required Programs of General Education in the Social Sciences at Columbia College, the College of the University of Chicago, and Harvard College," 57.

Depression years, Compton added social-scientific material to MIT's engineering curriculum in an attempt to combat his students' attraction to communism. He redoubled his support for that approach when Roosevelt and other administration officials charged that the nation's technical schools lacked social concern and fostered technological unemployment. But in 1942, when Stanford took up the task of humanizing its science and engineering students, it launched a School of Humanities rather than drawing on the social sciences. In fact, the School's initial hires included Lewis Mumford, giving him his first academic position after many years as an independent critic of the universities. The School's stated focus on "the study of man as a rational and artistic being" neatly captured the partnership of technical and humanistic knowledge that would dominate postwar curricula and conceptually marginalize the social sciences.[54]

Although Stanford president Ray Lyman Wilbur took this step in part because he worried that the American public would "lose the peace" after a successful war effort, he likely felt the pressure of political criticism as well. From his post at the Department of Agriculture, Henry A. Wallace took the lead in pressing the Roosevelt administration's case against exclusively scientific curricula, arguing that "the remorseless discipline of higher mathematics, physics and mechanics" required a strong leaven of "imaginative, non-mathematical studies, such as philosophy, literature, metaphysics, drama and poetry." The outbreak of war inspired political leaders from both parties to urge university administrators to widen the training of future scientists. Let the liberal arts languish, Republican presidential hopeful Wendell Willkie declared in 1943, "and you will lose freedom as surely as if you were to invite Hitler and his henchmen to rule over you." Engineering programs drew special attention as a source of antidemocratic attitudes during the war, even as several additional schools added humanistic and social-scientific material to keep their engineering students from becoming heartless calculators. Drawing on such models, and responding directly to criticism from Roosevelt's administration, a special committee of the Society for the Promotion of Engineering Education recommended in 1944 that engineers receive two roughly equal series of courses in technical fields and in "humanistic-social studies," the latter focusing on "the enduring ideas and aspirations which men have evolved as guides to ethical and moral values."[55]

[54] Richard M. Freeland, *Academia's Golden Age: Universities in Massachusetts, 1945–1970* (New York: Oxford University Press, 1992), 79; LeCuyer, "The Making of a Science Based Technological University"; Charles Dorn, "Promoting the 'Public Welfare' in Wartime: Stanford University during World War II," *American Journal of Education* 112, no. 1 (November 2005), 113, 116–117, 119 (quoted).

[55] Quoted in Dorn, "Promoting the 'Public Welfare' in Wartime," 117; Wallace, "The Social Advantages and Disadvantages of the Engineering-Scientific Approach to Civilization," 3–5; quoted in H. P. Hammond, "Liberalizing Technical Education," in *Proceedings of the Association of Land-Grant Colleges and Universities, Fifty-Eighth Annual Convention* (New Haven: Quinnipiack Press, 1945), 100. See also Clinton K. Judy, "Unexpected Light: Engineering and the Humanities," *Sewanee Review* 52 (1944): 199–202; and the symposium "How Can

Harvard's Redbook captured the transition from the "humanistic-social stud-ies" approach, with its lingering pluralism, to one focused solely on the human-ities, in keeping with the emerging two-cultures framework. The committee, led by the historian Paul H. Buck, did not ignore the social sciences entirely. At various points, it stressed the importance of teaching "the human past and human institutions," with the latter understood as "attempted embodiments of the good life." The authors also implied that social-scientific study helped forge "a mind capable of self-criticism." The "civilized man," they explained, was "a citizen of the entire universe" who had "overcome provincialism," becoming "objective" and "universal in his motives and sympathies." The report even invoked the interpersonal sphere of informal, nongovernmental control empha-sized by Progressive and post-Progressive scholars. By enabling "the sharing of meanings," it stated, this communicative sphere ensured that "human beings are welded into a society, both the living with the living and the living with the dead." Like many wartime texts, the Redbook identified leaders' use of persua-sion rather than force as the difference between democracy and totalitarianism. On this basis, it recommended a program of general education in which the "less gifted" – those exhibiting "lower facility with ideas" – would be fitted for "responsible private judgment" through concrete, relevant material appealing to their senses and emotions.[56]

In the end, however, the Redbook represented a major step away from the interwar emphasis on nongovernmental discipline embodied in a personal, knowledgeable commitment to the common good. It instead advanced an emerging, postwar view centered on deference to expertise and collaboration between natural scientists and humanists. Expertise figured centrally in the doc-ument, as the authors equated "the general art of the free man and the citizen" with the capacity to "distinguish the expert from the quack, and the better from the worse expert." Meanwhile, the committee divided the intellectual world into two distinct spheres: the factual and the normative, the scientific realm of truth and the humanistic realm of value. Social scientists, they explained awkwardly, simply straddled the line, borrowing from both spheres in keeping with the fact that "human ideals are somehow a part of nature." Like the Redbook's empha-sis on expert authority, its sharp separation of scientific and normative pursuits anticipated the postwar Western Civ and Great Books approaches. The com-mittee explained that its proposed curriculum centered on "the great writings of our culture," presented in a simplified, de-intellectualized form.[57]

Science Education Make Its Greatest Contribution in the Post-War Period?" *Science Education* 28 (1944): 231–238, 282–288. One beneficiary of this humanistic turn in the 1940s was the nascent discipline of history of science; George Sarton's introductory course at Harvard jumped dramatically in popularity with the start of the war, and the postwar years brought intense interest in the history of science among university leaders. Arnold Thackray and Robert K. Merton, "On Discipline Building: The Paradoxes of George Sarton," *Isis* 63 (1972), 489; Victor L. Hilts, "History of Science at the University of Wisconsin," *Isis* 75 (March 1984), 63.

[56] Buck et al., *General Education in a Free Society*, viii, 73, 53, 68, 94–95.
[57] *Ibid.*, 54, 73, 61–62, 59, 68, 94–95.

Thus, although the Redbook committee could hardly ignore the social sciences, it left those disciplines no distinctive role in the intellectual economy of the postwar university. What came through most clearly in the report was its image of a thoroughly value-neutral science and its insistence that Western civilization was defined, in part, by its use of this tool. The authors opened a wide conceptual gap between science and values. Repeatedly invoking "facts" and "truth," they described scientists as disinterested individuals who trafficked in "direct observation and precision," making a "direct appeal to nature." This approach enabled the committee to claim that science not only produced material goods but also advanced "the spiritual values of humanism" by teaching citizens to appeal to facts against arbitrary authority and false *a priori* claims. Science, they wrote, instilled "the morals of thinking," namely "intellectual integrity, the suppression of all wishful thinking and the strictest regard for the claims of evidence." But the authors did not ultimately ground democracy in scientific principles or practices. Instead, they traced it to a set of "fixed beliefs," derived from Old World sources, which centered on "the dignity and mutual obligation of man" and operated as "principles above the state." The true role of science in Western history, according to the Redbook, was to have "implemented the humanism which classicism and Christianity have proclaimed."[58]

On the ground, Harvard's social scientists managed to craft a program of general education featuring both humanities and social science courses, as did the faculty at Chicago. But the Redbook's vision proved widely influential in postwar America. Conant sent the report to an array of national leaders, and the *New York Times* reviewed it. The educational leader Earl McGrath also plugged the Harvard model, both as editor of the *Journal of General Education* from 1946 to 1949 and as U.S. Commissioner of Education thereafter. Within five years, more than 40,000 copies had been sold. Through such avenues, the Redbook, which assigned a central political role to the humanities, powerfully advanced the distinctive American version of the two-cultures analysis.[59]

In its prevailing forms, the emerging Western Civ model tended to undermine the post-Progressive vision of many interwar scientific democrats. It reinforced the image of science as strictly value-neutral, while often obscuring the very existence of the social sciences. Science now appeared as a new and uniquely powerful means of advancing ethical and political ends that had been set many centuries earlier, in ancient Greece, Rome, and Palestine. Western Civ, and the broader two-cultures framework it embodied, undercut the previously widespread conviction that scientific methods, defined broadly and applied to both past and present experience, could produce, or at least reveal, a system of ethical values appropriate to the industrial age. Science, in the dominant postwar understanding, meant a body of facts or theories about the physical and

[58] *Ibid.*, 153, 157, 72, 50, 47, 53–54, 57, 77, 55.
[59] Cohen-Cole, *The Open Mind.*

biological worlds, along with a few small areas of similarly reliable, apolitical knowledge of human affairs. To the extent that social scientists would find a place for themselves in American intellectual culture in the 1950s, they would do so largely by wedging their work into the narrow vision of the scientific enterprise forged by physical scientists and humanistic and religious critics.

11

Accommodation

As the physical scientists more actively took up the role of shaping public understandings of science and its relation to society, scientific democracy's disciplinary center of gravity shifted away from the social sciences and philosophy. The efforts of Conant and other influential commentators marginalized both the post-Progressive vision of a thick culture of public-mindedness and the disciplines from which that outlook had long emanated. But in the human sciences, too, the 1940s and early 1950s brought a decisive shift in the overall tenor of scientific democracy, setting the stage for its dissolution in the 1960s. This shift went hand in hand with changes in American liberalism, as the Protestant-inflected successor project to ethical economics that had fired so many thinkers up to World War II ceased to be a dominant force in American intellectual life.

It is hazardous to generalize about what replaced it. A combination of rapid university growth and massive new funding sources led the academic disciplines to expand and diversify dramatically, producing a highly variegated set of scholarly initiatives. In general, however, the direction of change ran toward more individualistic and, in many circles, more rationalistic models of human behavior and social relations. The various forces pressing on postwar human scientists led most to follow natural scientists in ceding the realm of the social emotions to humanists. Thus, as unprecedented political conditions intersected with the arrival of fresh cohorts of theorists, the major casualty was the post-Progressive orientation of the interwar period. No longer would science signal a direct, individual concern with the consequences of one's actions for the flourishing of others.

The intellectual sensibilities that emerged after World War II typically described the sciences as sophisticated tools for implementing values drawn from nonscientific sources – not from human needs and contemporary social realities, as vouched for by social scientists, but rather from long-standing cultural traditions rooted in the deep past. Science could offer citizens certain kinds of knowledge, but not the robust set of skills and virtues stressed by many interwar scientific democrats. In the postwar social sciences, too, scholars loudly proclaimed their neutrality but committed themselves to advancing the national interest – defined, it seemed to many critics, in terms of the values of liberal, middle-class professionals.

A HOUSE DIVIDED

After the war, the dramatically increased visibility and cultural authority of the natural sciences and humanities drew individual scholars and even entire fields of study away from the post-Progressive version of scientific democracy and the broad "cultural sciences" formation that accompanied and sustained it. Now, most historians, psychologists, and philosophers threw in their lot with either the humanities or the natural sciences. Economists pushed toward mathematics, meanwhile, and biologists turned away from the social sciences as well. This left a stripped-down social-scientific enterprise centered on sociology, political science, and anthropology. Scholars in these fields increasingly referred to themselves as "behavioral scientists" and emphasized their disinterestedness.

From the side of the natural sciences, some biologists continued to seek an active commitment to human needs among members of the public. But rather than authorizing the social sciences by validating human freedom, as had the emergentists of the 1920s, these postwar figures looked to biology to answer the full range of questions about human behavior. With the "modern synthesis" of Darwinian evolution and Mendelian genetics under their belt, they crafted broad, genetically based formulations of democracy. The plant morphologist Edmund W. Sinnott sounded the notes of liberal Protestant arguments for human brotherhood. "Love is the climax of all goal-seeking, protoplasm's final consummation," he declared. "If the stars in their courses are not fighting for us, at least the genes within our chromosomes do so." Meanwhile, George Gaylord Simpson, a key player in the modern synthesis, argued that genetics pointed the way toward a "socialized democracy" featuring "controlled capitalism without improper exploitation" by revealing the absolute moral freedom and responsibility of the individual. Reframing earlier arguments for purposive behavior and free will, while adding new emphases on human diversity and equality, these biologists argued that human genetics offered a sufficient basis for morality and democratic theory.[1]

Some social scientists themselves turned to biology after the war, arguing that it firmly supported their post-Progressive ideals. "Man does not want to be independent, free, in the sense of functioning independently of the interests of his fellows," wrote the anthropologist Ashley Montagu, a student of Ruth Benedict at Columbia. "This kind of negative independence leads to lonesomeness, isolation, and fear. What man wants is that positive freedom

---

[1] Edmund W. Sinnott, *The Biology of the Spirit* (New York: Viking, 1955), 155; Sinnott, "The Biological Basis of Democracy," *Yale Review* 35, no. 1 (September 1945), 72–73; George Gaylord Simpson, *The Meaning of Evolution* (New Haven: Yale University Press, 1949), 322.

which follows the pattern of life as an infant within the family – dependent security, the feeling that one is a part of a group, accepted, wanted, loved and loving." The very "tissues of every organism," Montagu argued, "remember their dependency and interdependency" from the period of infancy and sought always to retain or recapture that biological state. Like Sinnott and Simpson, Montagu mobilized the authority of biology on behalf of the ethical principles underlying his cooperative model of social relations.[2]

Other postwar scholars disavowed more thoroughly the interwar vision of a broadly defined human-scientific enterprise that would fulfill a crucial social function by sensitizing citizens to their impact on others. Instead, they aligned themselves with either the humanities or the natural sciences, reflecting the increasingly powerful two-cultures model. Many historians, for example, came to define themselves as humanists after World War II. This was particularly true of those who embraced either the Western Civ idea or consensus liberalism – including its "American mind" variant, associated with the emerging field of American studies. These interpretive frameworks rendered the historian a full-fledged "conservator of moral values" and enabled what one called the discipline's postwar "declaration of intellectual independence" from the social sciences. This disconnection was by no means complete; Columbia's Richard Hofstadter, among others, retained important ties to the social sciences. But the development of the Western Civ, consensus liberalism, and American mind approaches rested heavily on the work of historians who self-identified as humanists.[3]

In fact, a number of leading postwar historians endorsed Reinhold Niebuhr's biting critique of scientism. These "atheists for Niebuhr" included Perry Miller – trained in literature and employed in the English Department at Harvard – and Arthur M. Schlesinger Jr. They saw in the theologian's emphasis on human sinfulness an antidote to what they portrayed as the arid rationalism of the interwar post-Progressives. It is likely relevant here that Charles Beard and Harry Elmer Barnes, the leading Progressive historians after James Harvey Robinson's death in 1936, not only dabbled in economic and technological determinism but also vocally opposed World War II and postwar militarism. Many of the emerging historians of the 1940s and 1950s instead followed Niebuhr in supporting overseas campaigns on behalf of democracy and defining that political system as a product of deep-rooted American or Western values, while equating science with a fatuous optimism about human

---

[2] Ashley Montagu, *On Being Human* (New York: H. Schuman, 1950), 80, 30.

[3] Peter Novick, *That Noble Dream: The "Objectivity Question" and the American Historical Profession* (New York: Cambridge University Press, 1988), 382, 387; Roy F. Nichols, "Postwar Reorientation of Historical Thinking," *American Historical Review* 54, no. 1 (October 1948), 89; John Higham, *History* (Englewood Cliffs: Prentice-Hall, 1965), 135; David S. Brown, *Richard Hofstadter: An Intellectual Biography* (Chicago: University of Chicago Press, 2006). On the professional benefits for historians, see Gilbert Allardyce, "The Rise and Fall of the Western Civilization Course," *American Historical Review* 87, no. 3 (2005): 695–725; and Thomas Bender, "Politics, Intellect, and the American University, 1945–1995," *Daedalus* 126, no. 1 (Winter 1997), 5.

perfectibility and a deterministic elimination of human freedom. Advocates of this new, humanistic orientation in the history discipline criticized the scientism of the social sciences from the outside, rather than advocating a broader definition of science that also included their own views and methods.[4]

Meanwhile, psychology and philosophy each split down the middle, as institutionally powerful factions oriented toward the natural sciences squared off against culturally inclined factions identified with the humanities. These developments completed the postwar reduction of the cultural sciences. Attempts by psychologists to gain the mantle of natural science dated to the late nineteenth century, but a shared commitment to the post-Progressive vision of a psychologically unified citizenry had aligned most interwar behaviorists with other human scientists. After World War II, however, younger figures such as Harvard's S. S. Stevens and B. F. Skinner, both of whom had been inspired early in their careers by logical empiricism, pushed the field much closer to the natural sciences. Skinner, for example, described psychology as a purely experimental field and banished all other methods from the domain of legitimate scientific inquiry, after the fashion of John B. Watson. The fierce ambition of postwar behaviorists to establish their scientific credentials reinforced the image of a sharp line between genuine sciences, modeled on physics, and what critics such as Skinner saw as the nonscientific guesswork and prophecy of the so-called social sciences.[5]

On the other side of the two-cultures divide, the "humanistic psychologists" Carl Rogers, Abraham Maslow, and Rollo May countered Skinner's scientific ambitions by consciously aligning themselves with the humanities, and often with religious traditions as well. These figures took inspiration from émigré

---

[4] Richard Wightman Fox, *Reinhold Niebuhr: A Biography* (Ithaca: Cornell University Press, 1996); Arthur M. Schlesinger Jr., "The Statistical Soldier," *Partisan Review* 16 (August 1949): 852–856. Other notable isolationists in 1939–1941 included William F. Ogburn, John Dewey, Franz Boas, and the Columbia historian James T. Shotwell: R. Alan Lawson, *The Failure of Independent Liberalism, 1930–1941* (New York: Putnam, 1971), 231–233, 244–245; Robert C. Bannister, *Sociology and Scientism: The American Quest for Objectivity, 1880–1940* (Chapel Hill: University of North Carolina Press, 1987), 225; Robert B. Westbrook, *John Dewey and American Democracy* (Ithaca: Cornell University Press, 1991), 510–513; Edward A. Purcell Jr., *The Crisis of Democratic Theory: Scientific Naturalism & the Problem of Value* (Lexington: University Press of Kentucky, 1973), 227; Novick, *That Noble Dream*, 247–248, 291–292, 308–309; Robert C. Bannister, "Principle, Politics, Profession: American Sociologists and Fascism, 1930–1950," in *Sociology Responds to Fascism*, ed. Stephen P. Turner and Dirk Käsler (New York: Routledge, 1992), 193. Most of the isolationist historians were affiliated with Beard in some fashion. Unlike the other critics of intervention, however, Beard and Barnes stood their ground through the war and after: Ernst Breisach, *American Progressive History: An Experiment in Modernization* (Chicago: University of Chicago Press, 1993), 190–203; Mark C. Smith, "A Tale of Two Charlies: Political Science, History, and Civic Reform, 1890–1940," in *Modern Political Science: Anglo-American Exchanges Since 1880*, ed. Robert Adcock, Mark Bevir, and Shannon C. Stimson (Princeton: Princeton University Press, 2007), 136.

[5] James H. Capshew, *Psychologists on the March: Science, Practice, and Professional Identity in America, 1929–1969* (New York: Cambridge University Press, 1999); Rebecca Lemov, *World as Laboratory: Experiments with Mice, Mazes, and Men* (New York: Hill and Wang, 2005).

psychologists such as Karen Horney, Erich Fromm, and Erik Erikson. Sharing with Gordon Allport a concern for the moral freedom of the individual and the centrality of personal integrity and development in the face of threatening social forces, these émigrés added a powerful new dose of Freudianism and existentialism. Their American interlocutors, enmeshed in the struggle against the behaviorists, proved much less concerned than Allport to claim the mantle of science. Instead, they aligned their work with the journeys of self-discovery characteristic of art, literature, and religion.[6]

Philosophers likewise split after World War II. Some inclined toward the natural sciences, while others presented themselves as humanists, and virtually none lined up with the social sciences, as had Columbia's naturalists. On the humanistic side, Ralph Barton Perry, once a great advocate of a scientifically grounded philosophy, now stressed that science and technology "have nothing to do with morals" and claimed that the humanities offered the basis for a satisfying modern culture. By contrast, Perry's Harvard colleague W. V. O. Quine saw the field as more closely aligned with the natural sciences than even Perry had earlier imagined. Inspired by logical empiricism, though critical of many of its specific tenets, Quine argued that philosophers should rest content with the task of clarifying the character and implications of a "naturalized epistemology" rooted in the natural sciences. Many other postwar American philosophers likewise identified their work as subsidiary to the natural sciences and toiled in fields such as epistemology, the philosophy of mind, and the philosophy of science itself.[7]

Still, philosophers continued to vigorously debate science's relationship to values, as did scholars in many other fields. To see postwar American philosophy as a univocal defense of socially detached science is to miss the fact that explicitly normative subfields found new life in the postwar years. In fact, Quine's "Two Dogmas of Empiricism" (1951), a landmark in the development of analytic philosophy, appeared as part of a symposium with two pieces that celebrated the renaissance of metaphysical speculation and moral theory. Even Quine, moreover, rejected the model of empirical verification often associated with defenses of strict value-neutrality. The easy identification of science with facts and the humanities with values in the American two-cultures discourse hardly became universal across the disciplines.[8]

[6] Ellen Herman, *The Romance of American Psychology: Political Culture in the Age of Experts* (Berkeley: University of California Press, 1995), 264–274.

[7] Ralph Barton Perry, *On All Fronts* (New York: Vanguard, 1941), 113; Perry et al., *The Meaning of the Humanities* (Princeton: Princeton University Press, 1938); Joel Isaac, "W. V. Quine and the Origins of Analytic Philosophy in the United States," *Modern Intellectual History* 2 (2005): 205–234; Bruce Kuklick, *A History of Philosophy in America, 1720–2000* (New York: Oxford University Press, 2001), 225–258.

[8] Grace A. De Laguna, "Main Trends in Recent Philosophy: Speculative Philosophy"; W. V. [O.] Quine, "Main Trends in Recent Philosophy: Two Dogmas of Empiricism"; and William K. Frankena, "Main Trends in Recent Philosophy: Moral Philosophy at Mid-Century," *Philosophical Review* 60, no. 1 (January 1951), 3–19, 20–43, 44–55; Guy W. Stroh, *American Ethical Thought* (Chicago: Nelson-Hall, 1979). Defying our image of postwar philosophy's domination by

Yet while Quine and other analytic thinkers acknowledged the problematic character of the traditional positivist stance, they adopted a much more technical, specialized mode of writing. Even in metaphysics and aesthetics, in fact, the leading philosophical offerings of the 1950s were highly formal and abstract, aimed largely at other scholars. Criticism of value-neutrality had itself become a mode of expertise for philosophers. Rather than accompanying direct appeals to a public audience, it emerged from attempts to clarify the epistemological character of scholarly practice by a small group of specialists concerned professionally with that technical question.

The two-cultures split in American intellectual life did not render social scientists invisible in the wider public culture, even though it made little room for them conceptually. In fact, many historians see the 1950s as a golden age of publicly engaged social science, a time when titles such as C. Wright Mills' *White Collar* and David Riesman's *The Lonely Crowd* – the most commercially successful work of American social science, with sales of more than 1.4 million copies as of 1995 – captured the public's imagination. But that imagination itself had changed in ways that made post-Progressive understandings of human behavior far less palatable. The ubiquitous misreading of Riesman's book as a straightforward lament for a lost, "inner-directed" Victorian self is highly revealing. After World War II, the educated middle class was less likely to see itself as the vanguard of a more cooperative society and more likely to defend individual freedom – if often expressive freedom rather than economic independence – against putatively authoritarian forms of social conformity and political pressure. Many social-scientific analyses of the era likewise took their shape and their appeal from this orientation, which flourished as readily among scholars as it did among other middle-class Americans. The idea that individual behavior is shaped by a combination of childhood experiences and horizontal, person-to-person influences seems to have felt safe to postwar thinkers and audiences only when it accompanied the claim that American culture inevitably formed children in the mold of democratic individualism.[9]

This new context profoundly reshaped the program of scientific democracy. Between the wars, most scientific democrats in the human sciences held

---

analysts, accounts from the time typically marveled at its diversity: e.g., Edgar Sheffield Brightman, "Philosophy in the United States 1939–1945," *Philosophical Review* 56, no. 4 (July 1947), 390; Philip Blair Rice, "The Philosopher's Commitment," *Proceedings and Addresses of the American Philosophical Association* 26 (1952–1953), 32; Ralph Barton Perry, "Is There a North American Philosophy?" *Philosophy and Phenomenological Research* 9, no. 3 (March 1949), 365. Cf. Hilary Putnam, "A Half Century of Philosophy, Viewed From Within," *Daedalus* 126 (1997), 176.

[9] C. Wright Mills, *White Collar: The American Middle Classes* (New York: Oxford University Press, 1951); David Riesman et al., *The Lonely Crowd: A Study of the Changing American Character* (New Haven: Yale University Press, 1950); Herbert J. Gans, "Best-Sellers by Sociologists: An Exploratory Study," *Contemporary Sociology* 26, no. 2 (March 1997), 134; David Paul Haney, *The Americanization of Social Science: Intellectuals and Public Responsibility in the Postwar United States* (Philadelphia: Temple University Press, 2008), 208–221. A helpful overview of postwar thought is Richard H. Pells, *The Liberal Mind in a Conservative Age: American Intellectuals in the 1940s and 1950s* (Middletown: Wesleyan University Press, 1989).

that science could further the democratic cause in three ways. First and foremost, as John Dewey wrote in 1927, it could help "to make the interest of the public a more supreme guide and criterion of governmental activity, and to enable the public to form and manifest its purposes still more authoritatively." Second, science could validate democracy over and against other forms of government, by portraying individuals and societies as capable of sustaining democracy's characteristic model of collective opinion formation and institutional responsiveness on a day-to-day basis. Finally, science could teach citizens that the term democracy referred to the broad ideal of matching public policies to public purposes, not to any specific policies or institutions adopted for that end.[10]

During the 1940s, however, much of this program fell away. The new versions of liberalism located the public good in a political tradition, a suite of rights, or a set of social structures inherited from the past, rather than entrusting that good to the choices of a scientifically informed citizenry. To their advocates, the purpose of everyday action seemed more specific and less ambitious than it had to interwar thinkers: individuals needed simply to promote and defend the preexisting embodiments of universal human values, which could be trusted to reliably address their needs and the needs of others. In short, Americans did not need to assess and potentially remake their core institutions; they simply needed to acknowledge their necessity and uphold them.

In fact, some postwar thinkers, like Sinnott, Simpson, and Montagu, rendered the public good even less precarious by rooting it in the biological nature of the human animal itself. Like the relegation of liberal values to a practice or institution handed down from the past, this biological argument expressed a broader tendency among postwar thinkers to "naturalize" their core commitments – to embed them in the very structure of reality itself. This naturalizing move rendered the commitments relatively immune to challenge but also fixed them permanently and absolved proponents of the need to cultivate in others the virtues, skills, and sentiments that sustained them. Rather than seeing their values as embodied in the practices of the scientific community and holding that they needed to flow into the wider public culture, postwar American thinkers tended to locate these values in realms that shaped individual behavior more automatically, with less margin for error – and less need for active effort by scientists themselves.[11]

As these theorists relocated the source of American values outside the community of scientists, they implicitly identified humanists, historians, biologists, or religious leaders as the authoritative spokespersons for those values. This left to the human sciences the simpler function of suggesting more effective

---

[10] John Dewey, *The Public and Its Problems* (1927), in Jo Ann Boydston, ed., *The Later Works, 1925–1953, Volume 2* (Carbondale: Southern Illinois University Press, 1981), 327. Cf. Westbrook, *John Dewey and American Democracy*, 361–366.

[11] Andrew Jewett, "Naturalizing Liberalism in the 1950s," in Neil Gross and Solon J. Simmons, eds., *Professors and Their Politics* (forthcoming).

and efficient ways to institutionalize liberal principles. Reinforcing the physical scientists' image of an autonomous scientific community, the naturalization strategies pursued by many postwar social scientists profoundly altered the program of scientific democracy. The scientific enterprise no longer appeared as a model for democratic discourse, but simply as a source of concrete knowledge – for the state, practical techniques, and for the citizen, the fact of the wisdom and necessity of the cultural practices, rights, or social structures that embodied America's liberal ideals. Social scientists could advance the cause of democracy simply by demonstrating that these pre-given resources existed and urging citizens not to tamper with them.

CAUSES AND COHORTS

Historians have long recognized the relatively uncritical, even politically quiescent, tenor of the postwar social sciences and philosophy. Two overlapping forces have figured prominently: the new funding structures put into place during World War II and the early Cold War, and the broad stream of anti-communist politics often shorthanded as "McCarthyism." These phenomena were extraordinarily important and consequential. During World War II, social scientists flocked to Washington, contributing their skills to the war effort in ways both large and small. From psychological warfare to domestic morale-building, from postwar planning for the reeducation of Germans to improving race relations and even administering food rations, social scientists were in the thick of many aspects of the conflict. After the war, they carried forward their view of federal power as a potent force for good. Government agencies, along with contractors and foundations, reciprocated by directing plentiful resources to those scholars comfortable with a nonpartisan self-presentation and working in strategically important fields, especially those relevant to military capacity. The new funding avenues enabled social scientists to believe that they could powerfully promote the public good simply by toiling in technical, specialized research fields. By 1960, the social sciences had, to varying degrees, pivoted from the public to the federal government as their ultimate audience. Meanwhile, anti-communism had rendered it dangerous for scientific democrats to publicly advocate a more cooperative society, although scholarly seniority and prestige provided some immunity from political repression.[12]

---

[12] For overviews of the period, see Herman, *The Romance of American Psychology*; Ron Robin, *The Making of the Cold War Enemy: Culture and Politics in the Military-Industrial Complex* (Princeton: Princeton University Press, 2001); Robert Adcock and Mark Bevir, "The Remaking of Political Theory," in *Modern Political Science: Anglo-American Exchanges Since 1880*, ed. Adcock et al. (Princeton: Princeton University Press, 2007): 209–233; and Mark Solovey and Hamilton Cravens, eds. *Cold War Social Science: Knowledge Production, Liberal Democracy, and Human Nature* (New York: Palgrave Macmillan, 2012). The institutional relationship to government that physical scientists negotiated during the war and codified in the postwar National Science Foundation protected the formal autonomy of investigators by leaving scientists relatively free to set their own specific agendas and work in university settings rather than federal

As important as these factors were, however, others also helped produce the postwar era's less critical, more celebratory social-scientific enterprise. Changes in national politics, wider geopolitical developments, demographic shifts in academia, the resurgence of powerful competing voices in American public culture, and the redirection of cultural authority toward new areas of science all contributed to the postwar narrowing of the social sciences' disciplinary scope and the emergence of new understandings of the individual, society, and science. Although it is hardly comprehensive, the rest of this chapter emphasizes how the interaction of demographic developments with the threat of totalitarianism and the resurgence of the American right shaped many characteristic projects in the slimmed-down social sciences and accelerated the turn away from post-Progressive versions of scientific democracy.[13]

Ideas are made by people, and people live specific lives, in specific settings. Viewing shifts in scientific democracy in light of the subjective experiences of concrete groups thus is crucial to, if perhaps not sufficient for, understanding the causes of those shifts. From the mid-1930s to the early 1950s, left-leaning scholars watched a number of major political developments unfold. The emergence of totalitarianism abroad led them to rally to democracy's defense, not least because the Nazis, and later the Soviets, proved quick to impose ideological shackles on scientists. At home, meanwhile, conservative opponents of the New Deal enjoyed steadily growing power, especially in Congress. In the late 1940s and early 1950s, the right-wing anti-communism that fueled Joseph McCarthy's meteoric rise heightened the sense of embattlement on the domestic front. Finally, military leaders and other governing elites seemed to have developed a healthy, progressive interest in economic regulation and social-scientific tools, whereas the citizenry seemed to have sanctioned the alarming power of right-wing politicians and religious traditionalists. More and more, social scientists concluded that they should work to ensure that the values of the elites trickled down to ordinary Americans, rather than empowering the latter against their leaders. The people, it appeared, were no longer the most progressive force in the American political system. At the very least, the elites

---

laboratories. But, of course, outside funding from the state and from foundations was only available for certain broad fields of work. On these administrative dynamics, see Rebecca S. Lowen, *Creating the Cold War University: The Transformation of Stanford* (Berkeley: University of California Press, 1997).

[13] Nor can one forget the contention of the new Keynesian economics that governments could smooth boom-and-bust cycles and ensure universal prosperity with a minimum of painful redistribution and institutional reform. On the dangers of using the Cold War as an all-encompassing explanation for postwar intellectual developments, see David A. Hollinger, "Religion, Ethnicity, and Politics in American Philosophy: Reflections on McCumber's 'Time in the Ditch,'" *Philosophical Studies* 108, no. 1–2 (March 2002): 173–181; David C. Engerman, "Rethinking Cold War Universities: Some Recent Histories," *Journal of Cold War Studies* 5, no. 3 (2003): 80–95; Engerman, "The Romance of Economic Development and New Histories of the Cold War," *Diplomatic History* 28, no. 1 (January 2004): 23–54; and Joel Isaac, "The Human Sciences in Cold War America," *Historical Journal* 50, no. 3 (2007): 725–746. Hollinger stresses some of the demographic shifts discussed here.

Science and Politics

spoke a language of real-world causes and consequences rather than seeing conspirators under every bed.[14]

Who watched these developments and drew these conclusions? Generational and demographic shifts in the social sciences played a key role in the transformation of scientific democracy, as new groups brought new sensibilities to the task of defining science's social role under the changed conditions. Simplifying dramatically, and keeping in mind the merely heuristic value of the categories, one can speak of three broad cohorts that shaped American social science by the 1950s. Two of these had forged their political identities in the fire of the struggle against totalitarianism: first, the illustrious array of émigrés driven out of Europe by fascism, and second, a younger generation of American-born scholars, disproportionately though not exclusively Jewish, whose formative intellectual and political experiences had been the ideological conflicts of the late 1930s rather than the rampant consumerism of the 1920s or the economic slump of the early 1930s. Distrusting the public, members of these two groups tended to view the interpersonal sphere of communication and influence, not merely as a source of cultural inertia, but also as a powerful threat to the freedom and even the lives of individuals. When they thought about the prospect of noncognitive forces shaping human behavior, it did not trigger associations with the liberal Protestants' mutual uplift projects, but rather with childhood traumas and personality disorders, or even with jackbooted thugs and concentration camps. By contrast, an older group of culturally Protestant thinkers who had begun their careers in the 1920s or early 1930s placed far more trust in ordinary Americans and continued to see the main threat to democracy as a misguided, *laissez-faire* reading of individualism rather than a frightening loss of individual autonomy. Despite the many commonalities between them, each cohort responded in a distinctive manner to the developments of the World War II and early Cold War periods.

The thousand-plus European scholars who emigrated to the United States after 1933 were even more likely to be Jewish than the members of the 1930s generation, although Catholics figured increasingly prominently after France fell to Hitler in 1940. Many viewed the intellectual as an expositor of universal values and truths to a public that was relatively passive but potentially dangerous if provoked. Most of the émigrés had enjoyed considerable prestige in Europe, and none were steeped in the low-church ideal of informal, horizontal social control that shaped American political culture. Even many of those who denied the possibility of neutral knowledge and expected their influence to operate through their writings rather than through the state had in mind a

---

[14] Alan Brinkley, *Liberalism and Its Discontents* (Cambridge, MA: Harvard University Press, 1998); Elizabeth A. Fones-Wolf, *Selling Free Enterprise: The Business Assault on Labor and Liberalism, 1945–60* (Urbana: University of Illinois Press, 1994); Jonathan Bell, *The Liberal State On Trial: The Cold War and American Politics in the Truman Years* (New York: Columbia University Press, 2004).

polity in which scholars helped shape social and institutional change from the top down.[15]

This orientation, while hardly universal among the émigrés, bridged profound theoretical disagreements within their ranks. These figures brought with them a polarized, supercharged debate over epistemology and politics, dividing roughly into today's "analytic" and "Continental" camps. On each side of the science and values question, the émigrés typically defined their positions in much sharper terms than did their American equivalents. Moreover, they routinely accused one another of aiding in Hitler's victory. Whereas William F. Ogburn deemed the entanglement of science with emotional commitments an obstacle to correct understanding and a hindrance to reform efforts, Rudolf Carnap saw it as the source of fascism. Carnap's "metaphysics" referred to the views of Catholic monarchists and blood-and-soil fascists, not Ogburn's fuzzy-headed, liberal Protestant do-gooders – with whom, of course, Ogburn shared virtually all of his social and political values. For Carnap and his analytically minded allies, the question of science and values signaled a pitched battle between modernity and its critics, not a strategic dispute within a relatively homogeneous, reform-minded academic mainstream. Most of these analytic thinkers supported what they took to be the characteristically modern political form: a powerful welfare state armed with technical expertise.[16]

Meanwhile, those émigrés who rejected the concept of value-neutrality outdid even the earlier-arriving *Gestalt* theorists in declaring it frankly authoritarian. Associating the claim of value-neutrality with the cultural dominance of an instrumental reason that eradicated human values, critics as different as the left-wing social theorist Theodor Adorno and the conservative political theorist Leo Strauss blamed it for Nazism's rise out of the ashes of German liberalism. Rather than the domestic tussles over economic regulation that fired Protestant and post-Protestant figures such as Charles Ellwood and Robert Lynd, this group's critique centered on the far larger questions about the nature of the modern state that had preoccupied earlier émigrés such as Robert MacIver, Pitirim Sorokin, and the *Gestaltists*. Some of them, including Hans Morgenthau in political science, joined Reinhold Niebuhr in attacking scientism on religious grounds. All deemed scientism an utterly inadequate and highly dangerous account of the knowing process.[17]

Though the older cohort of American-bred thinkers had, like the émigrés, launched their careers in the 1920s and early 1930s, they did not have to grapple

---

[15] A classic overview is Laura Fermi, *Illustrious Immigrants: The Intellectual Migration from Europe, 1930–41* (Chicago: University of Chicago, 1971).

[16] Peter Galison, "The Americanization of Unity," *Daedalus* 127 (1998), 65; George A. Reisch, *How the Cold War Transformed Philosophy of Science: To the Icy Slopes of Logic* (New York: Cambridge University Press, 2005).

[17] Max Horkheimer and Theodor Adorno, *Dialectic of Enlightenment*, trans. John Cumming (New York: Seabury Press, 1972 [1947]); Leo Strauss, *Natural Right and History* (Chicago: University of Chicago Press, 1953); Hans J. Morgenthau, *Scientific Man vs. Power Politics* (Chicago: University of Chicago Press, 1946).

with totalitarian movements until the late 1930s. Prior to that, the leading public issues were economic in nature: first prosperity and consumerism, then stagnation and poverty. Figures such as Margaret Mead, Talcott Parsons, Gordon Allport, Gardner Murphy, Carl Rogers, and the anthropologist Robert Redfield remained sanguine about horizontal influences between citizens and optimistic about the prospect for cultural change. They carried forward the post-Progressive vision of a grand cultural transformation. After 1940, however, they typically reframed that vision in more gradualist, meliorist terms, speaking of a progressive improvement in Americans' self-understanding rather than a sharp break in the nation's history. Many of these theorists adopted the "postcapitalist" view that the best elements of socialism would emerge peacefully within the existing contours of American democracy. Roosevelt, they reasoned, had put into place a set of political structures adequate to the dislocations of industrialization. The dynamic of cultural lag now operated at the level of individual personalities, where psychologically informed scholars needed to build a base of legitimacy for the welfare state by helping citizens grow a more cooperative culture. When these thinkers embraced consensus liberalism, as did Mead, it often reflected a fairly shallow and largely rhetorical shift away from their earlier interventionist stance toward American public culture.[18]

Some scholars of Protestant extraction deviated from this sensibility, and younger Protestant thinkers sometimes embraced it. On the one side, a number of important figures such as the sociologist George A. Lundberg and the psychologist B. F. Skinner carried forward the rigorous, control-oriented mode of scientific democracy launched by the likes of Luther Lee Bernard and John B. Watson. Translating stern upbringings into equally stern calls for scientifically managed social change, they harbored expectations that individuals – or at least scientists – could become entirely selfless. Even after the war, they advanced a vision of total social transformation. But the more gradualist, less self-denying stream of thought proved more powerful culturally, if not institutionally. In fact, it appealed to many younger figures from Protestant backgrounds as well; Rollo May and Clyde Kluckhohn developed views of this kind in the context of World War II.[19]

---

[18] Charles Camic, "Introduction," in Talcott Parsons, *The Early Essays* (Chicago: University of Chicago Press, 1991); Uta Gerhardt, *Talcott Parsons: An Intellectual Biography* (New York: Cambridge University Press, 2002); Katherine Pandora, *Rebels Within the Ranks: Psychologists' Critiques of Scientific Authority and Democratic Realities in New Deal America* (New York: Cambridge University Press, 1997); Ian A. M. Nicholson, *Inventing Personality: Gordon Allport and the Science of Selfhood* (Washington: American Psychological Association, 2003); Lois Barclay Murphy, *Gardner Murphy: Integrating, Expanding, and Humanizing Psychology* (Jefferson, NC: McFarland, 1990); David Cohen, *Carl Rogers: A Critical Biography* (London: Constable, 1997); Clifford Wilcox, *Robert Redfield and the Development of American Anthropology* (Lanham: Lexington Books, 2004); Howard Brick, *Transcending Capitalism: Visions of a New Society in Modern American Thought* (Ithaca: Cornell University Press, 2006).

[19] Robert C. Bannister, *Sociology and Scientism: The American Quest for Objectivity, 1880–1940* (Chapel Hill: University of North Carolina Press, 1987); Daniel W. Bjork, *B. F. Skinner: A Life* (New York: BasicBooks, 1993); Roy J. deCarvalho, "Rollo R. May (1909–1994): A Biographical

But many of those who came of age during the campaign against totalitarianism rejected efforts to fundamentally alter American institutions. During this period, significant numbers of Jews found employment in sociology, political science, and philosophy, whose earlier practitioners had sought to advance a barely secularized Christian ethic. In earlier years, most Jewish scholars had made careers either in anthropology, with its anti-racist edge, or in the natural sciences, economics, law, and other comparatively technical fields that were not associated in the public mind with the conservation of American values. That changed in the late 1930s, when a much larger group of second-generation Jews made their way into academia just as the totalitarian threat rendered anti-Semitism suspect and dampened enthusiasm for a cooperative commonwealth. Robert K. Merton was hardly alone among the members of his cohort – including many non-Jewish figures as well – in seeing systematic interaction with the wider culture as a direct threat to scholarly communities. These figures lacked the confidence of most of their elders that ethical truths would prevail if simply placed before the public.[20]

These Jewish scholars generally compared the America of their day to Nazi Germany, not to an idealized vision of what it could become. The contrast could not have been clearer. Although powerful currents of anti-Semitism persisted in the United States, Jews could join the professions and speak to large public audiences, provided they did not unduly emphasize their backgrounds. Moreover, during the war, the United States committed itself to defeating European Jewry's persecutors and fostering equality at home, where interfaith initiatives proliferated. Finally, the massive projects – economic recovery, World War II, and the Cold War – that structured American politics in the mid-twentieth century tended to favor skill and efficiency over religiously tinged character traits in any realm of endeavor that could conceivably advance the nation's cause. Far from requiring a total overhaul, in the eyes of many young Jewish scholars, the United States primarily needed to preserve its existing institutions against right-wing forces within. Having witnessed, at a particularly impressionable moment, the political and human wreckage left by the sharp ideological warfare of the 1930s, these thinkers hoped to conserve the good in American political culture.[21]

Sketch," *Journal of Humanistic Psychology* 36, no. 2 (Spring 1996): 8–16; Walter W. Taylor et al., eds., *Culture and Life: Essays in Memory of Clyde Kluckhohn* (Carbondale: Southern Illinois University Press, 1973); Arthur M. Schlesinger Jr., *A Life in the Twentieth Century: Innocent Beginnings, 1917–1950* (Boston: Houghton Mifflin, 2000). At least one Jewish scholar, the psychologist Abraham Maslow, seems to fit in the latter category as well: Edward Hoffman, *The Right to Be Human: A Biography of Abraham Maslow* (New York: McGraw-Hill, 1999).
[20] Susanne Klingenstein, *Jews in the American Academy, 1900–1940: The Dynamics of Intellectual Assimilation* (New Haven: Yale University Press, 1991); David A. Hollinger, "Jewish Intellectuals and the De-Christianization of American Public Culture in the Twentieth Century," in *Science, Jews, and Secular Culture: Studies in Mid-Twentieth-Century American Intellectual History* (Princeton: Princeton University Press, 1996): 17–41.
[21] Sam B. Girgus, "The New Covenant: The Jews and the Myth of America," in *The American Self*, ed. Girgus (Albuquerque: University of New Mexico Press, 1981): 105–123.

Above all, this meant the civil liberties that carved out a protected space for religious and political dissent in an overwhelmingly Christian and seemingly quite conservative polity. To be sure, the core institutions of the New Deal seem to have felt less fragile to many members of the 1930s generation than to their elders. Although the historian Richard Hofstadter, the sociologist Daniel Bell, and the political scientist Louis Hartz worried about the threat from the economic right, most Jewish thinkers of that era, especially – including Merton, the philosopher Ernest Nagel, the sociologists David Riesman and Edward Shils, and the political scientist Gabriel Almond – concerned themselves with intellectual freedom rather than economic cooperation. Hofstadter famously engaged the issue of intellectual freedom as well. It was less *laissez-faire* sentiment that worried these figures than the prospect of an openly reactionary, antidemocratic right. Many in this generation had embraced socialism during the Popular Front years. Even after they shifted toward liberalism, they often joined émigré Marxists such as Adorno in embracing the theory of "social fascism" – the idea that Nazism had stemmed from a decision by corporate titans to mobilize the public's latent xenophobia against parliamentary procedures and civil freedoms when these produced anti-business policies. The 1930s generation's seemingly uncritical embrace of existing institutions often reflected a pervasive fear that tinkering with the delicate balance of social power would lead to a revitalized and potentially fascist right-wing populism of the kind exemplified by both Adolf Hitler and Joseph McCarthy.[22]

EXPRESSIONS

The areas of commonality and difference between these three cohorts profoundly shaped the human sciences after World War II. In some instances, commitments shared by all three groups became dominant across most or all of the core disciplines. At other times, one or two cohorts poured their energies into intellectual projects that reflected their distinct sensibilities. These postwar developments had one element in common: in virtually every instance, they signaled a turn away from the characteristic desire of interwar scientific

---

[22] Brown, *Richard Hofstadter*; Howard Brick, *Daniel Bell and the Decline of Intellectual Radicalism: Social Theory and Political Reconciliation in the 1940s* (Madison: University of Wisconsin Press, 1986); Mark Hulliung, ed., *The American Liberal Tradition Reconsidered: The Contested Legacy of Louis Hartz* (Lawrence: University Press of Kansas, 2010); Robert K. Merton, "A Life of Learning," in *Sociological Visions*, ed. Kai Erikson (Lanham, MD: Rowman & Littlefield, 1997): 275–295; Sidney Morgenbesser et al., eds., *Philosophy, Science, and Method: Essays in Honor of Ernest Nagel* (New York: St. Martin's, 1969); Irving Louis Horowitz, "Reflections on Riesman," *American Sociologist* 3, supp. (Summer 2002): 118–122; Michael A. Baer et al., eds., *Political Science in America: Oral Histories of a Discipline* (Lexington: University Press of Kentucky, 1991); Edward Shils, *A Fragment of a Sociological Autobiography: The History of My Pursuit of a Few Ideas* (New Brunswick, NJ: Transaction, 2006). Classic expressions of this postwar worry about a resurgent populist right include Daniel Bell, ed., *The New American Right* (New York: Criterion Books, 1955); and Richard Hofstadter, *Anti-Intellectualism in American Life* (New York: Knopf, 1963).

democrats to directly and substantially alter American culture through acts of publicly engaged scholarship.

For example, all of these cohorts could align themselves fairly easily with the new, rights-based form of American liberalism that emerged after the late 1930s, although that discourse's center of gravity lay well outside the human sciences. Most Progressives had flatly rejected the language of rights, which they identified with claims for the absolute sanctity of private property and freedom of contract. In the 1920s, many of these thinkers took up the new call for "civil liberties," seeking to create a free marketplace of ideas in which religious diversity could flourish and social-scientific concepts could spread widely. It was easy enough, however, to justify these liberties on thick, utilitarian grounds, as John Stuart Mill had done. One can best promote the needs of others, the argument ran, by leaving them entirely free in certain key areas of behavior. Yet interwar post-Progressives balked at the concept of defending economic rights in this manner. But as the Nazi atrocities came to light, American liberals tended to pivot toward a broader defense of "human rights." The apparent capacity of utilitarian frameworks to sanction the outright murder of the "unfit" gave a tremendous push to this alternative political language. Simultaneously, domestic political struggles encouraged liberals to reframe their program in the terms of individual rights rather than mutual obligations or the common good, in order to capitalize on that language's powerful hold over the public. The wartime New Dealers redefined even widespread material prosperity as individual "freedom from want," in President Roosevelt's famous formulation. Thus, both domestic and international developments pushed many liberals toward a rights-based framework in the World War II years.[23]

With their shared support for social diversity, civil liberties, and economic regulation, all three of the cohorts of human scientists could support the language of rights. As it turned out, however, political rights played a fairly minor role in the conceptions of society that they developed. Rights-based liberalism fit better with the historical orientation of the Western civilization discourse employed by humanists. Many defenders portrayed the Western tradition as the source of both political rights and, standing behind these rights, an ethical conception of the person as possessing intrinsic dignity and worth. Still, the rights-based approach also comported easily enough with the human scientists' postulate of a liberal consensus embodied in a set of political freedoms that Americans had embraced as the very ground of their collective being since the colonial period.

A liberalism grounded in individual rights represented a fundamental break with key tenets of interwar scientific democracy and left a straitened role for

---

[23] David M. Rabban, "Free Speech in Progressive Social Thought," *Texas Law Review* 74 (1996): 951–1038; Samuel Walker, *In Defense of American Liberties: A History of the ACLU* (New York: Oxford University Press, 1990); David Plotke, *Building a Democratic Political Order: Reshaping American Liberalism in the 1930s and 1940s* (New York: Cambridge University Press, 1996). Arthur M. Schlesinger Jr. used an historicist definition of human rights to square an emerging, individualistic liberalism with the older, utilitarian emphasis: *The Vital Center: The Politics of Freedom* (Boston: Houghton Mifflin, 1949), 8.

the public. Those who employed the new language relocated to the regime
of rights many of the tasks assigned to citizens by interwar thinkers. In this
understanding, citizens fulfilled one another's needs – provided the system
worked properly – by simply refraining from violating the rights of others,
not by cultivating and acting on a positive commitment to their flourishing.
Science, widely conceived by interwar post-Progressives as a tool for enabling
citizens to understand the consequences of their actions for others, occupied
a substantially diminished space in the new rights-oriented liberalism. As it
had for William Graham Sumner back in the 1870s, science functioned under
a regime of rights primarily to offer empirical proof of the social benefits of
faithful adherence to that regime.[24]

Even when the new language of rights was employed largely as window
dressing for a post-Progressive outlook, however, it pointed to a larger tendency among postwar liberals to assert that what was most fundamentally
good in the American project already existed, having been handed down from
the past, and did not need to be created through the voluntary acts of informed
and ethically committed individuals in the present. Whether through ingenuity or luck, it now seemed, Americans had long since solved the basic problem
of reconciling individual action with social consequences. This left only the
task of refraining from questioning the rights or other structures that connected actions to their effects. From this standpoint, any substantive change to
American institutions could only be for the worse.

Versions of consensus liberalism centered on broad political values rather
than specific political rights circulated much more widely, especially among the
non-émigré human scientists. Many members of these cohorts converged on the
idea that a liberal tradition kept their fellow citizens from deviating substantially from the mainstream. But the two groups framed that liberal tradition
differently. Margaret Mead and others of her generation confidently proclaimed
that Americans shared a dense web of foundational normative commitments
and were far better than their stated opinions, needing only to be released from
the sway of conventional beliefs. These figures were genuinely impressed by the
capacity of the American political system to weather the storms of depression
and war and to come out of them with an unprecedented burst of economic
growth to boot. They were also proud of their nation's expanding commitment
to spreading democracy around the world. At the same time, many believed
that core American values would continue to gradually remake the economy
along thoroughly collectivistic lines, if liberals could only protect the gains
of the 1930s against the *laissez-faire* diehards. By contrast, members of the
1930s generation leaned toward much thinner formulations of the liberal tradition. The Jewish thinkers in the group, especially, could hardly follow their
Protestant counterparts in assuming that all Americans shared their values.
Many of them denied that a thick national culture existed or could even be

[24] For a different take on the turn to rights, see David Ciepley, *Liberalism in the Shadow of
Totalitarianism* (Cambridge, MA: Harvard University Press, 2006).

created in the United States. Merton, adopting the most extreme version of this skepticism, identified the content of the liberal tradition as little more than an agreement to preserve institutional differences.

But the idea of a liberal consensus, however thin, proved highly attractive to postwar social scientists. For one thing, consensus liberalism bridged epistemological divides, as it had in the late 1930s. It allowed social scientists to see themselves as promoting democracy and other universal values rather than shrinking from the ethical struggles of the day. Thus, Gabriel Almond enthusiastically seconded the Christian political theorist John Hallowell's attack on the "scientificism" of most scholarly liberalism. Although advocates of scholarly neutrality deemed it "the only genuine scientific manliness," Almond asserted, that stance actually shirked "the scientist's responsibility as a man." But consensus liberalism enabled Almond to fulfill that ethical duty while still describing scientific knowledge itself as nonnormative. "Science cannot create values," he contended, but these emerged naturally from "the needs and aspirations of the people," and scientists could "demonstrate how and to what extent alternative public policies contribute to the realization of public values and aspirations." A coherent and consensual framework of shared values held society together, in Almond's view, and social scientists could both discover that framework and demonstrate how to implement it under current conditions. Farther along the epistemological spectrum, even George Lundberg, as fierce a critic of normative orientations as the mid-twentieth century produced, used the consensus assumption to link empirical and logical reasoning to ethical values. Conversely, other scholars could adopt that approach and describe their research as thoroughly normative, openly committed to the public good.[25]

Consensus liberalism also offered important rhetorical benefits in the age of McCarthy. Few social scientists possessed even the physical scientists' limited resources for parrying attacks from the right. In fact, the Ford Foundation and other philanthropies endured multiple Congressional investigations in the early 1950s that led them to scale back their funding for the social sciences. In this climate, the postulation of a liberal consensus enabled scientific scholars to position themselves as defenders of the public good without calling for systematic change or even violating the imperative of political nonpartisanship. For example, scholars could set themselves firmly against racism and religious prejudice by arguing, after the fashion of Gunnar Myrdal, that these phenomena violated age-old American ideals embodied in the nation's basic structure.[26]

In a similar fashion, consensus liberalism helped postwar social scientists deal with the sometimes uncomfortable new patterns of government funding

[25] Gabriel A. Almond, "Politics, Science, and Ethics," *American Political Science Review* 40, no. 2 (April 1946), 285, 293; George Lundberg, *Can Science Save Us?* (New York: Longmans, Green, 1947).

[26] Mark Solovey, "The Politics of Intellectual Identity and American Social Science, 1945–1970" (PhD dissertation, University of Wisconsin-Madison, 1996), 105–166; Gunnar Myrdal, *An American Dilemma: The Negro Problem and Modern Democracy* (New York: Harper & Row, 1944).

and even government employment by enabling them to believe that they could drive social progress without adopting an explicitly critical stance or engaging a public audience at all. It allowed them to see themselves as speaking for the public rather than operating as mere adjuncts of the state, even if the latter paid their bills. Consensus liberalism identified the scientific scholar as a faithful servant of the public, not a mercenary employed by political elites. It allowed critical social scientists to square their progressive views with their support for the current institutional order, by suggesting that scientific work would automatically foster social change in the long run.

Finally, at a time when business leaders and market enthusiasts worked furiously to roll back key elements of Roosevelt's program, consensus liberalism portrayed conservatism as a product of cultural maladjustment or even a symptom of psychological illness rather than a principled form of dissent. Conversely, it identified the New Deal as a natural outgrowth of America's liberal tradition, an application of long-approved values to a new economic environment rather than a radical departure from prevailing practices. Instead of trumpeting a grand historical transition from competition to cooperation, postwar theorists of consensus liberalism worked to convince Americans that their own deepest values dictated the defense and extension of New Deal policies. By telling Americans that their history and culture implicitly supported the reforms of the 1930s, the consensus approach celebrated the cultural power of liberalism rather than lamenting the cultural power of conservatism. In fact, it led social scientists to believe that the American political tradition dictated even more progressive policies than those currently in place. Meanwhile, consensus liberalism gave implicitly critical scholarship an affirmative tone that was well suited to the task of sustaining national morale during a time of crisis.[27]

Yet the body of work from the early 1950s called "consensus history," which featured the era's most explicit articulations of consensus liberalism, revealed the tensions as well as the commonalities between the non-émigré cohorts. These writings also marked the distance that the culturally Protestant thinkers, especially, had traveled since the days of Mead's *And Keep Your Powder Dry*. Two competing images of the consensus appeared in postwar histories. A celebratory version, offered by the centrists Daniel Boorstin and David Potter and the liberal Arthur M. Schlesinger Jr., identified economic prosperity as the backdrop to Americans' laudable embrace of pragmatic realism rather than romantic or utopian political views. More critical were Hartz and Hofstadter, former socialists who identified the consensus as a stultifying climate of support for individual freedom and private enterprise that squelched calls for substantive reform, even as it kept the ideological temperature of American politics comfortably low. While carrying forward the basic framework of

[27] Of course, the consensus approach also portrayed radical critics of postwar liberalism, like its conservative critics, as a tiny minority with views utterly alien to American tradition.

consensus liberalism, both of these postwar approaches departed substantially from Mead's formulation.[28]

So, too, did a third view of the liberal consensus that appeared in the 1950s. Embedded in the theories of Talcott Parsons and many other self-identified behavioral scientists, rather than explicitly articulated as a narrative of American history, this approach held that Americans harbored an egalitarian bent rather than the competitive drive that Mead had discerned. They embraced universalism rather than individualism, cosmopolitanism rather than Puritanism, and tolerance rather than moralism. This version of the consensus account identified Americans as open-minded citizens of the world, not the inhabitants of a distinctive and even parochial culture.[29]

Open-mindedness also keyed a body of postwar work exploring the relationship between democratic citizenship and psychological health. Here, émigrés versed in Freudianism reinforced the personality-centered theory of cultural lag offered by Mead and other culturally Protestant theorists. In the best-known example, the German sociologist Theodor Adorno and three collaborators implied in *The Authoritarian Personality* (1950) that the existing institutions of democracy stood in a precarious position because few Americans possessed the character traits required for responsible political behavior. "Conventionality, rigidity, repressive denial, and the ensuing break-through of one's weakness, fear and dependency" led to totalitarianism, in the authors' view, by producing "a desperate clinging to what appears to be strong and a disdainful rejection of whatever is relegated to the bottom." By contrast, democratic individuals had fully internalized "religious and social values" and sustained "affectionate, basically equalitarian, and permissive interpersonal relationships." Suggesting that scientific thinkers should work actively to alter childrearing practices, this psychologized analysis came closer to interwar scientific democracy than did most postwar theories. It suggested that scientific thinkers needed to help

[28] Daniel J. Boorstin, *The Genius of American Politics* (Chicago: University of Chicago Press, 1953); David M. Potter, *People of Plenty: Economic Abundance and the American Character* (Chicago: University of Chicago Press, 1954); Schlesinger, *The Vital Center*; Louis Hartz, *The Liberal Tradition in America: An Interpretation of American Political Thought Since the Revolution* (New York: Harcourt, Brace, 1955); Richard Hofstadter, *The American Political Tradition and the Men Who Made It* (New York: Vintage, 1948). Hofstadter's account was sufficiently critical that he worried about a negative public response: Novick, *That Noble Dream*, 323. On these studies, see also Philip Gleason, "The Study of American Culture," in *Speaking of Diversity: Language and Ethnicity in Twentieth-Century America* (Baltimore: Johns Hopkins University Press, 1992), 188–206; Nils Gilman, *Mandarins of the Future: Modernization Theory in Cold War America* (Baltimore: Johns Hopkins University Press, 2003), 62–68; and Wendy Wall, *Inventing the "American Way": The Politics of Consensus from the New Deal to the Civil Rights Movement* (New York: Oxford University Press, 2008), 87–100. The consensus approach had its challengers even in the 1950s, however: Ellen F. Fitzpatrick, *History's Memory: Writing America's Past, 1880–1980* (Cambridge, MA: Harvard University Press, 2002), 188–238.

[29] E.g., Talcott Parsons, *The Social System* (Glencoe, IL: Free Press, 1951).

create a democratic culture by demonstrating to parents how to form the personalities that constituted it.[30]

A related body of psychological research tied personality traits to scientific progress as well as to political stability. Milton Rokeach and many other postwar psychologists identified the autonomy and creativity of the working researcher as the core virtues of the democratic citizen as well. This view of the knowing process harmonized reasonably well with Merton, Conant, and Oppenheimer's understanding of science as a communal practice in which interpersonal commitments shared by tightly knit groups of inquirers enabled the production of new knowledge. But the psychologists' model did not include the genuine, generative originality that Conant and Oppenheimer built into their conceptions of the scientist as artist. Facing a deficit of authority, social scientists could hardly afford to describe their theories as equivalent to works of art. They instead made the link to creativity through open-mindedness. Only those able to buck the trend of dominant thinking, they argued, could discern the truth, whether scientific or political. All others remained caught in the web of prevailing cultural assumptions, clinging to false beliefs rather than heeding the evidence in front of them. Thus, while personality researchers adopted a thinner conception of creativity than did Conant and Oppenheimer, they too assumed that the survival of democracy depended on the creation of knowledge, knowers, and perhaps audiences that were constitutionally immune to political conformity. For these psychologists, too, the interpersonal sphere of culture represented a threat – to truth, and thus to social order – rather than a resource, and science served the polity by enabling individuals to resist cultural influences and think autonomously.[31]

Other thinkers of the 1930s generation joined analytically oriented émigrés in developing an even more individualistic model of cognitive behavior in the cluster of highly mathematical fields – operations research, cybernetics, game theory, and rational choice theory – developed under the auspices of the RAND Corporation in the 1950s. A joint venture launched by the Air Force, RAND lavishly funded research at the nexus of social science, natural science, and mathematics. In this fertile, interdisciplinary soil, theorists systematized and extended the classic image of a rational, relentlessly goal-directed *homo economicus* to cover a vast range of human interactions, most involving stakes other than money. They applied the model of a purely strategic individual to actors ranging from nations locked in a military standoff to voters and candidates attempting to maximize their gains. These RAND-sponsored fields promised policymakers and leaders of organizations new means of predicting

---

[30] Theodor Adorno et al., *The Authoritarian Personality* (New York: Harper, 1950), 971. Cf. Richard Hofstadter, *The Paranoid Style in American Politics* (New York: Knopf, 1965).

[31] Jamie Cohen-Cole, "The Creative American: Cold War Salons, Social Science, and the Cure for Modern Society," *Isis* 100, no. 2 (June 2009): 219–262; Cohen-Cole, *The Open Mind: Cold War Politics and the Sciences of Human Nature* (Chicago: University of Chicago Press, forthcoming).

and controlling human behavior, based on the presumption that individuals relentlessly pursued their self-interest and were essentially devoid of irrational, interpersonal commitments.[32]

Margaret Mead, a familiar public presence – a visual presence, especially – in the 1950s, put a reassuringly maternal face on key sectors of this body of work. After the war, she and Gregory Bateson reframed their anthropological concepts in terms of physics and biology, having latched onto the cybernetics of the MIT electrical engineer Norbert Wiener and the systems theory of the Austrian biologist Ludwig von Bertalanffy and the British-born economist Kenneth E. Boulding. Mead domesticated the starkly rationalistic and individualistic "Manichean sciences" associated with RAND by integrating them into an expansive scientific humanism that featured strong Christian overtones and an optimistic reading of historical change. Appearing frequently in the popular media, Mead's writings and her very persona identified science as a source of limitless and relatively painless progress along the road paved by traditional Western values.[33]

Not all social scientists claimed the rigorous objectivity of the RAND crowd. Because so many value-neutralists viewed themselves as spokespersons for a normative consensus, frank expressions of political disengagement by social scientists still proved controversial at times, especially before McCarthy's hammer fell. George A. Lundberg's *Can Science Save Us?* (1948), a forthright defense of value-neutral operationalism written in a hard-nosed debunking tone, generated a welter of outraged responses from other scientific thinkers. In fact, none of the major social science journals gave Lundberg's book an entirely favorable review. Only the *New York Times'* Waldemar Kaempffert, then touting the social sciences' scientific credentials in the context of the NSF struggle, eschewed criticism of the book altogether. Moreover, Lundberg himself employed the framework of consensus liberalism to square his emphasis

[32] Philip Mirowski, *Machine Dreams: Economics Becomes a Cyborg Science* (New York: Cambridge University Press, 2002); S. M. Amadae, *Rationalizing Capitalist Democracy: The Cold War Origins of Rational Choice Liberalism* (Chicago: University of Chicago Press, 2003); Hunter Crowther-Heyck, *Herbert A. Simon: The Bounds of Reason in Modern America* (Baltimore: Johns Hopkins University Press, 2005); Robert Leonard, *Von Neumann, Morgenstern, and the Creation of Game Theory: From Chess to Social Science, 1900–1960* (New York: Cambridge University Press, 2010), 293–343.
[33] Margaret Mead, *People and Places* (Cleveland: World, 1959); Mary Catherine Bateson, *With a Daughter's Eye: A Memoir of Margaret Mead and Gregory Bateson* (New York: HarperPerennial, 1994); Deborah Ruth Hammond, "Toward a Science of Synthesis: The Heritage of General Systems Theory" (PhD dissertation, University of California-Berkeley, 1997); Nancy Lutkehaus, *Margaret Mead: The Making of an American Icon* (Princeton: Princeton University Press, 2008). Unlike the combative Mills or the meditative Riesman, Mead did not pen a bestseller in the 1950s, though she did write a children's book. But Mead was a public presence in a way that they were not. At the same time, however, Mead became increasingly marginal to the postwar academic establishment: Peter Mandler, "One World, Many Cultures: Margaret Mead and the Limits to Cold War Anthropology," *History Workshop Journal* 68 (Autumn 2009): 149–172. "Manichean sciences" is Peter Galison's term: "The Ontology of the Enemy: Norbert Wiener and the Cybernetic Vision," *Critical Inquiry* 21, no. 1 (Autumn 1994): 228–266.

on the personal disinterestedness of the researcher with the claim that science fueled moral progress in the wider society.[34]

Even after McCarthyism chilled political speech on the left, postwar human scientists vigorously debated epistemological questions. The question of science and values generated sufficient heat to fill a 342-page bibliography of recent writings on the topic by 1959. As in philosophy, however, this epistemological discourse mostly appeared in highly specialized forums, disconnected from actual attempts to engage normative concerns and public audiences. Its products barely registered outside the academy, as explicitly contextual views of knowledge and the related projects of cultural change languished in the early Cold War years. Even the staunchest opponents of value-neutrality tended to take the existing disciplinary matrix for granted and to pursue their epistemological agendas in the company of specialists, on a high theoretical plane.[35]

The glaring exception to this generalization is C. Wright Mills' 1959 blast *The Sociological Imagination*. A Texan ex-Catholic who was inspired by Deweyan instrumentalism and the sociology of knowledge, the Columbia sociologist worried about individual conformity as an aid to the economic right rather than would-be dictators. He thus stood apart from the main cohorts shaping the postwar social sciences. Mills famously excoriated both quantifiers and grand theorists in those fields for the political sterility of their work. Yet he, too, followed many of the era's leading academic trends, even as he adopted the role of public intellectual in books such as *White Collar* and *The Power Elite*. Most importantly, Mills equated science with a fatuous pursuit of value-neutrality. Although he harbored expansive hopes for the democratic potential of intersubjective, empirical reasoning, Mills felt compelled to distance himself from the term "social science," rather than advocate a broader meaning. Sociology, in his portrayal, represented a species of humanistic interpretation rather than a scientific discipline.[36]

[34] George Lundberg, *Can Science Save Us?* (New York: Longmans, Green, 1947); Frank H. Hankins, review of Lundberg, *Can Science Save Us? Southern Economic Journal* 14, no. 1 (July 1947): 73–74; James W. Woodard, review of Lundberg, *Can Science Save Us? Annals of the American Academy of Political and Social Science* 252 (July 1947): 158–159; Oscar Lewis, "The Problem of Our Times," *Scientific Monthly* 65, no. 1 (July 1947): 84–85; J. O. Hertzler, review of Lundberg, *Can Science Save Us? Social Forces* 26, no. 1 (October 1947): 100–101; Frank H. Knight, "Salvation by Science: The Gospel According to Professor Lundberg," *Journal of Political Economy* 55, no. 6 (December 1947): 537–552; Ronald Freedman, review of Lundberg, *Can Science Save Us? American Sociological Review* 13, no. 2 (April 1948): 226–228; Howard Becker, review of Lundberg, *Can Science Save Us? American Journal of Sociology* 54, no. 2 (September 1948): 170–171; Elgin Williams, "Can We Save Science?" *Philosophy of Science* 15, no. 4 (October 1948): 333–341; B. F. Skinner, "Science of Society," *New York Times* (November 14, 1948): BR20, BR22; Waldemar Kaempffert, "A New Role for the Social Sciences," *New York Times* (June 8, 1947): BR28.
[35] Ethel M. Albert and Clyde Kluckhohn, *A Selected Bibliography on Values, Ethics, and Esthetics* (Glencoe, IL: Free Press, 1959).
[36] C. Wright Mills, *The Sociological Imagination* (New York: Oxford University Press, 1959); Daniel Geary, *Radical Ambition: C. Wright Mills, the Left, and American Social Thought* (Berkeley: University of California Press, 2009), 143–178. Before Mills published his attack,

Most members of the 1930s generation and analytically minded émigrés advanced the claims of value-neutrality that Mills decried. The experience of ideological conflict propelled these figures toward a purely technical conception of scientific inquiry, even when they viewed it as a means of reliably translating public values into concrete actions. Having seen the intersubjective sphere of culture serve as a site of sharp contest rather than a powerful integrative force, they tended to view themselves as cutting through it with precise analytical tools rather than improving it through the slow, organic force of mutual enlightenment. Even those, like Merton, who deemed scientific knowledge a collective rather than individual product argued that science could reconcile the competing claims of human agents because it revealed or reflected the structure of reality, not because it aligned wills, habits, and consequences in a thick, discursive culture connecting democratic citizens. These émigrés and young American scholars found a neutral, managerial conception of expertise personally comfortable and politically promising. Their fear of the political danger posed by intense, socialized emotions made attractive an image of science – of the scientific method as well as the scientific community – that conceptually isolated it from such psychological forces. Such forces could be allowed to operate in science only at the local level of the subdisciplinary tradition, to which Merton, no less than Conant, banished them. In this view, scientists were socialized, in the deep sense, only to the norms of the guild.[37]

A number of young sociologists of science joined Merton in ratifying Conant and Oppenheimer's emphasis on scientific autonomy. In his postwar writings, Merton dropped the argument that science flourished only in a democracy and focused exclusively on theorizing the internal workings of that unique and precious subsystem. Merton's student Bernard Barber offered a more accessible version of the analysis in the 1952 book *Science and the Social Order*. Edward Shils also took up the task of defining science as a communal enterprise. His take on modern society was hardly uncritical. In fact, Shils, though a fervent anti-communist, deemed the era's mission-oriented and largely secret "big science" a potent internal threat to the "open society" of the West. But Shils did not describe scientists as openly and actively engaged in building a free culture. With Merton and Barber, he instead echoed the physical scientists' twin emphases on the communal character of scientific knowing and the corresponding importance of professional autonomy. These theorists described the scientific community as potentially an unparalleled servant of human values, but only if

the most visible and controversial contribution to postwar epistemological debates was Pitirim Sorokin's *Fads and Foibles in Modern Sociology and Related Sciences* (Chicago: H. Regnery, 1956).

[37] For overviews of postwar intellectual shifts, see Thomas Bender and Carl Schorske, eds., *American Academic Culture in Transformation: Fifty Years, Four Disciplines* (Princeton: Princeton University Press, 1998) and Roger E. Backhouse and Philippe Fontaine, eds., *The History of the Social Sciences Since 1945* (New York: Cambridge University Press, 2010). Cf. Thomas L. Akehurst, *The Cultural Politics of Analytic Philosophy: Britishness and the Spectre of Europe* (London: Continuum, 2010).

it were insulated from the influence of those values and left to its own devices. Like Rokeach's open-minded individuals, in other words, scientists served the public good by eschewing the presuppositions of their culture and reasoning entirely from evidence and logic.[38]

As in the natural sciences, these images of near-absolute disciplinary autonomy accompanied a steadily expanding web of ties between scholars, government agencies, foundations, and private contractors. The self-consciously empiricist "behavioral science" orientation, advanced primarily by members of the 1930s generation, flourished at the point of convergence between those epistemological and institutional developments. Despite its similar name, postwar behavioralism departed from John B. Watson's behaviorism. That earlier approach had entailed a sharp interpretive stricture prohibiting the inclusion of immaterial entities, such as minds, in the explanation of behavior. By contrast, the behavioralists of the 1950s and early 1960s followed the looser methodological imperative of interwar neobehaviorists such as Edward C. Tolman: to rely solely on sensory evidence. Like Tolman, in fact, the postwar behavioralists denied that the observable patterns of human action could be explained without postulating minds, values, and purposes. By adopting an empiricist methodology, however, behavioral scientists claimed the ability to study value-laden phenomena – social structures, cultural influences, and personality development, for example – in a disinterested fashion.[39]

A profoundly anti-populist strain ran through much behavioralist work, especially in political science. Intimately familiar with the public's views, thanks to advances in polling techniques, many postwar political scientists expressed profound doubt about the average citizen's capacity to take an active political role, even through the limited avenue of informed voting. These figures viewed the intersubjective realm of public opinion as innately dangerous rather than reassuringly effective or even simply promising. Alexis de Tocqueville's analysis of "democratic despotism" gained new adherents in the 1950s, as scholars, finding the horizontal, person-to-person influence of citizens working directly against what they took to be the public good, responded by seeking to inoculate American institutions against the influence of public opinion.[40]

Fortunately, an emerging "pluralist" vision of democracy taught that the public hardly influenced political decisions, which were controlled by small

[38] Robert K. Merton, *Social Theory and Social Structure* (Glencoe: Free Press, 1957); Bernard Barber, *Science and the Social Order* (Glencoe: Free Press, 1952); Edward Shils, *The Torment of Secrecy: The Background and Consequences of American Security Policies* (Glencoe: Free Press, 1956); David A. Hollinger, "Free Enterprise and Free Inquiry: The Emergence of Laissez-Faire Communitarianism in the Ideology of Science in the United States," in *Science, Jews, and Secular Culture*: 97–120. On polling, see Sarah E. Igo, *The Averaged American: Surveys, Citizens, and the Making of a Mass Public* (Cambridge, MA: Harvard University Press, 2007).

[39] A classic statement of this approach is David Easton, "Traditional and Behavioral Research in Political Science," *Administrative Science Quarterly* 2, no. 1 (June 1957): 110–115.

[40] Matthew Mancini, *Alexis de Tocqueville and American Intellectuals: From His Times to Ours* (Lanham, MD: Rowman and Littlefield, 2006).

groups of insiders who negotiated policies on behalf of pressure groups and power blocs. Two Protestant members of the 1930s generation, Yale's Robert A. Dahl and the Canadian-born David Easton, of the University of Chicago, led the way in developing this model. Picking up themes from interwar explorations by Harold Lasswell and the neglected Progressive Era work of Arthur F. Bentley, they redefined a democracy as a polity ruled by several competing interest groups of roughly equal power. Dahl and Easton viewed the intrusion of emotionally charged ideologies into the public arena as a grave danger to the negotiated balance of organized interest groups. They looked to these groups, rather than the public itself, to promote the interests of the people. To the extent that political scientists in this vein engaged the public in their own writings, they worked primarily to instill trust in existing processes and the experts who kept them running smoothly.[41]

Political scientists, however, could not get rid of democratic legitimacy entirely. Led by Almond, a student of comparative politics, pluralists developed the concept of "political culture." Like the narrower ideal of a liberal consensus – said to constitute the distinctive political culture of the United States – this concept complemented the wartime "national character" approach developed by Margaret Mead and Ruth Benedict. In Almond's definition, a nation's political culture provided its citizens' understanding of the proper scope and purpose of its public institutions. Processes of socialization, education, and communication replicated this understanding through the generations, as with all elements of culture. Most users of the concept further held that a truly modern political culture identified government's role as that of promoting secular, human values and understood that legitimate political debate meant disputes over the adequacy of various proposed means to this end. A modern political culture sustained the practice of rational negotiation between individuals or interest groups, rather than dominance by a single faction or a bitter clash between irrational, ideological programs. Theorists in this vein rejected the interwar emphasis on public opinion and tasked political culture with keeping citizens out of the work of political elites.[42]

In sociology, meanwhile, scholars from the 1930s generation implicitly denied that a broad, consensual framework of values alone could guarantee

[41] Gilman, *Mandarins of the Future*, 47–56; Ciepley, *Liberalism in the Shadow of Totalitarianism*, 206–212; John G. Gunnell, "Making Democracy Safe for the World: Political Science Between the Wars," in *Modern Political Science*, ed. Robert Adcock et al., 137–157. The consummate statement is Robert A. Dahl, *Who Governs? Democracy and Power in An American City* (New Haven: Yale University Press, 1961). On the critical edge of Dahl's thought, see Tom Hoffman, "The Quiet Desperation of Robert Dahl's (Quiet) Radicalism," *Critical Review* 15, nos. 1–2 (December 2003): 87–122; cf. Tracy B. Strong, "David Easton: Reflections on an American Scholar," *Political Theory* 26, no. 3 (June 1998): 267–280. Lasswell's own postwar work aimed at developing autonomous "policy sciences" to serve a pluralist regime: Daniel Lerner and Lasswell, eds., *The Policy Sciences* (Stanford: Stanford University Press, 1951).

[42] Gabriel Almond, "Comparative Political Systems," *Journal of Politics* 18, no. 3 (1956): 391–409; Gilman, *Mandarins of the Future*, 47–56.

that individual behavior conduced to the good of all. They also looked to a set of more specific, formal mechanisms that channeled behavior into certain pathways. In a revealing linguistic shift, these sociologists spoke of social "structures" rather than "institutions." This usage emphasized the relative solidity, rather than the conventional character, of the social patterns that mediated between individual behavior and its social consequences and thus obviated the need for individuals to think constantly about the effects of each of their actions on others. At the same time, it implied that a highly organized and differentiated system of formalized practices, political rights, and professional roles reliably translated the core values of the liberal consensus into the specific behaviors most conducive to the general welfare under current conditions. These sociologists looked to what the philosopher John Rawls would later call the "basic structure" of a society, rather than to a culture comprised of knowledgeable and sympathetic individuals, to harmonize formally voluntary action with mutually beneficial consequences. And if the existing structures represented the path to the good, it stood to reason that scholars should teach citizens to defend and even defer to them – to play their appointed roles – rather than equip them to criticize any and all such constructs from the standpoint of human needs. Experts armed with technical knowledge could fine-tune American institutions more reliably than could ordinary citizens.[43]

A widely shared portrait of the basic structure of a modern society emerged among sociologists by the mid-1950s. Conant and Oppenheimer had ignored virtually everything except for the professions in thinking about the structure of society. But social scientists needed to find room alongside these bodies of vocational specialists for culture, personality formation, families, the state, and other features of collective existence. In doing so, they extrapolated from the central structures of the post-New Deal United States to an ideal-typical characterization of a modern society. Such a society, in the emerging portrait, featured a system of large-scale industrial production; a career structure that aligned neatly with the distribution of talents across the human species and thus allowed all individuals to find positions (including political offices) suited to their innate abilities; and a political system sufficiently decentralized and fluid to allow each occupational group to pursue its interests peacefully, through settled practices of negotiation rather than ideological crusades or outright violence. Completing the picture was a deep, emotional commitment to preserving these arrangements, transmitted across the generations by the family and the school. This commitment found embodiment in both a cluster of political rights and a carefully delineated, functionally interrelated set of personal and vocational roles.[44]

[43] E.g., Bernard Barber, *Social Stratification: A Comparative Analysis of Structure and Process* (New York: Harcourt, Brace, 1957).
[44] Edward A. Purcell Jr., *The Crisis of Democratic Theory: Scientific Naturalism and the Problem of Value* (Lexington: University Press of Kentucky, 1973), 235–272; Gilman, *Mandarins of the Future*, 12–20, 47–68.

Talcott Parsons codified this image of a modern society by building it into an influential, architectonic vision of the relationships between the behavioral sciences. Chronologically, Parsons stood midway between his culturally Protestant elders and the 1930s generation. His perspectivalist conception of science, described in Chapter 9, combined elements of both outlooks. He identified scientific conclusions as the product of pre-chosen interpretive lenses, yet at the same time he portrayed these lenses, and the scientific community itself, as standing apart from the surrounding culture. In other words, Parsons, like Conant and many other Harvard thinkers, described scientific disciplines and subdisciplines as craft guilds that employed perceptual tools handed down for generations. Unlike Conant, however, he thought that even the most abstract modes of scientific work could inspire detailed, empirical research and fuel practical advances. Like other theorists of behavioralism, Parsons went forward with a tremendous sense of optimism about its social potential.[45]

Parsons sought to advance the behavioral sciences by integrating the social-structural orientation with the culture and personality approach. As he did so, he too incorporated the tenet of a liberal consensus, firmly lodged in the pre-ideological sphere of political culture. The result of Parsons' labor was a comprehensive and influential, if extraordinarily abstract, model for the value-free analysis of value-laden behavior. Writing in a dry, technical style, he produced a pair of forbidding works in 1951 that described a society as a functionally integrated system knit together by a shared normative framework – a "moral consensus," as he and his co-authors termed it. For Parsons, as for so many postwar scholars, this normative framework provided stability but did not serve as a direct source of political decisions, except in a highly indirect sense. In his portrait, the United States was not a deliberative democracy. Yet the liberal consensus underlying American social structures required them to implement the public good, and the behavioral sciences offered the needed tools to assess, fine-tune, and, where necessary, reconcile those structures.[46]

Parsons' theoretical writings inspired a raft of scholars involved in the articulation of "modernization theory." Spanning the disciplines of political science, sociology, economics, and anthropology, this interdisciplinary field, which took shape at the end of the 1950s and grew rapidly in the early 1960s, emerged from a problem-oriented focus on international development. Faced with a rapidly decolonizing world of new nations seeking rapid growth, its practitioners offered comprehensive models of development to help nonindustrial nations navigate the transition to industrial production and the associated set of "modern" institutions without falling prey to the seduction of communism. Modernization theorists sought to connect basic, theoretical work such

[45] Joel Isaac, *Working Knowledge: Making the Human Sciences from Parsons to Kuhn* (Cambridge, MA: Harvard University Press, 2012), 160–190.

[46] Gilman, *Mandarins of the Future*, 74–94 (quote on 85). Parsons remained a thinker of the left: Brick, *Transcending Capitalism*, 145–151; Keith Doubt, "The Untold Friendship of Kenneth Burke and Talcott Parsons," *Social Science Journal* 34, no. 4 (October 1997): 527–537.

as Parsons' to the challenges facing the developing world, simultaneously fulfilling a felt moral obligation and working to pull "Third World" nations into the American sphere of influence. Inclined toward relatively inexpensive, noncontroversial forms of economic and technical aid, many such thinkers drew on the concept of political culture to argue that certain highly focused forms of intervention would trigger a cultural shift and lead citizens to voluntarily choose democratic governance. Here, as in so many other realms of postwar thought and practice, American leaders proved far more willing to countenance projects for systematic cultural change abroad than similar projects at home.[47]

Modernization theory overlapped substantially with postwar thinking about American institutions. Social scientists felt comfortable addressing the question of how to build modern social structures and political cultures abroad because they believed that their studies of the United States, and to a lesser extent Europe, had accurately revealed the contours of those structures and cultures. In particular, leading modernization theorists drew on the model of the liberal consensus put forth by Parsons, which focused on a thin, universalistic set of values as the centerpiece of American political culture. They also tended to see the specific social structures and modes of action outlined by Parsons as central elements of a modern society, wherever it emerged.[48]

While modernization theorists targeted the political cultures of "non-modern" nations as strategic sites for change, they identified key groups of elites as the immediate agents of that change. Modernization theorists were even more attuned to the role of elites in governance than were students of American politics. Scholars of many different political persuasions could agree that, in the short term, vanguard elites committed to modern, scientific values would need to spearhead the modernization process. They could also agree that ordinary citizens in the modernizing nations needed above all to understand what they could expect from their governments – or rather, what they could not expect, namely emotional fulfillment, utopian outcomes, or a substantive equalization of resources.[49]

A significant counter-strain in the development discourse, however, provided evidence of a lingering commitment to a deliberative understanding of democracy among members of the older Protestant cohort. Thwarted in their efforts to promote political change at home, many thinkers now called for thoroughly

[47] Gilman, *Mandarins of the Future*; Michael E. Latham, *Modernization as Ideology: American Social Science and "Nation Building" in the Kennedy Era* (Chapel Hill: University of North Carolina Press, 2000); David C. Engerman et al., eds., *Staging Growth: Modernization, Development, and the Global Cold War* (Amherst: University of Massachusetts Press, 2003). In this context, a number of prominent anthropologists declared cultural relativism dead: e.g., Clyde Kluckhohn, "Manners and Morals: A.D. 1950," *New Republic* (June 12, 1950), 14; Kluckhohn; "Ethical Relativity: Sic et Non," *Journal of Philosophy* 52, no. 23 (November 10, 1955), 663.
[48] Brick, *Transcending Capitalism*, 145–151; Gilman, *Mandarins of the Future*, 74–94.
[49] Gilman, *Mandarins of the Future*.

bottom-up modes of economic and political development abroad. In the most striking example, a group of left-leaning scientific democrats who had worked in Henry A. Wallace's Department of Agriculture in the late 1930s redirected their attention to the developing world after conservatives forced them out of the agency in the late 1940s and early 1950s. They promoted a "community development" model that emphasized local and informal leadership, collective effort, and the preservation of shared values amid growing prosperity. Carl C. Taylor, Douglas Ensminger, M. L. Wilson, and other rural sociologists and agricultural economists applied themselves to the task of building wells, roads, and community centers rather than massive factories and hydroelectric dams. They eventually took the community development program to every village in India, as well as innumerable others around the world. Like the massive comparative projects undertaken by the anthropologists Robert Redfield at Chicago and Clyde Kluckhohn at Harvard, the community development model identified local traditions as the avenues through which individuals and groups would come to grasp universal values.[50]

Overall, though, the institutional and intellectual changes of the crucial period stretching from the mid-1930s to the early 1950s brought an expansion of state power without the sense of interpersonal and collective responsibility that interwar scientific democrats had sought to instill in their fellow citizens. Moreover, important groups of postwar thinkers declared that a widely distributed, scientifically informed sense of responsibility was impossible. Many others simply deemed it unnecessary. Although postwar human scientists followed Dewey in describing the goal of their efforts as a "free society," they did not think that creating such a society would require the cultivation of direct, personal relations of mutuality in the intersubjective realm of culture. Instead, they typically identified the source of a free society as some combination of an inherited value set and a cluster of social structures and roles, including the role of citizen, with its accompanying rights.

---

[50] Olaf F. Larson and Julie N. Zimmerman, *Sociology in Government: The Galpin-Taylor Years in the U.S. Department of Agriculture* (University Park: Pennsylvania State University Press, 2003), 266–268; Daniel Immerwahr, "Quests for Community: The United States, Community Development, and the World, 1935–1970" (PhD dissertation, University of California-Berkeley, 2011); Andrew Sartori, "Robert Redfield's Comparative Civilizations Project and the Political Imagination of Postwar America," *Positions* 6, no. 1 (Spring 1998): 33–65; Wilcox, *Robert Redfield and the Development of American Anthropology*; Nicole Sackley, "Passage to Modernity: American Social Scientists, India, and the Pursuit of Development, 1945–1961" (PhD dissertation, Princeton University, 2004); Willow Roberts Powers, "The Harvard Study of Values: Mirror for Postwar Anthropology," *Journal of the History of the Behavioral Sciences* 36, no. 1 (Winter 2000): 15–29; John S. Gilkeson, "Clyde Kluckhohn and the New Anthropology: From Culture and Personality to the Scientific Study of Values," *Pacific Studies* 32, nos. 2–3 (June-September 2009): 251–272. On the tendency for progressive energies to be redirected outward as domestic hopes evaporated, see also Gilman, *Mandarins of the Future* and Elizabeth Borgwardt, *A New Deal for the World: America's Vision for Human Rights* (Cambridge, MA: Belknap, 2005).

By comparison to earlier, post-Progressive versions of scientific democracy, this vision of a scientific society entailed a dramatic contraction of the political tasks assigned to citizens and a substantial shift in the tasks assigned to scientific scholars. The new approach portrayed a one-way, top-down movement of knowledge from scholars to the masses, rather than employing the horizontal image of scholars enmeshed in a network of two-way communicative exchanges with their fellow citizens. It also posited a sharp distinction between the motives driving everyday behavior and those fueling research in the scientific disciplines. Readers will disagree on whether to call these postwar frameworks chastened and realistic or anemic and irresponsible. There is little doubt, however, that they differed substantially from their interwar predecessors.

Conclusion

Science and Democracy in a New Century

Many of today's thorniest political questions, from climate change to bio-engineering to economic regulation, involve contests over scientific expertise and authority. Much contemporary religious conflict also turns on competing responses to the challenge of modern science. To grapple with these problems effectively, we need a better account of past interactions between science and American politics than scholars have thus far provided. This book speaks to that interpretive shortfall, while calling into question a powerful complex of ideas that has hamstrung much American thinking – particularly that of academic progressives – on the political meanings of science since the 1960s.

The roots of that complex of ideas are easy to spot. The postwar convergence of the politics of autonomy with an apparently servile scientific establishment triggered a sharp reaction against expressions of scientism in the human sciences and in political debate. By the mid-1960s, a new generation of critics viewed with alarm a series of interrelated phenomena: the increasing interpenetration of science, industry, foundations, and the state, aimed at both military superiority and consumption-driven economic growth; the vigorous claims of value-neutrality and political detachment by researchers who worked to meet the instrumental needs of the governing complex; and the image – and to some extent the reality – of democracy as a process in which experts and bureaucrats simply handed down decisions and material benefits to passive, naïvely trusting citizens. Science lay at the heart of the "system" targeted by 1960s critics, who grasped all too well the ideological and practical centrality of science to postwar American governance. The epistemological claims and political ties of the postwar scientific establishment led many on the left to conclude that a stance of self-professed detachment was hypocritical and pernicious in its social and political effects.[1]

---

[1] Revealingly, when the young radicals of the 1960s took up the ideas of postwar gadflies such as C. Wright Mills, they often reframed them as attacks on science itself, fueling phenomena such as the anti-psychiatry movement, the feminist critique of Freudianism, and a counterculture that looked upon all forms of rationality with deep suspicion. Meanwhile, many scientists of the 1960s

Since the 1960s, one of the central threads of scientific democracy, namely the impulse toward public engagement, has been resurgent in the American universities. Yet that impulse is rarely linked to scientific ambitions these days. To the extent that many 1960s-era critics had difficulty disentangling science itself from postwar scientism's extravagant claims of value-neutrality, their political critique bore on both. Little in their lived experience prepared them to draw a distinction between science and scientism, which recent history seemed to prove were inseparable. As witnesses to the merger of a putatively value-neutral science with a militarized, bureaucratized, corporate-liberal polity whose advocates deemed it the epitome of individual freedom, critics often assumed that the links between these phenomena were logical and intrinsic – constitutive of modernity itself – rather than historical and contingent.

This experiential matrix fueled the rapid uptake and development of a series of "post-positivist" orientations in the American universities. For a time, critics of modern technology such as Lewis Mumford and the French philosopher Jacques Ellul held considerable influence. Thomas S. Kuhn's *The Structure of Scientific Revolutions* (1962), which proposed that extrascientific factors accounted for fundamental shifts in scientific theorizing, also fired the imaginations of many critical scholars in the early days of post-positivism. More recently, various forms of interpretive social science, critical theory, and post-structuralism have proliferated as alternatives to the mainstream behavioral science model and its rational-choice successor. All of these post-positivist schools share a deep-seated suspicion of "scientificity," which is said to have caused dire political consequences as well as epistemological confusion.[2]

Ironically, these critiques have flourished across virtually the same range of disciplinary niches – what I have called the cultural sciences – that once gave rise to an intellectual-political project centered on science rather than opposed

actively reinforced the sense that postwar epistemologies, ideologies, and political structures went hand in hand. "What I'm designing may one day be used to kill millions of people," acknowledged an MIT graduate student asked about the moral implications of his work during a November 1969 demonstration against military-funded research on campus. "I don't care. That's not my responsibility. I'm given an interesting technological problem and I get enjoyment out of solving it." Quoted in Stuart W. Leslie, *The Cold War and American Science: The Military-Industrial-Academic Complex at MIT and Stanford* (New York: Columbia University Press, 1993), 238. On the importance of postwar institutions in shaping epistemological arguments in the 1960s, see Mark Solovey, "Project Camelot and the 1960s Epistemological Revolution: Rethinking the Politics-Patronage-Social Science Nexus," *Social Studies of Science* 31, no. 2 (April 2001): 171–206 and, more generally, Howard Brick, *Age of Contradiction: American Thought and Culture in the 1960s* (New York: Twayne, 1998).

[2] Helpful overviews of post-positivism include Quentin Skinner, *The Return of Grand Theory in the Human Sciences* (New York: Cambridge University Press, 1990); Joan Wallach Scott and Debra Keates, eds., *Schools of Thought: Twenty-Five Years of Interpretive Social Science* (Princeton: Princeton University Press, 2001); John H. Zammito, *A Nice Derangement of Epistemes: Post-Positivism in the Study of Science from Quine to Latour* (Chicago: University of Chicago Press, 2004); François Cusset, *French Theory: How Foucault, Derrida, Deleuze, & Co. Transformed the Intellectual Life of the United States* (Minneapolis: University of Minnesota Press, 2008); and Daniel T. Rodgers, *Age of Fracture* (Cambridge: Belknap, 2011).

to it.[3] The intellectual sites and social purposes that generated the cultural sciences framework remain with us, but their exponents no longer seek to connect their work to public concerns by widening the epistemological and methodological scope of the term science. Quite the opposite: they typically strive to keep scientific methods at bay in their disciplines. Since the 1960s, large groups of human scientists, fearing any dalliance with a putatively monolithic "scientific rationality," have reinvented themselves as humanists, following the trail blazed earlier by many postwar historians and philosophers. Although the hardened administrative structure of the modern university kept specific disciplines frozen in place as "natural sciences," "social sciences," or "humanities," the secession from the cultural sciences framework that began during World War II accelerated in the 1960s and 1970s.

Today, a powerful aversion to pretensions of scientific status represents one of the strongest intellectual currents not only in the humanities, but also in what are sometimes euphemistically called the "humanistic social sciences." Several decades earlier, scholars in the latter fields would have followed the cultural scientists chronicled in this book by arguing that the label science also covered hermeneutic and interpretive modes of scholarship. Now, they simply validate a narrower definition of science, effectively ceding that potent term to the natural sciences and those social scientists that claim to model their work on them. Since the 1960s, in short, leading critics in the American universities have powerfully ratified the narrow, value-neutral conception of science crafted by their postwar predecessors. In part because of that interpretive shift, science and democracy came to seem opposed rather than mutually reinforcing. Echoes of scientific democracy can still be heard in certain corners today, but in the wake of the 1960s, a new set of arguments took hold that stressed science's complicity in a wide range of social and political ills.[4]

Such an approach leaves progressive scholars and activists interpretively impoverished amid a massive resurgence of theistic modes of conservatism. It obscures our understanding of the powerful alliance of free-market and traditionalist conservatives that shapes American political debate; of the tension between the universities' growing importance as credentialing devices and generators of new technology and their increasing marginality as sources of social criticism; and of many other related phenomena. At a time when widespread fears about secularization and battles over scientific, technological, and medical questions shape party politics and policy formation in the United States, the instinctive revulsion with which many critical scholars greet the prospect of

---

[3] The exception is biology, which has become as technical, specialized, and reductive as the other natural sciences.

[4] For the many 1960s critics who sought more genuine forms of community than seemed available at the time, science stood further condemned by its association with a postwar liberalism that centered on the creative, autonomous individual, perhaps moving within social structures but never shaped by those structures, let alone by relationships with others. Of course, this ideological link was even more contingent – or at least newer – than that between science and scientism.

allowing science a prominent role in political discourse has become a hindrance to developing an adequate theory and praxis for the twenty-first century.

In important respects, of course, scientific authority continues to shape American politics. For example, today's religious conservatives are often surprisingly friendly to technology, if not to Darwinism. But the current political scene differs substantially from that of the postwar decades. The forms of scientific liberalism that played such a crucial ideological role in American politics during that era have waned considerably in their public influence, even as scientific expertise has continued to hold a privileged place in legal proceedings and most federal agencies. Confounding theories of a tightly knit "modernity" in which science, technology, capitalism, and bureaucratic administration marched shoulder to shoulder, corporate interests have proven quite willing to ally themselves with religious conservatives who pin the alarming cultural effects of commerce and consumption on the secularity of modern liberalism.[5]

Given that the political meanings of science and religion are intertwined, we cannot fully understand the meteoric rise of the religious right without an adequate account of the science-oriented alternatives against which that movement defined itself. I have sought to show that a revised view of how scholars in the late nineteenth and early twentieth centuries described the role of science in American public life can help us rethink the sources and functions of the powerful modes of scientific liberalism that structured postwar American politics.

Above all, a new historical narrative can help us see that those forms of liberalism did not flow inexorably from science, or even from an epistemology that sharply differentiated scientific conclusions from normative claims. Rather, the roots of those political orientations lay elsewhere, in the combination of technical resources and ideological cover that science offered powerful elites outside the universities. Academic scholars certainly collaborated with these elites in many instances, but one can hardly pin this fact on their commitment to science, or even on their epistemological self-descriptions. Historical actors have never had trouble finding cultural resources with which to justify self-interested action. Moreover, even if modern science were uniquely useful for this ideological task, that fact would not foreclose the possibility of it playing other cultural roles as well. We would simply need to take additional care to embed the scientific enterprise in adequate frameworks of social value. In the end, we have not escaped what remains of the tyranny of modern scientism if we let science's most narrow-minded spokespersons and critics define its cultural and political meanings for us.

Today, we stand as far removed from the apogee of scientific liberalism in the mid-1960s as the theorists of the later Progressive Era stood from the defeat of the Confederacy when they watched Woodrow Wilson best William Howard

---

[5] Secularization does not necessarily mean the triumph of science, because there are many secular discourses (for example, political discourses centered on free markets and consumption) that are not in themselves scientific, even though they tend to shape particular cultures of science. The claim that we live in a "scientific culture" or "scientific age" is quite literally meaningless.

Taft, Theodore Roosevelt, and Eugene V. Debs in the freewheeling election of 1912. Surely it is time to craft interpretations of science and American politics that account for the nearly half-century of dramatic changes since Robert McNamara's "best and brightest" ruled the roost, rather than presuming that we still live in the world of the postwar military-industrial complex and its associated brands of scientific liberalism, which never reigned uncontested in the first place. This book does not recount the history of the troubled decades since the 1960s. But it does argue that we can cast new light on earlier developments we thought we understood well by questioning the view of science that many post-1960s critics carried forward from the postwar behavioral sciences – an inflexible, ahistorical understanding that presents science as a powerful solvent of ethical commitments, and thus of progressive politics.

My hope is that the story of the scientific democrats of the late nineteenth and early twentieth centuries can help us move toward a broader conception of science's political meanings and possibilities today. Like "religion" or any other linguistic construct, "science" expands or contracts over time – and we could, if we desired, adopt a more expansive meaning than has prevailed in recent decades. Science need not perform only the disciplinary function of marking off the normal from the abnormal, and its advocates need not grant religion an exclusive tie to social emancipation, civic commitment, and other goods. Science, like religion, offers a mix of promise and danger.[6]

I would like to see post-positivist scholars trade in the stark, blocky politics of anti-scientism that has flourished since the 1960s for a much more nuanced, flexible understanding of the relationship between intellectual practices, epistemological claims, and political discourses. Such an understanding would acknowledge that the prevailing caricature of science fits the natural sciences almost as poorly as it does the human sciences.[7] It would inspire scholars to undertake careful historical analyses of the cultural work that goes into constructing links between methodologies, epistemologies, and political understandings, rather than assuming that the links are pre-ordained. In short, it would thoroughly historicize the politics of epistemology.[8]

---

[6] To be perfectly clear, my argument is not that the scientific democrats I treat, or any subset of them, got the political import of science right. Nor is it just that it would be useful for us to recapture one or another of their understandings of science *in toto*; we need an understanding of science that is appropriate to our day and our situation. Nor, finally, is my intention simply to demonstrate that the scientific democrats held their view of science as legitimately and genuinely as did those who understood science differently, and that it too produced historical effects worthy of note. This is true, but to my mind the key lessons of the study lie elsewhere.

[7] As self-aware natural scientists have always known, and as recent studies of science have shown, knowledge claims in the natural sciences are as inescapably interpretive – if perhaps more reliable, on average – as those in the human sciences.

[8] Good examples include Anne Harrington, *Reenchanted Science: Holism in German Culture from Wilhelm II to Hitler* (Princeton: Princeton University Press, 1996); Sheila Jasanoff, *Designs on Nature: Science and Democracy in Europe and the United States* (Princeton: Princeton University Press, 2005); and John Carson, *The Measure of Merit: Talents, Intelligence, and Inequality in the French and American Republics, 1750–1940* (Princeton: Princeton University Press, 2007).

This book contributes to that task by recapturing the intense preoccupation with shaping public culture in American scientific thought during the late nineteenth and early twentieth centuries. In the original version of the project, I called the main characters "scientific republicans." Although the applicability of the label "republican" to periods after the early nineteenth century is a matter of fierce scholarly debate, my early formulation had the benefit of capturing the profound concern of so many scientific thinkers with the state of their nation, and especially the state of its inhabitants – their virtues, skills, and moral character as well as their beliefs. Most of the figures discussed in this book thought the spread of science would counterbalance the excesses of market society by inculcating democratic virtues and a commitment to the good of others.[9]

To the extent that republicanism has remained a viable counterpoint to a mode of liberalism rooted in the balancing of individual interests, it certainly reverberated through the sensibilities I have described. Scientific democrats of all stripes assumed that citizens had a stake in how their fellows thought – not just their cognitive beliefs, but also, and more importantly, their broader intellectual and ethical practices. "No man and no mind was ever emancipated by being left alone," John Dewey wrote bluntly in 1927.[10] He and other scientific democrats rejected a *laissez-faire* approach in the formation of public opinion, though they rarely sought an active role for the state in that regard. They held that Americans could reconcile industrial conditions with self-governance only by changing many of their core commitments, in keeping with the results and methods of modern science. In their ideal modern polity, virtue would be guaranteed by science rather than by the market, the frontier, or Christianity.

Of course, the shortcomings of scientific democracy were legion. The central image of the United States as a broadly deliberative polity in which the public already held the reins of power led many scientific democrats to neglect the procedural task of empowering citizens to voice their will effectively. Their portrait of the nation also inclined them to assume that citizens would inevitably reject any scholarly claims that did not reflect their own experiences – in short, to assume that science could never become an authoritarian force. Meanwhile, their tendency to connect ethics to science overstated both the distance of scientific claims from theological tenets and the logical relations between scientific practices and ethical values. Furthermore, even the post-Progressive scientific

---

[9] For suggestive continuities between republican tenets and American pragmatism, see Daniel S. Malachuk, "'Loyal to a Dream Country': Republicanism and the Pragmatism of William James and Richard Rorty," *Journal of American Studies* 34 (2000): 89–113; and Filipe Carreira da Silva, "Bringing Republican Ideas Back Home: The Dewey-Laski Connection," *History of European Ideas* 35 (2009): 360–368. If Steven Shapin is right about its origins, the modern scientific enterprise shares with much of American political thought a common root in the figure of the English gentleman: *A Social History of Truth: Civility and Science in Seventeenth-Century England* (Chicago: University of Chicago Press, 1994).

[10] John Dewey, *The Public and Its Problems*, in *John Dewey: The Later Works, 1925–1953*, vol. 2, ed. Jo Ann Boydston (Carbondale: Southern Illinois University Press, 1988), 340.

democrats vastly overestimated the base of shared human experience, in part because of their fervent belief that their intellectual models or representations could ground a form of solidarity adequate to a polyglot industrial society. They expected to knit together a contentious polity by producing a universally applicable – if, in some renderings, quite spare – system of mental constructs that could be shared across all lines of difference.

That goal has proven elusive, and the universalisms of the past, whether robust or spare, now stand exposed as mere projections of Western, middle-class values. Yet the faults characteristic of scientific democracy actually flowed from its particular, politically charged rendering of science, not from an abstract, timeless science. For example, the core assumption that citizens could harmonize modern economic and social forms with democratic political control by drawing on scientific resources was not a logical deduction from scientific practice, but rather a political tenet. The same holds for the assumption that a democratic public culture must be grounded in secular rather than religious truths. And scientific democracy, rather than science, is likewise to blame for the circular reasoning it fostered, in which advocates projected an idealized image of democracy onto the day-to-day workings of the scientific community and then held up that community as a model for democratic deliberation.[11]

Moreover, scientific thinkers are by no means alone in having built such dubious assumptions into their cultural practices. Equivalent faults can be found in abundance among the defenders of all other cultural programs, past and present. Many of these shortcomings ultimately stem from deep-seated human weaknesses. Others can be traced to the need to employ language in our social relations. Still others – including the tendency, widespread among the scientific democrats, to equate one's own vision of democracy with the essence of democracy itself – relate to the central, unsolvable epistemological problem that lies at the heart of democratic practice: how to know the needs, desires, and values of others.[12] The flaws of the scientific democrats were substantial, but they were not unique or even rare.[13]

---

[11] This act of projection had the additional effect of explaining away the tension between endorsing science as a model of open discourse and expecting voluntary compliance with the latest scientific findings. It allowed scientific democrats to believe that such findings represented the outcome of a process that was transparent, non-coercive, and deliberative, if not actually inclusive of all citizens. Deweyan views of science, which took scientific claims to be more like negotiated policy solutions than judicial decisions, produced this effect particularly strongly.

[12] In the waning days of scientific democracy, both public opinion polling, which seemed to provide a transparent window into the minds of citizens, and consensus liberalism, which posited that investigators shared in a homogeneous national culture, appeared to offer easy solutions to the epistemological problem of democracy and made attempts to empower citizens even less likely. For a long historical view of the main types of science-centered responses to that epistemological problem, see Yaron Ezrahi, *The Descent of Icarus: Science and the Transformation of Contemporary Democracy* (Cambridge, MA: Harvard University Press, 1990).

[13] Similarly, none of the other faults exhibited by the various groups of scientific democrats – faults that were highly consequential for both the fortunes of scientific democracy and the

Those who want to put science's powerful resources into the service of more progressive ends might begin by carefully parsing the distinctions and tensions between the putatively interlinked elements of what critics often shorthand – and thereby reify – as "modernity": science, positivism, capitalism, the administrative state, political secularity, and technocratic or managerial forms of governance. The various problems with these closely intertwined Western phenomena do not prove that scientific methods are useless. Nor do they prove that other intellectual methods are superior. It is hardly repressive or antidemocratic, in itself, to assume that the inhabitants of modern societies stand to gain valuable insight by ensuring that knowledgeable individuals spend their time thinking about natural and social phenomena in a careful, empirically informed, and collectively disciplined manner. At the end of the day, it is the institutional details of the instantiation of scientific inquiry – the funding structures, the choice of problems, the manner of decision-making, and the associated renderings of science's character, scope, and prospective contributions – that make the difference.

To say that the meaning of science is fluid, contingent, and contested is not to say that every cultural practice and truth claim could be labeled science. Nor is it to deny that specific scientific conclusions and methods often prove particularly reliable – nor even that we might be well advised to grant a certain degree of authority to those who spend their lives toiling in the pursuit of systematic, evidentially grounded knowledge claims. One need not posit the quixotic possibility of absolutely disinterested knowledge in order to recognize the capacity of scientific inquiry to inform and enrich all sorts of pursuits – not least the pursuit of a more just social and political order.

---

lives of others – represented a logical concomitant of science. These shortcomings included an unproblematic "methodological nationalism" that assumed the nation-state was the relevant locus of political action and defined cultures as coextensive with national polities; a propensity to reduce complex questions of institutional practice to abstract ethical principles; a tendency to exempt the overall framework of industrial production and private enterprise from critical scrutiny; a belief that solving problems of knowledge is necessary and perhaps even sufficient for solving problems of social order; an inclination to draw the distinction between established facts and exploded speculations in a manner that reflected their political values; and a dramatic overestimation of the historical power of academic ideas. Scientific democrats also tended to assume that the public had gained its cultural bearings when national policies appeared to match their sense of the public's needs – or when large audiences bought their books. Meanwhile, those scientific democrats of a deliberative bent rarely grasped how their program differed from – and often conflicted with – the more expertise-centered political projects pursued under the banner of science by other groups. Deliberatively minded scientific democrats were overly optimistic about the capacity of middle-class professionals to serve as vectors of progressive social change. Finally, one could conceivably charge these figures with professional self-interestedness by suggesting that those who identified their work as a contribution to strengthening democratic deliberation were simply unable to find buyers for their forms of knowledge and therefore made a virtue out of necessity. At the very least, a strong inverse relationship often prevailed between the existence of socially valued technical applications for a given field of inquiry and the commitment of its practitioners to improving public culture.

This book seeks to aid in that pursuit by explaining how and why an important group of American scholars came to see science as a form of critical discourse that challenged a technology-centered, market-based industrial order created by more powerful elites. It urges us to take seriously the fact that the likes of John Dewey, Charles Ellwood, and Gordon Allport saw themselves as scientists. Unlike many of their successors today, these figures recognized that the various methods employed in the natural and human sciences share intrinsic similarities that distinguish those fields from other cultural practices, such as art and religion. The commonality flows from a pair of simple facts: we must employ partial, abstract models of concrete experience in order to think or talk about the world, and certain procedures centered on empirical investigation have often made those models more reliable or helpful.

The twin scientific tenets – that we must simplify and abstract in order to think, and that we stand to gain by comparing our abstractions to comparatively concrete forms of experience – can be detached from the characteristic errors of the scientific democrats and reattached to a recognition that our abstractions must be extremely complex and open to challenge by others before we can safely rely on them. However, this is also what the scientific democrats, at their best, sought to do. Understanding that the unavoidable process of mental abstraction involves the projection of one's own values, beliefs, and interests onto the world, they sought by various methodological and social means to make their abstractions more responsive to the expressed and imagined needs of others. In the succeeding decades, we have gained valuable perspective on the limitations of both their methodological strategies and their conceptions of the human person. In the end, however, the scientific democrats were much more like contemporary scholars than most post-positivist critics have been willing to admit.

It is only by recognizing the contingency of the relationships between science, epistemology, and politics that we can see the flaws and failures of the scientific democrats as instructive lessons for present practice, rather than simply concluding that they stood worlds apart from us and took a dead-end road that we can congratulate ourselves on having avoided. From a political standpoint, science, like religion, is what we make of it.[14] In other words, cultural work is required to turn any of the phenomena currently assigned to the category of science – or religion – into a specific, politically salient claim.

Acknowledging the cultural and political contingency of science's current meanings offers the hope of producing new kinds – new cultures – of science. If critical scholars today continue to employ the epistemological and methodological frameworks of the nuclear era, wrought as they were by the confluence of behavioral scientists' participation in the military-industrial complex and the New Left's revolt against that work, we will have little chance of developing a political understanding of science capable of meeting the challenges of

[14] One should properly say "sciences" and "religions," because each category contains multitudes.

the twenty-first century. A caricature ostensibly modeled on physics and only tenuously related to the concerns of the human sciences can hardly buttress a progressive project.

The historical record, however, offers comparatively little guidance for crafting visions of science adequate to our own day. Looking back at the book of history, we cannot discern on any of its pages a pristine, unvarnished science with an easily legible political meaning. We can see only the specific, historical formations to which the term science became attached through dedicated cultural work, at the expense of competing interpretations. In the end, a history of science's past political meanings can demonstrate only that its present political import remains up to us.

Index

AAAE. *See* American Association for Adult
 Education
AAAS. *See* American Association for the
 Advancement of Science
AAUP. *See* American Association of University
 Professors
About Ourselves (Overstreet), 209
academic freedom, 46, 200, 214
academic left, 2, 4, 5, 365–367, 369
ACDIF. *See* American Committee for
 Democracy and Intellectual Freedom
ACLS. *See* American Council of Learned
 Societies
Adams, Charles Kendall, 34, 48, 49
Adams, Henry, 42n26, 206
Adams, Henry Carter, 77, 78, 80, 80n61,
 118n3. *See also* economics, ethical
Addams, Jane, 7, 98, 125n16
Adler, Felix, 125n16
Adler, Mortimer, 205, 221–222, 322–323
administration, 3–4n3, 4–7, 10–14, 25,
 71–73, 79, 119–120, 123–124, 137, 145,
 177–178, 181, 234, 303–304, 307, 368,
 372. *See also* bureaucracy
Adorno, Theodor, 301, 345, 348, 353–354
advertising, 281
AEA. *See* American Economic Association
agrarianism, 23, 328
agricultural society, 78
agriculture, 31–32, 33, 140, 207, 254, 274
Agriculture, U.S. Department of (USDA),
 13n15, 227, 272n1, 305, 331, 363
agnosticism, 59–61, 63–64, 65, 85–86
Allied Expeditionary Force, 292
Allport, Gordon, 227n2, 285–288, 339, 346,
 373
Almond, Gabriel, 348, 351, 359

altruism, 66, 85, 125, 127, 133,
 184–185, 190, 199, 200, 306, 321.
 See also humanitarianism
An American Dilemma, 299–300, 351
American Association for Adult Education
 (AAAE), 215
American Association for the Advancement
 of Science (AAAS), 101n30, 274n4, 304,
 305, 306–307
American Association of University Professors
 (AAUP), 200
American Committee for Democracy and
 Intellectual Freedom (ACDIF), 324
American Council of Learned Societies
 (ACLS), 292
American Creed, 299–300
American Economic Association (AEA), 79–80
American Federation of Labor, 281
American Legion, 281
American mind, 337. *See also* American
 studies; consensus liberalism
American Philosophical Association (APA), 26,
 98, 100, 101n31, 102–107, 257, 273n2
American Social Science Association (ASSA),
 73
American Sociological Association (ASA), 274
American studies, 271, 337. *See also* history
Amherst College, 202, 203n14, 205
And Keep Your Powder Dry (Margaret Mead),
 294, 295–296, 352
Angell, James, 42, 52, 53–54
Anshen, Ruth Nanda, 291
anthropology, 48, 118, 169, 264n60, 269n72,
 319, 346, 362n47, 363; Benedict, 8,
 165, 269n72, 284–285, 286, 287, 293,
 294, 296n48, 298, 300–301, 336, 359;
 Boas, 8, 110, 117, 118, 120, 132–133,

anthropology *(cont.)*
135–137, 142, 155, 183–185, 187, 188, 238, 240, 259–261, 262, 263, 266–267, 268, 297, 311, 324, 338n4; Boasian, 8, 165, 183–187, 208, 215, 234, 238, 262, 269n72, 270, 277, 280, 284–285, 286, 287, 293–296, 298, 300–301, 326, 336–337, 341, 346, 350, 352–353, 355, 359; comparative method, 137, 259, 260–261, 363; and consensus liberalism, 293–294, 359; cultural relativism, 8, 120, 142, 183–187, 259–264, 270, 319, 362n47; and cultural sciences, 13, 16, 111, 114–115; culture concept, 117, 118, 120, 132, 137, 183; culture at a distance, 300–301; as democratic theory, 120, 132–133, 136, 137, 183; Lowie, 184, 186–187; Mead, 8, 165, 234, 293–296, 300, 326, 346, 350, 352–353, 355, 359; national character, 293–296, 300–301, 359; and pragmatism, 136, 136n39; Sapir, 165, 185–186, 262; and scientific democracy, 8, 13, 16, 111, 114–115, 117–118, 120, 132–133, 135–137, 165, 172, 183–187, 188, 215, 238, 258–259, 270, 283–285, 293–296, 297, 300–301, 336–337, 346, 355, 363. *See also* culture and personality; linguistics
Anthropology and Modern Life (Boas), 184–185, 240
anti-communism, 229, 234, 291, 303, 309, 324–326, 342, 343, 357. *See also* communism, McCarthy, Joseph
anti-evolutionism, 149–150, 156, 159, 210, 216
anti-fascism, 304–305, 324. *See also* fascism
Antioch College, 212, 213
anti-Semitism, 196, 219n45, 347
APA. *See* American Philosophical Association
Aquinas, Thomas, 217
Are We Civilized? (Lowie), 186–187
argument from design, 34, 37, 39, 47
Aristotle, 202, 252, 262, 269
Arno, Peter, 251
Arnold, Matthew, 198
Arnold, Thomas, 58
Arnold, Thurman, 269n72
artisans, 30, 318
arts. *See* humanistic tradition
ASA. *See* American Sociological Association
ASSA. *See* American Social Science Association
associationalism, 303. *See also* Hoover, Herbert; science administrators
astronomy, 70, 208, 312, 322. *See also* physical scientists

atheism, 49, 59–60, 64, 226, 229, 323n38, 324, 325, 337
Atkinson, William P., 33–34, 51, 53
atomic bomb. *See* Manhattan Project; nuclear weapons
Atlantic Monthly (magazine), 156
The Authoritarian Personality (Adorno et al.), 353–354
auxiliary language movement, 268. *See also* linguistics
Ayer, A. J., 242n15
Ayres, Clarence E., 182

Babbitt, Irving, 198–199, 200, 217, 219, 222, 328. *See also* New Humanism
Bacon, Francis, 56, 252
Bagehot, Walter, 107
Bain, Read, 179
Baltimore Sun (newspaper), 306
Barber, Bernard, 357–358, 360n43
Bard College, 213
Barnes, Harry Elmer, 337, 338n4
Barry, Frederick, 254
Barzun, Jacques, 257n44, 328
Bascom, John, 43, 46
Basic English, 268
basic science, 46, 71, 308
Bateson, Gregory, 293, 355
Beard, Charles A., 8, 110, 187, 188–189, 214, 218, 267, 337, 338n4. *See also* New History
Becker, Howard, 246, 356n34
Beecher, Henry Ward, 35–36
Beecher, Lyman, 36
behavioral sciences, 16, 234, 288, 290, 301, 336, 353, 358–362, 366, 369, 373. *See also* human sciences
behaviorism. *See* psychology
Bell, Daniel, 348
Bemis, Edward, 80
Benedict, Ruth, 8, 165, 269n72, 294, 336; *The Chrysanthemum and the Sword*, 300–301; conformity to individuality, 298; and consensus liberalism, 296n48, 301; national character approach, 293, 300–301, 359; need for self-determination, 284–285, 286, 287, 296n48, 298; *Patterns of Culture*, 269n72, 284. *See also* anthropology, Boasian
Bennington College, 213
Bennion, Milton, 205
Bentham, Jeremy, 66, 69, 191
Bentley, Arthur F., 270n72, 359

Bergson, Henri, 153
Berlin, Isaiah, 163
Bernal, J. D., 244–245
Bernalists, 244–245, 247, 248, 255, 276n9, 306, 311
Bernard, Luther Lee, 123–124, 126, 133, 145, 146, 150, 176, 178, 227n2, 280n17, 346
Bertalanffy, Ludwig von, 355
Bible, 34, 35, 38, 39n22, 43, 50, 54. *See also* Christianity
big science, 270, 304, 357
biology, 97, 128, 136, 171, 194, 211, 216, 245, 255, 285, 286, 306, 311, 316n23, 318, 322, 323, 324n39, 341, 355, 365, 367n3; anatomy, 160–161; botany, 47, 76, 336; Child, 160–161; Coghill, 161, 269n72; Conklin, 149, 155, 157–158, 163, 169, 240n12, 306–307; as cultural science, 13, 16, 111, 114, 183; in curricula, 206–208; embryology, 152, 160; entomology, 156, 157; and eugenics, 169, 170n54, 207; and general semantics, 269n72; genetics, 148, 149, 156, 169, 207, 245, 336; Herrick, 149–150, 151n7, 159–162, 162–163n36, 163–164, 208; holism, 157–158, 161, 170; Lamarckism, 156, 156n20; natural history, 59; neurology, 160, 269; physiology, 26, 160–161, 166–168, 242, 286, 289, 316n24; and scientific democracy, 13, 16, 111, 114, 148–152, 169–170, 189, 191–192, 209, 238, 336–337; Simpson, 336, 337, 341; Sinnott, 324n39, 336, 337, 341; and sociology, 66, 69, 77, 117–118, 118n3, 120–121, 128, 149, 162–163n36, 289–290, 336–337; vitalism, 152–153, 264; and World War I, 155–156; zoology, 47, 156, 206. *See also* emergentism; eugenics; evolution; psychobiology
Bismarck, Otto von 77
Black Mountain College, 213
Blanshard, Brand, 241n13
Bliss, W. D. P., 75
Bloomfield, Leonard, 265–266, 267, 268
Boas, Franz, 110, 135, 155, 188, 238, 240, 338n4; *Anthropology and Modern Life*, 184–185, 240; comparative method, 137, 259, 260–261; cultural diffusion, 260; cultural relativism, 8, 120, 142; culture concept, 117, 118, 120, 132, 137, 183; as democratic theorist, 120, 132–133, 136, 137, 183; epistemology, 136–137, 259; introduction to *Handbook of American Indian Languages*, 259–261, 266–267; on

language, 259–262, 263, 266–267, 268; materialism, 184, 185, 259–260, 263; political activities, 136, 311, 324, 338n4; on social role of scholar, 135, 137, 184–185, 187, 297. *See also* anthropology, Boasian; linguistics, Boasian
Bode, Boyd H., 216, 239
Bogardus, Emory S., 179
Book-of-the-Month Club, 229
Boorstin, Daniel, 352
Boring, Edwin G., 165–166
Boston University, 193
Boulding, Kenneth E., 355
Bowen, Francis, 37–38
Bowman, Isaiah, 312
Bowne, Borden Parker, 193
Brain Trust, 212, 228
Brains of Rats and Men (Herrick), 160
Brasch, Frederick E., 252–253
Brick, Howard, 298n53
Bridgman, Percy W., 236, 238, 242, 316
Brill, Abraham A., 269n72
Brinton, Crane, 282n21
Brookings Institution, 307
Brother Chrysostom (John Conlon), 101n31
Brown, Elmer Ellsworth, 220
Brown, Harold Chapman, 253
Brown University, 29
Brown v. Board of Education, 283
Brownell, Baker, 208–209, 254–255
Bryn Mawr College, 105
Bryson, Lyman, 215, 290, 322, 325
Buchler, Justus, 299n54
Buck, Paul H., 332
Buckley, William F., Jr., 326
Bulletin of the Atomic Scientists (newsletter), 309
Burbank, Luther, 210
bureaucracy, 1, 3, 4, 6, 13, 19, 71, 76, 119, 138, 172, 200, 227, 249, 286, 315, 319, 328, 365, 366, 368. *See also* administration
Burgess, John W., 49
Burke, Kenneth, 246, 247
Burtt, Edwin A., 192
Bush, Vannevar, 230, 302–303, 304, 307–308, 309–310, 312–313, 315, 320, 324n39. *See also* physical scientists; science administrators
Bushnell, Horace, 35–36, 297
business, 1, 25, 29, 30, 31, 34, 51, 74, 80, 81, 112, 201, 202, 204, 206, 222, 268, 295, 315, 320, 348. *See also* capitalism; corporations; industry

business cycles, 6, 52, 65, 74, 80, 182, 343n13.
  See also capitalism; Depression
business leaders, 6, 12, 25, 29, 31, 32, 33, 52,
  73, 76, 77, 96, 139, 146, 174, 210, 227,
  249, 297, 299, 300, 305, 308, 348, 352,
  368. *See also* capitalism; corporations;
  industry
Butler, Nicholas Murray, 203, 214
Butts, R. Freeman, 239–240

Calder, Ritchie, 306
Caltech, 305
Cambridge University, 193
Can Science Save Us? (Lundberg), 351n25,
  355–356
Canby, Henry Seidel, 219
Cannon, Walter B., 324n39
Cantril, Hadley, 267
capitalism, 3–4n3, 6, 17, 19, 24, 25–26, 29,
  35, 52, 68, 69, 74–75, 77, 78, 95, 111,
  112, 142, 188, 189, 193, 210–211, 223,
  244, 248, 249, 250, 252, 263, 281, 289,
  294, 296, 303, 306, 325, 336, 352, 368,
  372. *See also* business; business leaders;
  corporations; industry; postcapitalism
Carnap, Rudolf, 241–242, 242n15, 326, 345.
  See also philosophy, logical empir
Carnegie, Andrew, 33
Carnegie Corporation, 215
Carnegie Hall, 219
Carnegie Institution of Washington, 253
Catholicism, 101n31, 217–218, 221, 222,
  223, 226, 229, 290, 326–327, 344, 345,
  356. *See also* Christianity; philosophy,
  neo-Thomist; religious leaders
CC. *See* Contemporary Civilization
CCF. *See* Committee for Cultural Freedom
Chadbourne, P. A., 42
Chamberlin, Rollin T., 208
Chapin, F. Stuart, 178
Chase, Stuart, 172n1, 258n48, 269, 312n16
chemistry, 12, 30, 32, 47, 51, 61, 155,
  211, 230, 254, 256, 269, 307, 323.
  See also physical scientists; science
  administrators
Child, Charles Manning, 160–161
Childs, John L., 216
Christian Century (magazine), 306
Christianity, 2, 12, 22, 23, 24–26, 29, 37, 40,
  41, 44, 48, 49–50, 52, 58, 68, 99, 199,
  212, 218, 219, 239, 240, 250, 277–278,
  305, 306, 348, 351, 370; and Darwinism,
  59–65, 98, 156, 159, 210, 216; as source
  of democratic culture, 23, 24, 210, 229,

230, 321–323, 324n39, 326–327, 333;
  and economic policy, 65–66, 73–76, 77,
  79–82, 94; ethics, 43, 65, 73–74, 80,
  88–89, 94, 95–96, 97, 98, 112, 193, 323,
  347. *See also* Catholicism; ministers;
  Protestantism; religious leaders
Christianity and the Social Crisis
  (Rauschenbusch), 76
The Chrysanthemum and the Sword
  (Benedict), 300–301
"The City" (Park), 181–182
City College of New York, 103, 273n2,
  323n38
civil liberties, 243, 296n48, 304, 324–325,
  348, 349
civil service reform, 49, 71–73
civil society, 37, 41
Civil War, 8, 11, 15, 18, 21, 23, 28, 77, 79;
  impact on scientific democracy, 2, 3, 22,
  24–25, 30, 33–34, 39, 99
Civilian Conservation Corps, 276
civilization, 65, 135, 139, 141, 143, 186–187,
  203, 205, 206, 278, 290, 294, 332;
  ancient, 29, 47–48, 175, 206, 255, 265,
  299, 328–330, 333; Christian, 24, 37,
  50; modern, 47–48, 145, 208, 211, 253,
  256, 297, 317, 322. *See also* collegiate
  curricula, Western Civ
Clark, John Bates, 8, 80, 82, 95n17.
  See also economics, ethical
Clark, John M., 110
Clark University, 323
classical curriculum, 29–30, 31, 34, 37,
  42, 43, 44n32, 46, 47, 48, 49–50, 51,
  54, 71
classics. *See* languages and literature
Clifford, William Kingdon, 61n12
Coffin, Henry Sloane, 220
Coghill, George E., 161, 269n72
Cohen, Morris R., 103, 199–200, 238, 243
Cohen, Nancy, 71–72
Cold War, 5n6, 8, 11, 16, 21, 52, 138, 233,
  243, 303, 309, 314–315, 336, 344, 347;
  impact on scientific democracy, 14, 16,
  225–226, 342, 343n13, 356
The College of William and Mary, 29
colleges and universities: administrators, 203,
  212–213, 215, 221, 257n44, 331; classical
  colleges, 29–30, 34, 35, 54, 67n25;
  denominational colleges, 28, 30, 34,
  39–46, 54; Eastern, 13, 254, 256, 256n43,
  258; extension programs, 212–213;
  founded, 14, 15, 21, 28–34, 45–54, 213–
  214; graduate education, 33, 47, 48–49,

197, 198, 200–201, 221, 251, 255n41, 366n1; land-grant institutions, 13, 30–33, 55, 74n44, 254–255; metropolitan universities, 13, 254; Midwestern, 13, 73, 238–239, 254–256, 258; Morrill Act, 30–31, 74n44; presidents, 32–33, 37, 40, 42, 43, 49–50, 72n40, 198, 202, 203, 205, 212, 213, 214, 220, 221, 255, 256, 268n71, 302, 306, 312, 315–318, 330–331; professional schools, 48–49, 72, 201, 206, 215–216; progressive colleges, 213; public universities, 13, 49, 54, 238; research function, 2, 11, 12, 24, 28, 32–33, 38–39, 46, 47, 51–52, 54, 71, 102–103, 115, 117, 119, 124, 176–178, 180, 196–197, 200–201, 214, 217, 218, 220, 226, 230, 238–239, 249, 253, 257–258, 268, 270–271, 272, 274–277, 281–282, 302–320, 342, 351, 354–355, 361; Southern, 13, 238–239; student discipline, 45; summer courses, 212–213; teacher-student relations, 45–46; trustees, 31, 32, 74, 80, 202, 213, 329n51; tuition, 53. *See also* collegiate curricula; names of individual institutions

collegiate curricula: classical curriculum, 29–30, 31, 34, 37, 42, 43, 44n32, 46, 47, 48, 49–50, 51, 54, 71; elective system, 14, 30, 47, 55, 196, 197–198, 200, 253; engineering, 305–306, 330–331; general education, 14, 197, 201–209, 219, 220, 315–316, 318, 328–333; general semantics, 269; Great Books, 201, 205, 219, 229, 327n48, 332; history of science, 253–256, 305n2, 332n55; liberal arts, 254, 315, 331; liberal education, 13, 31–32, 33, 51, 71n36, 205; major-minor system, 14; modern subjects, 30, 46, 47–48, 55, 197–198; moral philosophy, 29, 35, 37–38, 39, 43, 46, 48, 49–50, 57, 58, 74, 98, 103, 273; nonsectarianism, 21, 49–50; parallel tracks, 31–32, 47; pedagogy, 42–43; preprofessional, 200–201; sciences, 37, 45, 47, 55; substitutions allowed, 30; survey/orientation courses, 175, 197, 201–209, 216, 219, 229, 327n48, 328–330, 332–333, 337; textbooks, 42–43, 43n28, 92, 98, 123, 191, 203–204, 208, 210n28, 215, 262, 279, 286–288, 316n23, 329–330; vocational education, 13, 30–32, 33, 46, 47, 52–53, 196–197, 203, 213; War Issues, 202–203; Western Civ, 201, 229, 232, 328–330, 332–333, 337, 349.

See also colleges and universities; names of individual institutions

Columbia University, 29, 104, 177, 182, 209–210, 215, 221, 273, 273n2, 279, 306, 328, 337; Beard leaves, 187, 214; Benedict at, 336; Boas at, 132, 136, 188; CC course, 203–204, 205, 205n17, 212, 213, 219, 228, 268n71, 329n51, 330, 330n53; commitment to public service, 13, 254; commitment to liberal education, 13; commitment to teaching, 258; and consensus liberalism, 330n53; contextual theories of science at, 188, 190–191, 203–204, 238, 239–240, 243, 250–252, 253–254, 256n43, 257; curricular reform, 30, 202–204, 219, 220; Dewey at, 98, 124, 158, 188, 251, 253; Erskine at, 205, 219; and federal government, 212, 228; General Honors, 219; and Great Books, 205, 219; history of science at, 251–252, 254, 257; intellectual history at, 250–252, 257; isolationism at, 338n4; Lynd at, 273, 281; Merton at, 243, 254; Mills at, 246, 356; T. H. Morgan at, 149; naturalistic philosophers, 190–191, 203–204, 238, 243, 251–252, 257, 329–330, 339; New College, 216; New History, 8, 187–189, 212, 251–252, 337, 338n4; Ogburn at, 178; Readability Laboratory, 215; Robinson at, 188, 251, 252; Robinson leaves, 187, 212, 214; School of Political Science, 49; and scientific democracy, 13, 213, 215; Teachers College, 216, 240, 253; War Issues, 202–203; Western Civ at, 329–330

Coming of Age in Samoa (Mead), 294
Commager, Henry Steele, 299
Committee for Cultural Freedom (CCF), 324
Committee of Ten (NEA), 32
Committee on Food Habits, 294
Commons, John R., 80, 182, 228
communication research, 267–268
communism, 225, 226, 230, 234, 245, 291, 296, 303, 309, 324–326, 329, 329n51, 331, 361. *See also* anti-communism; totalitarianism
Communist Party, 245, 325
community development, 363
competition, 17, 23, 25–26, 29, 36–37, 68, 73, 74n44, 75, 76–77, 78, 90, 91, 108, 118, 123, 133–134, 139, 155–156, 204, 233, 244, 284–285, 294, 296, 297, 298, 305n2, 352, 353. *See also* capitalism
Compton, Arthur Holly, 305, 322, 323, 324n39

Compton, Karl T., 268n71, 302–303, 306,
    312–313, 330–331
Comte, Auguste, 61–62, 65, 77, 85, 86, 87,
    122, 140, 280, 328
Conant, James B., 335, 360–361; as Harvard
    president, 256, 302, 315, 317, 319,
    330; and history of science, 256, 316;
    laissez-faire communitarianism, 231–
    232, 302–303, 357; and meritocracy,
    315; *Modern Science and Modern Man*,
    316–318, 320–321, 330; and NAS
    establishment, 302–303; vs. naturalistic
    ethics, 321, 330; and neo-orthodoxy,
    320–321; and operationalism, 316; and
    Redbook, 330, 333; science akin to
    art, 231, 314, 317, 320, 354; science as
    model of democratic freedom, 231, 320,
    354; vs. social sciences, 312, 317–318,
    320; and Western Civ courses, 330;
    writings, 256, 315–318, 320–321, 330.
    See also physical scientists; science
    administrators
Condon, Edward U., 239n8
Conference on Science, Philosophy and
    Religion in Their Relation to the
    Democratic Way of Life (CSPR), 296,
    320, 322–323, 324n39
Congress, 212, 226, 229, 304, 307–308, 312,
    343, 351
Conklin, Edwin Grant, 149, 155, 157–158,
    163, 169, 240n12, 306–307, 324n39
Conlon, John (Brother Chrysostom), 101n31
consensus liberalism, 304; benefits, 16,
    233–234, 273, 282–283, 292, 299–300,
    329, 346, 351–352, 355–356, 371n12;
    consensus history, 352–353; contents of
    consensus, 350–351, 352–353; defined,
    232, 283; expressions, 232–234, 271,
    282–283, 288–290, 292–301, 329,
    330n53, 337, 346–353, 355–356, 357,
    359–362, 371n12; and postcapitalism,
    346, 352; roots, 283, 297–298, 301, 359;
    and value-neutrality, 16, 273, 282–283,
    329, 351, 355–356; and Western Civ,
    327n48, 328
conservatism, 16, 65–74, 80, 120, 135, 142,
    178n15, 184, 188, 189, 226–227, 230,
    233, 244, 250, 282, 287, 291, 298, 301,
    312, 324–326, 328, 343, 345, 348, 352,
    363, 367–368. *See also laissez-faire*
consumer goods, 14, 17, 36, 304, 315
consumerism, 150, 190, 195, 278, 344, 346,
    368, 368n5
consumption, 74, 81, 138–140,
    214, 365

Contemporary Civilization (CC), 203–204,
    205, 205n17, 212, 213, 219, 228,
    268n71, 329n51, 330, 330n53
context of internalization, 270–271
contexts of discovery and justification,
    270–271, 311
contextualism. *See* philosophy of science
conversion, 35–36, 41, 45, 105, 297, 323
Cooley, Charles Horton, 8, 119–120, 124–128,
    128n22, 130, 131, 131n28, 132, 133,
    155, 182, 183n25, 292
Cornell, Ezra, 31, 32n5
Cornell University, 31–32, 33, 40, 45, 49, 54,
    55, 77, 99, 100, 105, 252, 256
corporate liberalism, 3–4n3, 6–7, 139, 146,
    227, 366
corporations, 1, 74, 118n3, 140, 183, 307.
    See also business; business leaders;
    capitalism; industry
Coss, John J., 213
Counts, George S., 216
Creative Experience (Follett), 173–174
Creighton, J. E., 100–101, 102, 103, 105, 237,
    288, 289n35
The Crisis of Our Age (Sorokin), 281
Croly, Herbert, 8, 137–138, 140–142, 143,
    173, 214
cross-cultural comparison. *See* anthropology,
    comparative method
CSPR. *See* Conference on Science, Philosophy
    and Religion in Their Relation to the
    Democratic Way of Life
cultural lag. *See* sociology
cultural sciences: defined, 16, 111; give way
    to behavioral sciences, 16, 234, 336–342;
    and post-positivism, 366–367; and
    scientific democracy, 16, 111, 114–115,
    196–197, 209, 239n8; three wings, 183.
    See also biology, human sciences, names
    of individual disciplines
culture. *See* anthropology
culture and personality, 8, 283–288,
    292–296, 298, 299–301, 361.
    See also anthropology; psychiatry;
    psychology
The Culture Demanded by Modern Life
    (Youmans), 63, 66–67
curricula. *See* collegiate curricula
Curti, Merle, 271, 299
Cushing, Harvey, 313
cybernetics, 354–355

Dahl, Robert A., 359
Darrow, Clarence, 150, 163, 208, 210
Dartmouth College, 29, 156, 206–207

Darwinism. *See* evolution
Davies, Arthur Ernest, 103n34
de Kruif, Paul, 211
de Laguna, Theodore, 105, 168
de Tocqueville, Alexis, 21, 109, 225, 358
Debs, Eugene V., 368–369
deism, 35
deliberative democracy, 7, 9n11, 11, 19,
   106, 107, 119, 129–130, 146, 172–
   175, 225, 227, 282, 309, 313, 361,
   362–363, 370, 371, 371n11, 372n13.
   See also participation, political; public
   opinion
democracy, epistemological problem of, 371,
   371n12
Democracy and Education (Dewey), 130–132
Democratic Party, 212
democratic realism, 7n9, 11, 175–176,
   358–359, 362
Department of Agriculture. *See* Agriculture,
   U.S. Department of
Department of Defense, 316n23
Department of Education. *See* Education, U.S.
   Department of
Depression, 16, 115, 195, 214–215, 220,
   305–306, 331, 344, 346, 350
Descartes, Rene, 252
determinism, 99, 210, 327n48; social, 150,
   244, 327; biological, 136, 149, 150,
   286, 290; criticized, 78–79, 102, 136,
   149–170, 172, 192, 222, 240, 262, 279,
   289–290, 324, 327, 337–338; cultural,
   134, 286; economic, 289–290, 337;
   technological, 178–179, 187n34, 189,
   275–276, 337. *See also* free will
Dewey, John, 19, 133, 210, 214, 222,
   253, 256, 257, 303, 363; as
   anti-communist, 324–325;
   anti-communists target, 326; and biology,
   94, 97, 154, 158, 162, 181, 236, 247; and
   Boasian anthropology, 136; called threat
   to democracy, 218, 321, 327; at Chicago,
   95, 131n28; and consensus liberalism,
   297–299; *Democracy and Education*,
   130–132; as democratic theorist, 6, 7n9,
   8, 18, 19, 90, 95, 119–120, 131, 172–173,
   177, 191, 203, 297–299, 341, 370n9; on
   education, 6, 11, 98, 124, 130–132, 216;
   emergentism, 154, 158, 162; and ethical
   economics, 90, 94–95, 183n25, 190,
   191, 298; ethics, 89, 92, 94–95, 96, 97,
   98, 102, 110, 120, 189, 190, 191, 193,
   203, 243, 321; *Experience and Nature*,
   158, 172, 250n32; fallibilism, 12, 17,
   89, 95, 96, 108, 110, 192; on freedom,
   163, 297, 298, 370; and guild socialism,
   141n47, 227n2; Hegelian roots, 99;
   hegemony assumption, 96, 191, 250;
   historical treatments, 6–8, 11; and history
   of science, 250–251; *How We Think*,
   239n10; *Human Nature and Conduct*,
   158, 172, 250n32; vs. Hutchins and Adler,
   221, 221n49, 223; *Individualism, Old
   and New*, 251n32, 298; influence, 6, 7–8,
   11–12, 142, 172, 182, 188, 189, 190, 191,
   192, 194, 203, 216, 225, 236, 238–240,
   250–251, 254, 270–271, 298–299, 330,
   356; instrumentalism, 97, 148, 203, 218,
   236n2, 238–240, 356; and intellectual
   history, 250–251, 271, 299; *Intelligence
   in the Modern World*, 239n10; vs. Kant,
   190; on language, 258; vs. Locke,
   190–191; vs. logical empiricism,
   240–242; and Meiklejohn, 202, 206; as
   metaphysician, 158n26, 172, 194n46;
   vs. metaphysics, 92; at Michigan, 42, 95,
   131n28, 158; mind as natural fact, 158,
   192; naturalism, 7n10, 158, 159, 162,
   192, 194, 218, 236n2, 278, 321; vs. New
   Deal, 227, 227n2; on platform idea, 105;
   Outlines of a Critical Theory of Ethics,
   163n38; pragmatism, 8, 26, 89–90, 92,
   94–98, 101–102, 108, 132, 136, 154, 192,
   202, 218, 236n2, 243, 273n2, 298–299,
   370n9; pragmatism criticized, 101–102,
   202, 206, 218, 221; *The Public and Its
   Problems*, 172–173, 239, 250n32, 341,
   370; *The Quest for Certainty*, 172, 250–
   251n32; *Reconstruction in Philosophy*,
   172, 250n32; and religion, 89, 92, 94,
   95–96, 97, 190, 223, 297n52; *The
   School and Society*, 98; on science, 7–8,
   10–11, 12, 17, 89–90, 94–98, 110–111,
   112, 119–120, 130–132, 154, 191, 192,
   236–237, 238–240, 254, 262, 270, 278,
   311, 321, 371n11, 373; social organism
   model, 94, 131, 247; on social role of
   scholar, 6, 11, 89, 94–98, 110, 119–120,
   124, 125, 130, 172–173, 188, 189, 190,
   191, 248, 254, 270–271; and social
   sciences, 188, 189–190, 191, 238, 271,
   341; social self, 124–125, 130–131, 297;
   students, 105, 158, 189–190, 191, 193,
   203, 216, 239n10, 243, 250–251, 271,
   297, 298–299, 330; *Theory of Valuation*,
   242n15; "Thought News," 98, 180; on
   values, 98, 110, 112, 179, 188, 189, 190,
   191, 192, 236, 239–240, 273n2, 321;
   and World War II, 292, 338n4; *See also*
   philosophy of science, Deweyan

disciplinary associations. *See* scholarly
  organizations
discussion groups, 227, 309
disengagement thesis, 3–4
Dodson, George R., 103n35
Dorsey, George A., 209
Drake, Durant, 192
Draper, John W., 58, 64–65, 85
Driesch, Hans, 152
Drift and Mastery (Lippmann), 142–145
Du Bois, W. E. B., 7
Duke University, 277
Durant, Will, 209–210, 210n28
Durkheim, Émile, 152n10, 186, 298

Easton, David, 358n39, 359
*An Economic Interpretation of the
  Constitution of the United States* (Beard),
  188–189
economics, 70, 110, 114, 183, 199, 207,
  273, 287, 290, 318n37, 347, 355; of
  abundance, 138–139; agricultural, 363;
  American Economic Association, 79–80;
  classical, 36, 66, 69, 72, 72n40, 74, 78,
  79, 250; and cultural sciences, 13, 16,
  111, 114; in curricula, 203, 204, 255;
  economists' debate, 26, 78–79; ethical,
  8, 26, 65, 74–75, 77–82, 86, 90, 94–95,
  95n17, 111, 115, 118n3, 154, 182, 190,
  298, 335; Ely, 8, 26, 74–75, 77–81, 86,
  94–95, 111, 118n3, 120, 125, 131, 163,
  182, 208; evangelical, 77n53; and federal
  government, 227–228, 272, 273, 312n16;
  historical, 77–78, 81n63; institutional,
  182, 202, 227–228; Keynesian,
  343n13; marginalist, 74, 81–82, 138;
  mathematical, 336; Mitchell, 182, 214,
  228, 306–307, 312; and modernization
  theory, 361; and scientific democracy,
  8, 13, 16, 26, 74–82, 85–86, 111, 114,
  182–183, 188, 190, 226–228, 272, 289–
  290, 306–307, 335–336, 343n13, 363.
  See also laissez-faire; political economy
Eddington, Arthur, 245
Edison, Thomas, 210, 299
Edman, Irwin, 203, 299, 329–330
education, 12, 13, 28, 37, 58, 70–71, 78, 88n6,
  112, 122, 124, 135, 139, 140, 150, 156,
  161, 178, 179, 180, 182, 198, 201, 203,
  206, 211, 265–266, 274, 276, 278, 294,
  306, 315, 316n23, 333, 359, 360; adult,
  151, 197, 209, 214–215, 267; American
  Association for Adult Education, 215;
  Dewey, 6, 11, 98, 124, 130–132, 216; vs.

eugenics, 169; history of, 207; National
  Education Association, 32, 321n34;
  progressive, 6, 98, 124, 130–132,
  213, 215, 294; public, 48; schools of,
  215–216, 250, 253; teacher-training
  programs, 206, 216; theory of, 216, 239;
  workers', 202; Youmans, 62, 63, 66,
  67–68. *See also* colleges and universities;
  collegiate curricula; socialization
Education, U.S. Department of, 227, 333
efficiency, 10n14, 12, 70, 138, 142, 188, 199,
  203, 204, 211, 265, 266, 295, 341, 347.
  See also administration; bureaucracy;
  managerial liberalism
egalitarianism, 2, 3, 7, 11, 45, 75, 77, 131,
  134, 138–139, 191, 198, 205, 233, 254,
  291, 322, 336, 347, 353, 362
Einstein, Albert, 210, 263, 269, 307, 309, 323
Eisenhower, Dwight D., 327
Eldridge, Seba, 174–175, 180
elective system, 14, 30, 47, 55, 196, 197–198,
  200, 253
Eliot, Charles W., 8, 30, 32, 43–44, 45, 46n37,
  47–48, 51, 55, 58n6, 72n40, 74n44, 94,
  197, 198, 300
Eliot, T. S., 219
Elkus, Savilla Alice, 105
Ellul, Jacques, 366
Ellwood, Charles A., 128n21, 165, 166, 227n2,
  228, 273, 277–278, 279, 282, 345, 373
Ely, Richard T., 8, 26, 74–75, 77–81, 86,
  94–95, 111, 118n3, 120, 125, 131, 163,
  182, 208
emergentism, 9, 101, 122, 148, 152–155,
  156–163, 164–169, 170, 172, 181, 218,
  273, 289n35, 336. *See also* biology
émigrés, 86, 136, 159, 165–166, 209,
  237–238, 240–243, 246n23, 249, 269,
  273, 279, 280–281, 293, 294, 301,
  319–320, 328, 338–339, 344–345, 348,
  353–355
Encyclopedia of Social Reforms (Bliss), 75
Engels, Friedrich, 244
engineering, 12, 32, 201; and
  associationalism, 210, 230, 302–303;
  bioengineering, 365; Bush, 230, 302–303,
  304, 307–308, 309–310, 312–313,
  315, 320, 324n39; chemical, 269;
  civil, 213; criticized, 305–306, 331;
  curricula, 305–306, 330–331; electrical,
  302, 355; mechanical, 79, 125; and
  military-industrial complex, 230,
  302–304, 315; mining, 210; and
  religion, 22, 210, 230, 315; and social

sciences, 226, 302, 304. *See also* physical scientists; science administrators
Enlightenment, 10n12, 110, 140, 186, 191, 191n40. *See also* Scottish Enlightenment; sentimental Enlightenment
Ensminger, Douglas, 363
eponymy, 249
Erikson, Erik, 339
Erskine, John, 205, 219
Eskimo, 185, 260
Esperanto, 268
Ethical Culture, 125n16
ethos of science, 247–249
eugenics, 13, 87, 169, 170n54, 207, 220
Europe, 1–2, 3n2, 29, 32, 104n37, 113n5, 192, 237, 245, 246n23, 249, 263–264, 271, 290, 292–294, 307, 328, 333, 344, 347, 362. *See also* émigrés
Everson v. Board of Education, 326
evolution, 9, 26, 28, 40, 55, 59–64, 65, 66, 69, 73n42, 75, 76, 77, 83, 85, 91, 97, 98, 101, 117, 148–150, 152–170, 172, 177n14, 191–192, 206–207, 210, 216, 218, 245, 275, 289n35, 306, 324n39, 336, 368; cultural, 152, 161, 169; social, 127, 157. *See also* biology; emergentism
Experience and Nature (Dewey), 158, 172, 250n32
Experimental College (University of Wisconsin), 205–206, 213, 215
experimentation, 32, 88, 92, 109, 112–113, 151, 158, 161, 164–167, 168, 215, 226, 237, 286–288, 313, 338
experts, 4, 6, 10, 11, 16, 17, 19, 73, 91, 113, 116, 117, 122, 123–124, 143, 147, 171, 172–178, 183, 191, 220, 227–228, 234, 270, 272, 274–275, 282, 294, 300, 306–307, 313, 315, 332, 336, 345, 357, 359–360, 365, 368, 372n13

Fairchild, Henry Pratt, 220
fallibilism, 12, 40, 62, 66, 91, 95, 97, 110, 135, 143–145, 158, 240, 259
farmers, 30
fascism, 16, 226, 244, 290–291, 296, 304–305, 324, 329, 344, 345, 348. *See also* Hitler, Adolf; Nazism; totalitarianism
Federation of American Scientists (FAS), 309, 313–314
Feigl, Herbert, 242. *See also* philosophy, logical empiricist
feminism, 9, 63, 87, 143, 146, 208, 365n1
Fichte, Johann Gottlieb, 155
Finkelstein, Louis, 323

first scientists' movement, 274n4, 302, 304–307, 310–311, 313, 323
Fiske, John, 40, 42–43, 48, 60
Fite, Werner, 193
Flexner, Abraham, 200–201, 308
Florida State College, 33
folkways, 120, 132, 133–135, 136, 146
Folkways (Sumner), 120, 132–135, 136, 248n27, 260
Follett, Mary Parker, 173–174, 175, 176
football, 29
Ford Foundation, 351
Ford, Henry, 210, 299
Fosdick, Raymond B., 209
Foucault, Michel, 121
foundationalism, 82
foundations, 176–178, 180, 215, 228, 230, 268, 284–285, 292, 342, 343n12, 351, 358, 365. *See also* Social Science Research Council
Fountainheads of Freedom (Edman), 299
Frank, Glenn, 205
Frank, Lawrence K., 284, 285, 286, 287, 322
Frank, Philipp, 314, 319–320
Frankfurter, Felix, 214
fraternities, 29
free will, 40, 56, 92–93, 172, 199, 240, 326; reconciled with naturalism, 99–101, 149–170, 172, 189–192, 236–237, 240, 273, 289–290, 336, 358. *See also* determinism
Freedman, Ronald, 356n34
Freud, Sigmund, 142, 150, 175, 211, 245, 301, 339, 353, 365n1
Friley, Charles E., 255
Fromm, Erich, 301, 339
frustration of science, 244

game theory, 354–355
general education, 14, 197, 201–209, 219, 220, 315–316, 318, 328–333
General Education in a Free Society (Harvard faculty report), 330, 332–333
general semantics, 269
General Theory of Value (Perry), 164, 193, 222, 273–274
genetics. *See* biology
geography, 46, 260, 280
geology, 60, 208, 267, 291
George, Henry, 32
Gestalt, 164–167, 168, 170, 345
Gestalt Psychology (Köhler), 166
Giddings, Franklin H., 118n3
Gilfillan, S. C., 246, 247
Gilman, Charlotte Perkins, 208

Gilman, Daniel Coit, 8, 30, 32–33, 42, 49, 50, 51, 59, 73
Gladden, Washington, 79
Goddard College, 213
Godkin, E. L., 71–73
Godwin, Parke, 62n15
Goethe, Johann Wolfgang von, 262, 286
Goldenweiser, Alexander, 277.
    See also anthropology, Boasian
Goodnow, Frank, 177
Gorer, Geoffrey, 293
Goucher College, 252
governmentality, 121
The Grammar of Science (Pearson), 86–87
Gramsci, Antonio, 300
Grangers, 32
Graubard, Mark, 323
Great Books, 201, 205, 219, 229, 327n48, 332
The Great Chain of Being (Lovejoy), 257
Great Depression. *See* Depression
Growing Up in New Guinea (Mead), 294
The Growth of the Mind (Koffka), 165
Gruenberg, Benjamin C., 306
Guerlac, Henry, 256

Hadley, Arthur T., 79, 87n5
Haeckel, Ernst, 155
Hall, G. Stanley, 42n26, 88
Hallowell, John, 351
Hamilton, Walton H., 182, 202, 236n2, 251n32
Hand, Learned, 214
Handbook of American Indian Languages (Boas introduction), 259–261, 266–267
Hankins, Frank H., 356n34
Harding, Warren G., 149
Harper's (magazine), 245
Harris, William Torrey, 99
Hart, Hornell, 174n6
Hartz, Louis, 348, 352
Harvard University, 29, 31, 34, 37, 54, 90, 157, 158, 159, 165, 192, 214, 266n66, 291, 312, 338, 363; Allport at, 286n27; ascendant after World War II, 330n53; Babbitt at, 198–199; and civil service reform, 72n40; Conant as president, 256, 302, 315, 317, 319, 330, 361; curricular debates, 30, 32, 102, 197–199, 221n49, 271, 315–316, 327, 330, 332–333; elective system, 47, 197–198; Eliot as president, 32, 47, 72, 197–198; Fifth International Congress for the Unity of Science, 242; Fiske appointment scotched, 60; graduate education, 33; history of

science at, 242, 247, 253, 254, 256, 316, 332n55; humanities defended at, 102, 198–199, 222, 271, 327, 330, 332–333, 337, 339; Lowell as president, 198; Merton trained at, 247; Norton at, 102; operationalism at, 236, 316, 316n24; Pareto Circle, 289; Parsons at, 242, 288–290, 319, 361; Perry at, 101–102, 104, 166–167, 193, 339; perspectivalism at, 237n2; Quine at, 242, 316n24, 339–340; Redbook, 330, 332–333; Royce at, 189; Sarton at, 242, 247, 253, 254, 256, 332n55; Sorokin at, 247, 280–281; Western Civ model, 330, 332–333; Whitehead at, 193–194, 221n49
Hawkes, Herbert E., 203, 204
Hayes, Carlton J. H., 251
Headquarters Nights (Kellogg), 156
Hegel, G. W. F., 26, 99, 127, 155.
    See also philosophy, idealism
Henderson, Lawrence J., 242, 256, 289, 316n24
Henry, Joseph, 41, 53
Herrick, C. Judson, 149–150, 151n7, 159–162, 162–163n36, 163–164, 208
Herron, George D., 75
Hertzler, J. O., 356n34
Hessen, Boris, 244, 247, 252, 254n38
Higham, John C., 257n45
higher criticism, 35, 50
Hill, Thomas, 32
history, 60, 191n40, 287, 332, 341, 348; in adult education, 215; Beard, 8, 110, 187, 188–189, 214, 218, 267, 337, 338n4; book of, 38, 43, 47, 50; consensus, 352–353; corporate-liberal school, 4n3, 6–7, 139; cultural, 230, 283n23; and cultural sciences, 13, 16, 110, 111, 114–115, 183, 277, 280; in curricula, 30, 46, 48–49, 62, 65, 203–204, 205, 205n17, 207, 252–256, 305n2, 330n53, 332n55; History of Science Society, 252, 255–256, 313; and humanities, 102, 299, 336, 337–338, 367; intellectual, 48, 191, 193–194, 209–210, 228–229, 235–237, 250–252, 257–258, 271, 299, 337–338; Lovejoy, 250, 257–258, 271; of mathematics, 252n35, 256; of medicine, 256n43; modern, 48–49; as modern subject, 30, 46, 48–49, 62, 65; New History, 8, 187–189, 212, 251–252, 337, 338n4; popular, 209; postwar, 228–229, 235–237, 250, 256, 330n53, 332n55, 336, 337–338, 341, 348, 349, 352–353, 367; and pragmatism, 271, 299;

Progressive, 189, 337, 338n4; and religion, 327, 337; Robinson, 8, 187–188, 212, 214, 251, 252, 257n45, 264, 337; Sarton, 242, 247, 250, 253, 254, 256, 257, 332n55; of science, 9, 64–65, 166, 228–229, 235–237, 242, 244, 247, 250–256, 257n44, 270–271, 289, 303, 305n2, 313, 316, 317n24, 332n55, 366; and scientific democracy, 8, 9, 13, 16, 30, 46, 48–49, 62, 110, 111, 114–115, 183, 188, 189, 212, 238, 336, 337–338; social, 310; of technology, 246, 247; and two-cultures model, 310, 337; and Western Civ, 329, 337, 349; and World War II, 290, 329, 330n53, 337–338

A History of Experimental Psychology (Boring), 166

A History of Modern Culture (Preserved Smith), 252

History of Science Society (HSS), 252, 255–256, 313

History of the Conflict Between Religion and Science (Draper), 64

Hitler, Adolf, 160, 249, 290–291, 293, 323, 331, 344, 345, 348. *See also* fascism; Nazism; totalitarianism

Hocking, William Ernest, 107, 193, 194–195

Hofstadter, Richard, 63n17, 337, 348, 352, 353n28, 354n30

Hogben, Lancelot, 245, 255

holism, 165, 170, 213, 279, 281; biology, 157–158, 161, 170; linguistics, 261–262; psychology, 164–165, 168–169, 170, 287n29

Holmes, Oliver Wendell, Jr., 123

Holmes, Oliver Wendell, Sr., 68

Holt, Edwin B., 104, 165, 167, 168, 173

Hook, Sidney, 162, 266, 270n72, 324–325

Hooton, Earnest A., 269n72

Hoover, Herbert, 210, 274–275, 302–303

Hoover, J. Edgar, 325–326

Hopi, 263

Hopkins, Johns, 32

Horney, Karen, 301, 339

How We Think (Dewey), 239n10

HSS. *See* History of Science Society

Hull House, 98

Hull, Clark L., 242

human nature, 36, 37, 82, 97, 125–128, 136, 158, 172, 177, 191, 200, 203, 250n32, 298, 300, 327

Human Nature and Conduct (Dewey), 158, 172, 250n32

Human Nature and Enduring Peace (Murphy), 300

Human Nature and the Social Order (Cooley), 125–128

human sciences: and scientific democracy, 3, 11–12, 13, 16, 110–116, 146, 170, 172, 188, 195, 206, 225–234, 238–239, 335–342, 363–364; defined, 3n2; demographic shifts, 14, 343–348; emergentism authorizes, 152, 154–155, 159, 162, 170, 336; and federal government, 76, 115–116, 119, 202n12, 225–228, 272–273, 274–275, 276, 292, 290–294, 300–301, 312, 312n16, 342; liberal Protestantism authorizes, 38; value-neutrality, 7, 8, 66, 73, 81, 87–88, 88n6, 109–115, 176–177, 178, 199, 228, 232–233, 235, 236, 271, 272–273, 273–282, 307n7, 311, 312, 327, 329, 330n53, 333, 345, 346, 351, 355–356, 361; value-neutrality ascendant after 1930s, 9, 225–226, 232, 312, 333–334, 357–358, 365; value-neutrality criticized, 8, 102, 109–111, 114–115, 166, 198–200, 216–223, 226, 228–230, 232–233, 235, 250, 272–273, 277–282, 283, 321–323, 326–327, 329, 337–340, 345, 351, 355, 356, 356n34, 365–367. *See also* cultural sciences, names of individual disciplines

humanism: naturalistic faith, 143, 194, 209, 247n26; science-centered liberal education, 253, 254, 314, 315–320; New Humanism, 198–199, 200, 201, 216, 217, 217n40, 218, 219, 222, 231, 328; Western ethical tradition, 333, 355

"Humanist Manifesto" of 1933, 194

humanistic tradition, 4n3, 12, 102, 198–199, 217, 218–219, 230, 231, 310, 314, 321, 322, 327–333

humanists, 216, 222, 223, 226, 229–230, 231–232, 234, 319, 321–322, 327–328, 334, 335, 341, 349, 356, 367; academic, 102, 162, 196–197, 198–199, 200, 217, 219, 220, 222, 229–230, 231, 322, 327–328, 336–339; non-academic, 198, 199, 218–219, 222, 229, 231, 290, 327–328, 331, 366

humanitarianism, 15, 24, 26, 34, 36–37, 39, 40, 45, 46, 51, 70, 74, 76, 77, 114, 115, 133–134, 139, 222, 227, 232. *See also* altruism

humanities, 3, 102–103, 196, 198, 201, 229, 231, 253, 310, 318, 320, 327–328, 330–333, 336–339, 367. *See also* languages and literature

The Humanizing of Knowledge (Robinson), 212

Hume, David, 101
Hutchins, Robert M., 205, 221–222, 223, 326
Huxley, Julian, 245
Huxley, Thomas Henry, 61n12, 245

IALA. *See* International Auxiliary Language
    Association
IAS. *See* Institute for Advanced Study
Ideology and Utopia (Mannheim), 245–246
immigration, 12, 82, 166, 171, 196, 285.
    See also émigrés
Indiana University, 254
individualism, 104, 112, 133, 154, 183, 199,
    204, 211, 233, 251n32, 284–285, 288,
    294–295, 296, 297, 298, 340, 344, 353.
    See also laissez-faire
Individualism, Old and New (Dewey),
    251n32, 298
industrial democracy, 197, 250, 291
industrial society, 12, 39, 78, 94, 98, 99,
    108, 115, 125, 126, 127, 130, 131, 137,
    141, 154, 181, 185–186, 189, 197, 198,
    202, 209, 210, 219, 229, 328, 333, 360,
    370–371, 373. *See also* capitalism
industrialization, 12, 15, 17, 22, 23, 24,
    25–26, 28, 29, 34, 36–37, 39, 65, 71n36,
    74, 78, 95, 178–179, 183, 190, 203, 260,
    283, 298, 346, 361–362; crisis of, 98,
    146, 210
industry, 13, 24, 36, 37, 39, 47, 48, 50, 52,
    70–71, 74, 75, 139–140, 141, 143, 179,
    188, 190, 200, 213, 218, 219, 274, 360,
    372n13; science and, 1, 5, 12, 25, 36, 46,
    51, 69–70, 111, 115, 178, 210, 211–212,
    230, 305–306, 307, 308n9, 311, 315,
    318, 365, 367, 369. *See also* business;
    business leaders; capitalism; corporations
Ingersoll, Robert, 60
The Inquiry. *See* National Conference on the
    Christian Way of Life
Institute for Propaganda Analysis (IPA),
    267–268, 268n70
Institute for Social and Religious Research
    (Rockefeller), 180
Institute for Advanced Study (IAS), 308
intellectual freedom, 25, 27, 32, 41, 46, 57,
    58, 94, 104, 105, 106, 108, 184, 202,
    212, 229, 231, 248, 262, 298, 304, 314,
    324–325, 339, 340, 348
Intelligence in the Modern World (Ratner
    introduction), 239n10
interfaith initiatives, 347
International Auxiliary Language Association
    (IALA), 268

International Congress of the History of
    Science and Technology, 244
International Encyclopedia of Unified Science
    (Neurath et al.), 242. *See also* philosophy,
    logical empiricist
interpretive pluralism, 93, 99–101, 103,
    105–106, 126, 152, 155, 159, 162, 167,
    170, 192, 263, 284, 286–290, 319, 322,
    336, 339–340n8, 361
Interstate Commerce Commission, 125
Introduction to the History of Science
    (Sarton), 256
inventors, 210, 299, 308, 317.
    See also technology
Iowa State College, 254–255
IPA. *See* Institute for Propaganda Analysis
The Irony of American History (Niebuhr),
    321n34, 327
Isis (journal), 253

James, William, 98, 164, 190, 222, 250n31,
    299; on Bergson, 153; compared to Dewey,
    90, 92, 94, 96; compared to Peirce, 90,
    92; conception of experience, 92–93, 100,
    104, 167, 208, 259; as democratic hero,
    7; emergentism, 154, 158; vs. materialism,
    92–93, 96, 149; vs. metaphysics, 92; mind
    as a natural fact, 158–159, 192; and
    the New Psychology, 26, 88, 89, 92–93;
    politics, 94, 94n14, 370n9; pragmatism, 8,
    26, 89–90, 92–93, 95, 101–102, 108, 136,
    154, 254; *The Principles of Psychology*,
    215; vs. professionalization, 102–103;
    religious hypothesis, 93; students, 104,
    164, 167, 189, 214; theory of learning,
    183, 215, 297
Japanese national character, 300–301
Japanese internment, 292
Jeans, James, 245
Jefferson, Thomas, 30, 298, 326
Jelliffe, Smith Ely, 269n72
Jennings, Herbert Spencer, 149, 150n3, 151n7,
    158, 162, 170n54
Jensen, Howard, 227n2
Jewett, Frank B., 302–303, 308n10
Johns Hopkins University, 31, 32–33, 41,
    51n51, 54, 55, 60, 70, 106, 151, 256n43,
    257, 312
Johnson, F. Ernest, 216, 323
Jones, Howard Mumford, 271, 327
Journal of General Education, 333
Journal of Philosophy, 238
*Journal of Philosophy, Psychology and
    Scientific Methods*, 238

Journal of Speculative Philosophy, 99
Journal of Symbolic Logic, 238n7
Journal of the History of Ideas, 257
journalism, 13, 31, 71, 137–138, 173, 180–
    181, 229, 276; science, 61, 63–64, 211,
    245, 306, 308n9, 355. *See also* periodicals
Judaism, 12, 77, 103, 136, 186, 196–197,
    247n26, 323, 327, 344–345, 347–348,
    350–351
Judeo-Christianity, 229, 320, 323
jurisprudence. *See* law

Kaempffert, Waldemar, 306, 307, 308n9, 355
Kaempffert-Calder Plan, 306
Kallen, Horace, 214, 219n45, 242, 299, 326
Kant, Immanuel, 26, 99–100, 155, 162.
    See also philosophy, idealism
Karier, Clarence, 6
Kellogg, Vernon L., 156
Kemp Smith, Norman, 103, 105–106
Keppel, Frederick P., 215
Kettering, Charles P., 305
Keynesianism, 343n13
KGB, 304
Kilgore, Harley M., 307, 308n9, 312
Killian, James R., Jr., 330
Kilpatrick, William H., 216
Kluckhohn, Clyde, 319, 346, 362n47, 363
Knies, Karl, 77
Knight, Frank H., 356n34
Knights of Labor, 29
Knowledge for What? (Lynd), 273, 281–282
Koffka, Kurt, 164, 165, 167.
    See also psychology, *Gestalt*
Köhler, Wolfgang, 164, 165–166.
    See also psychology, *Gestalt*
Korean War, 302, 308n11
Korzybski, Alfred, 269
Krutch, Joseph Wood, 155n16, 322n35, 327
Kuhn, Thomas S., 256, 317n24, 366

labor relations, 17, 29, 52, 65, 73, 74, 77, 80,
    83, 94, 119, 127, 139, 140–141, 185,
    188, 193, 209–210, 254, 274, 281, 304,
    305. *See also* workers
Labor Temple, 209–210
Laboratory School, 98
Ladd-Franklin, Christine, 103–104
LaFarge, John, 323
laissez-faire, 25–26, 36, 48–49, 72, 72n37,
    115, 120, 127, 196, 200; criticized,
    15, 17, 74–82, 95, 96, 115–116, 139,
    144–145, 183, 190–191, 233, 250, 275,
    278, 289, 291, 295, 296, 298, 344,

348, 350; economists' debate, 78–79;
    justification contested, 65–66, 73–74;
    Newcomb, 70–71, 73–74, 78–79; Sumner,
    68–70, 71, 73–74, 87–88, 96, 133–134;
    Youmans, 66–68, 69, 70, 71, 73–74, 96.
    See also conservatism
laissez-faire communitarianism, 52, 229,
    232, 235–236, 243, 247–249, 270–271,
    302–304, 310–311, 312–313, 314–316,
    319, 321, 342, 357–358; defined, 310.
    See also scientific community
Lamarck, Jean-Baptiste, 156, 156n20
Language (Bloomfield), 265–266
languages and literature; classical, 29–30,
    33, 34, 47–48, 51, 54, 67; as cultural
    products, 77, 276; in curricula, 31,
    47–48, 51, 103, 198–199, 207, 215, 266,
    269, 330, 331; and history, 271, 337;
    modern, 30, 46, 47–48, 267; literary
    criticism, 75, 110, 114, 219, 222, 230,
    246, 258, 264, 266, 327; and science,
    310, 314–319, 339. *See also* humanities;
    humanistic tradition; humanists;
    philology; poetry
Lashley, Karl, 159–160, 165, 168
Lasswell, Harold D., 163, 175–176, 177,
    178n15, 322, 359, 359n41
Laura Spelman Rockefeller Memorial, 285.
    See also Rockefeller Foundation
law, 29, 62, 67, 73, 80, 86, 109, 125, 127, 139,
    146, 163, 181, 201, 205, 218, 297, 347,
    368; legal realism, 123, 214, 221, 269n72
LeConte, Joseph, 60
Lee, Dorothy Demetracopoulou, 264n60
Leigh, Robert Devore, 213
Leopold and Loeb, 150
Lewin, Kurt, 167, 294
Lewis, C. I., 192, 193
Lewis, Oscar, 356n34
liberal arts, 254, 315, 331
liberal consensus. *See* consensus liberalism
liberal education, 13, 31–32, 33, 51, 71n36,
    205
liberal tradition. *See* consensus liberalism
liberalism, 1, 4n3, 5, 6–7, 7n9, 8, 17, 25, 41,
    72n37, 72n39, 106, 126, 138, 139, 142,
    190–191, 205–206, 214, 225–228, 247,
    247n26, 249, 297, 298n53, 324–325,
    327, 329, 330n53, 335, 336, 341–342,
    345, 348, 349–350, 351, 366, 367n4,
    368–369, 370. *See also* consensus
    liberalism; managerial liberalism;
    Protestantism, liberal
Library of Congress, 229

Lincoln's Birthday Committee for Democracy and Intellectual Freedom, 304
Lindeman, Eduard C., 214, 246n23, 267
Linguistic Society of America, 267
linguistics, 207; and auxiliary languages, 268; behaviorist, 265–266, 266n67, 267, 268; Bloomfield, 265–266, 267, 268; Boas, 259–261; Boasian, 259–264, 267, 270; and communication research, 268; comparative method, 258–259; and cultural sciences, 13, 16, 111, 114–115, 226, 235–236, 280; emotive-referential distinction, 264–269; and general semantics movement, 269, 269–270n72; holism, 261–262; interwar flowering, 108, 226, 235–236, 258–270; and logical empiricism, 265; as moral science, 265; Ogden and Richards, 264, 266, 267, 267n68, 268, 269; pragmatics, 266–267; and propaganda analysis, 267–268; and science, 259, 261, 263–264, 266, 270; Sapir, 259, 261–263, 267, 270; Sapir-Whorf hypothesis, 259, 261; and scientific democracy, 13, 16, 111, 114–115, 228, 235–236, 258–259, 263–264, 267–270; structuralist, 261–264; Whorf, 259, 261, 263–264, 270. *See also* anthropology
Lippmann, Walter, 175, 214; as democratic realist, 8, 18, 172; *Drift and Mastery*, 142–145; as New Deal critic, 229; and *New Republic*, 8, 137–138; *A Preface to Politics*, 142–144; as Progressive theorist, 137–138, 142–146; on science, 142, 143–146; vs. social sciences, 220, 222, 229
literature. *See* humanistic tradition; languages and literature
Literature and the American College (Babbitt), 198–199, 200
Lloyd Morgan, C., 152–153, 154, 157, 162, 289n35
Locke, Alain, 214, 322
Locy, W. A., 255
logic, 29, 71–72, 90, 91, 104n36, 104n37, 105, 107, 108, 154, 168, 192, 207, 238n7, 241, 255, 257, 275, 288, 351, 357, 358. *See also* mathematics; philosophy
London School of Economics, 245
The Lonely Crowd (Riesman), 340
Los Alamos. *See* Manhattan Project; nuclear weapons
Lovejoy, Arthur O.; anti-communism, 325; on emergentism, 162, 162n35; on philosophy

and science, 106–107; and intellectual history, 250, 257–258, 271
Lowell, A. Lawrence, 198, 222
Lowie, Robert H., 184, 186–187. *See also* anthropology, Boasian
Lundberg, George A., 346, 351, 355–356
Lustig, R. Jeffrey, 6
Lynd, Helen M., 180
Lynd, Robert S., 267, 326, 345; *Knowledge for What?*, 273, 281–282, 283; *Middletown*, 180; on social role of scholar, 282, 283, 313; vs. value-neutrality, 228, 273, 281–282, 283
Lysenko, Trofim, 245

Mach, Ernst, 86, 87, 92, 93n13, 187
MacIver, Robert M., 228, 246n23, 261n55, 273, 279, 280, 281, 291, 322, 345
MacLeish, Archibald, 229
The Making of the Modern Mind (Randall), 191, 204, 251, 252n35, 330n52
Malisoff, William M., 304, 305n2
Malthus, Thomas 36, 134, 138
management theory, 174n4
managerial liberalism, 3, 5, 6–7, 120, 137, 138, 144n53, 172, 225, 272, 306–307, 313, 345, 346, 357, 372. *See also* liberalism
Manhattan College, 101n31
Manhattan Project, 307, 309, 325
Manichean sciences, 355
Mann, Charles R., 253, 255
Mannheim, Karl, 244, 245–246, 247, 327n47
marginalism, 74, 81–82, 138
Marvin, Walter T., 104–105, 106n43, 153, 154. *See also* philosophy, New Realist
Marx, Karl, 298
Marxism, 229, 244–245, 246, 248, 254, 276, 287, 321, 348. *See also* communism
Maslow, Abraham, 338–339, 347n19
Massachusetts Institute of Technology (MIT), 32, 34, 51, 55, 72n40, 268n71, 302, 306, 307, 330–331, 355, 366n1
mathematics, 29, 63, 90, 103, 104n37, 105, 108, 207, 252n35, 253, 255, 256, 258, 262, 269, 281, 331, 336, 354
Mathematics for the Millions (Hogben), 255
Mather, Kirtley F., 267, 291, 326
Matthews, J. B., 326. *See also* anti-communism
May, Henry, 299
May, Rollo, 338–339, 346
McCarthy, Joseph, 16, 225, 325–326, 342, 343, 348, 351, 355, 356. *See also* anti-communism

McCollum v. Board of Education, 326
McCosh, James, 49–50, 217
McDougall, William, 159
McGrath, Earl, 333
McNamara, Robert, 369
Mead, George Herbert, 119–120, 124, 125, 128–130, 131, 131n28, 133, 174, 203, 243, 246, 255
Mead, Margaret, 8, 165, 234, 293–296, 300, 326, 346, 350, 352–353, 355, 359
And Keep Your Powder Dry, 294, 295–296, 352. *See also* anthropology, Boasian
The Meaning of Adult Education (Lindeman), 214
The Meaning of Meaning (Ogden and Richards), 264, 267, 269
medicine, 12, 29, 68, 73, 92, 200, 211, 256n43, 269n72, 276, 285, 313, 318, 367. *See also* professionals
Meiklejohn, Alexander, 202, 203n14, 205–206, 215, 222
mental hygiene, 269n72
merchants, 30, 31
Mercier, Louis J. A., 222
meritocracy, 26, 52–53, 69, 140, 315, 360
Merriam, Charles E., 176, 177–178, 201
Merton, Robert K., 245, 347, 348; eponymy, 249; ethos of science, 247–248, 249; and history of science, 247, 254, 254n38; laissez-faire communitarianism, 243, 247–249, 310, 316, 316n24, 354, 357; "A Note on Science and Democracy," 247–248; postwar writings, 249, 357–358; underdetermination theory, 248; wartime writings, 247–249, 351
Methodism, 31, 326
Methods in Social Science (Rice), 280
Methods in Sociology (Ellwood), 278
Meyer, Adolf, 269n72
Michels, John, 58n6, 63–64, 65n21, 85
Microbe Hunters (de Kruif), 211
middle class, 1, 6, 29, 80, 94, 96, 112, 123, 133–134, 138, 142–143, 190–191, 204, 244, 247, 251–252, 258, 269, 336, 340, 372n13. *See also* professionals
Middletown (Lynds), 180, 281
military-industrial complex, 226, 230, 234, 270–271, 303, 304, 308n11, 316n23, 320, 342, 343, 365, 369, 373
Mill, John Stuart, 61n12, 86, 114, 349
millennialism, 24
Miller, Perry, 337–338
Millikan, Robert A., 210, 211, 304–305, 313n8, 324n39

Mills, C. Wright, 234, 246, 246n23, 340, 355n33, 356, 357, 365n1
Mind and the World-Order (Lewis), 192
The Mind in the Making (Robinson), 187–188, 212, 264
mining, 32, 210
ministers, 25, 29, 43–44, 44n32, 54, 73, 75–76, 79, 201, 326. *See also* Protestantism; religious leaders
Mirowski, Philip, 270n73
missionaries, 13, 77, 195
MIT. *See* Massachusetts Institute of Technology
Mitchell, Wesley C., 182, 214, 228, 306–307, 307n7, 312
Mizruchi, Susan L., 75
Modern Arms and Free Men (Bush), 309
A Modern College and a Modern School (Flexner), 200–201
Modern Science and Modern Man (Conant), 316–318, 320–321, 330
modernism, 142, 229–230
modernity, 181, 328, 345, 366, 368
modernization theory, 361–362
The Monist (journal), 99
Montagu, Ashley, 336–337, 341. *See also* anthropology, Boasian
Montague, W. P., 104–105
moral government, 37, 40, 58, 67, 217
More, Paul Elmer, 199, 222, 328. *See also* New Humanism
mores, 120, 132, 134–135, 136, 146, 181
Morgan, Arthur E., 212, 213, 268n71
Morgan, C. Lloyd. *See* Lloyd Morgan, C.
Morgan, T. H., 148, 149
Morgenthau, Hans, 345
Morrill Act, 30–31, 74n44. *See also* land-grant institutions
Morris, Charles W., 242–243, 255
Morris, George Sylvester, 42
Moulton, Forest Ray, 208, 307
Moulton, Harold G., 307
Mumford, Lewis, 218–219, 222, 229, 290, 331, 366
Muncie, Indiana, 180
Munsterberg, Hugo, 273
Murphy, Gardner, 285–288, 300, 327n47, 346
Murphy, Lois Barclay, 285
Murray, Henry A., 286n27
Myrdal, Gunnar, 299–300, 351

Nagel, Ernest, 162, 243, 251–252, 257n44, 270n72, 348
NAS. *See* National Academy of Sciences

Nashville Agrarians, 328
The Nation (magazine), 71
National Academy of Sciences (NAS), 302–303, 304. *See also* science administrators
National Association of Manufacturers, 281
national character. *See* anthropology
National Conference on the Christian Way of Life (The Inquiry), 173
National Education Association (NEA), 32, 321n34
National Institutes of Health, 304
National Research Council (NRC), 292, 294
National Review (magazine), 326
National Science Foundation (NSF), 304, 307–308, 308n11, 311–313, 316n23, 342–343n12, 355
national security state, 8
nationalism, 202–203, 372n13
nativism, 111, 171, 286
natural law, 217–218, 221
natural laws, 68, 88, 95, 207–208, 209, 240, 248n27
natural theology, 34, 37, 47, 50, 61
Nazism, 228, 248, 249, 301, 324, 343, 345, 347, 348, 349. *See also* fascism; Hitler, Adolf; totalitarianism
NEA. *See* National Education Association
Needham, Joseph, 245
neo-positivism. *See* philosophy of science
neo-Thomism. *See* philosophy
Neurath, Otto, 306. *See also* philosophy, logical empiricist
New Aspects of Politics (Merriam), 176, 177–178
New Deal, 5, 72n37, 115, 119, 138, 182, 208, 229, 232, 233, 235, 236, 249, 275, 283, 303, 304, 306, 343, 348, 349, 352, 360; impact on scientific democracy, 111, 212, 225–227, 272–273
New Deal order, 5
The New Democracy (Weyl), 138–140
New Humanism, 198–199, 200, 201, 216, 217, 217n40, 218, 219, 222, 231, 328
New Left, 7, 16, 234, 336, 365–366, 367n4, 373
The New Republic (magazine), 8, 120, 137–138, 214, 308n9
New School for Social Research, 165, 213–214
The New State (Follett), 173–174
The New Universe (Brownell), 208–209
New York (city), 65, 68, 73, 76, 186, 209, 212, 213, 242, 268, 304
New York (state), 31

New York Times, 306, 333, 355
New York University (NYU), 220, 266
New York World's Fair, 304
New Yorker (magazine), 251
Newcomb, Simon, 70–71, 72, 74, 77, 78–79, 79n58
Newcomb, Theodore M., 179–180
Newman, Horatio Hackett, 207–208
Newton, Isaac, 244, 263
Niebuhr, Reinhold, 217n40, 218, 222, 223, 226, 229, 290, 321, 327, 337–338, 345
Nietzsche, Friedrich, 155
1930s generation, 344, 347–351, 358–362
North American Review (magazine), 60
Northwestern University, 207, 208–209, 254–255
Norton, Charles Eliot, 102, 198, 217
"A Note on Science and Democracy" (Merton), 247–248
NRC. *See* National Research Council
NSF. *See* National Science Foundation
nuclear weapons, 230–231, 307, 308–309, 313–314, 315, 325–326
NYU. *See* New York University

Oberlin College, 207
Office of Facts and Figures, 292
Office of Scientific Research and Development (OSRD), 302, 307
Office of Strategic Services (OSS), 292
Office of War Information, 292
Ogburn, William F., 8, 238, 313, 321, 345; anti-communists target, 326; criticized, 228, 273, 277–279, 281; cultural lag, 178–179, 189, 215, 275, 305, 311; epistemology, 275–276, 307n7, 311; and federal government, 274–275, 276, 312; "The Folk-Ways of a Scientific Sociology," 109, 274, 275–277; isolationism, 338n4; neo-positivism, 109, 235, 236, 274, 275–278; in NSF debate, 312; and SSRC, 178, 312; *Social Change, With Respect to Culture and Original Nature*, 178–179; on social role of scholar, 179, 274–275, 276–277, 282; technological determinism, 178–179, 189, 275
Ogden, C. K., 264, 267, 268, 269
Ohio State University, 216, 239, 265
The Old Savage in the New Civilization (Fosdick), 209
One World or None (various), 309
Open Court Publishing, 99
operationalism. *See* philosophy of science
operations research, 354

Oppenheimer, J. Robert, 231–232, 314–315, 318–319, 320, 321, 325–326, 354, 357, 360
organic analogy. *See* social organism
organization theory, 273
OSRD. *See* Office of Scientific Research and Development
OSS. *See* Office of Strategic Services
Otto, Max C., 192, 193, 240
The Outline of History (Wells), 209
Outlines of a Critical Theory of Ethics (Dewey), 163n38
Overstreet, Harry A., 209, 273n2

Paine, Thomas, 299
Palestine, 333
pan-Protestant establishment, 2, 21, 22–23, 24–25, 98. *See also* Christianity; Protestantism
Pareto Circle, 289
Park, Robert E., 180–182
Parrington, Vernon L., 271
Parsons, Talcott, 242, 319, 346; anti-communists target, 326; and behavioral sciences, 288, 353, 361; and consensus liberalism, 288, 290, 353, 361; and culture and personality, 288, 361; vs. determinism, 289–290; laissez-faire communitarianism, 361; and modernization theory, 361–362; in NSF debate, 312, 312n16; perspectivalism, 237, 288–290, 361; postwar writings, 353, 361; on social role of scholar, 289–290; and SSRC, 312, 312n16; *The Structure of Social Action*, 288–290; voluntaristic theory of action, 289–290; wartime activities, 292, 300
participation, political, 6, 7n9, 9–10, 10n14, 17, 72, 124, 172–175, 177, 178, 191, 227, 282, 298, 311, 313. *See also* deliberative democracy; public opinion
Paton, Stewart, 269n72
Patten, Simon, 138–139
Patten, William, 156, 169, 206–207, 208
Patterns of Culture (Benedict), 269n72, 284
Pearl, Raymond, 155, 170n54
Pearson, Karl, 22–23n2, 86–87, 92, 112, 113, 124
Peirce, Benjamin, 90
Peirce, Charles S., 8, 26, 38, 89–92, 95, 97, 98, 100, 156n20, 187, 316
People's Institute of New York, 151, 209
periodicals: *The New Republic*, 8, 120, 137–138, 202, 214, 308n9; newsletters,

267, 309; newspapers, 306, 333, 355; other magazines, 60, 71, 156, 245, 251, 267n68, 269, 282n19, 306, 326. *See also* scholarly journals
Perry, Ralph Barton, 114n8, 165n44, 167, 168, 174; *General Theory of Value*, 164, 193, 222, 273–274; and *Gestalt*, 165, 166; and emergentism, 163, 164; and humanities, 339; and New Realism, 101–102, 104–105, 106, 107; and World War II, 290–291, 339
perspectivalism. *See* philosophy of science
philanthropy, 3, 29, 31, 32, 67, 68, 71, 95. *See also* foundations
philology, 46, 102. *See also* languages and literature
The Philosopher of the Common Man (various), 299
Philosophical Review, 99
philosophy, 134, 135, 212, 213, 219n45, 225, 261, 265, 266, 269–270n72, 280, 302, 311, 323n38, 347, 348, 360; aesthetics, 37, 104n36, 193, 273–274, 329, 340; American Philosophical Association, 26, 98, 100, 101n31, 102–107, 257, 273n2; analytic, 193, 242, 243, 339–340, 345, 357n37; and anti-communism, 324–325, 326; Aristotle, 202, 252, 262, 269; behaviorism, 165, 165n44, 167, 168, 173, 236; and biology, 13, 16, 26, 59–60, 63–64, 83, 85, 92, 94, 97, 114, 128, 131, 148–149, 151–155, 156, 156n20, 157, 158–159, 160, 162, 163, 164, 170, 171, 183, 189–190, 191–192, 194, 236, 238, 242, 247, 289n35; common-sense, 15, 26, 56–58, 82, 83, 84, 85, 88, 89, 93, 107, 217n41; and consensus liberalism, 297–299; Creighton, 100–101, 102, 103, 105, 237, 288, 289n35; critical realism, 241; in curricula, 26, 29, 35, 37–38, 46, 48, 49, 83–84, 98, 102–103, 198–200, 202–206, 207, 208–210, 251, 255, 330, 331; dualism, 158, 169, 199, 200, 219; of education, 6, 11, 98, 124, 130–132, 216; and emergentism, 148, 151–155, 157, 158–159, 162, 163, 164, 170, 172; empiricism, 56, 60, 96, 101, 104, 190, 250, 280, 339; ethical transcendentalism, 94, 95, 96, 129, 188, 190, 250, 278; ethics, 10, 16, 19, 22, 25, 26, 28, 34–36, 37, 39, 43, 50, 54, 58, 73–74, 77, 79, 87, 88–89, 92, 94, 95, 96, 97, 98, 108, 112–114, 134, 135, 154n13, 156, 164,

philosophy (*cont.*)
171, 188, 189, 193, 194, 200, 201, 202, 207, 216, 219, 227, 232, 273–274, 276, 278, 321, 323, 324n39, 327, 329–331, 333, 337, 349, 351, 370, 372n13; ethics of communication, 24, 41; ethics, consequentialist, 38, 113–114, 128, 129, 131, 132, 203–204, 205; ethics, emotivist, 241, 242n15; ethics, utilitarian, 114, 191; ethics, virtue, 23–24, 73–74, 112; Frank, 314, 319–320; and freedom, 27, 99–100, 103, 105–108, 151–152, 154, 156, 163, 164, 189–190, 192, 240, 297, 298, 370; and *Gestalt*, 164, 165, 166, 167, 170; history of, 251; and history of science, 228–229, 235–237, 250–252, 254–255, 257n44; and humanities, 102–103, 193, 198–200, 201, 205, 216–217, 218–219, 221–223, 229–230, 271, 320, 331, 336, 338, 339–340; idealism, 73, 76, 86, 99, 101, 105, 106, 107, 126, 153, 173, 189, 193–195, 241, 242, 245, 264, 273; idealism, British, 99, 104; idealism, Hegelian, 26, 99; idealism, Kantian, 26, 99–101, 202, 288; instrumentalism, 97, 148, 203, 218, 236n2, 238–240, 356; and intellectual history, 191, 193–194, 228–229, 235–237, 250–252, 257–258, 271, 299; James, 7, 8, 26, 88–90, 92–93, 94, 95, 96, 98, 100, 101–102, 102–103, 104, 108, 136, 149, 153, 154, 158–159, 164, 167, 183, 189–190, 192, 208, 214, 215, 222, 250n31, 254, 259, 297, 299, 370n9; logical empiricism, 237–238, 240–243, 265, 270, 306, 319–320, 338, 339; materialism, 51, 59, 60, 63–64, 75, 85–86, 91–93, 96, 99, 101, 151, 155, 156, 159, 162, 192, 197, 199, 200, 208, 217, 229, 230, 241, 245, 251–252, 264, 297, 320, 321; and mathematics, 103, 104n37, 105, 108; metaphysics, 2, 29, 31, 51, 59, 61, 85, 86–87, 91–93, 94, 99, 102, 104, 107, 155, 158n26, 169, 172, 189, 191, 192n42, 193–194, 208, 221, 263, 274, 323, 326, 331, 339–340, 345; moral, 29, 35, 37–38, 39, 43, 46, 48, 49–50, 57, 58, 74, 98, 103, 273; natural, 29, 244; naturalism, 4n3, 7n10, 9, 60, 61n12, 68, 76, 96n19, 106, 118n3, 148, 151, 152, 158–159, 160, 161–162, 166, 189–192, 194, 203–204, 208, 216–217, 218, 219, 229, 236n2, 238, 241, 242, 243, 251–252, 254, 257, 278, 297, 320–322, 323, 326–327, 329–330, 339; naturalism,

Columbia, 158, 162, 190–191, 194, 203–204, 238, 251–252, 254, 257, 257n44, 297, 329–330, 339; neo-Kantianism, 237n2; neo-Thomism, 101n31, 217–218, 221, 222, 223, 229, 290, 327; vs. New Deal, 227, 227n2; New Realism, 101–102, 103, 104–106, 153–154, 158, 173, 193, 241; Peirce, 8, 26, 38, 89–92, 95, 97, 98, 100, 156n20, 187, 316; Perry, Ralph Barton, 101–102, 104–105, 106, 107, 114n8, 163, 164, 165, 165n44, 166, 167, 168, 174, 193, 222, 273, 274, 290–291, 339; personalism, 193; and physical sciences, 27, 84, 86, 108, 191, 194, 199, 236, 239n8, 240, 241, 243, 289, 320, 338, 339–340; platform idea, 84, 103–108; Plato, 95, 188, 205, 206, 221; popular, 209–210, 210n28; pragmatism, 8, 26, 38, 84–85, 88–98, 99, 101–102, 104, 108, 113, 117, 128–132, 136, 136n39, 142, 154, 167–168, 189, 191, 192, 194, 199, 202, 203, 221, 236n2, 238, 241, 243, 254, 273, 298–299, 370n9; process, 194n46; professionalization efforts, 9, 26–27, 83–84, 98–108; Randall, 190–191, 194, 204, 204n15, 251–252, 257n44, 297, 330n52; and religion, 26, 29, 35, 37–38, 39, 59–65, 85, 87, 89, 91, 92, 93–94, 95–96, 97, 98, 101n31, 104n36, 114, 173, 190, 193–195, 199, 202n11, 205, 216–218, 219, 221–223, 229–230, 239, 240, 297n52; as a science, 35, 37–38, 43, 46, 85, 100–101, 103–108, 336; and scientific democracy, 6–9, 11–12, 13, 16, 18, 19, 26–27, 38, 39, 43, 46, 48, 49, 56–65, 74, 83–108, 111, 113–115, 119–120, 124–125, 131, 148–149, 151–155, 162, 170, 172–174, 177, 183, 188, 189–195, 198–200, 202–206, 208–210, 214, 216–218, 219, 221–223, 228–230, 235–243, 248, 250–252, 257–258, 273–274, 278, 288–290, 297–299, 316, 335, 341, 342, 356, 370n9; Sellars, 157, 159, 164, 191–192, 193, 194; and social sciences, 3n3, 13, 16, 46, 85, 94–96, 111, 114–115, 124–125, 126, 128–132, 148–149, 151–155, 159, 164, 165, 166, 167, 169, 170, 171, 172–174, 183, 188, 189–191, 193, 198–200, 202–206, 208–210, 214, 221, 228–230, 235–237, 238, 240, 242, 246, 248, 257–258, 271, 273–274, 276–277, 278, 288–290, 297–298, 312,

317, 319–320, 326–327, 335, 341, 342, 367; social self, 124–125, 128–132, 203–204, 297; truth, coherence theory of, 241; truth, correspondence theory of, 26, 38, 82, 88, 89; truth, pragmatic criterion of, 26, 38, 89–90, 104, 113, 167–168; Whitehead, 193–194, 221n49, 263; and World War II, 290–291, 292, 338n4, 339; value theory, 164, 193, 273–274; vitalism, 152–153, 264. *See also* Dewey, John; emergentism; logic; philosophy of science; social organism

philosophy of science, 15, 16, 22–23n2, 26, 38, 50n48, 56–65, 82, 83–98, 99–101, 103–108, 109, 110, 112, 113, 117, 129, 140, 143–145, 166, 169, 187, 191, 192, 199, 206, 208, 216, 221, 225, 228–229, 235–236, 236n2, 237, 237n3, 238–240, 241, 242, 243–244, 247, 248, 250–252, 253–256, 257, 258, 262, 262n57, 269, 270–271, 274–277, 282, 288–290, 303, 308, 311, 316, 323, 326, 355, 356, 361, 366, 369, 371, 373; contextualist, 9, 16, 93, 110, 191, 192, 216, 225, 228–229, 235–236, 237, 238–240, 242, 243–244, 247, 248, 250–252, 253–256, 257, 258, 259–260, 262, 268, 270–271, 303, 308, 311, 356, 371; Deweyan, 191, 192, 216, 225, 236, 237, 238–240, 242, 243–244, 247, 248, 250–252, 253–256, 257, 258, 262, 270–271, 303, 311, 371; neo-positivist, 109, 112, 235, 236, 236n2, 237, 240, 241, 248, 274–277, 311, 326; operationalist, 236, 237, 316, 355; perspectivalist, 99–101, 237, 237n3, 262n57, 288–290, 361; positivist, 22–23n2, 26, 61–62, 65, 84–88, 91, 92, 93n13, 94, 113, 113n6, 117, 140, 166, 169, 187, 199, 208, 282, 323, 340; post-positivist, 4, 5n6, 366, 369, 373; underdeterminationist, 248, 316. *See also* philosophy; sociology of knowledge; sociology of science

Philosophy of Science (Philipp Frank), 319–320

Philosophy of Science (journal), 238

physical scientists: and anti-communism, 309, 325–326, 351; associationalism, 230, 302–303; Christian commitments, 22, 210, 230, 305n2, 320–323, 324n39; Conant, 231–232, 256, 302–303, 309, 312, 314–318, 319, 320–321, 330, 333, 335, 354, 357, 360, 361; and humanists, 231–232, 304, 310, 314–321,

323; laissez-faire communitarianism, 302–303, 310–311, 321, 342, 342n12, 357; and military-industrial complex, 230, 302–304, 307–308, 311–313; vs. New Deal, 115, 230, 304–305, 305n2; political activism, 274n4, 302, 304–309, 310–311, 313–314, 323; vs. scientific democracy, 210–211, 237, 302, 310–314, 317–318, 320–321, 322–323, 334, 335; vs. social sciences, 226, 230–231, 302, 310–314, 317–318, 320–321, 322–323; and World War I, 155. *See also* names of individual disciplines; science administrators

physics, 41–42, 53, 63, 88, 155, 167, 194, 218, 241, 244, 260, 261, 262, 277, 289, 338, 355, 374; and art, 230–232, 314, 318–319; atomic theory, 60, 86, 159, 194, 244; criticized, 331; crystallography, 244; cultural authority, 210, 230, 302–303, 310, 322; in curricula, 316n23; Einstein, 210, 263, 269, 307, 309, 323; and federal government, 230, 302–303, 312–313, 325–326, 331; Frank, 314, 319–320; and idealism, 86, 245; and industry, 12, 230, 305; laissez-faire communitarianism, 230–232; modern, 191, 193–194, 320; Newtonian, 105, 244, 263; nuclear, 309, 314; and operationalism, 236; Oppenheimer, 231–232, 314–315, 318–319, 320, 321, 325–326, 354, 357, 360; practical relevance, 155, 230, 305, 318–319; quantum theory, 314; relativity, 263, 314; and religion, 22, 210, 230, 305n2, 322–323, 324n39; and scientific democracy, 12, 149, 162, 170, 210, 230, 239n8, 302–303, 310, 312–313, 322; and two-cultures model, 231, 310, 319–320. *See also* physical scientists; science administrators

Pitkin, Walter B., 104–105, 106n3, 273n2

The Place of Magic in the Intellectual History of Europe (Lynn Thorndike), 252

planning, 181–182, 249, 256n43, 273, 275, 281, 306. *See also* social engineering

plant breeding, 210

plutocracy, 138–140, 145

Plato, 95, 188, 205, 206, 221

poetry, 50, 68, 91, 185, 219, 229, 267, 327, 331. *See also* humanities; humanistic tradition; languages and literature

policy sciences, 359n41

political economy, 30, 37–38, 46, 48–49, 65, 66, 70, 79, 82, 87–88, 125, 188, 190;

political economy (*cont.*)
  classical, 36, 69, 72, 74, 78, 79, 199, 250.
    See also economics, political science
political science, 188, 347, 348; Almond,
  348, 351, 359; antidote to social conflict,
  33–34; behavioralist, 336, 358–359,
  361–362; comparative, 359; and cultural
  sciences, 13, 16, 111, 114–115, 183, 188;
  in curricula, 203, 204–205; democratic
  realism, 7n9, 11, 175–176, 358–359,
  362; empiricist, 109–110, 176–178,
  198; and emergentism, 163; and federal
  government, 13, 227–228, 272; graduate
  programs for civil servants, 49, 72;
  Lasswell, 163, 175–176, 177, 178n15,
  322, 359, 359n41; Merriam, 176, 177–
  178, 201; as modern subject, 33–34; and
  modernization theory, 361–362; opinion
  polling, 267, 268n70, 283n22, 358;
  pluralism, 141, 358–359; political culture,
  138, 232, 359, 361, 362; political theory,
  18, 271, 348, 351, 352, 360; politics
  and administration, 177; and scientific
  democracy, 13, 16, 111, 114–115,
  176–178, 183, 188, 198, 272, 322–323,
  336, 352–353, 358–359, 361–362; vs.
  scientism, 345; and SSRC, 176–177, 201,
  228, 272. *See also* political economy
Popper, Karl, 61, 162
Popular Front, 325, 348
Popular Science Monthly, 60, 61n13, 63, 68,
  73
popular sovereignty, 9–10, 15, 72, 111, 234
popularization, 15, 17, 40, 60, 61n13, 68, 98,
  112, 150–151, 157n22, 159, 173, 184,
  192, 195, 196, 207, 209–212, 219, 256,
  267n68, 269, 273n2, 276, 281, 294–295,
  309, 312n16, 316–318, 320–321, 355
populism, 2, 3, 53, 265, 304, 307, 318, 348,
  358
Porter, Noah, 49–50, 64, 217, 217n40
positive freedom, 77, 94, 163–164, 297–298,
  336–337
positivism. *See* philosophy of science;
  sociology
postcapitalism, 249, 263, 346, 352
post-positivism. *See* philosophy of science
Potter, David, 352
Pound, Roscoe, 123, 269n72
The Power Elite (Mills), 356
A Preface to Politics (Lippmann), 142–143
Primitive Society (Lowie), 186–187
Princeton University, 29, 49–50, 54, 103, 104,
  157, 193, 267, 273n2

The Principles of Psychology (James), 215
problem choice and theory choice, 244,
  244n20, 254, 276, 281, 308, 311, 314
process theology, 194n46
professional journals. *See* scholarly journals
professional organizations. *See* scholarly
  organizations
professionalism, 1, 3, 4, 12, 15, 337
professionalization, 1, 4, 9, 57, 86n2,
  109–110, 127–128, 128n21, 141–142,
  146, 201–202, 210–211, 267; humanities,
  102, 198; philosophy, 9, 26–27, 83–84,
  98–108; science studies fields, 228–229,
  237–238, 242–243, 247, 250, 253,
  256–258, 270–271, 340
professionals, 1, 3, 12, 21n1, 25–26, 29,
  30, 44n32, 65–66, 68, 73, 76, 96, 123,
  201, 232, 254, 269, 319, 336, 347, 360,
  372n13. *See also* administration; experts;
  managerial liberalism; middle class
Progressive Democracy (Croly), 141
Progressivism, 1, 6–7, 8, 17, 26, 72n37, 74–82,
  98, 111, 115, 117–132, 133, 135–147,
  149, 163, 171, 173, 177, 178–179, 181,
  189, 203, 208, 226, 227, 232, 250, 258,
  264, 279, 292, 297, 300, 310, 332, 335,
  349, 359, 368
progressives, 218, 222, 249, 293, 304, 305,
  309, 310–311, 312, 325, 336, 363n50,
  365
The Promise of American Life (Croly), 141–142
propaganda, 8, 105, 111, 176, 258, 264, 267,
  276, 278, 282, 292
Propaganda Analysis (newsletter), 267–268
Protestantism, 91; evangelical, 2, 5, 23, 25,
  30, 34, 36n16, 41, 43, 75, 210, 226;
  liberal, 1, 3, 15, 21, 24, 25, 28, 34–39,
  58, 64, 65, 75–76, 77, 88–89, 95–96,
  105, 117, 119, 149, 190, 210, 230, 273,
  297, 305n2, 306, 336, 344, 345; low
  church, 3, 25, 344; neo-orthodox, 218,
  321; pan-Protestant establishment, 2,
  21, 22–23, 24–25, 98; continuities with
  scientific democracy, 21, 24–26, 30,
  34–39, 44, 45, 46, 50, 54, 59, 88–89,
  94, 95–96, 97, 112–113, 173, 193,
  202n11, 297, 347, 355; Social Gospel, 26,
  75–76, 77, 79–81, 94, 115, 202n11, 277.
  See also Christianity; ministers; religious
  leaders
psychiatry, 8, 12, 16, 245, 269n72, 283,
  284, 292, 301, 365. *See also* culture and
  personality
psychoanalysis. *See* psychiatry

psychobiology, 159–162, 169, 181.
See also biology; psychology
Psychological Warfare Division, 292
psychology, 12, 16, 35n14, 100, 171, 219,
246n23, 248, 264, 267, 268n71, 274n4,
298, 311, 317, 318n27; adult learning,
215; Allport, 227n2, 285–288, 339,
346, 373; animal studies, 160, 168;
attitude concept, 113n7, 179–180, 206;
authoritarian personality, 353–354; and
behavioral sciences, 290; behaviorism,
148, 150–151, 159–160, 161, 164,
165–169, 173, 236n2, 242, 258, 264,
265–266, 266n67, 285–287, 327,
338–339, 346, 356n34, 358; Bentham,
66, 69, 191; common-sense, 15, 26, 56,
82, 83, 93; creativity, 354; and cultural
sciences, 13, 16, 111, 114–115, 164,
165, 166, 169, 172, 175–176, 183, 189,
209, 215, 226, 258, 283–288, 292–293,
319, 336, 338–339, 352; and culture and
personality, 8, 165, 283–288, 292–293,
301; in curricula, 207; and emergentism,
148, 152–153, 158, 159–162, 164–169;
experimentation, 88, 92, 109, 112–113,
151, 165–167, 168, 215, 286–288,
338; faculty, 56, 83, 93; Freudian, 142,
150, 175, 211, 245, 269n72, 301, 339,
353, 365n1; *Gestalt*, 164–167, 168,
170, 345; holism, 164–165, 168–169,
170, 287n29; humanistic, 338–339,
346, 347n19; industrial, 12; instincts,
118, 121, 122; introspection, 151, 159;
Lloyd Morgan, 152–153, 154, 157,
162, 289n35; and moral philosophy, 37;
Murphy, 285–288, 300, 327n47, 346;
as natural science, 88, 151, 165–169,
287–288; neobehaviorism, 166–169,
269, 358; and New Deal, 227n2;
New Psychology, 15, 26, 82, 88, 89,
93, 112, 113n7, 122, 124, 125, 191,
215; open-mindedness, 353–354, 358;
operationalism, 236, 237, 242, 338;
personality, 285–288, 317, 353–354;
popular, 150, 209–210, 211; and
pragmatism, 26, 88, 89, 92–93, 95, 154,
167–168, 259; and scientific democracy,
8, 13, 16, 26, 82, 83, 88, 89, 109, 111,
113, 113n7, 114–115, 118, 132–133,
136–137, 142–143, 146, 148–149,
150–151, 152, 166, 169, 172, 179–180,
183, 189, 209–210, 212, 226, 231, 232,
234, 258, 283–288, 300–301, 311, 336,
338–339, 346, 353–354; social, 172,

179–180, 191, 206, 236, 285–288, 324;
Society for the Psychological Study of
Social Issues, 287, 300, 324; Tolman,
166–169, 170, 179, 242, 269, 275–276,
286, 326, 358; Watson, 148, 150–151,
159, 160, 164, 165, 166, 167, 168,
210, 265, 346, 358; and World War II,
292–294, 300, 342. See also culture and
personality; psychobiology
Psychopathology and Politics (Lasswell),
175–176
The Public and Its Problems (Dewey),
172–173, 239, 250n32, 341, 370
public good, 1, 12, 13, 23, 75, 78, 108, 114,
115–116, 121, 122, 123, 124, 128, 129,
139, 171, 176, 225, 227, 233, 234, 272,
281, 282, 300, 304, 306, 308, 308n9,
313, 315, 341, 342, 349, 351, 358, 360,
361, 370
public opinion, 9, 11, 14, 19, 24, 44, 68, 71,
119–120, 122–123, 124, 127, 133, 146,
148, 172, 177, 181, 183, 186, 189, 204,
234, 267, 297, 300, 313, 358–359, 370,
371n12. See also deliberative democracy;
participation, political
Public Opinion Quarterly, 267–268

Quakerism, 31, 32
quantification, 5n6, 109, 178–180, 206, 235,
236, 238, 240, 277–281, 336, 354, 356.
See also statistics
The Quest for Certainty (Dewey), 172,
250–251n32
Quine, W. V. O., 242, 243, 316n24, 339–340

Rabi, I. I., 312
racism, 118, 120, 136, 160, 233, 293,
299–300, 326, 351. See also eugenics
radicalism, 3, 4, 6, 16, 32, 80–81, 94, 111,
139, 141, 142–143, 171, 178n15, 180,
226, 244–245, 246, 256n43, 276, 281,
304, 306, 352n27, 365
RAND Corporation, 354–355
Randall, John Herman Jr., 190–191, 194, 204,
204n15, 251–252, 257n44, 297, 330n52
rational choice theory, 354, 366
Ratner, Joseph, 239n10
Rauschenbusch, Walter, 76, 202n11
Rawls, John, 360
Raytheon Company, 315
Readability Laboratory (Columbia), 215
Recent Social Trends in the United States
(government report), 228, 274–275
Reconstruction, 29, 34, 63

Reconstruction in Philosophy (Dewey), 172, 250n32

Redbook. *See General Education in a Free Society*

Redfield, Robert, 346, 363

Reformation, 211

Reichenbach, Hans, 270, 270n73, 311. *See also* philosophy, logical empiricist

Reid, Thomas, 56

religious leaders, 2, 9n11, 43–44, 49, 54, 65, 95, 217–218, 222–223, 226, 229, 290, 321–323, 326–327, 341. *See also* Catholicism; Christianity; Judaism; ministers; Protestantism

Rensselaer Polytechnic Institute, 30

Report of the Committee on Organization (White), 31–32, 45

The Republic (Plato), 205, 206

Republican Party, 24, 29, 331

republicanism, 1, 23, 94n14, 112, 175, 205, 315, 370

research moratorium, 305, 313

revivals, 35–36

rhetoric, 29

Rice, Stuart A., 280

Richards, I. A., 110, 264, 266, 267, 267n68, 268, 269

Riesman, David, 38, 234, 340, 348, 355n33

rights, 123, 130n26, 134, 341, 342, 349–350, 360, 363; property, 82, 133

Rignano, Eugenio, 156

Riley, Isaac Woodbridge, 250n31

Ritter, William E., 158

Robinson, James Harvey, 8, 187–188, 212, 214, 251, 252, 257n45, 264, 337. *See also* New History

Robischon, Thomas, 252n34

Rockefeller, John D., 95, 140

Rockefeller Foundation, 176–178, 180, 230, 268, 285. *See also* Social Science Research Council

Rogers, Carl, 338–339, 346

Rokeach, Milton, 354, 358

Romanticism, 126

Roosevelt, Eleanor, 293

Roosevelt, Franklin Delano, 16, 182, 226–228, 233, 272, 274–275, 297n51, 302–303, 305–306, 307, 331, 346, 349, 352

Roosevelt, Theodore, 123, 143, 146, 368–369

Ross, Earle D., 254

Ross, Edward A., 8, 119–123, 124, 125, 125n15, 126, 127, 131, 133, 142, 144n54, 155, 171, 202n10, 255, 297

Rousseau, Jean-Jacques, 199

Rowland, Henry A., 41–42, 60, 61

Royce, Josiah, 101, 189, 190, 192, 193, 208

Ruml, Beardsley, 176

Russell, Bertrand, 104, 208, 323n38

Russell, Doug, 311n15

Rutgers University, 29, 104, 204

St. Louis Hegelians, 99

San Francisco School of Social Studies, 215

Santayana, George, 158–159, 192, 208

Sapir, Edward, 184, 187, 208, 280; auxiliary language, 268; genuine and spurious cultures, 185–186, 262; and *Gestalt*, 165; linguistics, 259, 261–263, 267, 270; Sapir-Whorf hypothesis, 259, 261. *See also* anthropology, Boasian

Sarah Lawrence College, 213

Sarton, George, 242, 247, 250, 253, 254, 256, 257, 332n55

Schelting, Alexander von, 246n23

Schilpp, Paul Arthur, 222, 246n23

Schlesinger, Arthur M., Jr., 337–338, 349n23, 352

Schmidt, Karl, 103–104, 105, 106

Schmoller, Gustav von, 77

Schneider, Herbert W., 190, 251, 271

scholarly journals, 48, 53, 63–64, 99, 238, 238n7, 245, 253, 254n38, 257, 267–268, 274n4, 287, 333

scholarly organizations, 101n30, 176–178, 200, 201, 228, 267–268, 272, 274, 274n4, 280, 287, 290, 292, 296, 300, 302–303, 304, 305, 306–307, 309, 312, 313–314, 320, 322–323, 324, 324n39; disciplinary associations, 26, 32, 57, 73, 79–80, 98, 100, 101n31, 102–107, 127–128, 200, 252, 255–256, 257, 267, 273n2, 274, 313, 321n34

The School and Society (Dewey), 98

science: and administration, 3–5, 6–7, 10, 12–14, 25, 71–72, 119–120, 124, 137, 178, 227–228, 272, 342; and art, 106–107, 169, 173, 207, 218, 230–231, 279, 310, 314–330, 373; as communicative practice, 15, 23–24, 26, 40–46, 90–91, 96, 101, 108, 145–146, 240, 247–248; produces consensus, 26, 54, 55–59, 83–98, 100, 103–108, 110, 129, 247–248, 287–288, 357, 371; denaturalizing tendency, 12, 17, 95, 209; as ethical practice, 2–3, 10, 13, 15, 23–24, 26, 28, 34–35, 39–46, 73–74, 89, 96; ethos of, 247–249; fallible, 12, 40, 62, 66, 91, 95, 97, 110, 135, 143–145, 158, 240, 259; as foundation for ethics,

89, 91–92, 94, 108, 113–114, 151, 194, 333; frustration of, 244; and health, 12, 211; as humanitarian aid, 40, 46, 51–52; and individual adjustment, 211; and individual autonomy, 231, 318–319, 354; and industrial progress, 1, 5, 10, 12, 14, 23, 25, 31, 36, 39, 46, 51, 53, 69–70, 111, 115, 178, 210–212, 230, 305–306, 307, 308n9, 311, 315, 318, 333, 365, 369; and language, 258–271; naturalizing tendency, 3, 12, 94–95, 95n17, 177, 178–179, 206–207, 209, 341–342; as personal orientation, 10, 22n2, 23–24, 39–46, 58n6, 73–74, 96, 129, 131, 144, 204; and religion, 1, 2–3, 4n3, 5, 15, 22, 25, 26, 30, 34–46, 50, 50n48, 59–65, 75, 77, 81–82, 85–87, 93–94, 98, 157, 193–195, 210, 218, 240, 320–321, 323, 323–324n39; vs. social hierarchy, 52–53, 75–82, 131, 205, 254, 265–266, 278, 293; as social process, 90–91, 101, 104, 108, 145–146, 187, 240, 247–248, 319, 354; as spiritual resource, 320–321; as storehouse of knowledge, 212; and technical standards, 141; virtues associated with, 22n2, 23–24, 39–46, 51, 73–74, 75, 84–85, 129, 132, 247, 253, 271, 275–275, 308, 353–354. *See also* cultural sciences; human sciences; names of individual disciplines; physical scientists; two-cultures model

Science (journal), 48, 53, 63–64

science administrators, 230–231, 302–304, 307–308, 309, 310, 312–313, 314–315

Science & Society (journal), 245, 254n38

science and technology studies, 5n6, 369n7

Science and the Modern World (Whitehead), 193–194

Science and the Social Order (Barber), 357

Science for the Citizen (Hogben), 255

Science Service, 211

science studies fields. *See* history of science; philosophy of science; sociology of science

Science – The Endless Frontier (Bush), 308, 320

scientific attitude, 132, 166, 172, 180, 196–197, 201, 204, 206, 227, 248, 252; defined, 113–114

scientific community, 90–91, 230, 235–236, 243, 247–249, 302, 310, 315, 316n24, 319, 320, 341–342, 357–358, 361, 371. *See also* laissez-faire communitarianism

scientific democracy: impact of Cold War on, 14, 16, 225–226, 342, 343n13, 356; criticized, 102, 194–195, 198–200,

216–223, 226, 229–231, 272, 302, 310–314, 320–324, 326–327, 337–338; defined, 9–10, 12, 17–18, 340–341; distribution, 13, 17, 238–239, 254–256, 258; emergence, 15, 22–25, 28, 30, 34–39, 99; globalized, 290–293, 300–301, 350, 361–363; impact of New Deal on, 111, 212, 225–227, 272–273; postwar attenuation, 14, 16, 225–234, 235–237, 270–271, 335–343, 349–350, 363–364; shortcomings, 370–371, 371–372n13; impact of World War I on, 14, 98, 111, 146–147, 155–156, 171–172, 189, 197, 200, 202–203; impact of World War II on, 14, 16, 201, 225–227, 229–232, 235–236, 237, 257–258, 267, 271, 273, 290–297, 299–301, 302–304, 307–308, 321–323, 328, 330n53, 331, 335, 342, 343, 347, 349–350, 367. *See also* names of individual disciplines

scientific management, 138. *See also* efficiency

scientific method, 23n2, 52, 54, 71n36, 83, 84, 87, 93, 101, 104, 105, 107, 141, 145–146, 159, 187, 226, 237, 238, 243, 260–261, 270, 357, 370, 372; applied to human behavior, 15, 16, 27, 61–62, 65, 85–86, 92, 95, 96, 109, 110–111, 112–113, 122, 137, 151, 155, 160, 162, 165, 167–169, 178, 179–180, 216, 228, 277–281, 286, 287–288, 310, 317, 333, 338, 358, 367, 373; as personal orientation, 10, 22n2, 40–43, 58n6, 73–74, 96, 129, 131, 144, 204

Scientific Revolution, 2, 56, 211, 254n38

scientific spirit, 25, 29, 39–46, 47, 55–56, 59, 63, 65, 71n36, 74, 83, 84–85, 88–89, 90, 112, 113, 132, 176; defined, 23–24, 41–43

scientism, 3–5, 10n12, 109–110, 367n4, 368, 369; criticized, 102, 216–223, 272, 320–324, 326–327, 337–338, 345, 365–367

Scopes Trial, 159

Scottish Enlightenment, 126. *See also* Enlightenment

Scripps Institution, 158

Scudder, Samuel H., 53, 64

Second Great Awakening, 30, 35

second scientists' movement, 308–309, 310–311, 313–314, 323

secularization, 21–22n1, 28, 326–327, 367–368, 368n5, 372

Seligman, E. R. A., 77, 80, 134n35. *See also* economics, ethical

Sellars, Roy Wood, 157, 159, 164, 191–192, 193, 194
semantics, 264. *See also* general semantics
sentimental Enlightenment, 118.
    See also Enlightenment
Shapin, Steven, 370n9
Shapley, Harlow, 312, 322, 326
Sheen, Fulton J., 217, 218n42, 326, 327
Shils, Edward, 288n33, 348, 357
Shotwell, James T., 338n4
Shuster, George N., 323
Sikorsky, Igor, 324n39
Simpson, George Gaylord, 336, 337, 341
Sinnott, Edmund W., 324n39, 336, 337, 341
Skinner, B. F., 242, 327, 338, 346, 356n34
slavery, 29, 33
Slosson, Edwin E., 211
Small, Albion W., 75, 118n3
Smith, Adam, 126, 127
Smith, Christian, 21n1
Smith, Norman Kemp. *See* Kemp Smith, Norman
Smith, Preserved, 252
Smith, T. V., 114n8, 212, 245–246, 269n72, 325
Smith College, 165
Snow, C. P., 231, 310. *See also* two-cultures model
Social and Cultural Dynamics (Sorokin), 281
Social Change, With Respect to Culture and Original Nature (Ogburn), 178–179
social control, 7, 120–124, 127, 146, 171, 181
Social Control (Ross), 120–123, 125n15, 144n54
social engineering, 13, 123–124, 138, 143–145, 150–151, 175–176, 186, 285–286. *See also* planning
social fascism, 348. *See also* fascism
Social Gospel. *See* Protestantism
social organism, 66, 76, 78–80, 82, 94–95, 118n3, 123–124, 126, 128, 131, 146, 161, 163, 178–179, 247, 265, 285, 298
Social Organization (Cooley), 125–128
social responsibility of scientists, 244–245, 302, 304–307, 310–311, 313–314
Social Science Research Council (SSRC), 176–178, 201, 228, 272, 274, 280, 292, 312
social sciences. *See* cultural sciences; human sciences; names of individual disciplines
social self. *See* philosophy; sociology
social structures, 359–361
social work, 214, 267, 273, 277
socialism, 17, 66, 80, 87, 116, 119n4, 121n7, 123, 139–141, 141n7, 143, 144, 186,
203, 226, 227n2, 230, 241, 245, 249, 306, 312, 320, 336, 346, 348, 352
socialization, 124–132, 135, 178, 266, 284–285, 286, 294, 298, 340, 353–354, 359. *See also* education
Society (MacIver), 279
Society for the Psychological Study of Social Issues (SPSSI), 287, 300, 324
The Sociological Imagination (Mills), 356
sociology, 174n6, 287, 297, 347, 348; and adult education, 214, 267; American Creed, 299–300; American Sociological Association, 274; anti-communists target, 326; and behavioral sciences, 234, 288–290, 301, 336, 353, 359–362; Bernard, 123–124, 126, 133, 145, 146, 150, 176, 178, 227n2, 280n17, 346; and biology, 66, 69, 77, 117–118, 118n3, 120–121, 126, 128, 152n10, 159, 162–163n36, 178–179, 181, 183, 289–290, 336–337; Chicago school, 180–182, 246; and consensus liberalism, 232–233, 234, 273, 282–283, 288–290, 292, 299–301, 351, 353, 355–356, 359–362; Cooley, 8, 119–120, 124–128, 128n22, 130, 131, 131n28, 132, 133, 155, 182, 183n25, 292; cultural lag, 178–179, 187–188, 189, 215, 275, 305, 311; and cultural sciences, 8, 13, 16, 109–111, 114–115, 159, 165, 178–179, 183, 188, 226–227, 228, 234, 235, 236, 237, 242, 272–273, 278, 280, 282–283, 288–290, 297, 301, 336; and culture and personality, 288, 290, 292, 361; in curricula, 203; Eldridge, 174–175, 180; Ellwood, 128n21, 165, 166, 227n2, 228, 273, 277–278, 279, 282, 345, 373; and emergentism, 152n10, 159, 165, 181; empiricism, 16, 85, 109, 178, 277, 279, 280, 282; and eugenics, 220; and federal government, 76, 115–116, 227–228, 272, 274–275, 276, 282, 312, 312n16; and *Gestalt*, 165; of knowledge, 134, 135, 191, 203–204, 245–246, 248n27, 250, 263, 327n47; and *laissez-faire*, 26, 66, 68–70, 115–116, 133; Lynd, 180, 228, 267, 273, 281–282, 283, 313, 326, 345; MacIver, 246n23, 261n55, 273, 279, 280, 281, 291, 322, 345; Mannheim, 244, 245–246, 247, 327n47; Mead, 119–120, 124, 125, 128–130, 131, 131n28, 133, 174, 203, 243, 246, 255; of medicine, 276; Merton, 243, 245, 247–249, 254, 254n38, 310, 316, 316n24, 347, 348, 351, 354, 357–358; methodology, 85, 109, 178, 180, 277–281; Mills, 234, 246,

246n23, 340, 355n33, 356, 357, 365n1;
and modernization theory, 361–362;
Myrdal, 299–300, 351; neo-positivism,
109–110, 112, 178, 199, 228, 235, 236,
236n2, 237, 240, 241, 248, 274–282,
307n7, 311, 326; and New Deal, 226–
227, 227n2, 227–228, 249, 272–273;
Ogburn, 8, 109, 178–179, 189, 215,
228, 235, 236, 238, 273, 274–278, 279,
281, 282, 305, 307n7, 311, 312, 313,
321, 326, 338n4, 345; operationalism,
355; Park, 180–182; Parsons, 237, 242,
288–290, 292, 300, 312, 312n16, 319,
326, 346, 353, 361–362; positivism, 85;
of professions, 246; quantification, 109,
178, 277–279, 281, 356; and religion,
68, 75–76, 122–123, 175, 277–278;
Ross, 8, 119–123, 124, 125, 125n15,
126, 127, 131, 133, 142, 144n54, 155,
171, 202n10, 255, 297; rural, 280, 363;
of science, 228–229, 235–237, 243–249,
250, 251–252, 254, 255, 257–258,
270–271, 303, 357–358; and scientific
democracy, 8, 13, 16, 26, 109–111,
114–115, 117–130, 132–135, 174–175,
176–177, 178–179, 180–182, 183, 188,
226–227, 228, 234, 235, 236, 237, 242,
243–249, 257–258, 272–273, 278–279,
282–283, 289–290, 297, 300–301, 336,
340, 346, 348–349, 355–356, 357–358,
359–362; social control, 7, 120–124, 127,
146, 171, 181; social laws, 66, 181–182,
277, 279, 281; social self, 87–88, 124–
131, 146, 297; social structures, 359–361;
Sorokin, 228, 247, 273, 280–281,
287n29, 326, 345, 356–357n36; Spencer,
60, 61, 63–64, 65, 66, 69, 76, 77, 85,
118n3, 157, 205, 328; Sumner, 26, 68–70,
71, 72n40, 73, 74, 76, 79, 87–88, 96, 120,
132–135, 136, 137, 138, 142, 145, 146,
155, 181, 248n27, 260n52, 297n51, 350;
and SSRC, 176–177, 178–179, 180, 228,
272, 274, 312, 312n16; of technology,
246; value-neutrality advocated, 8, 87–88,
88n6, 109–111, 114–115, 176–177,
178, 199, 228, 235, 236, 272–273,
273–282, 307n7, 311, 312, 327, 346,
351, 355–356, 361; value-neutrality
criticized, 8, 109–111, 114–115, 166,
199, 228, 235, 272–273, 277–282, 283,
327, 355, 356, 356n34; Ward, 76–77, 87,
118n3, 121–122, 152, 164; and World
War II, 226–227, 282, 291, 292, 299–300,
338n4. *See also laissez-faire*; philosophy
of science; social organism

Sociometry (journal), 287
Sorokin, Pitirim A., 228, 247, 273, 280–281,
287n29, 326, 345, 356–357n36
South, 29, 34
Soviet Union, 16, 244, 245, 256n43, 305n2,
308n11, 315, 324–325, 326, 343
Spanish Civil War, 302, 304, 324
Spaulding, E. G., 104–105, 153–154, 273n2
specialization, 6, 47, 55, 57, 59, 86n2, 98,
99–101, 103, 149, 152, 155, 159, 162,
196, 199, 201, 207–208, 213, 220, 238,
250, 253, 276, 282, 288–290, 319, 335,
340, 342, 356, 361, 367
Speier, Hans, 246n23, 248n27
Spencer, Herbert, 60, 61, 63–64, 65, 66, 69,
76, 77, 85, 118n3, 157, 205, 328
SPSSI. *See* Society for the Psychological Study
of Social Issues
Sputnik, 308n11, 315
SSRC. *See* Social Science Research Council
Stallo, J. B., 60n11, 86
Standard Average European, 263
Stanford University, 54, 156, 204,
252–253, 331
statistics, 70, 75, 78, 79, 86, 178, 179–182,
199, 240, 262, 266n67, 274, 277, 280.
See also quantification
Stern, Bernhard J., 254, 256n43, 276
Stevens, S. S., 242, 338
Stevenson, C. L., 242n15
Stimson, Dorothy, 252
Storck, John, 203–204
The Story of Mankind (van Loon), 209
The Story of Philosophy (Durant), 209–210,
210n28
Strauss, Leo, 123, 345
The Structure of Scientific Revolutions (Kuhn),
317n24, 366
The Structure of Social Action (Parsons),
289–290
Studies in the History of Ideas (Columbia
Department of Philosophy), 251
The Study of the History of Mathematics
(Sarton), 256
The Study of the History of Science (Sarton),
256
suffrage, 9
Sumner, William Graham, 74, 76, 79, 138,
142, 146, 155, 181, 297n51; and civil
service reform, 72n40; compared to
Youmans, 69; *Folkways*, 120, 132–135,
136, 248n27, 260n52; on *laissez-faire*,
26, 68–70, 71, 96, 133–134; on science,
68–70, 73, 87–88, 137, 145, 350; on
social role of scholar, 133–135, 137

symbolic logic, 192, 238n7. *See also* logic
symbolism, 90n10, 192, 235, 258, 262, 264.
    See also linguistics
systems theory, 355

Taft, Robert A., 312
Taft, William Howard, 137, 146, 368–369
Taylor, Carl C., 363
Taylor, Charles, 298
Taylor, Frederick Winslow, 138
technocracy, 5, 7, 10, 10n14, 110, 123n13,
    124, 191, 227, 294, 306, 327, 372.
    See also bureaucracy; administration
technological unemployment, 305–306, 331
Technological Trends and National Policy
    (government report), 228
technology, 94, 127, 155, 187, 193, 201,
    228, 266, 281, 294, 339, 366, 368, 373;
    history of, 246, 247; and moral progress,
    15, 24, 28, 31, 34, 35, 36–37, 38, 39, 48,
    53, 70, 125–126, 127–128, 131, 133–134,
    185, 292, 324n39; and science, 9, 10, 13,
    34, 39, 46, 51–52, 52n52, 53, 70, 131,
    146, 210–211, 230, 244, 303, 304–306,
    307, 308, 308n9, 310–311, 314, 315,
    316–317, 318, 324n39, 330–331, 367.
    See also inventors
Teller, Edward, 309, 325
Tennessee Valley Authority (TVA), 212, 213
Theory of Valuation (Dewey), 242n15
Thilly, Frank, 105
Thomas, Norman, 227n2
Thomas, William I., 182
Thomism. *See* philosophy, neo-Thomist
Thorndike, Edward, 215–216, 268n71, 274n4
Thorndike, Lynn, 252
"Thought News," 98, 180
Thurstone, L. L., 179
Time (magazine), 269, 282n19
Tocqueville, Alexis de. *See* de Tocqueville,
    Alexis
Tolman, Edward C., 166–169, 170, 179, 242,
    269, 275–276, 286, 326, 358
Toomer, Jean, 208
totalitarianism, 16, 223, 225, 229, 233, 247,
    258, 270, 271, 273, 289, 291, 296,
    324–325, 328, 332, 343, 344, 346, 347,
    353. *See also* communism, fascism, Hitler,
    Nazism
*The Transition to an Objective Standard of
    Social Control* (Bernard), 123–124
Treadway, Walter L., 269n72
Trilling, Lionel, 230, 328
Truman, Harry S., 309, 312, 327

Tufts, James H., 98, 203
Tugwell, Rexford G., 212, 228
Tulane University, 246
Turner, Frank M., 17n18
TVA. *See* Tennessee Valley Authority
*The Two Cultures and the Scientific
    Revolution* (Snow), 231
two-cultures model, 231, 310, 319–320,
    328–329, 332–334, 337–340
"Two Dogmas of Empiricism" (Quine), 339
The Tyranny of Words (Chase), 172n1,
    258n48, 269

UCPIP. *See* Universities Committee on
    Post-War International Problems
underdetermination, 248, 316
Union College, 30
Union Theological Seminary, 220
unions. *See* labor relations
Unity of Science movement, 242, 319.
    See also philosophy, logical empiricist
universalism, 24, 37, 57–59, 78, 82, 83, 85,
    86–87, 88–89, 90, 93, 107, 110, 134, 137,
    184–185, 205, 220, 247–249, 259, 281,
    287, 321, 321n33, 328–329, 332, 341,
    344, 351, 353, 362, 363, 371
universities. *See* colleges and universities;
    names of individual institutions
Universities: American, English, German
    (Flexner), 200
Universities Committee on Post-War
    International Problems (UCPIP), 290, 292
University of California, Berkeley, 32, 54, 166,
    184, 307. *See also* Scripps Institution
University of Chicago, 54, 114n8, 123, 185,
    212, 242, 305, 307, 359, 363; Bloomfield
    at, 265; Chicago School of sociology,
    180–182, 246; commitment to public
    service, 13; commitment to liberal
    education, 13; curriculum, 207–208,
    210n28, 219, 333; Dewey at, 95, 98,
    131n28; history of science at, 253,
    254–255; Hutchins and Adler at, 205,
    221–222; Lasswell at, 163, 175–176;
    Mead at, 124, 131n28, 243; Merriam
    at, 176–177; Ogburn at, 178–179;
    psychobiology at, 159–162; and scientific
    democracy, 13; and SSRC, 177–178
University of Giessen, 167
University of Illinois, 254, 255
University of Kansas, 174
University of Michigan, 34, 42, 49, 52, 53–54,
    77, 95, 124, 125, 131n28, 158, 159, 254
University of Minnesota, 242, 255

University of Pennsylvania, 29, 138, 280
University of Texas, 182
University of Utah, 205
University of Virginia, 30
University of Washington, 271
University of Wisconsin, 42, 43, 46, 74, 120, 182, 192, 246, 254–255, 271; Experimental College, 205–206, 213, 215
Urban, Wilbur Marshall, 273
urbanization, 12, 75, 181
Urey, Harold C., 307, 309, 312, 323
USDA. *See* Agriculture, U.S. Department of
utilitarianism, 69, 73, 80n61, 114, 139, 191, 349

Van Doren, Carl, 219
van Loon, Hendrik Willem, 209
Vassar College, 264n60
Veblen, Thorstein, 139n44, 182, 188, 214, 228
vitalism. *See* biology
vocational education, 13, 30–32, 33, 46, 47, 52–53, 196–197, 203, 213
Voltaire, 280

Walker, Francis Amasa, 72n40
Wallace, Henry A., 305, 309, 331, 363
War Issues, 202–203
Ward, Lester Frank, 76–77, 87, 118n3, 121–122, 152, 164
Ware, Caroline, 283n23
The Warfare of Science (White), 52n53, 64–65
Washington, D.C., 76, 159, 227, 253, 294, 342
Washington Post (newspaper), 306
Watson, Goodwin, 324
Watson, John B., 148, 150–151, 159, 160, 164, 165, 166, 167, 168, 210, 265, 346, 358. *See also* psychology, behaviorist
Weaver, Warren, 230
Weber, Max, 1, 4
Weiss, Albert Paul, 168, 265
welfare state, 77, 138, 233, 345–346
Wells, H. G., 209
Weltfish, Gene, 293, 326
Wertheimer, Max, 164, 165, 166n46. *See also* psychology, *Gestalt*
Westbrook, Robert B., 6n8, 7n9, 145n55, 194n46, 297n52, 341n10
Western Civ, 201, 229, 232, 328–330, 332–333, 337, 349
Western tradition. *See* humanistic tradition
Weyl, Walter, 8, 137, 138–140, 142, 143, 227
Wheeler, William Morton, 157–158, 159, 162, 164, 269n72
Whewell, William, 60–61, 62, 63

Whigs, 15, 24, 26, 28, 35, 36–39, 48, 53, 69, 73–74
White Collar (Mills), 340, 356
White, Andrew Dickson, 8, 48, 72n40, 73n43, 74, 79, 300; at Cornell, 30, 31–32, 32n7, 34n12, 43n30, 45, 48–49, 52, 59; on religion, 40, 49, 50, 52n53, 64–65; *The Warfare of Science*, 52n53, 64–65
White, Morton, 299
Whitehead, Alfred North, 193–194, 221n49, 263
Whorf, Benjamin Lee, 259, 261, 263–264, 270
Why We Behave Like Human Beings (Dorsey), 209
Wiener, Norbert, 355
Wilbur, Ray Lyman, 331
William and Mary, 29
Williams, Elgin, 356n34
Willkie, Wendell, 331
Wilson, Logan, 246
Wilson, M. L., 363
Wilson, Woodrow, 106, 137, 146, 368
Wirth, Louis B., 242, 246n23
Wistar Institute of Anatomy and Biology, 161
Woodard, James W., 356n34
Woodbridge, Frederick J. E., 102n32, 158–159, 192, 199–200, 203
Woodhull, John F., 253
Wooster, Harvey A., 207, 269n72
workers, 12, 31, 34, 51, 52, 53n54, 94, 134, 140–141, 143, 146, 184, 190, 191, 202. *See also* labor relations
World Congress of Philosophy (1926), 162
world organization, 157, 174, 207, 290, 292, 300
World War I, 7, 7n9, 17, 118, 152, 187, 188, 198, 199, 209, 212, 214, 252, 253, 255, 261, 290, 307; impact on scientific democracy, 14, 98, 111, 146–147, 155–156, 171–172, 189, 197, 200, 202–203
World War II, 5n6, 114, 115, 214, 230, 234, 243, 249, 256, 258, 267, 280, 282, 298–299, 309, 315, 320, 332, 336, 337, 338, 338n4, 339, 340, 344, 346, 348, 355, 359; impact on scientific democracy, 14, 16, 201, 225–227, 229–232, 235–236, 237, 257–258, 267, 271, 273, 290–297, 299–301, 302–304, 307–308, 321–323, 328, 330n53, 331, 335, 342, 343, 347, 349–350, 367
The World's Need of Christ (Ellwood), 277
Wright, Chauncey, 41
Wundt, Wilhelm, 88

Yale University, 29, 30, 31, 32, 49–50, 54, 68, 72n40, 73, 132, 182, 185, 221, 250n31, 313, 359
Yale Report, 30, 34
Yerkes, Robert M., 312
Youmans, E. L., 43, 44, 45, 63n17, 107n46; compared to Spencer, 66; *The Culture Demanded by Modern Life*, 63, 66–67; educational proposals, 47, 63, 67–68, 70, 71; fallibilism, 62–63, 66, 82, 85; *laissez-faire*, 66–68, 69, 70, 71, 73–74, 96; *Popular Science Monthly*, 60, 61n13, 63, 68, 73; positivism, 61–62, 65, 66, 85, 89; on religion, 40n23, 62, 63, 64–65

Youmans, William Jay, 43, 44–45
Young, Kimball, 179

Zipf, George K., 266n67
Znaniecki, Florian, 182, 246